FIBROBLAST SURFACE PROTEIN

ANNALS OF THE NEW YORK ACADEMY OF SCIENCES
Volume 312

FIBROBLAST SURFACE PROTEIN

Edited by Antti Vaheri, Erkki Ruoslahti,
and Deane F. Mosher

The New York Academy of Sciences
New York, New York
1978

Library of Congress Cataloging in Publication Data
Main entry under title:

Fibroblast surface protein.

(Annals of the New York Academy of Sciences; v. 312)
Papers from a conference held by the New York Academy of Sciences, Nov. 30 and Dec. 1–2, 1977.
1. Fibroblasts—Congresses. 2. Plasma membranes—Congresses. 3. Glycoproteins—Congresses. I. Vaheri, Antti. II. Ruoslahti, Erkki. III. Mosher, Deane F. IV. New York Academy of Sciences. V. Series: New York Academy of Sciences. Annals; v. 312. [DNLM: 1. Fibroblasts—Analysis—Congresses. 2. Membrane proteins—Analysis—Congresses. W1 AN626YL v. 312 / QS532.5.C7 F443 1977]
Q11.N5 vol. 312 [QP88.23] 508'.1s [591.8'2]
ISBN 0-89072-068-1 78-15899

SP

Printed in the United States of America

ISBN 0-89072-068-1

ANNALS OF THE NEW YORK ACADEMY OF SCIENCES
VOLUME 312
June 20, 1978

FIBROBLAST SURFACE PROTEIN*

Editors and Conference Chairmen
ANTTI VAHERI, ERKKI RUOSLAHTI, AND DEANE F. MOSHER

CONTENTS

Some Early History of Cold-Insoluble Globulin. *By* JOHN T. EDSALL 1

Part I. Cold-Insoluble Globulin: Structure and Possible Functions

Structure of Human Plasma Cold-Insoluble Globulin and the Mechanism of Its Precipitation in the Cold with Heparin or Fibrin-Fibrinogen Complexes. *By* MICHAEL W. MOSESSON 11

Are Lobster Fibrinogen and Cold-Insoluble Globulin Related Molecules? *By* GERALD M. FULLER 31

Cross-linking of Plasma and Cellular Fibronectin by Plasma Transglutaminase. *By* DEANE F. MOSHER 38

Physiologic Role of Cold-Insoluble Globulin in Systemic Host Defense: Implications of Its Characterization as the Opsonic α_2-Surface-Binding Glycoprotein. *By* THOMAS M. SABA, FRANK A. BLUMENSTOCK, PETER WEBER, AND JOHN E. KAPLAN 43

Distribution of Cold-Insoluble Globulin in Plasma and Tissues. *By* MICHIO MATSUDA, NOBUHIKO YOSHIDA, NOBUO AOKI, AND KUNIO WAKABAYASHI 56

Bovine Plasma Cold-Insoluble Globulin: Gross Structure and Function. *By* SADAAKI IWANAGA, KOJI SUZUKI, AND SENICHIRO HASHIMOTO 74

Part II. Characteristics of the High-Molecular-Weight Glycoprotein in Connective Tissue and Basement Membranes

Organization of Extracellular Proteins on the Connective Tissue Cell Surface: Relevance to Cell-Matrix Interactions *in Vitro* and *in Vivo*. *By* PAUL BORNSTEIN, DAN DUKSIN, GARY BALIAN, JEFFREY M. DAVIDSON, AND ED CROUCH 93

Nature of the Glycoprotein Components of Basement Membranes. *By* ROBERT G. SPIRO 106

Synthesis of Fibronectin by Cultured Human Endothelial Cells. *By* ERIC A. JAFFE AND DEANE F. MOSHER 122

*This series of papers is the result of a conference entitled Fibroblast Surface Protein, held by The New York Academy of Sciences on November 30 and December 1 and 2, 1977.

Cell Surface and Extracellular Matrix Glycoprotein Fibronectin: Expression in Embryogenesis and in Teratocarcinoma Differentiation. *By* Jorma Wartiovaara, Ilmo Leivo, Ismo Virtanen, Antti Vaheri, and Christopher F. Graham 132

Presence of Fibronectin in Basement Membranes and Acidic Structural Glycoproteins from Human Placenta and Lung. *By* Bonnie Anderson Bray 142

Distribution of Fibronectin in Human Tissues and Relationship to Other Connective Tissue Components. *By* E. Linder, S. Stenman, V.-P. Lehto, and A. Vaheri 151

Part III. Characteristics of the High-Molecular-Weight Fibroblast Surface-Associated Glycoprotein

Chemical Composition, Gross Structure, and Organization of Transformation-Sensitive Glycoproteins. *By* William G. Carter, Minoru Fukuda, Clifford Lingwood, and Sen-itiroh Hakomori 160

Immunochemical and Collagen-Binding Properties of Fibronectin. *By* Erkki Ruoslahti and Eva Engvall 178

Labeling, Solubilization, and Peptide Mapping of Fibroblast Surface Glycoprotein. *By* R. Colin Hughes and Roderick Nairn 192

Interaction of an L-Cell Surface Protein with the Cytoskeleton. *By* Richard C. Hunt and Jay C. Brown 207

Cell Surface Coat of Hamster Fibroblasts. *By* John M. Graham, Richard O. Hynes, Charles Rowlatt, and Jane K. Sandall 221

Identification and Characterization of Normal and Malignant Human Blood Leukocytes by Surface Glycoprotein Patterns. *By* Carl G. Gahmberg and Leif C. Andersson 240

Part IV. Role of the High-Molecular-Weight Glycoprotein in Cellular Morphology and Adhesion

Transformation-Sensitive Cell Surface Protein: Isolation, Characterization, and Role in Cellular Morphology and Adhesion. *By* Kenneth M. Yamada, Kenneth Olden, and Ira Pastan 256

High-Molecular-Weight Glycoprotein as a Mediator of Cellular Adhesion. *By* Edward Pearlstein and Leslie I. Gold 278

Are High-Molecular-Weight Glycoproteins Regulators of Cellular Growth. *By* John M. Buchanan, Lan Bo Chen, and Bruce R. Zetter 293

Role of the High-Molecular-Weight Glycoprotein in Cellular Morphology, Adhesion, and Differentiation. *By* Bruce R. Zetter, Gail R. Martin, Charles R. Birdwell, and Denis Gospodarowicz 299

Part V. Role of the High-Molecular-Weight Glycoprotein in Cellular Growth and Malignant Transformation

A Large Glycoprotein Lost from the Surfaces of Transformed Cells. *By* R. O. Hynes, I. U. Ali, A. T. Destree, V. Mautner, M. E. Perkins, D. R. Senger, D. D. Wagner, and K. K. Smith 317

Fibronectin and the Pericellular Matrix of Normal and Transformed Adherent Cells. *By* Antti Vaheri, Kari Alitalo, Klaus Hedman, Jorma Keski-Oja, Markku Kurkinen, and Jorma Wartiovaara 343

Cellular Transglutaminase, Growth, and Transformation. *By* Paul J. Birckbichler and M. K. Patterson, Jr. 354

Modulation of Cell Surface Glycocalyx: Studies on Large, External, Transformation-Sensitive Protein. *By* Lan Bo Chen, Keith Burridge, Andrew Murray, Marcia L. Walsh, Christine D. Copple, Anne Bushnell, James K. McDougall, and Phillip H. Gallimore 366

Glycoprotein-Containing Factor That Mediates Contact Inhibition of Growth. *By* G. Lipkin, M. E. Knecht, and M. Rosenberg 382

Synthesis and Function of Carbohydrate Units of Cell Surface Glycoproteins. *By* S. J. Turco and P. W. Robbins 392

Part VI. Poster Session

Cold-Insoluble Globulin: Transamidation-Sensitive Site and Affinity Behavior. *By* Helmut Hörmann and Franz Jilek 399

Properties of Fibronectin Isolated from Culture Medium of Human Fibroblasts and from Human Plasma. *By* Matti Vuento 401

Role of Fibronectin and Collagen Types I and II in Chondrocytic Differentiation *in Vitro*. *By* W. Dessau, J. Sasse, R. Timpl, and K. von der Mark 404

Changes in Cell Surface Proteins During Normal and Vitamin A-Inhibited Chondrogenesis *in Vitro*. *By* J. R. Hassell, J. P. Pennypacker, K. M. Yamada, and R. M. Pratt 406

Amniotic Fluid Fibronectin and the Synthesis of Fibronectin by Amniotic Fluid Cells in Culture. *By* E. Crouch, G. Balian, K. Holbrook, H. Hoehn, and P. Bornstein 410

Transmembrane Linkage of Fibronectin to Intracellular Actin-Containing Filaments in Cultured Human Fibroblasts. *By* Michael H. Heggeness, J. F. Ash, and S. J. Singer 414

Subunit Structure of Surface and Shed Large, External, Transformation-Sensitive Protein of Chick Embryo Fibroblasts. *By* Michael R. McConnell and Peter M. Blumberg 418

Attachment of Chicken Myoblasts to a Gelatinized Substratum Is Mediated by Cell Attachment Protein from Horse Serum and by Chicken Fibroblast Cell Surface Protein. *By* Matthias Chiquet, Emilio C. Puri, and David C. Turner 420

Bioautography of Serum- and Cell-Derived Cell Attachment Proteins. *By* W. Daryl Dickey, James R. Hall, Cary M. Seals, and Robert J. Klebe 423

Differences in the Ultrastructural Localization of Fibronectin on Normal and Transformed Cells. *By* Leo T. Furcht, Deane F. Mosher, and Gwen Wendelschafer-Crabb 426

Thrombin-Induced Lectin Activity of Human Platelets: A Possible Determinant of Platelet Aggregation. *By* T. KENT GARTNER, DANNY C. WILLIAMS, AND DAVID R. PHILLIPS 432

Cold-Insoluble Globulin Is Not Absolutely Required for Cellular Adhesion to Native Collagen Substrata. *By* FREDERICK GRINNELL AND DIANNE MINTER 434

Binding of Cell Attachment Protein to Collagen: Effect of Chemical Modifications. *By* HYNDA K. KLEINMAN, ERMONA B. MCGOODWIN, GEORGE R. MARTIN, AND ROBERT J. KLEBE 436

Fibroblasts Release Surface Proteins That Function as Concanavalin A Receptors. *By* ALBERT J. T. MILLIS, MARIAN HOYLE, AND PAULETTE J. MCCORMICK 439

Models of Cell Surface Events During Normal and Transformed Human Cellular Adhesion *in Vitro*. *By* R. RAJARAMAN, R. A. FOX, AND J. M. MACSWEEN 441

Immunofluorescence Studies on Fibronectin Distribution Patterns During Adhesion, Deformation, and Spreading of Human Glial and Glioma Cells. *By* R. RAJARAMAN, B. WESTERMARK, A. VAHERI, AND J. PONTÉN 444

A Collagenous Component of the Microexudate Carpet Secreted by Attaching Human Fibroblasts. *By* C. E. SCHWARTZ, C. G. HELLERQVIST, AND L. W. CUNNINGHAM 450

Serum Glycoproteins in Fibroblast Spreading on Glass. *By* DAVID THOM, ANDREW J. POWELL, R. ANDREW BADLEY, ANNE WOODS, COLIN G. SMITH, AND DAVID A. REES 453

Financial assistance was received from:

- ABBOTT LABORATORIES
- AMERICAN HOECHST CORPORATION
- BURROUGHS WELCOME CO.
- ELI LILLY AND COMPANY
- MERCK SHARP AND DOHME POSTGRADUATE PROGRAM
- NATIONAL CANCER INSTITUTE

SOME EARLY HISTORY OF COLD-INSOLUBLE GLOBULIN*

John T. Edsall

Department of Biochemistry and Molecular Biology
The Biological Laboratories
Harvard University
Cambridge, Massachusetts 02138

Cold-insoluble globulin, like many other plasma proteins, was a mystery to us when we first came across it in the plasma fractionation program that started in World War II. It remained a mystery for many years thereafter, and none of us, then, could have foreseen the future intense interest in the subject, so impressively manifested in the papers in this monograph. I hope that a glimpse of the events of a generation ago will be of some interest to you. For some of us, it is fun to look back at the past; whether such a historical look will provide any useful lessons for the future is something I will leave for you to judge.

Let me say at once that our laboratory, the Department of Physical Chemistry at Harvard Medical School, as it was then, was by no means the only one that studied cold-insoluble globulin in those early days. We cited significant work in two or three other laboratories when we published our own studies, and there were probably more. I make no claims for priority; I merely want to give my recollections of the story as I remember it.

The Department of Physical Chemistry was unique; no other medical school in the United States or (I think) elsewhere had anything like it, either in name or function. It had been originally set up in 1920, for Professor L. J. Henderson; but almost from the beginning, the operating head of the laboratory was Edwin J. Cohn, who centered its research on the physical chemistry of proteins.

Cohn was an extraordinary man, a dynamic leader, with a remarkable capacity for drawing other people into projects that he saw as significant. He could be very aggressive, and he made many enemies; but he had vision, and he provided understanding help and encouragement to many younger people, of whom I was one. Like his teachers, L. J. Henderson, T. B. Osborne, and S. P. L. Sörensen, he looked on proteins as definite chemical molecules and rejected the then widely held opinion that all proteins were rather heterogeneous mixtures of colloidal particles of different sizes. I have written about him in more detail elsewhere,[1] and FIGURE 1 shows a portrait of him.

After Cohn's early work on proteins in the 1920s, there was a period of about a decade, from 1930 to 1940, when we turned back from proteins to study their constituents, the amino acids and peptides. The aim was to relate their physical properties to their structure, to study the influence of the groups carrying electric charges, and of polar and nonpolar groups, on the solubility, Raman spectra, dielectric properties, molal volumes, heat capacities, and other properties of these molecules in solution. All this effort seemed terribly abstract to many people, who wondered what an outfit like ours was doing in a medical school. It all looked quite irrelevant to many of our medical colleagues, but it was a necessary prelude to a deeper study of proteins.

*Supported by National Science Foundation Grant SOC 72-05516.

0077-8923/78/0312-0001 $01.75/0 © 1978, The New York Academy of Sciences

FIGURE 1. Edwin J. Cohn (1892–1953).

Our situation changed dramatically after war broke out in Europe in 1939. As the probability of United States' involvement became increasingly clear, the need for blood plasma, on a large scale, was clearly going to be urgent. It was Edwin Cohn, with his long previous experience in protein fractionation, who had the vision to see that plasma could be separated, on a large scale, into a variety of constituents, each of which could then be used for purposes appropriate to its specific function. Given Cohn's scientific authority, and his great persuasive powers, money from Washington was forthcoming, in previously unimaginable amounts, and plasma fractionation got under way. The pilot plant at Harvard Medical School received each week 60 liters of human plasma, derived from Red Cross donors, and the successive steps in fractionation in the cold room took just about a week. The resulting fractions were studied by chemical analysis, in the ultracentrifuge, and in the Tiselius electrophoresis apparatus. Fractions that might be of clinical use were tested in animals and eventually, if all went well, in clinical trials on patients, carefully monitored. For a year or two, the pilot plant operated 24 hours per day, the staff working on three 8-hour shifts. Later, we cut it down to 16 hours per day, which proved just about as effective. There was searching appraisal each week of the results of the fractionation procedure; as a result of the appraisal, debated at a group seminar, the fractionation procedure was modified for the following week. Eventually, acceptable procedures emerged, and the fractionation went into action on an industrial scale, in the seven firms that had undertaken its large-scale production.

Consider the state of knowledge about blood plasma at that time. Cohn, as the fractionation program was beginning, summarized the situation.[2] The known constituents were few, and fewer still had been obtained in purified form. Serum albumin had been crystallized nearly a half century before, but its chemical purity was still uncertain. Fibrinogen had been prepared by Olof Hammarsten in the late nineteenth century. Immense research had gone into the study of the plasma globulins, which had been fractionated into euglobulins and pseudoglobulins by different criteria. In any case, none of them appeared to meet rigorous criteria of chemical purity. The electrophoretic technique of Tiselius, from 1937 on, became a powerful tool of separation, but it resolved the globulins of plasma into only three major fractions, denoted α, β, and γ, and there was strong evidence that all of them were rather complicated mixtures. Apart from fibrinogen, no component of the blood-clotting system had been prepared in even moderate purity. Pillemer and Ecker had recently split complement into four components, and that seemed complicated enough; we could not foresee the almost unimaginable complexities of the complement system and the blood-clotting system as they are known today. It was known that many kinds of antibody activity were to be found in the γ-globulin fraction, but, of course, nothing like a pure antibody was yet known or was indeed to be seen for many years to come. That is about where we stood in 1941. Five years later, the situation was phenomenally different, as may be seen from Cohn's account in 1947[3] or a review that I published about the same time.[4] The increase in the number of known components of plasma, in that short interval, was phenomenal. Since that time, of course, our knowledge has increased by more than another order of magnitude, and the state of affairs 30 years ago surely must look rather primitive by comparison. But it represented an immense change from the state of knowledge in 1940.

The use of ethanol fractionation at low temperature, which was the central feature of the Cohn procedure, was not novel. As far back as 1871, a monograph by Preyer,[5] on the crystallization of hemoglobins from many species of animals, had described the use of ethanol as one among other suitable reagents for precipitating and crystallizing proteins; and Preyer was not original in this respect. Cohn's procedure, however, unlike its predecessors, involved rigorous control of ethanol concentration, pH, ionic strength, and temperature at each stage of the process. Temperature was never allowed to rise above 0°C; it was lowered to −3°C during the separation of the first fraction and was set at −5°C thereafter to minimize the risk of denaturation. The purified fractions were taken up in solution and freeze-dried.

I need not go into details about the various steps in the process; serum albu-

TABLE 1

SEPARATION OF FRACTION I FROM HUMAN PLASMA (METHOD 6)*

Cool citrated plasma to 0°C
Add acetate buffer in 53% ethanol to pH 7.2
Cool to −2.5 to −3°C, 8% ethanol by volume
Remove precipitate by centrifuging
Dissolve precipitate in citrate buffer pH 6.1, I/2 0.3
Freeze dry (~60% of protein in clottable)

*From Cohn *et al.*[6]

min, which was the most urgently needed material for treatment of shock, was obtained in the last major step in the procedure. What concerns us here is the very first step, which yielded fraction I. This fraction contained the much greater part of the fibrinogen, as well as the antihemophilic factor (factor VIII in modern terms), and there also, in time, we found the cold-insoluble globulin. TABLE 1 shows, in simplified form, the procedure (method 6) that was finally adopted for separation of fraction I from plasma.[6]

Our interest in that fraction naturally centered first on fibrinogen and on the fibrin clot that formed in the presence of thrombin. We obtained thrombin, impure but adequate for our purposes, by conversion from prothrombin that was present in another of the plasma fractions. Fibrinogen and thrombin turned out to be of great value in surgery, especially to neurosurgeons. Fibrin film, brilliantly developed by John Ferry and Peter Morrison,[7] proved to be excellent as a replacement for the dural membrane after brain operations; it did not produce any of the adverse reactions that were common with the artificial replacements that had been used earlier. Likewise, fibrin foam, with thrombin, developed by Edgar Bering, Peter Morrison, and myself, found widespread use among neurosurgeons as a hemostatic.

With all of the demand for those products in the war years, we had no time to think much about other constituents of fraction I, except for the antihemophilic globulin, which often proved immensely valuable for hemophiliacs. However, we wanted to obtain and characterize pure fibrinogen, and it was in the course of that work that we became particularly aware of cold-insoluble globulin. Peter Morrison, Susan Miller, and I approached the problem.[8] The procedure we worked out was fairly simple and is shown in outline form in TABLE 2. We precipitated fraction I at low pH (4.9–5.5) and low ethanol concentration (0–4% by volume), separated the resulting precipitate, redissolved it at pH 6.2–6.4, and brought the

TABLE 2

SEPARATION OF FRACTION I-1 (COLD-INSOLUBLE GLOBULIN) AND PURIFIED FIBRINOGEN*

Adjust fraction I to pH 4.9–5.5, ethanol 0–4°C by volume
Separate precipitate in centrifuge (I-A)
Redissolve at pH 6.2–6.4, 0.5% ethanol
Fraction I-1 precipitates (cold-insoluble globulin)
Bring supernatant to 8% ethanol; fraction I-2 is nearly pure fibrinogen (98% clottable)

*From Morrison *et al.*[8]

solution to 0.5% ethanol at 0°C. Under these conditions, a fraction, which we called I-1, separated out, containing the constituents of lower solubility than that of fibrinogen. After removing this fraction and raising the ethanol to 8%, the resulting precipitate was nearly pure fibrinogen. FIGURE 2 shows the solubility of fibrinogen as a function of pH and ethanol concentration, at 0.3 ionic strength and 0°C. We determined similar solubility patterns at two lower ionic strengths; the solubility, of course, decreased markedly as the ionic strength fell, but the general pattern was similar to that of FIGURE 2.

The center of interest on this occasion, however, is in fraction I-1, because it contained the cold-insoluble globulin. The yield of this fraction was variable; in different runs, it ranged from 6 to 17% of the total protein in fraction I, suggesting that it might be quite variable in amount from one individual to another. There

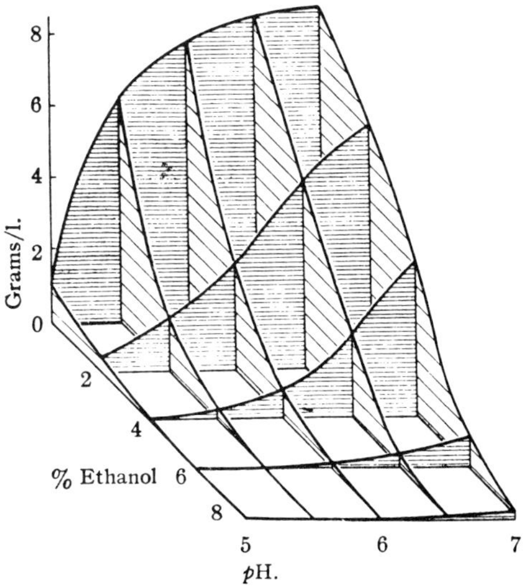

FIGURE 2. Solubility of fibrinogen (from fraction I) as a function of pH and ethanol concentration; ionic strength, 0.30. Ordinate, fibrinogen concentration (grams/liter); abscissa, pH; Z axis, ethanol concentration in percent by volume; temperature, 0°C. The values in this Figure were interpolated from two-dimensional charts of solubility as a function of pH at constant ethanol concentration.

was still a large amount of fibrinogen in fraction I-1. Of the total protein in fraction I-1, about 50–60% was clottable with thrombin; it was evident that there was extensive complex formation between fibrinogen and cold- insoluble globulin.

On cooling a preparation of fraction I-1 to 0°C without stirring, the precipitate appeared like a coarse-structured fibrin clot, but the gelation was completely reversible on rewarming. Obviously, the problem needed more work.

In 1947, two promising postdoctoral fellows arrived in the laboratory. Geoffrey Gilbert came from England and Harold Scheraga from Duke University, where he had just received his Ph.D. in physical chemistry. The three of us together decided to push ahead further with the study of cold-insoluble globulin.[9] Every attempt to separate it from fibrinogen by varying the usual conditions for fractionation was a failure; on adding thrombin, about half of the protein in the prep-

TABLE 3
SEDIMENTATION COEFFICIENTS OF FRACTION I-1 AND THE NONCLOTTING PROTEIN*

Fraction I-1 (50% clottable)
s_{20} = 8S (16%), 11S (12%), 14S (63%), 19S (8%)
Nonclotting protein (cold-insoluble globulin)
s_{20} = 8S (12%), 15S (75%), 22S (13%)
Pure fibrinogen gave 8S

*From Edsall *et al.*[9]

aration was always found in the clot. It seemed clear that some of the material in the clot was not ordinary fibrinogen. The system behaved as if a stoichiometric complex was being formed between fibrinogen and the nonclotting protein. Only about one sixth of the protein in fraction I-1 sedimented, like pure fibrinogen, at 8S; most of it sedimented close to 14S, and about 8% at 19S. Clearly, most of the material in the clot came from soluble proteins that sedimented more rapidly than did pure fibrinogen. TABLE 3 summarizes the sedimentation patterns.

The only way we were able to obtain cold-insoluble globulin free of fibrinogen was to clot with thrombin, remove the clot, and precipitate the residual nonclotting protein, under the same conditions as those used to obtain the original fraction I-1, namely, at pH 6.3, 0°C, and an ethanol concentration of 5 ml/liter. Under these conditions, the solubility of the nonclotting protein, like that of fraction I-1, was close to 0.2 mg/ml. At room temperature, its solubility was about 100 times as great.

We studied these preparations by flow birefringence, observing the double refraction of the flowing solution between concentric cylinders, as we had done previously with fibrinogen, γ-globulin, and serum albumin.[10,11] FIGURES 3 and 4 show the results for the nonclotting protein. In both cases, the abscissa shows the product $G\eta$, where G is the velocity gradient in the flowing solution, and η is its

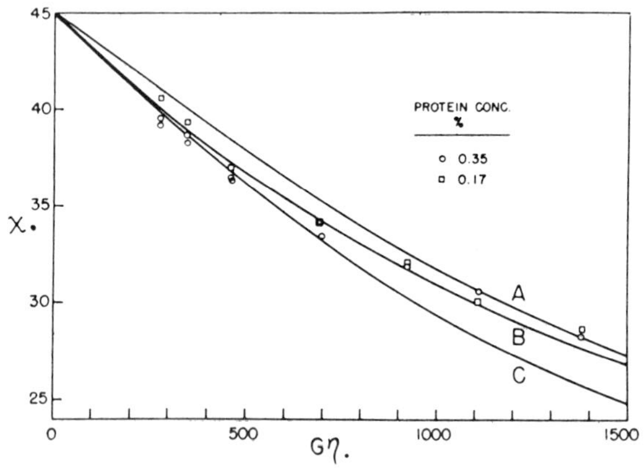

FIGURE 3. Extinction angle χ as a function of $G\eta$ for the nonclotting protein. The symbols are experimental data, while the curves are theoretical ones. Curve A is calculated for particles with θ = 312; curve B is calculated for a mixture of particles, 98% of which have θ = 312 and 2% of which have θ = 53; curve C is for particles with θ = 252. Theta is the rotary diffusion coefficient (sec^{-1}), G is the velocity gradient, and η is the viscosity.

viscosity. The solvent was a glycerol-water mixture that contained about 53% glycerol by weight; the glycerol increased the viscosity sufficiently to assure adequate orientation of the protein molecules in the velocity gradient. It did not apparently cause any significant change in the size or shape of the protein. The extinction angle χ in FIGURE 3 is the mean angle of orientation between the long axes of the protein molecules and the stream lines; it is 45° at very low gradients and shifts over gradually toward parallelism with the stream lines as the gradient becomes larger than the rotary diffusion coefficient of the molecules. From the data, assuming the molecule to be a moderately hydrated prolate ellipsoid of revolution, we calculated its length to be about 700 Å. Of course, you must remember that we were not dealing with a single pure protein; about 75% of it sedimented at 15S, about 13% at 22S, and the remainder at 8S, like fibrinogen (see TABLE 3). Also, we are not at all sure that the prolate ellipsoid model is correct; there are good reasons for doubting it in the case of fibrinogen, for which a variety of models have been proposed.[12]

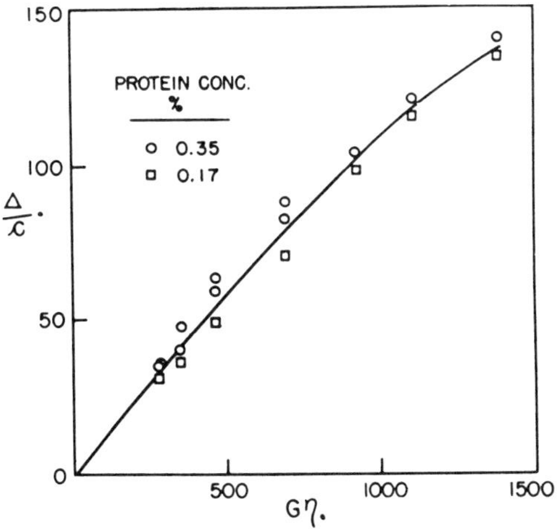

FIGURE 4. Birefringence data for the nonclotting protein. The ordinate gives the ratio of birefringence (Δ) to protein concentration (c).

FIGURE 4 shows that the birefringence/concentration ratio is a nearly linear function of $G\eta$. At the highest $G\eta$ values, it displays signs of leveling off as the fibrinogen molecules become better oriented in the gradient.

We never obtained cold-insoluble globulin that was really pure. The system was so complicated, and so difficult to interpret, that we hesitated to publish our findings. Finally, after seven years, we decided that it was such an interesting protein that we should definitely report the work; our paper appeared in 1955. One very valuable result of the research was that it stimulated Dr. Gilbert to develop an important theory of the behavior of macromolecules interacting in electrophoresis and ultracentrifugation; this theory has had wide application in protein chemistry. Perhaps he would have developed the theory anyway, but I am inclined to believe that puzzling over the unusual interactions of cold-insoluble globulin with fibrinogen was a major stimulus to the development. Indeed, in a recent letter, he has told me that my assumption was correct.

FIGURE 5. Geoffrey A. Gilbert and Mrs. Gilbert in Italy, June 1976.

Finally, a few words as to what happened later to some of the people in my story. Edwin Cohn, who had suffered for more than 20 years from high blood pressure, but had driven ahead with unflagging energy, died in 1953. Peter Morrison is a zoologist and physiologist, with a particular interest in the physiology of small mammals; to study them, his travels have taken him to South America, Australia, and other places. He has been for years a Professor at the University of Alaska and loves it. Geoffrey Gilbert (shown in FIGURE 5 with Mrs. Gilbert, at a meeting in Italy in 1976) is Professor of Biochemistry at the University of Birmingham in England and a Fellow of the Royal Society. I have already spoken of his work. Harold Scheraga (FIGURE 6) is Professor of Chemistry at Cornell University and is also associated with the Weizmann Institute of Science in Rehovot, Israel. He had performed much work on fibrinogen, fibrin, and throm-

FIGURE 6. Harold A. Scheraga.

bin; his major efforts are now devoted to understanding the three-dimensional structure of proteins and why the peptide chains fold up the way they do, a formidable problem for anyone to tackle but one in which he is making real progress. We were fortunate, in that unusual laboratory at Harvard Medical School, that these and other remarkedly gifted people became involved in the research on plasma proteins during that critical period.

References

1. Edsall, J. T. 1961. Edwin J. Cohn (1892–1953). Biogr. Mem. Nat. Acad. Sci. **35:** 47.
2. Cohn, E. J. 1941. Chem. Rev. **28:** 395.
3. Cohn, E. J. 1947. Experientia **3:** 125.
4. Edsall, J. T. 1946. Adv. Protein Chem. **3:** 383.
5. Preyer, W. 1871. Die Blutkristalle. Mauker's Verlag. Jena, Germany.
6. Cohn, E. J., L. E. Strong, W. L. Hughes, Jr., D. J. Mulford, J. N. Ashworth, M. Melin & H. L. Taylor. 1946. J. Amer. Chem. Soc. **68:** 459.
7. Ferry, J. D. & P. R. Morrison. 1947. J. Amer. Chem. Soc. **69:** 388, 400.
8. Morrison, P. R., J. T. Edsall & S. G. Miller. 1948. J. Amer. Chem. Soc. **70:** 3103.
9. Edsall, J. T., G. A. Gilbert & H. A. Scheraga. 1955. J. Amer. Chem. Soc. **77:** 157.
10. Edsall, J. T., J. F. Foster & H. Scheinberg. 1947. J. Amer. Chem. Soc. **69:** 2731.
11. Edsall, J. T. & J. F. Foster. 1948. J. Amer. Chem. Soc. **70:** 1860.
12. Doolittle, R. F. 1973. Adv. Protein Chem. **27:** 1.

STRUCTURE OF HUMAN PLASMA COLD-INSOLUBLE GLOBULIN AND THE MECHANISM OF ITS PRECIPITATION IN THE COLD WITH HEPARIN OR FIBRIN-FIBRINOGEN COMPLEXES*

Michael W. Mosesson

Department of Medicine
State University of New York
Downstate Medical Center
Brooklyn, New York 11203

Thirty years ago Morrison *et al.*[1] reported studies of the purification of human plasma fibrinogen. They characterized a component of their fibrinogen-rich fractions that was cold insoluble and, unlike fibrinogen, not thrombin coagulable. It also displayed a more rapid electrophoretic migration rate and higher sedimentation coefficient than did fibrinogen. This protein was named "cold-insoluble globulin." Several years later, Smith and von Korff[2] called attention to a protein with properties similar to those of cold-insoluble globulin, that they had found in a heparin-induced cold precipitate of normal plasma. The importance of this protein as a component of the heparin-precipitable fraction (HPF) of plasma has only recently been clarified.[3]

The first detailed physicochemical investigations of cold-insoluble globulin,[4] now abbreviated CIg,[5] led to the suggestion that it might be a dimer of fibrinogen modified in a way that had rendered it insusceptible to the action of thrombin (i.e., noncoagulable) yet somehow had left it able to form complexes with the fibrinogen in that fraction. Subsequently, Pollara,[6] mainly on the basis of NH_2 terminal analyses, suggested that CIg and fibrinogen were not different forms of the same protein.

More definitive clarrification of the nature of CIg began about 10 years ago with observations stemming from clinical investigation of a patient with a chronic intravascular coagulation syndrome,[7] whose illness was characterized by the persistent occurrence of a pathologic cold-induced plasma precipitate term "cryofibrinogen." The fact that the solubilized precipitate was coagulable with thrombin indicated the presence of fibrinogen; this presence was confirmed also by immunoelectrophoresis. The analysis also revealed another component in the fraction that migrated as a fast β-globulin and that was immunochemically related to or identical with a protein of normal serum. When immunoelectrophoresis was performed in the cold, most of the fibrinogen precipitated in the starting well with the β-globulin. However, precipitation of the β-globulin did not occur under these conditions if the fibrinogen had first been removed by clotting with thrombin. Review of the literature at that time suggested that the known properties of this globulin were like those of CIg, and this possibility stimulated several other investigations, the results of which form the basis for this review. In particular, I wish to summarize our present knowledge of the structure of CIg and discuss studies of its interactions with fibrin-fibrinogen complexes, fibrinogen, and hepa-

*Supported by National Heart and Lung Institute Grant HL-17419.

0077-8923/78/0312-0011 $01.75/0

rin. In so doing, I hope to draw the information together in a way that will connect early observations with more recent ones. The information should also provide a useful background for comparing the properties of CIg with those of a fibroblast surface protein (fibronectin, LETS) that has recently been shown to be antigenically identical with it.[8]

Isolation and Characterization of CIg

Investigations launched soon after the clinical study cited above[7] led to a method for obtaining CIg in a highly purified form,[9] virtually free of fibrinogen and other plasma proteins. The percentage recoveries of CIg after each of the main steps in the original purification procedure and in a recent modification[10] designed to improve the yield are summarized in Table 1. A thrombin defibrination step, which had been used previously[4,6] to separate fibrinogen from CIg, had certain disadvantages and was avoided. Immunochemical analyses with rabbit antiserum to the purified human protein[9] showed that the antigenic determinant(s) of CIg was distinct from those of fibrinogen and other plasma proteins; these and other physical and biochemical data, contrary to earlier speculations,[4] provided unequivocal evidence that CIg was not a modified form of fibrinogen.

Other data permitted an outline of its basic features. Carbohydrate analyses plus periodic acid-Schiff staining of CIg-containing polyacrylamide gels indicated that CIg was a glycoprotein[11] (Table 2). There was no sulfhydryl group detectable[13]; thus, virtually all of the cysteine residues in the molecule are involved in disulfide bridges. Moreover, the integrity of at least some of these bridges was necessary for maintaining solubility in physiologic buffers, because the addition of dithiothreitol (DTT) to such solutions resulted in precipitation.[11]

Table 1

Recovery of CIg During Purification from Plasma*†

Plasma Fraction or Step	Recovery (%) Original Method[9]	Modified Method[10]
ACD plasma	100	100
After barium absorption (step 1)	—	93
Cryoprecipitate	25‡	—
Cohn fraction I (step 2)	—	77
Cohn fraction I–O (step 3)	—	68
Extracted cryoprecipitate	18	—
Precipitate from the glycine supernatant solution (step 5)§	11	41
After chromatography (steps 6 & 7)	3	19¶

*Data from Reference 10.

†In each case, the data presented are the measurements from a single preparative experiment.

‡The amount precipitated with this fraction is approximately 60% of that in plasma. However, it is usually difficult to achieve complete dissolution of the fraction, an effect that results in considerable reduction in recovery.

§CIg is a minor constituent of the fractions prior to this step, which serves to remove most of the major contaminant protein, fibrinogen.

¶The mean recovery of CIg from five separate preparative experiments, including this one, was 15% (range, 11–19%).

TABLE 2

AMINO ACID AND CARBOHYDRATE COMPOSITION OF CIg*

Residue	Moles per 10^5 Grams of Protein†‡
Aspartic acid	85.3
Serine	62.6
Threonine	89.1
Glutamic acid	107.0
Proline	70.1
Glycine	73.5
Alanine	39.5
Half-cystine	24.2
Valine	74.3
Methionine	10.3
Isoleucine	40.6
Leucine	52.6
Tyrosine	41.8
Phenylalanine	25.0
Lysine	33.6
Histidine	19.1
Arginine	47.6
Tryptophan	25.5§
Amide-N	128.0
Sialic acid	3.8
Hexose	10.0
Hexosamine	11.8

*Data from Reference 11.
†The amino acid composition of CIg reported by Mosher[12] is similar to that given here.
‡Mean of duplicate analyses.
§Calculated from previous determination[9] of the tyrosine/tryptophan ratio.

Amino terminal analysis revealed that the NH_2 terminal group could not be labeled with dansyl chloride[11] but could be removed by pyrrolidone carboxylyl peptidase, a bacterial enzyme specific for NH_2 terminal pyroglutamic acid residues.[14] Following this step, a single NH_2 terminal acid was found, alanine, thus suggesting that all chains in the preparation began with the sequence pyroglutamyl-alanyl. Acidic urea or alkaline urea gel electrophoretic analyses revealed no evidence of major charge differences among the reduced components of the preparation,[11] although other information summarized in the next two sections clearly indicates that the chains are not alike in all respects.

Studies of the dichroic spectrum[11] at 25°C (FIGURE 1) revealed three distinct bands at 229, 215, and 201 nm. The band pattern provided no indication of the negative doublet characteristic of an α-helix.[15-17] The presence of a positive band at 201 nm and a negative band at 215 nm suggested that a considerable proportion of the molecule possessed β-structure. The complexities of such secondary structure in proteins (cf. References 17 and 18) precludes a more accurate assessment of the data without more information. Similarly, elucidation of the nature of the bathychromic shift of the band at 201 nm relative to that described for other polypeptides possessing β-structure such as poly-L-lysine,[15] poly-L-serine,[18] or silk fibroin,[19] awaits further study.

The band at 229 nm was asymmetric and manifested a "shoulder" at 235 nm. The nature of the band in this region is unclear. The peak seems to be positioned too far toward the longer wavelengths to be attributable to the presence of random

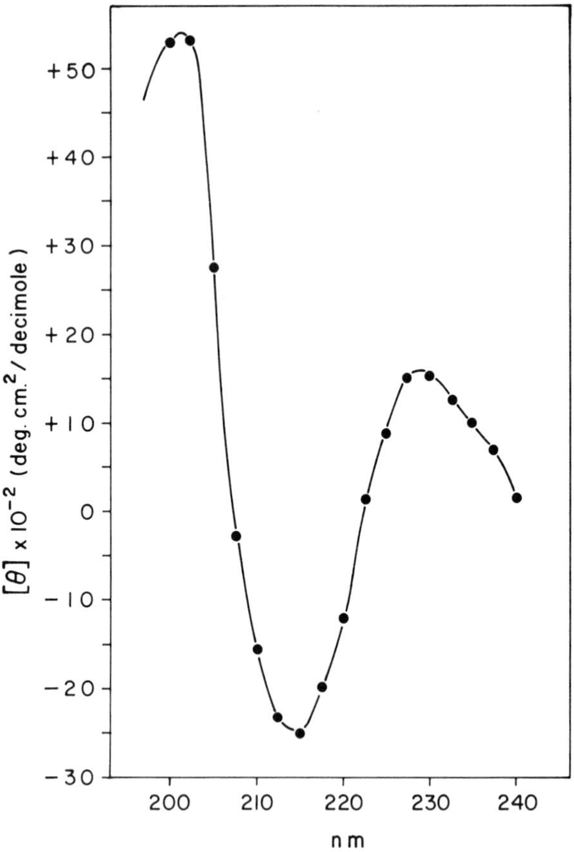

FIGURE 1. Circular dichroic spectral analysis of CIg at 25°C in 0.1 M Tris-phosphate buffer (pH 7.0). Ellipticity (θ) is plotted against wavelength (nm); the assumed mean residue weight was 104. Recent unpublished dichroic spectral analyses, performed in collaboration with Dr. Nancy Tooney, have confirmed the qualitative nature of the features shown in this Figure but indicate that the magnitudes of the bands are only approximately 1/10 of that reported. Drs. Ken Yamada and Steve Alexander[60] have made similar observations with respect to the band magnitude. Thus, the previous calculations probably were in error by a factor of about 10. Accordingly, the ellipticity scale on the ordinate has been changed from 10^{-3} to 10^{-2}. (Adapted from Reference 11.)

coil. It may reflect the presence of certain aromatic side chain chromophores,[17,20,21] may be due to asymmetric disulfide chromophores,[17,22,23] or both; some contribution may also arise from β-structure itself.[17,24]

COVALENT STRUCTURE OF CIg

As had been the case with plasma fibrinogen (cf. References 25 and 26), analyses of even the best CIg preparations were complicated by evidence of inhomogeneity. Despite such problems, a definitive outline of the molecular architecture of CIg can be sketched, and a preliminary understanding of the nature of the inhomogeneities has also emerged.

Sedimentation equilibrium centrifugation experiments with CIg always indicated the presence of a major large component. Computation of molecular weight from such experiments was made more difficult by the nonlinearity of the log J vs r^2 plot, the slope of which is proportional to the molecular weight (FIGURE 2). Such nonlinearity suggested the presence of small amounts of low-molecular-weight material (~1–5%). The data could be normalized for the

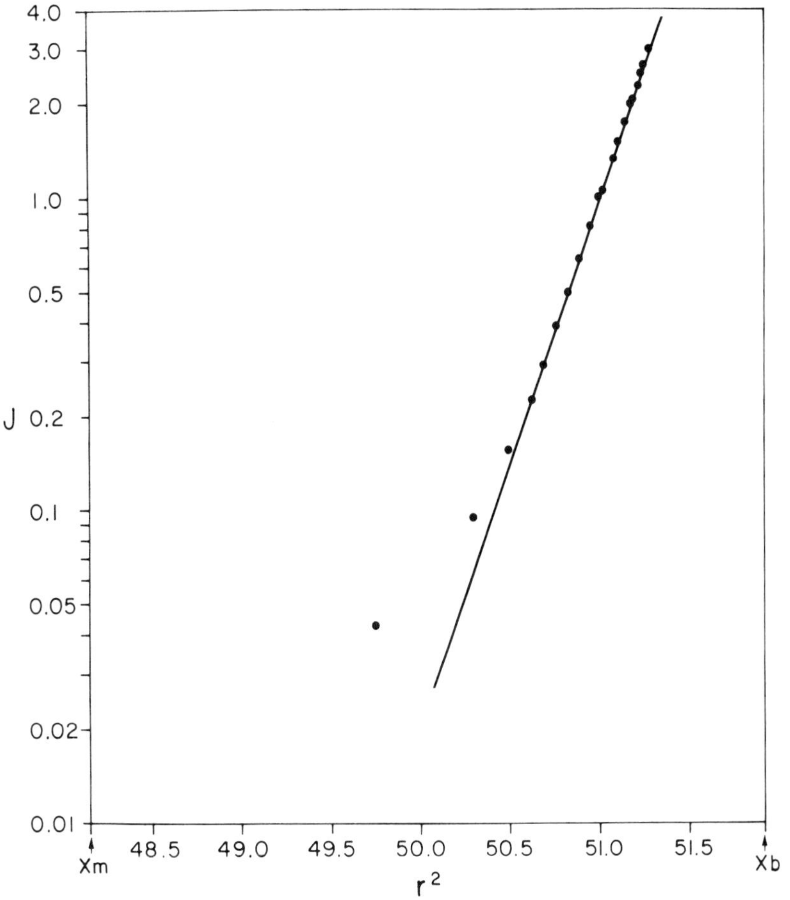

FIGURE 2. Sedimentation equilibrium analysis of CIg (1 mg/ml) at 20°C in 0.1 M Tris-phosphate buffer (pH 7.0) at 12,000 rpm. Fringe displacement (J) is plotted against the square of radial displacement (r^2). The meniscus (Xm) and cell bottom (Xb) are indicated. Even though corrections were not required in calculating the molecular weight in this particular experiment, the nonlinearity of the plot in the low-fringe regions is evident. The line drawn through the data points corresponds to a molecular weight of 454,000 (TABLE 3). (From Mosesson *et al.*[11] By permission of *Biochimica et Biophysica Acta.*)

TABLE 3

MOLECULAR WEIGHT OF CIg BY SEDIMENTATION EQUILIBRIUM ANALYSIS*

Experiment	Molecular Weight †‡
1	463,000
2	454,000
3	442,000
4	451,000

*Data from Reference 11.

†$\bar{v}$ = 0.72 cm^3/g (value calculated from amino acid and carbohydrate analysis or determined directly by pycnometry).

‡0.1 M Tris-phosphate buffer (pH 7.0).

presence of lower-molecular-weight components by performing analyses at two speeds and applying corrections in the manner used for the light and heavy components of skeletal myosin.[27] Following such manipulations, we concluded that the molecular weight of the major high-molecular-weight component was 450,000 ± 25,000 (TABLE 3).

Dodecyl sulfate gel electrophoresis of unreduced preparations (FIGURE 3) showed that the anodal migration rate of the main CIg band was less than that of unreduced fibrinogen fraction I-4 (mean mol wt 340,000; see Reference 28); this behavior was consistent with the molecular weight estimate of 450,000 from ultracentrifugal experimental (TABLE 3). At relatively high protein loads (15 μg or higher), a portion of the unreduced samples aggregated and remained at the top of the gel, even in the presence of 4–5 M urea or 1% sodium dodecyl sulfate or both. This phenomenon made it difficult to intensify minor bands simply by increasing the sample load (compare gels 1 and 2, FIGURE 3).

After reduction of disulfide bridges with DTT, the material no longer dis-

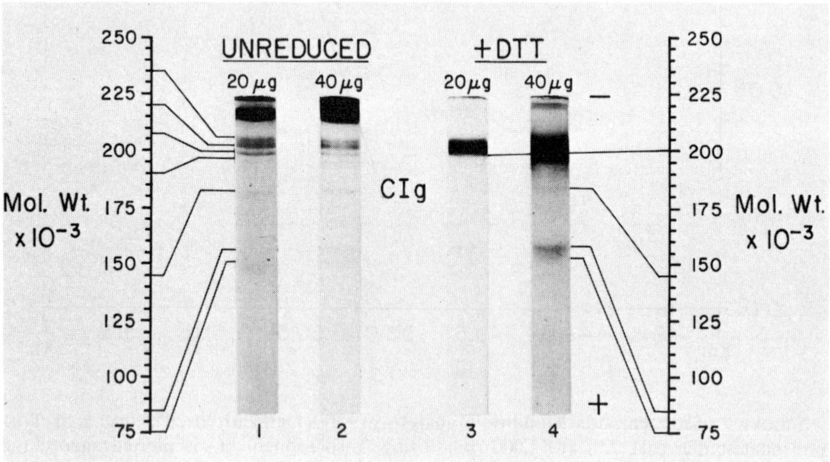

FIGURE 3. Dodecyl sulfate gel electrophoresis of unreduced and reduced CIg. The final acrylamide concentration in this and other such analyses was 5%, unless otherwise stated. The sample load is given above each gel. The molecular weights of certain bands are indicated by the arrows drawn from a given band to the scale on the right or left of the Figure.[61]

played a load-dependent aggregation tendency and migrated in a more anodal position, corresponding to a molecular weight of about 220,000 (FIGURE 3; see References 11 and 12). The latter observation indicated that each CIg molecule of approximate molecular weight 450,000 was composed of two disulfide-bridged chains of about the same size (FIGURE 4). A similar structure can be proposed for the bovine CIg molecule.[30,31] Additional support for this covalent structure in the case of human CIg was drawn from electrophoretic or sedimentation velocity experiments conducted before and after reduction of disulfide bridges.[11] Furthermore, data obtained from plasmic digests of CIg,[13,29] as presented in the next section, suggest that the disulfide bridging is within about 175 residues of the NH_2 or COOH terminus of the molecule. Of particular interest with regard to the

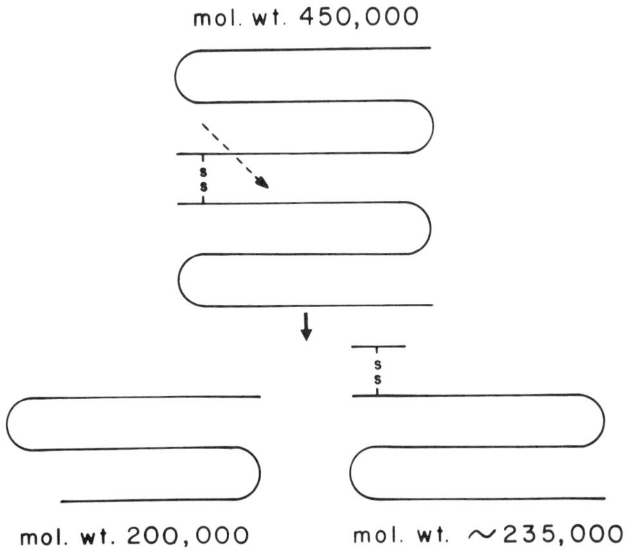

FIGURE 4. Proposed primary structure of human CIg. Most circulating molecules are composed of two chains of approximate molecular weight 220,000 each, linked near the NH_2 or COOH terminus by disulfide bridges (-ss-). The dashed arrow indicates an early site of *in vitro* plasmic attack [13,29]; the products of such attack on the two-chain molecule are shown in the lower portion of the diagram.

dimeric nature of CIg structure is the recent demonstration by dodecyl sulfate gel electrophoresis that the cell surface form of this protein also exists in a covalently bridged dimeric form.[32,33] However, in both of these studies, considerable amounts of material did not enter the gels, thus indicating the presence of aggregates of very high molecular weight. Whether this phenomenon reflects aggregation due to noncovalent forces, similar to that observed for CIg (FIGURE 3), or whether it reflects the presence of disulfide-bridge multimers, or both possibilities, remains to be determined.

The question of whether the subunits that comprise disulfide-bridged molecules are identical cannot be answered without introducing certain qualifying reservations, such as those covered in the next section. Nevertheless, the available evidence,[11] though meager, suggests that the two peptides that comprise any given dimeric molecule represent the same gene product and differ from one another only in minor ways.

Heterogeneity of CIg

Evidence that most CIg molecules have a two-chain covalent structure was summarized in the preceding section. Nevertheless, plasma CIg,[11,13] the material found in amniotic fluid,[5] and that from tissue cultures[8,33-37] have all given evidence of size heterogeneity. A summary of recent studies on the molecules found in plasma follows.

Gradient elution chromatography is the final step applied in purifying CIg (Table 1). Investigation of the CIg-containing peak by dodecyl sulfate gel electrophoretic analysis (Figure 5) of fractions that comprise the ascending and middle portions of the peak (fraction A) and of those that comprise the descending portion (fraction B) revealed a major band that corresponded to the two-chain disulfide-bridged molecule. This band was occasionally resolved into a closely spaced doublet. Several fainter, more anodal bands were also evident; the most prominent ones migrated in zone II.

Only minor differences were observed between the electrophoretic band patterns of unreduced fractions A and B, but differences became clearly evident when reduced samples were compared. Reduced fraction A displayed two prominent zone-II bands, the most intensely staining of which was that of molecular weight 220,000. Reduced fraction B also possessed two prominent zone-II bands, but the smaller one, of molecular weight 200,000, was considerably more intense in this fraction than in fraction A. These observations suggested that the smaller chains found in CIg preparations tend to be more anionic.

Previous immunochemical analyses of our CIg preparations had indicated a high degree of purity with respect to the presence of proteins other than those that possess CIg antigenic determinants.[9,10,38] This finding implies that virtually all bands observed in electrophoretic experiments reflect the presence of different forms of CIg. This conclusion was strengthened by crossed immunoelectrophoretic analyses that revealed a single biphasic CIg precipitin arc overlapping electrophoretic zones I–III.[13] Semiquantitative immunochemical analyses with antiserum absorbed with material from zone I or zones II and III strengthened this conclusion further, because, on a weight basis, material from zones II and III was shown to have at least as much antibody-binding potential as that from zone I.

Several converging lines of evidence indicate that material that migrates in electrophoretic zone I could not have given rise to zone-II material (or vice versa) by disulfide exchange prior to or during preparation for electrophoresis. First, there was no change in the electrophoretic band pattern of unreduced samples after they had been incubated prior to analysis for up to 18 hr at 37°C or for 2 hr at 56°C[13] at pH 7 in 0.1 M Tris-phosphate buffer. Second, unreduced samples of CIg that had been subjected to electrophoresis in dodecyl sulfate-containing gels, then sliced transversely and each slice (without reduction of disulfide bridges) again subjected to dodecyl sulfate gel electrophoresis, showed that material from zone I migrated in the same relative position during the second analysis. Similarly, slices from zone II did not reveal bands migrating in the zone-I position. Third, the fact that CIg preparations contain no detectable sulfhydryl group[13] minimizes the possibility that such an exchange might occur. Finally, results from gel slicing experiments with alkaline-urea gels or those obtained with CIg fractions separated

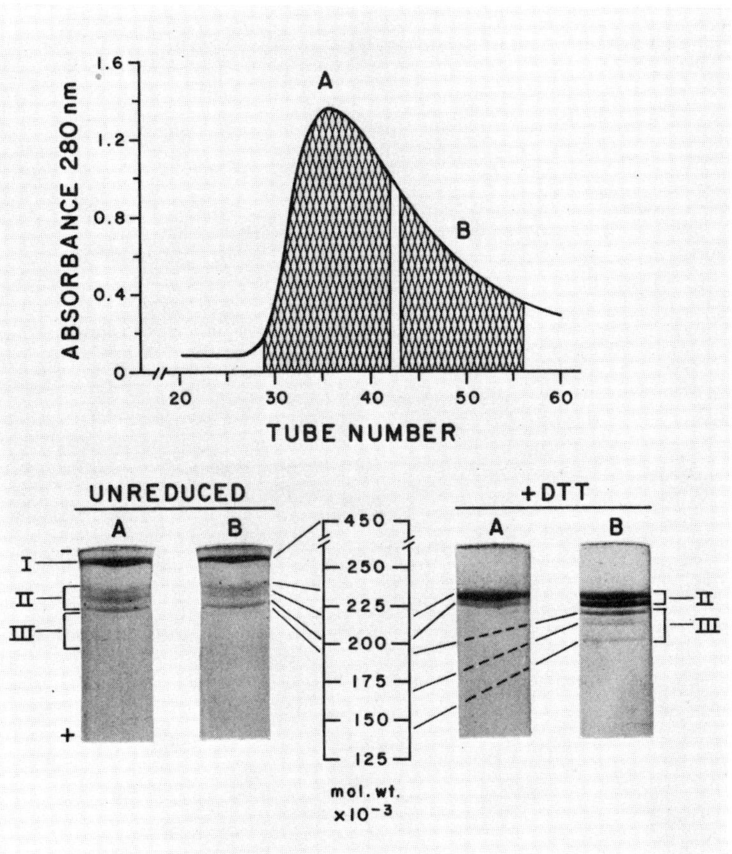

FIGURE 5. DEAE-cellulose linear gradient elution chromatography of CIg prepared by a modification[10] of the original method.[9] Pooling of eluate fractions (A and B) is shown by the shaded areas. Dodecyl sulfate gels of unreduced and reduced samples from each pool are shown in the lower portion of the Figure. For purposes of descriptive clarity, electrophoretic bands are grouped in three regions designated as zone I (two-chain disulfide-bridged molecules of approximate molecular weight 450,000), zone II (apparent mol wt, 200,000–235,000), and zone III (apparent mol wt, 146,000–190,000). The assignment of a molecular weight of 450,000 to material that migrates in zone I is based on determinations made by analytic ultracentrifugation.[11] Molecular weight assignments for bands in zones II and III are based on their anodal migration rates relative to reduced marker peptides. (From Chen *et al.*[13] By permission of *Biochimica et Biophysica Acta.*)

by molecular exclusion chromatography[13] indicate that the proteins that migrate in the different electrophoretic zones do not ordinarily interchange positions.

Available evidence also makes it unlikely that the size heterogeneity can be attributed to proteolysis occurring during collection or subsequent processing of blood.[10] The same type of heterogeneity as is found in purified material[10,13] could also be found in freshly drawn citrated blood (FIGURE 6) by analyzing the immunoprecipitate that resulted from the addition of anti-CIg serum. The electro-

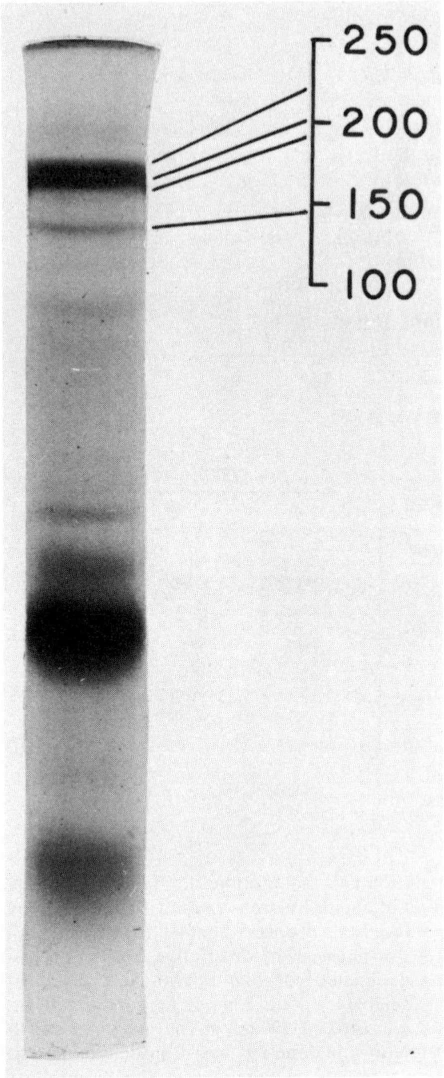

FIGURE 6. Dodecyl sulfate gel electrophoresis of reduced material from an immunoprecipitate of fresh single donor citrated plasma containing Kunitz' pancreatic trypsin inhibitor and hirudin. The reduced light and heavy chains of rabbit IgG appear as heavily stained bands that occupy more anodal positions relative to bands attributable to CIg; other faintly-staining unidentified bands are also present between the positions of the band of molecular weight 146,000 and the IgG heavy chain.[62]

phoretic pattern of the washed precipitate after reduction with DTT revealed the presence of the zone-II bands of molecular weights 220,000 and 200,000, respectively, plus zone-III bands (including that of molecular weight 146,000) that corresponded to those previously observed in purified CIg preparations (cf. FIGURES 3 & 5). Interestingly, unreduced immunoprecipitates of CIg from plasma became aggregated and entered the dodecyl sulfate gels poorly, as did a control immunoprecipitate prepared from purified CIg. These aggregated immunoprecipitates could not be satisfactorily dissociated, even after heating in a boiling water bath for several minutes in the presence of sodium dodecyl sulfate and

4–5 M urea. Tentatively, we attribute this aggregation phenomenon to noncovalent binding.

In vivo proteolysis at an extravascular site or within the circulation probably accounts for a significant proportion of the size heterogeneity we have observed. This question has been explored to some extent by determining whether digestion with plasmin can produce derivatives *in vitro* that resemble those found in plasma (FIGURE 7); such experiments have also permitted deduction of the location of the interchain disulfide bridges within the two-chain molecule.

Following the addition of plasmin, unreduced digest samples displayed an increase of material in the 200,000 molecular weight position (zone II) and of material that migrated in zone III. This was accompanied by depletion of zone I and the 235,000 and 220,000 molecular weight positions of zone II. No bands appeared between zones I and II at any phase of digestion.

The gels of reduced digest samples showed that material that occupied the 220,000 molecular weight position became progressively depleted, whereas that in the 200,000 molecular weight position increased during early and intermediate phases of the hydrolysis. There was, at most, a small increase in material that migrated in the 146,000 molecular weight position at any phase of hydrolysis.

After zone-I molecules had been consumed, the band patterns of reduced and unreduced samples became very similar. This digestion pattern suggests that proteolytic attack on two-chain molecules results in separation of a fragment that contains the interchain disulfide bridges that join the two chains. Since monomeric peptides released from the disulfide-bridged parent molecule can have a molecular weight as large as 200,000, this indicates that the disulfide bridging is near the NH_2 or COOH terminus of the molecule. Jilek and Hörmann[29] have drawn similar conclusions. The conclusion has been strengthened by the additional finding that during digestion, depletion of material from zone I is accompanied by the evolution of a disulfide-bridged fragment with an apparent molecular weight of about 23,000.[59]

Components of CIg preparations that have been characterized thus far, other than those of approximate molecular weight 450,000, range in size from 146,000 to 235,000. Even smaller components probably exist (FIGURE 3), but detailed information is not yet available. Polypeptides of unreduced preparations that migrate in electrophoretic zones II and III appear to be monomeric, with the important exception of the one of molecular weight 235,000. The fact that chains of molecular weight 220,000 are components of zone II and of zone I (FIGURES 3,5&8) suggests that chains of this size are either linked to a second chain and migrate in zone I or, less commonly, exist in a monomeric form and migrate in zone II.

The 235,000 molecular weight component most likely represents an *in vivo* proteolytic derivative formed from two-chain molecules and is comprised of an "intact" chain of molecular weight 220,000 covalently linked to a smaller remnant. This conclusion is based on the following reasoning: First, after gel slice analysis of reduced material from zone II, the largest electrophoretic component from the 235,000 molecular weight position has a size of only 220,000 (FIGURE 8). Second, results of *in vitro* plasmic digestion of CIg (FIGURE 7) suggests the mechanism whereby such derivatives might arise *in vivo*. That is, cleavage and release of a fragment of molecular weight 200,000 from a disulfide-bridged two-chain parent structure would result in a two-chain derivative species with an approximate molecular weight of 235,000 (FIGURE 4). Indeed, monomeric chains of molecular weight 200,000 are known to coexist in plasma with those of molecular weight 235,000 (FIGURES 3 & 5).

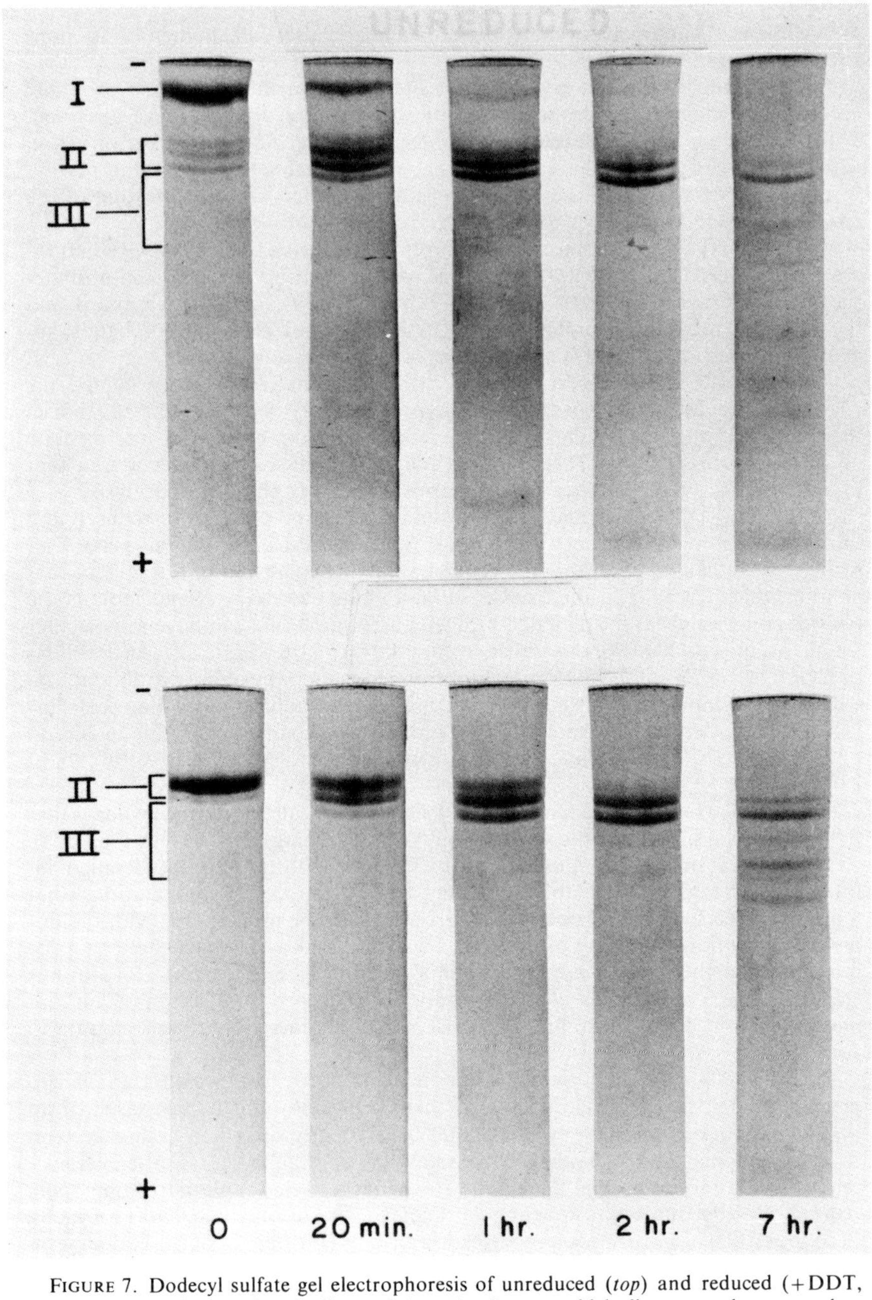

FIGURE 7. Dodecyl sulfate gel electrophoresis of unreduced (*top*) and reduced (+DDT, *bottom*) samples from a plasmic digest of CIg. The times at which digest samples were taken are indicated beneath the lower set of gels. (From Chen *et al.*[13] By permission of *Biochimica et Biophysica Acta.*)

A dimeric structure has also been found for the cell surface form of the protein,[32,33] but the possible presence of other molecular forms has not been rigorously investigated. It is useful to point out in this regard that a minor component of our plasma CIg preparations (mol wt 146,000) corresponds in size to a fibronectin-related peptide that was discovered in extracts of cells grown in culture.[8,34–37] Vaheri *et al.*[35,36] found that [^{35}S]methionine was incorporated into this peptide particularly well, suggesting that it might be synthesized independently from the larger chains and that it was not simply a proteolytic derivative of a larger chain. New information on this subject is needed to clarify these relationships.

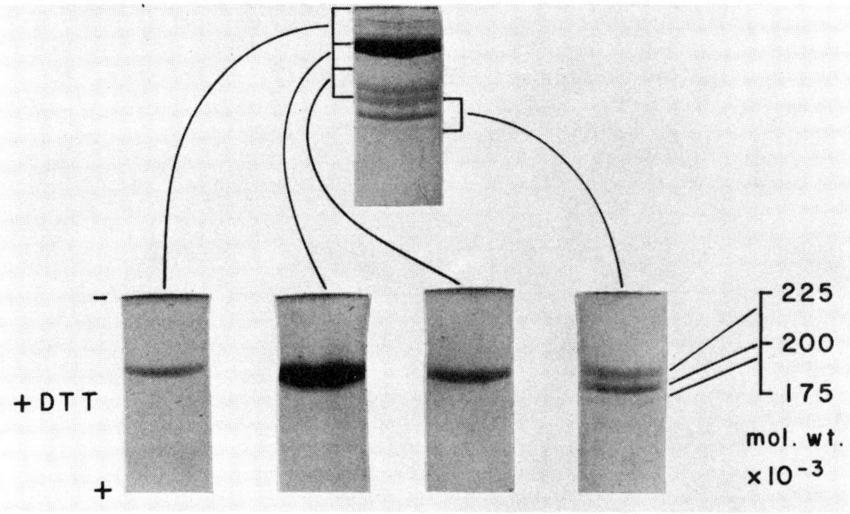

FIGURE 8. Dodecyl sulfate gel slice analysis of a sample of unreduced CIg (*top*); a duplicate gel was sectioned transversely, and the homogenized slices were reduced and again subjected to dodecyl sulfate gel electrophoresis (*bottom*); the region of the upper gel from which the slices were obtained is indicated.[63]

INTERACTIONS WITH FIBRIN, FIBRINOGEN, AND HEPARIN

Many investigations, beginning with those of Edsall *et al.*,[4] have shown that CIg has an affinity for fibrinogen and fibrin.[7–9,12,39–41] Thus, even though CIg is not itself thrombin coagulable, some of it is usually incorporated in the fibrin clot formed from plasma, and the resulting serum CIg level is somewhat lower than that in plasma.[8,9]

The recognition that CIg was a component of the HPF is attributable to Smith and Von Korff.[2] More contemporary studies[3,42,43] have provided even better evidence that this is the case. CIg is also a component[7] of another type of fibrinogen-rich cold-induced plasma precipitate termed "cryofibrinogen"; this type of precipitate is observed under certain pathologic circumstances and does not require heparin addition for induction. Until recently,[3,44] it was not known what role CIg played in the formation of HPF or "cryofibrinogen" and in what ways these two types of cold-induced plasma precipitates were alike or different. The following section summarizes investigations designed to resolve these questions.

The importance of CIg in the heparin-induced cyroprecipitation phenomenon has been examined.[3] Heparin was shown to form a cold-precipitable complex with purified CIg or with mixtures of CIg and fibrinogen but not with fibrinogen alone. Consistent with the earlier findings of Smith and von Korff,[2] cryoprecipitation could be augmented by addition of Ca^{2+} or by selection of optimal heparin levels; it could be reduced and even abolished by raising the ionic strength or pH or both or by raising the heparin concentration above that required for maximum precipitation of CIg. The presence of fibrinogen reduced the threshold for heparin-induced CIg cryoprecipitation and, by coprecipitating with heparin and CIg, increased the amount of precipitate that formed. In contrast to the heparin-precipitable fraction of normal plasma, which contained both fibrinogen and CIg, that from a patient with congenital afibrinogenemia contained CIg but lacked fibrinogen (FIGURE 9). Normal plasma depleted of CIg by immunoabsorption, but possessing a normal level of fibrinogen, failed to form a heparin-induced cryoprecipitate.

Based on the reported affinity of fibrinogen or fibrin for heparin,[45,46] it had been suggested previously[42] that the interaction between heparin and fibrinogen

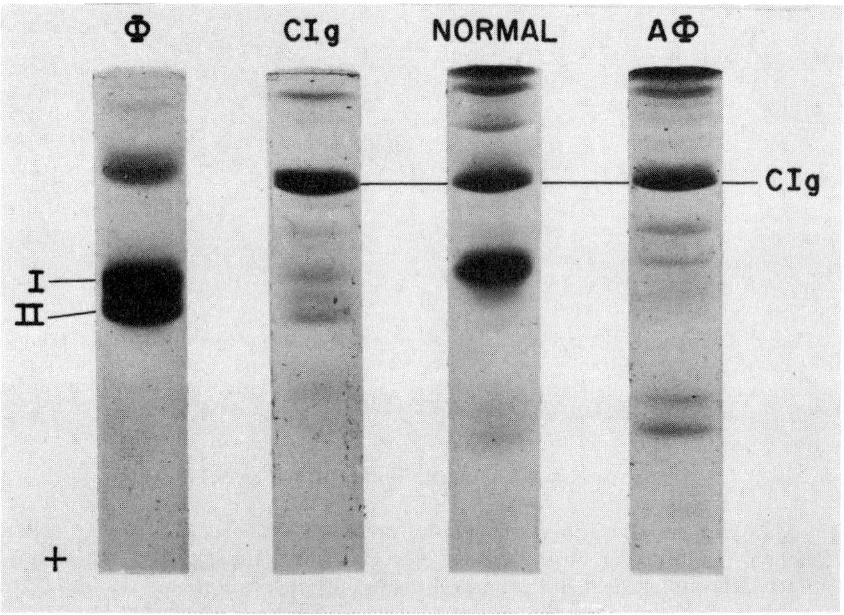

FIGURE 9. Dodecyl sulfate gel electrophoresis of unreduced (3.5% acrylamide) samples of fibrinogen fraction I-2(Φ), purified CIg (CIg), the HPF from normal plasma (normal) or from afibrinogenemic plasma (AΦ). Two major fibrinogen bands (I and II) are discernible in the fraction I-2 sample, in addition to small amounts of CIg. The fibrinogen bands differ with respect to their Aα chain populations.[26] Band I contains molecules with a very high proportion of intact Aα chains. In band II, a considerable proportion of the intact Aα chains (mol wt, 70,900) are replaced by a remnant chain termed Aα/$_4$ (mol wt, 57,700) that lacks a COOH terminal peptide, accounting for this difference in size.[26] (From Stathakis & Mosesson.[3] By permission of *Journal of Clinical Investigation.*)

TABLE 4

COMPARISON OF MINIMUM REQUIREMENTS FOR INDUCTION OF TWO TYPES OF FIBRINOGEN-RICH COLD-INSOLUBLE PLASMA PRECIPITATES

Component	Required for	
	HPF	"Cryofibrinogen"
CIg	yes	yes
Heparin	yes	no
Fibrin-fibrinogen complexes	no	yes

was crucial for formation of the HPF and that CIg was incorporated in the fraction by virtue of its affinity for fibrinogen. Nevertheless, the investigations summarized above provide compelling evidence for a different sequence of interactions. They indicate that CIg, not fibrinogen, plays the central role in mediating the formation of a heparin-induced plasma precipitate, as represented by the following formula:

$$\text{heparin} + \text{CIg} \underset{22^\circ\text{C}}{\rightleftarrows} \text{soluble heparin} \cdot \text{CIg complexes}$$
$$22^\circ\text{C} \uparrow \downarrow 2^\circ\text{C}$$
$$\text{heparin} \cdot \text{CIg precipitate}$$

Fibrinogen is neither essential for heparin-induced precipitation (FIGURE 9) nor does fibrinogen itself form a cold-precipitable complex with heparin under the conditions employed.[3] It appears that fibrinogen is incorporated in the HPF and augments the process mainly, if not entirely, by virtue of its affinity for CIg. The observations that the anticoagulant activity of heparin per se does not seem to be required for precipitate induction[3] and that such a precipitation phenomenon can even be produced by sulfated mucopolysaccharides that have little or no anticoagulant potential[2] indicate that it would be of considerable interest to explore the binding relationships between cellular glycosaminoglycans and cell surface proteins like CIg.

Fibrinogen is known to be a major constituent of the HPF from normal and of that from pathologic plasma.[2,43,47,48] Early observations led to the inference that the fibrinogen in the fraction might have been altered in some way,[2] perhaps by having been converted to a form of fibrin. Inasmuch as fibrin can circulate in a soluble form in blood under pathologic conditions[49,50] and, perhaps, to a limited extent, even in normal subjects,[51] it might be included as a significant component of the HPF under proper circumstances. Nevertheless, it is clear from recent studies[3,48] that fibrin is not an obligatory component of the HPF from normal plasma (TABLE 4). In fact, the fibrinogen in the HPF appears to be as close to the "native" form of the molecule as can be found in blood. That is, electrophoretic analyses indicate that the fibrinogen in this fraction is characterized by the presence of intact Aα chains (i.e., band I, FIGURE 9). Partially degraded circulating fibrinogen catabolites that lack the regions that contain the COOH terminal portion of the Aα chain, even though present in plasma[26,52,53] or in the mixtures of CIg and heparin, failed to participate in formation of such cold-precipitable complexes (FIGURES 9 & 10). The failure of fibrinogen catabolites that migrate in electrophoretic band II or beyond to participate in heparin-induced precipitation implies that the region of the fibrinogen Aα chain that accounts for precipitation with CIg is contained within a segment that represents the difference in size

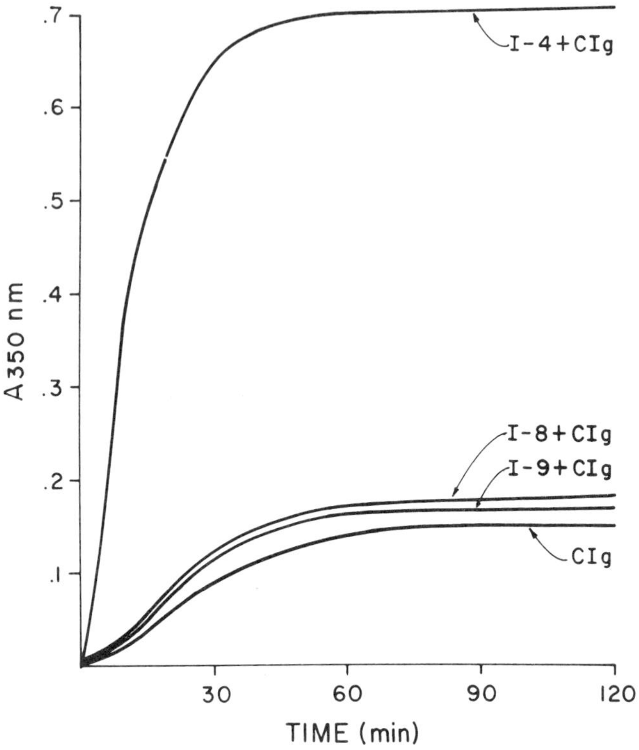

FIGURE 10. Heparin-induced cryoprecipitation of CIg in the absence or presence of various fibrinogen fractions. The rate and extent of precipitation at 2°C were monitored spectrophotometrically at 350 nm. In contrast to fibrinogen fraction I-4, which contains a high proportion of intact Aα chains, virtually all molecules that comprise fractions I-8 and I-9 lack the COOH terminal region of that chain.[26] (From Stathakis & Mosesson.[3] By permission of *Journal of Clinical Investigation.*)

(mol wt ~ 13,200) between the intact Aα chain and the principal Aα chain remnant that characterizes the fibrinogen that migrates in the band-II position (see the legend to FIGURE 9). These results do not, however, permit us to distinguish whether two such binding sites per molecule (or just one) are required for precipitate formation, whether the CIg-binding site is actually contained within this segment, or, alternatively, whether the binding site(s) is situated elsewhere in the molecule but is masked by conformational rearrangements that result from the loss of that region. It is useful to emphasize here that in addition to its importance for mediating binding between CIg and fibrinogen, the COOH terminal region of the Aα chain also contains a site(s) that is utilized during factor XIIIa-catalyzed cross-linking of fibrin α-chains[54-56] and, possibly, during cross-linking with CIg.[12]

This type of investigation has been extended to include studies of the mechanism by which soluble complexes of fibrin and fibrinogen [fibrin(ogen)] are rendered insoluble in the cold in the presence of CIg.[44] This type of precipitate is of particular interest because it resembles a pathologic cold-induced plasma precipitate termed "cryofibrinogen."[7] Precipitates like cryofibrinogen can be induced in normal plasma after the direct addition of small amounts of thrombin or thrombin-like enzymes or in plasma samples taken after the infusion of agents that enhance fibrin formation[44,49,50,57] (FIGURE 11).

It has been established for some time that in fibrin(ogen) solutions, whether in plasma or not, fibrinogen serves to maintain the fibrin in a soluble state.[49,50] Such fibrin(ogen) solutions failed to form cold-insoluble precipitates, even when "satu-

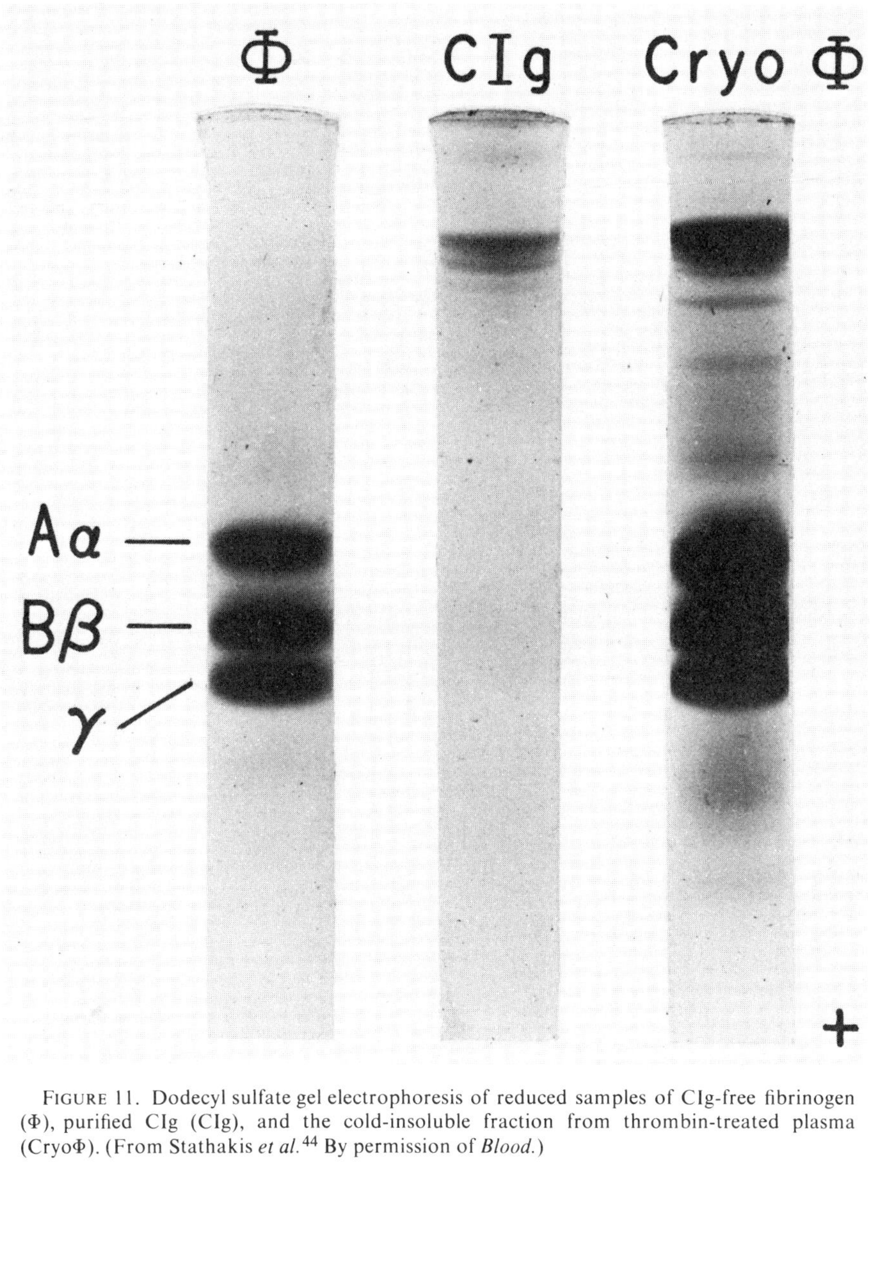

FIGURE 11. Dodecyl sulfate gel electrophoresis of reduced samples of CIg-free fibrinogen (Φ), purified CIg (CIg), and the cold-insoluble fraction from thrombin-treated plasma (CryoΦ). (From Stathakis *et al.*[44] By permission of *Blood.*)

rated" with fibrin, unless CIg was present or had been added to the mixture. Thus, CIg is indispensable (TABLE 4) for induction of cryoprecipitation, even though its presence is certainly not a requirement for fibrin-fibrinogen complex formation itself. Fibrin is also essential for producing this phenomenon, because precipitation of mixtures of CIg with fibrin-free fibrinogen did not occur under the same conditions (FIGURE 12). The absolute requirement for fibrin in this system stands in sharp contrast to the lack of such a requirement for forming the HPF.[3,48]

When precipitation in the cold does occur, fibrinogen presumably becomes incorporated in the fraction by virtue of the soluble complexes it had formed with fibrin. However, since fibrinogen has an affinity for CIg, one that is particularly

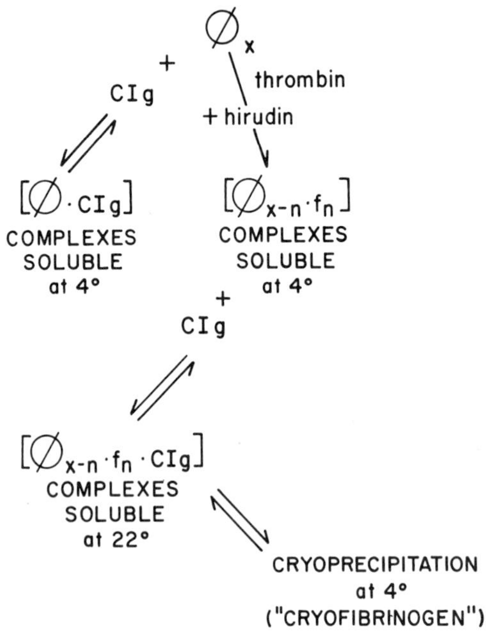

FIGURE 12. Schematic diagram that summarizes the interactions that occur among the three main components of "cryofibrinogen": fibrinogen (ϕ), fibrin (f_n), and CIg. Initial fibrinogen concentration, x; fibrin concentration, n. The ratio of $x - n/n$ under saturating conditions is in the range of 4:1.[50,58] The fibrin(ogen)/CIg ratio at the onset of CIg-induced cryoprecipitation is in the range of 20:1 or more,[44] suggesting that CIg behaves as a nucleus with multiple binding sites for participating fibrin-fibrinogen complexes.

evident in the cold,[44] it is probably incorporated in the complex on the basis of this interaction and the one with fibrin.

The interaction of CIg with thrombin-treated fibrinogen subfractions that lack the COOH terminal portion of the Aα chain (i.e., fractions I-8 and I-9) has also been examined. The results extend conclusions that relate to the importance of this region of the molecule for mediating binding between CIg and fibrinogen in the HPF.[3] Thrombin-treated fractions I-8 and I-9 were shown to form soluble fibrin-fibrinogen complexes,[44] but they failed to form cold-insoluble precipitates in the presence of CIg. Thus, molecules that contain the COOH terminal region of the Aα (or α) chain must be present for CIg and fibrin-fibrinogen complexes to precipitate together as cryofibrinogen.

Concluding Remarks

The foregoing presentation has summarized the present state of knowledge of the structure of plasma CIg and its interactions with heparin, fibrinogen, and fibrin-fibrinogen complexes. It is still difficult to gain from these results any clear indication of what function CIg might serve. Hopefully, the information presented will provide some useful clues for future studies designed to provide these answers.

Acknowledgments

I gratefully acknowledge the many contributions of my present and former collaborators, and I also thank Mrs. Lori Hubert for her invaluable secretarial assistance.

References

1. Morrison, P. R., J. T. Edsall & S. G. Miller. 1948. J. Amer. Chem. Soc. **70:** 3103–3108.
2. Smith, R. T. & R. W. von Korff. 1957. J. Clin. Invest. **36:** 596–604.
3. Stathakis, N. E. & M. W. Mosesson. 1977. J. Clin. Invest. **60:** 855–865.
4. Edsall, J. T., G. A. Gilbert & H. A. Scheraga. 1955. J. Amer. Chem. Soc. **77:** 157–161.
5. Chen, A. B., M. W. Mosesson & G. I. Solish. 1976. Amer. J. Obstet. Gynecol. **125:** 958–961.
6. Pollara, B. 1963. Thesis 63-7948. University of Minnesota. University Microfilms. Ann Arbor, Mich.
7. Mosesson, M. W., R. W. Colman & S. Sherry. 1968. N. Engl. J. Med. **278:** 815–821.
8. Ruoslahti, E. & A. Vaheri. 1975. J. Exp. Med. **141:** 497–501.
9. Mosesson, M. W. & R. A. Umfleet. 1970. J. Biol. Chem. **245:** 5728–5736.
10. Chen, A. B. & M. W. Mosesson. 1977. Anal. Biochem. **79:** 144–151.
11. Mosesson, M. W., A. B. Chen & R. M. Huseby. 1975. Biochim. Biophys. Acta **386:** 509–524.
12. Mosher, D. F. 1975. J. Biol. Chem. **250:** 6614–6621.
13. Chen, A. B., D. L. Amrani & M. W. Mosesson. 1977. Biochim. Biophys. Acta **493:** 310–322.
14. Doolittle, R. F. 1970. *In* Methods in Enzymology. G. E. Perlmann & L. Lorand, Eds. Vol. 19: 555–569. Academic Press, Inc. New York, N.Y.
15. Greenfield, N. & G. D. Fasman. 1969. Biochemistry **8:** 4108–4116.
16. Holzwarth, G. & P. Doty. 1965. J. Amer. Chem. Soc. **87:** 218–227.
17. Sears, D. W. & S. Beychok. 1973. Physical Principles and Techniques of Protein Chemistry. Part C. S. J. Leach, Ed.: 445–593. Academic Press, Inc. New York, N.Y.
18. Quadrifoglio, F. & D. W. Urry. 1968. J. Amer. Chem. Soc. **90:** 2760–2765.
19. Iizuka, E. & J. T. Yang. 1966. Proc. Nat. Acad. Sci. USA **55:** 1175–1182.
20. Green, N. M. & M. D. Melamed. 1966. Biochem. J. **100:** 614–621.
21. Beychok, S. & G. D. Fasman. 1964. Biochemistry **3:** 1675–1678.
22. Ludescher, U. & R. Schwyzer. 1971. Helv. Chim. Acta **54:** 1637–1644.

23. URRY, D. W., F. QUADRIFOGLIO, R. WALTER & I. L. SCHWARTZ. 1968. Proc. Nat. Acad. Sci. USA **60:** 967–974.
24. SAXENA, V. P. & D. B. WETLAUFER. 1971. Proc. Nat. Acad. Sci. USA **68:** 969–972.
25. MOSESSON, M. W., J. S. FINLAYSON, R. A. UMFLEET & D. K. GALANAKIS. 1972. J. Biol. Chem. **247:** 5210–5219.
26. MOSESSON, M. W., D. K. GALANAKIS & J. S. FINLAYSON. 1974. J. Biol. Chem. **249:** 4656–4664.
27. GERSHMAN, L. C., A. STRACHER & P. DREIZEN. 1969. J. Biol. Chem. **244:** 2726–2736.
28. MOSESSON, M. W. & J. S. FINLAYSON. 1976. *In* Progress in Hemostasis and Thrombosis. T. H. Spaet, Ed. Vol. 3:67–107. Grune & Stratton, Inc. New York, N.Y.
29. JILEK, F. & H. HÖRMANN. 1977. Hoppe-Seyler's Z. Physiol. Chem. **358:** 133–136.
30. SUZUKI, K., S. NAKAMURA, S. IWANAGA, S. HASHIMOTO & S. MATSUURA. 1976. 16th Congr. Int. Soc. Hematol. Abstr. 312.
31. WEINSTEIN, M. J., M. E. LEGAZ & C. M. HELDEBRANT. 1974. Fed. Proc. **33:** 1452 (Abst.).
32. KESKI-OJA, J., D. F. MOSHER & A. VAHERI. 1977. Biochem. Biophys. Res. Commun. **74:** 699–706.
33. HYNES, R. O. & A. DESTREE. 1977. Proc. Nat. Acad. Sci. USA **74:** 2855–2859.
34. RUOSLAHTI, E. & A. VAHERI. 1974. Nature (London) **248:** 790–791.
35. VAHERI, A. & E. RUOSLAHTI. 1974. Int. J. Cancer **13:** 579–586.
36. KUUSELA, P., E. RUOSLAHTI & A. VAHERI. 1975. Biochim. Biophys. Acta **379:** 295–303.
37. PEARLSTEIN, E. & J. SEAVER. 1976. Biochim. Biophys. Acta **426:** 589–597.
38. COOPER, H. A., R. H. Wagner & M. W. MOSESSON. 1974. J. Lab. Clin. Med. **84:** 258–263.
39. BANG, N. U., M. S. HANSEN, G. F. SMITH & M. W. MOSESSON. 1973. Thromb. Diath. Haemorrh. Suppl. **56:** 75–90.
40. MOSHER, D. F. 1976. J. Biol. Chem. **251:** 1639–1645.
41. STEMBERGER, A. & H. HÖRMANN. 1976. Hoppe-Seyler's Z. Physiol. Chem. **357:** 1003–1005.
42. FYRAND, O. & N. O. SOLUM. 1976. Thromb. Res. **8:** 659–672.
43. MATSUDA, M. T., T. SAIDA & R. HASAGAWA. 1976. Thromb. Res. **9:** 541–552.
44. STATHAKIS, N. E., M. W. MOSESSON, A. B. CHEN & D. K. GALANAKIS. 1978. Blood **51:** 1211–1222.
45. GODAL, H. C. 1960. Scand. J. Clin. Lab. Invest. **12:** 56–65.
46. GODAL, H. C. 1961. Scand. J. Clin. Lab. Invest. **13:** 550–552.
47. SMITH, R. T. 1957. J. Clin. Invest. **36:** 605–616.
48. FYRAND, O., N. O. SOLUM & P. KIERULF. 1976. Thromb. Res. **8:** 17–29.
49. SHAINOFF, J. R. & I. H. PAGE. 1960. Circ. Res. **8:** 1013–1021.
50. SHAINOFF, J. R. & I. H. PAGE. 1962. J. Exp. Med. **116:** 687–707.
51. NOSSEL, H. L., I. YUDELMAN, R. W. CANFIELD, V. P. BUTLER, JR., K. SPONANDIS, G. D. WILNER & G. D. QURESHI. 1974. J. Clin. Invest. **54:** 43–53.
52. LIPINSKA, I., B. LIPINSKI & V. GUREWICH. 1974. J. Lab. Clin. Med. **84:** 509–516.
53. MOSESSON, M. W. & J. S. FINLAYSON. 1974. Thromb. Res. **4:** 895–900.
54. FINLAYSON, J. S., M. W. MOSESSON, J. T. BRONZERT & J. J. PISANO. 1972. J. Biol. Chem. **247:** 5220–5222.
55. FINLAYSON, J. S. & M. W. MOSESSON. 1973. Thromb. Res. **2:** 467–478.
56. FOLK, J. E. & J. S. FINLAYSON. 1977. Adv. Protein Chem. **31:** 1–133.
57. HARDER, A. J. & P. W. STRAUB. 1972. Thromb. Diath. Haemorrh. **27:** 337–348.
58. LANDIS, W. J. & D. F. WAUGH. 1975. Arch. Biochem. Biophys. **168:** 498–511.
59. AMRANI, D., C. WOLFENSTEIN-TODEL & M. W. MOSESSON. 1976. Unpublished observations.
60. YAMADA, K. & S. ALEXANDER. 1977. Personal communication.
61. MOSESSON, M. W. 1975. Unpublished results.
62. MOSESSON, M. W. & A. B. CHEN. 1976. Unpublished results.
63. CHEN, A. B. & M. W. MOSESSON. 1976. Unpublished results.

ARE LOBSTER FIBRINOGEN AND COLD-INSOLUBLE GLOBULIN RELATED MOLECULES?*

Gerald M. Fuller

Department of Human Biological Chemistry and Genetics
The University of Texas Medical Branch
Galveston, Texas 77550

A major cell surface glycoprotein found on fibroblasts in culture has been the subject of intensive investigation for the past few years.[1] The principal reason for this interest has been the observation that cells that are virally or chemically transformed display a striking diminution of this cell surface marker.[2-4] Moreover, it has been shown[5,6] that when transformed cells are supplied with an exogenous source of the cell surface protein, the cells acquire a phenotype with characteristics that resemble those of normal morphology. It has also been shown that the cell surface protein is immunologically identical with a circulating plasma protein known as the cold-insoluble globulin (CIg).[7]

Mosher[8] has suggested that some molecular similarities exist between the invertebrate clotting molecule and CIg. Since the precise functional role of either CIg or cell surface protein (CSP, LETS, fibronectin†) has yet to be elucidated, the suggestion that they exhibit properties similar to those of to the lobster clotting molecule may be of interest. In this paper, a review of the molecular characteristics of these two proteins is presented. To gain some perspective on the functional role of the invertebrate fibrinogen, we shall present a description of the process of fibrin formation in the lobster. Finally, some speculation regarding the role of CIg is offered.

Molecular Properties of Lobster Fibrinogen

The hemolymph of most decapod crustaceans contains a protein that undergoes a gelation process when exposed to a tissue enzyme and calcium. The most thoroughly investigated invertebrate fibrinogen molecule is that from the California spiny lobster, *Panulirus interruptus.*[9-11] Although fibrinogen from other species has been investigated, these earlier studies placed more emphasis on elucidating sedimentation properties of the molecule rather than their chemical properties or subunit structure.[12-14]

Lobster fibrinogen can be separated from most proteins of the hemolymph by isoelectric precipitation.[9] A simple step of lowering the pH from neutrality (pH 7.5) to 4.8 causes the fibrinogen to precipitate from solution. Repeated cycles of precipitation and solubilization yield a preparation that is nearly 100% clottable. Isolation to a single molecular species can be achieved by preparative gel electrophoresis. Lobster fibrinogen shows no insolubility characteristics when stored at 4°C for short periods of time. Sedimentation velocity and equilibrium experiments

*Supported in part by National Heart and Lung Institute Grant DHEW 5R01 HL 00162.

†The protein found on the surface of fibroblasts is commonly referred to as CSP (cell surface protein), LETS (large, external, transformation-sensitive) protein, and fibronectin. All three terms appear to be synonyms for the same protein.

demonstrated that lobster fibrinogen has a Svedberg coefficient of 14.5S and a mass of about 420,000 daltons. The subunit composition of lobster fibrinogen has been determined by sodium dodecyl sulfate polyacrylamide gel electrophoresis.[11] It appears that the lobster molecule is composed of two high-molecular-weight polypeptides of approximately 220,000 daltons. Examination of the early stages of the clotting reaction by high-resolution electron microscopy has revealed that the fibrinogen is basically spherical and that polymerization proceeds by adherence of the spheres to one another like beads on a string (FIGURE 1).

Lobster fibrinogen has about 3% carbohydrate associated with it, including glucose, mannose, and glucosamine; thus, the molecule must be considered a glycoprotein. In addition, the molecule has an associated carotenoid pigment that causes preparations of pure fibrinogen to display a bright orange color. Quantitative amino terminal analysis reveals 2 mol of leucine per mole of protein, indicating the presence of two subunits per molecule.

FORMATION OF LOBSTER FIBRIN

Transformation of lobster fibrinogen into a fibrin gel is a phenomenon that has been studied by scores of investigators.[14] The conversion is catalyzed by a calcium-dependent enzyme found in certain lobster blood cells (hemocytes or coagulocytes) and muscle tissue. Vertebrate thrombins cannot substitute for the lobster clotting enzyme, and limited proteolysis, which is the basis of all vertebrate coagulation, is not involved. The production of lobster fibrin appears to be directly dependent on the formation of intermolecular covalent bonds between fibrinogen molecules, resulting in a clot that cannot be dispersed by solvent systems known to disrupt weaker intramolecular forces.

The discovery by Doolittle and Lorand[15] that lobster fibrin production is inhibited by glycine derivatives, which also prevent the formation of covalent cross-links in vertebrate fibrin,[16] led to renewed interest in the invertebrate coagulation scheme. It had been suggested that vertebrate fibrin stabilization occurred through a transpeptidation reaction that involves the glycine amino terminals of the α and β chains. Since lobster fibrin formation was inhibited by the same glycine derivatives, it was apparently a logical conclusion that the lobster fibrinogen amino terminals would be glycine residues and that they became incorporated into some molecular cross-link during the clotting reaction. When the amino terminals of lobster fibrinogen were shown to be leucine and not glycine, it was realized that the earlier notion could not be correct. We were able to demonstrate that the ϵ-amino groups of lysine were involved in both the vertebrate stabilization reaction and the invertebrate coagulation process. It was subsequently shown by Matacic and Loewy[17] that the acceptor moiety of the vertebrate cross-linking process was composed of glutaminyl side chains. Further analysis of lobster fibrin by use of procedures described by Pisano *et al.*[18,19] demonstrated that the isopeptide linkage was the mechanism by which lobster fibrin formation occurred. The molecular process of lobster cross-linking is shown in FIGURE 2, and its comparison to the vertebrate system is given in FIGURE 3.

COMPARISON BETWEEN COLD-INSOLUBLE GLOBULIN AND LOBSTER FIBRINOGEN

Isolation and determination of the molecular properties of human cold insoluble globulin have been performed primarily by Mosesson *et al.*[20-22] The

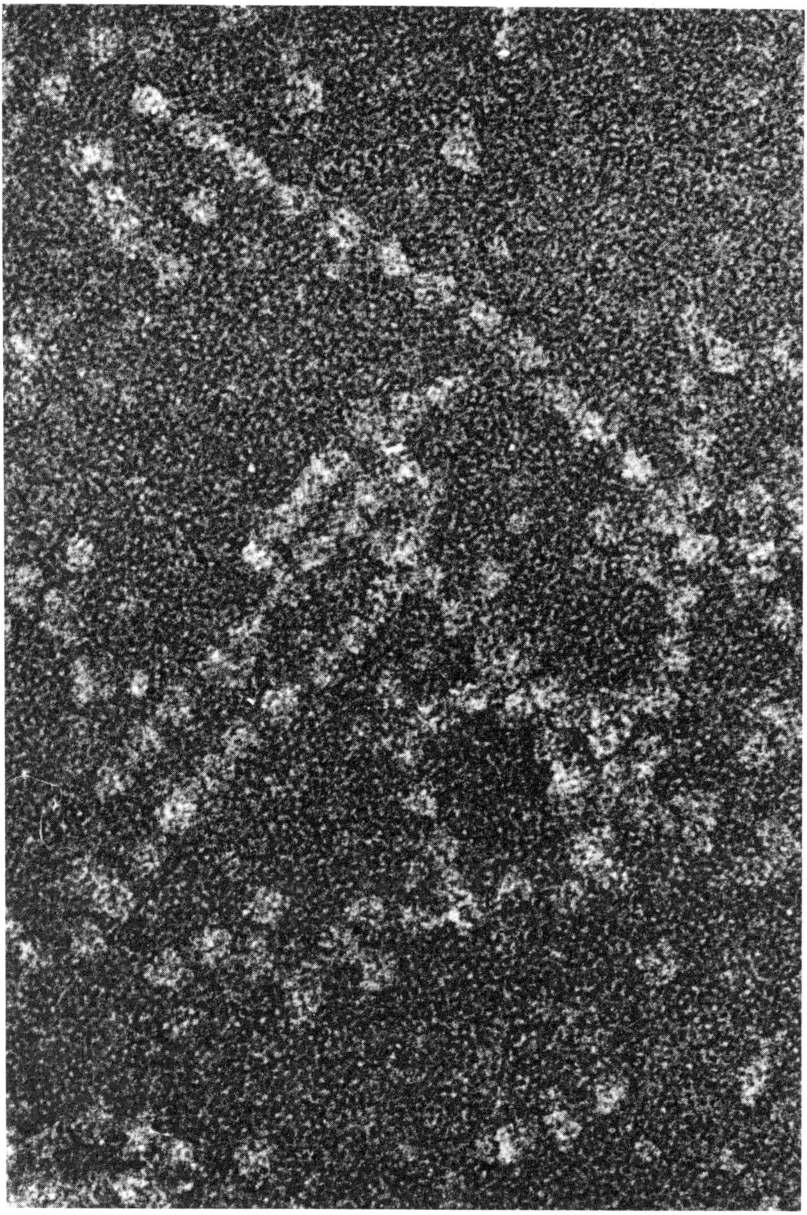

FIGURE 1. Negative-stained preparation of lobster fibrinogen-fibrin conversion. The carbon grid was dipped into a dilute solution of clotting reaction (fibrinogen, cell enzyme, calcium) × 210,000.

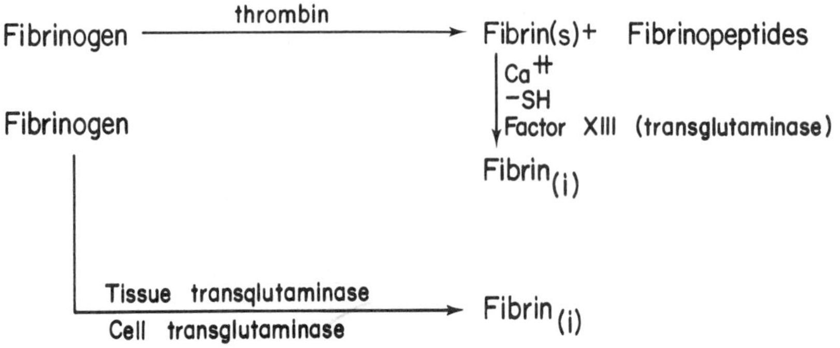

FIGURE 2. Molecular process of cross-linking in both the lobster and vertebrate clotting reactions.

molecule is the major nonclottable component of Cohn fraction I; it copurifies with fibrinogen in the cold, has a molecular weight of about 450,000 daltons, and is comprised of two high-molecular-weight subunits that have a mass of 220,000 daltons. Recently, CIg has been shown to be a substrate for plasma transglutaminase (clotting factor XIII); that is, CIg is capable of being cross-linked into large polymers by forming isopeptide linkages.[8] At first glance, these properties are strikingly similar to those described for the lobster fibrinogen molecule.

TABLE 1 shows the amino acid composition of the two proteins together with that of an "average" protein.[23] It is very difficult to draw any valid conclusions

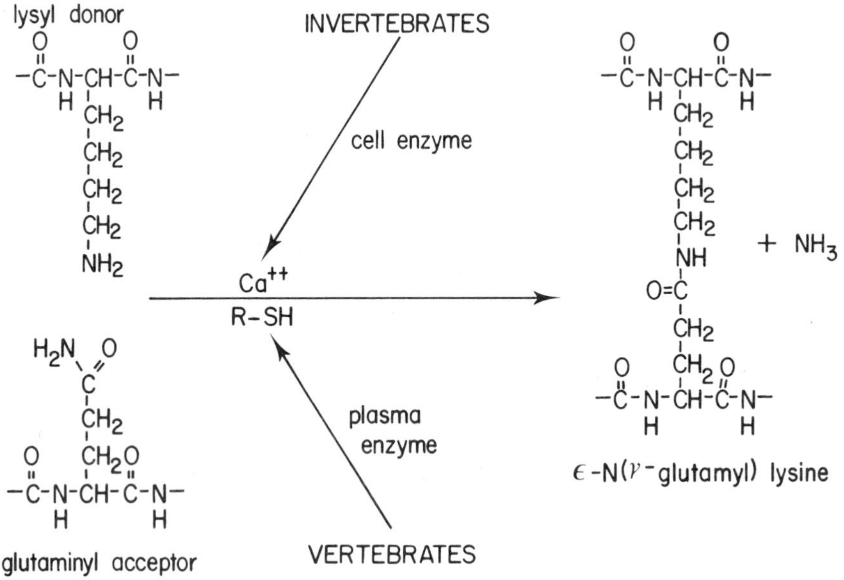

FIGURE 3. Comparisons of the lobster coagulation processes and the fibrin stabilization reactions of vertebrates.

about homology between rather dissimilar proteins based solely on amino acid composition. Computer analyses have been described, however, wherein amino acid composition can be used to develop a function termed "composition divergence," which gives some indication of the relatedness of molecules from their amino acid composition.[24]

TABLE 2 compares the molecular properties between the lobster molecule and those of CIg. This Table emphasizes the similarities between the molecules rather than their differences. Hemagglutination assays performed with antibodies directed against CIg (a generous gift of Dr. M. W. Mosesson) show no immunologic identity when reacted against lobster fibrinogen.

TABLE 1

COMPARISON OF AMINO ACID COMPOSITION BETWEEN LOBSTER FIBRINOGEN AND COLD-INSOLUBLE GLOBULIN

	Moles per 100 Grams of Protein		
Residue	Lobster Fibrinogen*	Cold-Insoluble Globulin†	"Average Protein"‡
Aspartic acid	9.90	8.53	10.4
Threonine	7.12	8.91	6.3
Serine	8.19	6.26	7.6
Glutamic acid	10.91	10.70	9.4
Proline	5.25	7.01	4.8
Glycine	6.17	7.35	7.7
Alanine	5.46	3.95	8.4
Half-cystine	1.31	2.42	3.4
Valine	6.90	7.43	6.9
Methionine	1.69	1.0	1.7
Isoleucine	5.12	4.0	4.8
Leucine	9.60	5.26	7.5
Tyrosine	3.18	4.18	3.6
Phenylalanine	4.10	2.50	3.6
Lysine	4.18	3.36	6.9
Histidine	4.16	1.91	2.2
Arginine	4.75	4.76	4.1
Tryptophan	1.94	2.55	1.3

*Data from Fuller and Doolittle.[9]
†Data from Mosesson *et al.*[21]
‡Data from Dayhoff and Hunt.[23]

An attempt has been made to compile and compare the available data on the molecular properties of lobster fibrinogen with those of CIg. The data available are tantalizing but certainly not definitive in establishing the relatedness between these two molecules. There is at present simply not enough information to make a strong case for their being evolutionarily related molecules. One of the original intents in the investigation of the lobster clotting mechanism was to determine if the lobster molecule was a primitive vertebrate fibrinogen. Even though these molecules performed a similar function and were substrates for the same enzymatic activity, their dissimilarities indicated that the lobster molecule is not an early prototype of the vertebrate fibrinogen. Clearly, the hemostatic process in invertebrates represents an example of parallel evolution.

Blood Coagulation in Invertebrates

The principal mechanism of blood coagulation in invertebrates appears to involve free-floating cells that are foreign contact sensitive. Cytologic description demonstrates that the cellular events of hemostasis in the invertebrates involve cell lysis, with the release of some "coagulum," causing cells to adhere to one another. In addition, cells appear to send out processes that entrap other floating cells, resulting in the formation of a cellular mass, or "plug." Quantitatively, of the species examined to date, only a few appear to possess extracellular proteins that function in the clotting process. It should be pointed out, however, that in only a few cases have modern biochemical methods been employed to study the events of clotting in the invertebrate phyla. It is not surprising that the principal clotting mechanism of invertebrates appears to be cell to cell adhesion, because the vast majority of these animals have an open circulatory system. There are sporadic reports, however, suggesting that the cellular plugs have a greater amount of protein associated with them than can be accounted for by the cells alone, implying the involvement of extracellular proteins in the plug.

Table 2

Comparison of Physical and Chemical Properties of Cold-Insoluble Globulin and Lobster Fibrinogen

	Cold-Insoluble Globulin	Lobster Fibrinogen
Molecular weight	$\simeq$450,000	$\simeq$440,000
Subunit structure	$(220{,}000)_2$	$(220{,}000)_2$
Amino terminal amino acid	blocked pyrrolidone (Glu-Ala)	leucine
Cross-linked by a transglutaminase	yes	yes
Sedimentation coefficient	11.8	14.5
Hydrodynamic shape	rod	sphere

One is tempted to suggest that there may be at least a functional correlation between CIg and the protein associated with cell surfaces in the vertebrate species in invertebrate hemostasis. It may be that there is not true physiologic function of CIg; rather, it may be remnant of a hemostatic system long since bypassed in vertebrate systems.

Summary

Several of the molecular properties of lobster fibrinogen and CIg have been compared. There are some significant similarities between these two proteins; however, to date, no direct experiments have demonstrated molecular homologies. Studies involving cyanogen bromide mapping, additional immunologic probing, and modifications of cell morphology effected by lobster fibrinogen are currently in progress. It is anticipated that these and other experimental approaches will provide direct evidence to answer the question, posed by the title of this paper.

REFERENCES

1. Hynes, R. O. 1976. Biochim. Biophys. Acta **458:** 73–107.
2. Hynes, R. O. 1973. Proc. Nat. Acad. Sci. USA **70:** 3170–3174.
3. Vaheri, A. & E. Ruoslahti. 1974. Int. J. Cancer **13:** 579–586.
4. Yamada, K. M. & J. A. Weston. 1974. Cell **5:** 75–81.
5. Yamada, K. M., S. S. Yamada & I. Pastan. 1976. Proc. Nat. Acad. Sci. USA **73:** 1217–1219.
6. Ali, U. I., V. Mautner, R. Lanza & R. O. Hynes. 1977. Cell **11:** 115–126.
7. Ruoslahti, E. & A. Vaheri. 1975. J. Exp. Med. **141:** 497–501.
8. Mosher, D. F. 1975. J. Biol. Chem. **250:** 6614–6621.
9. Fuller, G. M. & R. F. Doolittle. 1971. Biochemistry **10:** 1305–1311.
10. Fuller, G. M. & R. F. Doolittle. 1971. Biochemistry **10:** 1311–1315.
11. Doolittle, R. F. & G. M. Fuller. 1972. Biochim. Biophys. Acta **263:** 805–809.
12. Duchateau, G. & M. Florkin. 1954. Bull. Soc. Chem. Biol. **36**(2–3): 295–305.
13. Stewart, J. E., J. R. Dingle & P. H. Odense. 1966. Can. J. Biochem. **44:** 1447–1459.
14. Glavind, J. 1948. Studies on the Coagulation of Crustacean Blood. Arnold Buske. Copenhagen, Denmark.
15. Doolittle, R. F. & L. Lorand. 1962. Biol. Bull. **123:** 481.
16. Lorand, L., R. F. Doolittle, K. Konishi & S. K. Riggs. 1963. Arch. Biochem. Biophys. **102:** 171–179.
17. Matacic, S. & A. G. Loewy. 1966. Biochem. Biophys. Res. Commun. **24:** 858–866.
18. Pisano, J. J., J. S. Finlayson & M. P. Peyton. 1968. Science **160:** 892–893.
19. Pisano, J. J., J. S. Finlayson & M. P. Peyton. 1969. Biochemistry **8:** 871–876.
20. Mosesson, M. W. & R. A. Umfleet. 1970. J. Biol. Chem. **245:** 5728–5736.
21. Mosesson, M. W., A. B. Chen & R. M. Huseby. 1975. Biochim. Biophys. Acta **386:** 509–524.
22. Chen, A. B. & M. W. Mosesson. 1977. Anal. Biochem. **79:** 144–151.
23. Dayhoff, M. O. & L. T. Hunt. 1972. Atlas Protein Sequence Struct. **5:** D-355.
24. Harris, C. E. & D. C. Teller. 1973. J. Theor. Biol. **38:** 347–362.

CROSS-LINKING OF PLASMA AND CELLULAR FIBRONECTIN BY PLASMA TRANSGLUTAMINASE

Deane F. Mosher

Department of Medicine
University of Wisconsin
Madison, Wisconsin 53706

Transglutaminases are a class of calcium ion-dependent enzymes that catalyze an acyl transfer reaction that involves glutaminyl residues in proteins and peptides as acyl donors and a variety of primary amines, including lysyl residues, as acyl acceptors, thus resulting in mono-substituted γ-amide bonds.[1] By this reaction, transglutaminases can catalyze the incorporation of ϵ-(γ-glutamyl)lysine linkages between proteins. Inasmuch as specific glytaminyl residues are susceptible to the action of transglutaminases and accepting lysyl residues must be in favorable stereochemical juxtaposition, the cross-linking is a highly specific reaction. As recently reviewed by Folk and Finlayson,[1] transglutaminases are widespread and are thought to catalyze many biologically important reactions: the cross-linking of vertebrate fibrin by plasma transglutaminase (blood coagulation factor XIII_a, fibrinoligase), clotting of seminal plasma by transglutaminases of the anterior prostate gland of rodents, cross-linking of hair proteins by hair follicle transglutaminases, cross-linking of keratin by epidermal transglutaminase, and clotting of lobster fibrinogen. In the present paper, we review evidence that indicates that both plasma fibronectin (cold-insoluble globulin) and cell surface fibronectin of cultured fibroblasts (LETS protein, cell surface protein) are specific substrates for transglutaminases.

Materials and Methods

Fibronectin was purified from human plasma by precipitations with ammonium sulfate and ethanol, defibrination (either by clotting with purified thrombin or heat treatment), and chromatography on DEAE-cellulose.[2,3] When heat defibrination was performed, plasma protransglutaminase copurified with fibronectin and could be recovered in a pure state during the final chromatographic separation. Antisera to purified plasma fibronectin were prepared in rabbits or goats and used in Laurell's electroimmunoassy[4] or double-antibody radioimmunoassay[5] to measure the concentration of fibronectin. The antisera reacted identically with plasma fibronectin and with the fibronectin produced by cultured fibroblasts. Cell surface fibronectin of cultured fibroblasts was labeled by lactoperoxidase-catalyzed iodination.[6] Polyacrylamide gel electrophoresis in sodium dodecyl sulfate was performed in cylindrical gels[2] or on discontinuous slab gels.[6] Radioactivity was detected by autoradiography of dried slab gels.[6]

Results and Discussion

When purified plasma fibronectin was incubated with purified plasma transglutaminase and *N*-(5-aminopentyl)-5-dimethylaminonaphthalene-1-sulfonamide

0077-8923/78/0312-0038 $01.75/0

(dansylcadaverine), the fluorescent primary amine was incorporated into fibronectin.[2] Jilek and Hörmann found that under similar conditions, approximately 0.75 mol of the radioactive primary amine, [^{14}C]putrescine, was incorporated per 200,000–220,000-dalton subunit of plasma fibronectin. On cyanogen bromide or plasmin digestion, the labeled putrescine was localized to 27,000-dalton fragments.[7] On the basis of their experiments, Jilek and Hörmann concluded that there probably is a single transamidation-sensitive site per fibronectin subunit.[7]

When human plasma was clotted in the presence of dansylcadaverine, the fluorescent amine was incorporated into plasma fibronectin, as ascertained by gel electrophoresis of fibronectin purified from the serum by ammonium sulfate precipitation, gel exclusion chromatography, or immunoprecipitation.[8] Dansylcadaverine was not incorporated into fibronectin when plasma congenitally deficient in protransglutaminase was clotted or when transglutaminase was inhibited by chelation of calcium ions.[8] Only three other substrates for the transglutaminase were identified: fibrin, α_2-macroglobulin, and an unidentified protein with an apparent molecular weight greater than 150,000 and a subunit of 110,000 daltons.[8]

Plasma transglutaminase catalyzed the incorporation of dansylcadaverine into a 220,000 dalton polypeptide of cultured fibroblasts.[9] Inasmuch as the labeled polypeptide was extremely trypsin sensitive and not present in cell layers of transformed fibroblasts, it is likely that the labeled polypeptide is the large, external, transformation-sensitive subunit of cellular fibronectin.

In the presence of dithiothreitol or cysteine, plasma transglutaminase catalyzed the cross-linking of purified plasma fibronectin to itself; as monitored by polyacrylamide gel electrophoresis after reduction, progressively higher multimers appeared with time.[2] In support of this finding, Birckbichler and Patterson have identified ϵ-(γ-glutamyl)lysine in enzymatic hydrolysates of cross-linked plasma fibronectin.[10] No cross-linking was found if sulfhydryl reagents were omitted and purified plasma fibronectin was incubated with only plasma transglutaminase and calcium ion.[2] Sulfhydryl reagents are known to enhance the activity of transglutaminases,[11] but the effect of sulfhydryl reagents on fibronectin cross-linking seemed to be on the fibronectin rather than on plasma transglutaminase, inasmuch as increasing the concentration of plasma transglutaminase did not cause fibronectin-fibronectin cross-linking in the absence of sulfhydryl reagents[2] and since plasma transglutaminase cross-linked fibronectin to fibrin perfectly well in the presence of calcium ion alone (see below).

The large, external subunit of cell surface fibronectin, prelabeled by lactoperoxidase-catalyzed iodination, was readily cross-linked to a very high-molecular-weight complex by plasma transglutaminase, as ascertained by polyacrylamide gel electrophoresis in sodium dodecyl sulfate after reduction.[6] The intermediate multimers seen when plasma fibronectin was cross-linked could not be observed when cell surface fibronectin was cross-linked. It should be noted that the bulk of cell surface fibronectin was localized by immunofluorescence to fibrils[6] and was bound together by forces that could not be disrupted by sodium dodecyl sulfate in the absence of a reducing agent.[3] In older cell cultures, fibronectin-containing fibrils stain positively for collagen and for fibronectin.[12] Plasma fibronectin is known to bind to collagen, especially denatured collagen.[13] Thus, transglutaminase-catalyzed cross-linking of cell surface fibronectin may not occur among fibronectin molecules but rather between fibronectin and another constituent of the pericellular matrix, such as collagen.

A variety of techniques, including immunoelectrophoresis,[14] analysis of cryofibrinogen,[14,15] analysis of the heparin-precipitable fraction of plasma,[16,17] and absorption to insolubilized fibrinogen/fibrin,[18,19] have demonstrated an affinity of fibronectin for fibrinogen and fibrin, especially in the cold. When normal plasma was clotted at room temperature or above, the serum concentration was less than the plasma concentration.[4,8] When protransglutaminase-deficient plasma was clotted at 22° or 37°C or when normal plasma was clotted in the absence of calcium ions at these temperatures, the serum concentration of fibronectin was the same as the plasma concentration.[8] The serum concentration of α_2-macroglobulin was the same as the plasma concentration under all clotting conditions.[8] In various mixtures of protransglutaminase-free fibrinogen, purified plasma protransglutaminase, purified thrombin (needed to clot fibrinogen and activate protransglutaminase), calcium ion, and purified fibronectin, bands of 260,000 and 300,000 daltons, compatible with cross-linking of fibronectin to one of the chains of fibrin, were seen only if fibronectin, fibrin, and active transglutaminase were present.[2] In time course experiments, stainable protein was lost from the fibronectin band at the same rate as from the α-chain band of fibrin.[2] When α-chain polymer formation was inhibited, either by clotting at high sodium chloride concentration or at

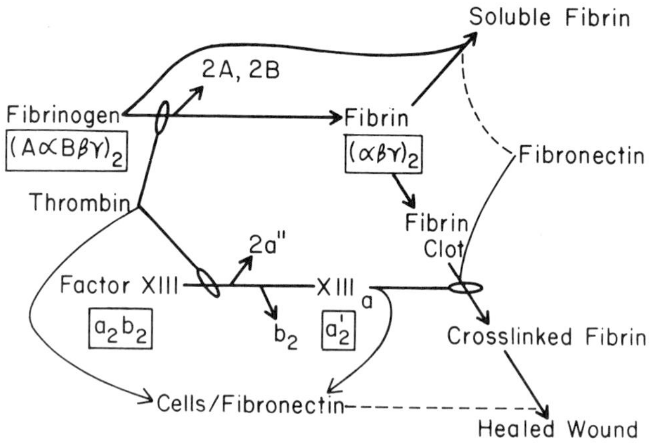

FIGURE 1. Interactions of fibronectin with components of the final stage of blood coagulation. Accepted blood coagulation pathways are shown by heavy lines, newly recognized interactions of fibronectin and thrombin are shown by light arrows, and speculative interactions are shown by dotted lines. The subunit compositions of fibrinogen and coagulation factor XIII (plasma protransglutaminase) are in boxes.

low temperature, the 260,000- and 300,000-dalton bands became more pronounced.[2] Similar 260,000- and 300,000-dalton bands were observed in gels of clots formed from normal plasma and washed free of serum proteins.[8] However, under no conditions could cross-linking of iodinated fibroblast cell surface fibronectin to fibrin be demonstrated.[6]

The experiments described above indicate that the various forms of fibronectin are specific substrates for plasma transglutaminase and that plasma fibronectin, by virtue of its affinity for fibrin, is cross-linked to the α-chain of fibrin under physiologic conditions. Fibronectin-fibrin cross-linking could involve a glutaminyl residue of the α-chain of fibrin and a lysyl residue of fibronectin. However, the cross-linking of cell surface fibronectin probably involves a transglutaminase-susceptible glutaminyl residue of fibronectin. A cyanogen bromide fragment of the α-chain of fibrin that is involved in α-polymer formation, CBI, has a very unusual amino acid composition (17.6% glycine, 16.4% serine, 12.5% threonine).[20] We are currently investigating whether this portion of the α-chain of fibrin is involved in fibronectin-fibrin cross-linking and whether the determinants of fibronectin-collagen interaction are identical to the determinants of the fibronectin-fibrin interaction.

The affinity of fibronectin for collagen and fibrin and the cross-linking of fibronectin by transglutaminases may be of considerable physiologic significance. As depicted by the thick lines in FIGURE 1, a cross-linked fibrin clot is thought to be important for wound healing and tissue repair by fibroblasts and endothelial cells. Fibroblasts and endothelial cells, in turn, are known to both produce and bind fibronectin.[21,22] Thrombin causes fibroblasts to proliferate[23-25] and to secrete more fibronectin.[5] The experiments described above indicate that plasma fibronectin is localized in a newly formed fibrin clot and that fibronectin secreted by proliferating cells is a potential substrate for plasma transglutaminase. People congenitally deficient in plasma protransglutaminase suffer from delayed bleeding and poor wound healing (reviewed in Reference 1). Thus, cross-linking of fibronectin may be important for the ordered growth of cells into a fibrin clot and the replacement of fibrin by the connective tissue matrix. Inasmuch as plasma fibronectin closely resembles the α_2-opsonic glycoprotein of serum,[26] it is possible also that fibronectin binds to soluble fibrinogen-fibrin complexes and mediates their clearance by the reticuloendothelial system.

REFERENCES

1. FOLK, J. E. & J. S. FINLAYSON. 1977. The ϵ-(γ-glutamyl) lysine cross-link and the catalytic role of transglutaminases. Adv. Protein Chem. **31:** 1–133.
2. MOSHER, D. F. 1975. Cross-linking of cold-insoluble globulin by fibrin-stabilizing factor. J. Biol. Chem. **250:** 6614–6621.
3. KESKI-OJA, J., D. F. MOSHER & A. VAHERI. 1977. Dimeric character of fibronectin, a major cell surface-associated glycoprotein. Biochem. Biophys. Res. Commun. **74:** 699–706.
4. MOSESSON, M. W. & R. A. UMFLEET. 1970. The cold-insoluble globulin of human plasma. I. Purification, primary characterization, and relationship to fibrinogen and other cold-insoluble fraction components. J. Biol. Chem. **245:** 5728–5736.
5. MOSHER, D. F. & A. VAHERI. 1978. Thrombin stimulates the production and release of a major surface-associated glycoprotein (fibronectin) in cultures of human fibroblasts. Exp. Cell Res. In press.
6. KESKI-OJA, J., D. F. MOSHER & A. VAHERI. 1976. Cross-linking of a major fibroblast surface-associated glycoprotein (fibronectin) catalyzed by blood coagulation factor XIII. Cell **9:** 29–35.
7. JILEK, F. & H. HÖRMANN. 1977. Cold-insoluble globulin, III. Cyanogen bromide and plasminolysis fragments containing a label introduced by transamidation. Hoppe-Seyler's Z. Physiol. Chem. **358:** 1165–1168.
8. MOSHER, D. F. 1976. Action of fibrin-stabilizing factor on cold-insoluble globulin and α_2-macroglobulin in clotting plasma. J. Biol. Chem. **251:** 1639–1645.
9. MOSHER, D. F. 1977. Labeling of a major fibroblast surface protein (fibronectin) catalyzed by blood coagulation factor XIII_a. Biochim. Biophys. Acta **491:** 205–210.
10. BIRCKBICHLER, P. J. & M. K. PATTERSON, JR. This monograph.
11. LORAND, L. & A. JACOBSEN. 1958. Studies on the polymerization of fibrin. The role of the globulin: fibrin-stabilizing factor. J. Biol. Chem. **230:** 421–434.

12. Bornstein, P. & J. F. Ash. 1977. Cell surface-associated structural proteins in connective tissue cells. Proc. Nat. Acad. Sci. USA **74:** 2480–2484.
13. Engvall, E. & E. Ruoslahti. 1977. Binding of soluble form of fibroblast surface protein, fibronectin, to collagen. Int. J. Cancer **20:** 1–5.
14. Mosesson, M. W., R. W. Colman & S. Sherry. 1968. Chronic intravascular coagulation syndrome: report of a case with special studies of an associated plasma cryoprecipitate ("cryofibrinogen"). N. Engl. J. Med. **278:** 815–821.
15. Matsuda, M., T. Saida & R. Hasegawa. 1976. Cryofibrinogen in the plasma of patients with skin ulcerative lesions on the legs: a complex of fibrinogen and cold-insoluble globulin. Thromb. Res. **9:** 541–552.
16. Fyrand, O. & N. D. Solum. 1976. Heparin precipitable fraction (HPF) from dermatological patients. II. Studies on the non-clottable proteins. Identification of cold-insoluble globulin as the major non-clottable component. Thromb. Res. **8:** 659–672.
17. Stathakis, N. E. & M. W. Mosesson. 1977. Interactions among heparin, cold-insoluble globulin and fibrinogen in formation of the heparin-precipitable fraction of plasma. J. Clin. Invest. **60:** 855–865.
18. Ruoslahti, E. & A. Vaheri. 1975. Interaction of soluble fibroblast surface protein with fibrinogen and fibrin. Identity with cold-insoluble globulin of human plasma. J. Exp. Med. **141:** 497–501.
19. Stemberger, A. & H. Hörmann. 1976. Affinity chromatography on immobilized fibrinogen and fibrin monomer. II. The behavior of cold-insoluble globulin. Hoppe-Seyler's Z. Physiol. Chem. **357:** 1003–1005.
20. Doolittle, R. F., K. G. Cassman, B. A. Cottrell & S. J. Friezner. 1977. Amino acid sequence studies on the α chain of human fibrinogen. Isolation and characterization of two linked α-chain cyanogen bromide fragments from fully cross-linked fibrin. Biochemistry **16:** 1715–1719.
21. Mosher, D. F., O. Saksela, J. Keski-Oja & A. Vaheri. 1977. Distribution of a major surface-associated glycoprotein, fibronectin, in cultures of adherent cells. J. Supramol. Struct. **6:** 551–557.
22. Jaffe, E. A. & D. F. Mosher. 1978. Synthesis of fibronectin by cultured human endothelial cells. J. Exp. Med. In press.
23. Chen, L. B. & J. M. Buchanan. 1975. Mitogenic activity of blood components. I. Thrombin and prothrombin. Proc. Nat. Acad. Sci. USA **72:** 131–135.
24. Blumberg, P. M. & Robbins, P. W. 1975. Effect of proteases on activation of resting chick embryo fibroblasts and on cell surface proteins. Cell **6:** 137–147.
25. Pohjanpelto, P. 1977. Proteases stimulate proliferation of human fibroblasts. J. Cell. Physiol. **91:** 387–392.
26. Blumenstock, F., P. Weber & T. M. Saba. 1977. Isolation and biochemical characterization of alpha-2-opsonic glycoprotein from rat serum. J. Biol. Chem. **252:** 7156–7162.

PHYSIOLOGIC ROLE OF COLD-INSOLUBLE GLOBULIN IN SYSTEMIC HOST DEFENSE: IMPLICATIONS OF ITS CHARACTERIZATION AS THE OPSONIC α_2-SURFACE-BINDING GLYCOPROTEIN*

Thomas M. Saba, Frank A. Blumenstock,†
Peter Weber, and John E. Kaplan

Departments of Physiology and Biochemistry
The Albany Medical College of Union University
Albany, New York 12208

The reticuloendothelial system (RES), or macrophage system, represents a diffuse collection of fixed and wandering mononuclear phagocytic cells that participate in host defense mechanisms.[1,2] The most apparent physiologic function of RE cells is the rapid phagocytic ingestion of foreign and effete particulate matter, a process described in the 1880s by Metchnikoff and fundamental to the concept of cellular immunity.[3] Macrophages phagocytize and process foreign particulate matter and are intimately involved in the immune response.[4] Early investigations of the participation of the RES in various pathophysiologic events utilized the technique of RE blockade to depress the macrophage system.[1,2] Reticuloendothelial blockade is induced by intravenous injection of a large dose of foreign particles, which are phagocytized primarily by fixed RE cells in the liver, spleen, and bone marrow. While the mechanisms of colloid-induced RE blockade was initially speculated to be due to a physical saturation of RE cells,[1] subsequent studies indicated that blockade was due to the depletion of a serum protein that interacts with the particle prior to phagocytic engulfment.[2,5] Such studies formed the basis for investigations into the nature of this protein, especially as it may influence liver Kupffer cell phagocytosis, which represents 80–90% of the total systemic clearance capacity of the RES.[1,2] These cells line the sinusoids of the liver and "filter," or remove, blood-borne particulate matter. Reticuloendothelial blockade decreases resistance to tumor challenge,[6] infection,[7] and experimental injury,[8] a fact that emphasizes the role of the RES in defense against septicemia, cancer, and trauma. Thus, a critical role for this specific protein in altered RES function and host defense was suggested.

Participation of the macrophage system in antitumor and antibacterial immunity has been adequately documented.[6,9] Mononuclear phagocytic cells and polymorphonuclear leukocytes are active in bacterial phagocytosis and killing, and macrophage processing of antigenic material has been described.[4,10] Similarly, the role of the RES in resistance to traumatic shock, hemorrhage, and burn injury has also been extensively investigated.[7,8,11–19] Reticuloendothelial system depression decreases resistance to various forms of experimental shock, and severe injury will compromise RES function.[8,11,13,14,16] Low-grade intravascular coagulation undermines RE phagocytic activity,[20] and experimentally induced RE blockade will exaggerate the cardiovascular and pulmonary consequences of intravas-

*Supported by National Institutes of Health Research Grants GM-21447 and CA-16011.
†National Institutes of Health Postdoctoral Fellow supported by Grant T32-GM-07033.

cular coagulation,[21,22] endotoxin challenge, and traumatic injury.[8,23] While a variety of mechanisms have been proposed to mediate such disturbances in the RES, recent data emphasize the role of a high-molecular-weight acid glycoprotein as a determinant of RE phagocytic activity.[2,24,25] This α_2-globulin has been referred to as opsonic protein, humoral recognition factor, aspecific opsonin, and α_2-surface-binding glycoprotein (α_2-SB-glycoprotein)[2,6,26] and has been shown to be the protein depleted during colloid-induced RE blockade.[5,25] Thus, the opsonic α_2-SB-glycoprotein appears to be of central importance in host defense.

Recent studies from our laboratory have led to the isolation and characterization of this protein from rat[24] and human serum.[26] The purified rat protein is a high-molecular-weight α_2-glycoprotein that is dependent on heparin for expression of biologic activity, as assayed by liver slice bioassay.[24,27] It is thermolabile at 60°C for 20 min,[28] distinct from the complement system,[29] and rapidly consumed during colloid clearance from the blood.[5,25,30] It has a molecular weight of 4.5×10^5 daltons at 25–30°C, a minimum molecular weight of 2.26×10^5, and appears to consist of two high-molecular-weight polypeptide chains held together by disulfide bridges.[27] The protein rapidly binds to foreign colloids,[5,25,30] leukemic leukocytes,[6] fibrin aggregates, and particulate collagen[44] and is sequestered in areas of tissue injury.[30] It is distinct from other characterized opsonic factors, such as immunoglobulins and C3b, which exert opsonic activity relative to bacterial phagocytosis.[10,28] On the basis of the amino acid analysis of the rat protein, its molecular weight, its cryoprecipitable nature from chilled heparinized plasma, and its serum concentration prior to and after surgery, as determined by electroimmunoassay, we speculated that the analogous human opsonic glycoprotein may be identical to human cold-insoluble globulin (CIg) or plasma fibronectin. The present study was designed to evaluate this concept by a comparative analysis of these two proteins by use of both immunologic and biochemical techniques.

To emphasize the potential clinical significance of this protein, the temporal pattern of immunoreactive serum α_2-SB-glycoprotein levels after surgery and traumatic injury were studied, because, theoretically, postoperative and posttraumatic opsonic deficiency should be reversible by intravenous cryoprecipitate infusion in view of its CIg content.

Materials and Methods

Purification of Human Opsonic Protein[26]

Serum Collection

Fresh human sera from normal healthy volunteers who ranged in age from 21 to 37 years were collected by withdrawing 100 ml of whole blood in sterile plastic syringes rinsed with sterile 0.1 M EDTA (ethylenediaminetetraacetate, sodium salt) to prevent fast clotting. The blood was allowed to clot for 1 hr at room temperature (25°C) and then centrifuged at 5000*g* for 20 min at 4°C. The collected serum was maintained at 4°C during all subsequent purification steps.

Ammonium Sulfate Precipitation

The serum was diluted with 1.6 vol of cold (4°C) phosphate-buffered saline (PBS, pH 7.4), and 1.28 vol of 100% saturated $(NH_4)_2SO_4$ in PBS (4°C) were added dropwise to yield a 33% saturated $(NH_4)_2SO_4$ solution, while stirring at 4°C. 2-Mercaptoethanol (ME) was then added to yield a final concentration of 0.02%, and the mixture was allowed to equilibrate while stirring slowly for 1 hr at 4°C. The suspension was then centrifuged for 20 min at 8000*g* at 4°C. The supernatant was discarded, except when being tested for biologic activity. If such tests were performed, it was dialyzed extensively against PBS at 4°C to remove $(NH_4)_2SO_4$ and assayed for its ability to augment Kupffer cell colloid uptake by the *in vitro* liver slice bioassay technique, as previously described.[31] The precipitated fraction was dissolved in cold distilled water to a volume equal to the original serum volume. The dissolved precipitate was dialyzed overnight against 100 vol of 0.054 M Tris-citrate buffer (pH 8.4) at 4°C with the addition of ME to a concentration of 0.01%.

Preparative Free-Flow Electrophoresis

The dialyzed biologically active ammonium sulfate fraction was further fractionated by means of preparative free-flow electrophoresis (FF-1, Brinkman Instruments) under conditions similar to those described previously.[26] The purified free-flow fractions were dialyzed individually for 24 hr against PBS to remove the Tris buffer, and each fraction was tested for biologic activity, while protein concentrations were determined in parallel by the Lowry method[32] to calculate specific activity (phagocytosis-promoting activity per milligram of protein). Fractions that manifested opsonic activity, as tested by bioassay, were pooled and dialyzed for 18 hr against 55% saturated $(NH_4)_2SO_4$ (0.01% ME) at 4°C. This pooled fraction was subsequently centrifuged at 8000*g* for 20 min, and the sedimented protein was stored at −78°C to preserve biologic activity.

Sepharose 4B Gel Filtration

The combined $(NH_4)_2SO_4$-precipitated biologically active fraction was dissolved in 4–6 ml of PBS (0.01% ME) and centrifuged for 10 min at 8000*g*. The clear supernatant was applied to a 72 × 2 cm Sepharose® 4B column (Pharmacia) previously equilibrated with PBS that contained 0.01% ME supplemented with 0.01% NaN_3 to prevent bacterial growth. Elution of the applied protein fraction was performed by pumping at a rate of 12.6 ml/hr, with 4.2-ml fractions being collected. The fractionation was performed at 4°C, and protein content was monitored by absorbancy at 280 nm.

When the eluted fractions were to be tested for biologic activity (i.e., ability to stimulate Kupffer cell phagocytosis), the column was washed extensively with PBS without sodium azide prior to fractionation, so that the eluted fractions could be tested directly for biologic activity by bioassay. With all samples, protein concentration was determined by the Lowry technique[32] to calculate specific activity units.

Once the position of the biologically active fractions had been repeatedly verified by bioassay, the protein fractions from subsequent repeated gel filtration separations were monitored by electroimmunoassay by use of polyspecific antiserum to the human opsonic α_2-SB-glycoprotein fractions. This immunochemical test for antigenic purity is well documented and was previously used for the rapid identification of gel filtration fractions with the highest content and greatest homogeneity of the α_2-SB-glycoprotein isolated from rat serum.[27] The fractions with the greatest homogeneity (first portion of the eluted protein peak) were pooled and dialyzed against 55% saturated $(NH_4)_2SO_4$ at 4°C. The collected protein precipitate was then rechromatographed on the Sepharose 4B column in a manner similar to the initial fractionation. This repeated gel filtration separation results in a further purification of the human protein comparable to that of the protein isolated from rat serum.[26] All fractions were tested by electroimmunoassay for antigenic purity and analyzed by gradient polyacrylamide gel electrophoresis (2.7–27% concave gradient, Isolabs, Inc.) for homogeneity.

Antiserum to Opsonic Protein and CIg

The rechromatographed protein fractions that exhibited homogeneity by electroimmunoassay and a single band by gradient gel electrophoresis were used to immunize male (3–5 kg) New Zealand white rabbits to develop antiserum, as previously described.[25] The protein solution (1 mg/ml) was heat denatured (100°C) in the presence of 1% sodium dodecyl sulfate (SDS) to increase the protein's antigenicity. Antiserum to CIg was obtained from Dr. Deane F. Mosher. The antiserum was monospecific for CIg, as previously reported.[33] The purified CIg was homogeneous when analyzed by SDS gel electrophoresis and gradient gel electrophoresis.

Immunochemical and Gel Electrophoresis Techniques

Ouchterlony double diffusion was performed at room temperature by means of standard techniques.[34] Antiserum to human haptoglobin was obtained from commercial sources (Behring Diagnostics). Sodium dodecyl sulfate polyacrylamide gel electrophoresis on 3.3% gels was performed according to the method of Weber and Osborn.[35] Samples were reduced by heating at 100°C for 10 min in the presence of 0.11 M ME and 1% SDS prior to electrophoresis. Gradient gel electrophoresis on a 2.7–27% concave gradient was accomplished as described previously.[24] Both gel systems were stained by the method of Fairbanks *et al.*[36]

Amino Acid Analysis

For amino acid analysis, the solutions of the respective isolated proteins (CIg and α_2-SB-glycoprotein) were dialyzed free of salt against 0.002 M triethylamine-acetate buffer (pH 8.4), and the salt-free dialysate was lyophilized. After standard hydrolysis with 6 N HCl at 110°C for 24 hr *in vacuo,* the hydrolysates were analyzed by use of an automatic amino acid analyzer.[37] For comparison, the moles of each amino acid per 100 mol of total amino acids (mol%) were calculated for both proteins. The values obtained were not corrected for incomplete release or partial destruction of some of the protein amino acids. Analysis of each protein was performed in duplicate.

Bioassay for Opsonic Activity

The standard *in vitro* liver slice assay was used to detect opsonic activity.[31] Liver slices were obtained from normal male adult Sprague-Dawley rats (200–

400 g). One milliliter of fresh human serum was added to 2.0 ml of Krebs-Ringer phosphate buffer (pH 7.4) and supplemented with 100 USP units of heparin (Upjohn) in a 25-ml Erlenmeyer flask. Thereafter, a fresh liver slice prepared by a Stadie-Riggs tissue slicer was added to each flask, and the phagocytic system was initiated by the addition of test particles to the medium. Opsonic activity of serum or plasma was evaluated by its ability to support the phagocytic uptake of an ^{131}I-labeled gelatinized "RE test lipid emulsion."[31] Liver Kupffer cell phagocytosis of the test particles was determined isotopically. Phagocytic uptake was expressed as the percentage of added counts phagocytized per 100 mg of wet liver slice weight (%ID/100 mg).

The assessment of the biologic activity of human cryoprecipitate and cryoprecipitate after extraction with 6.5% ethanol buffer was achieved by dissolving the protein precipitates in physiologic saline (0.9%) by use of 0.1 vol of the plasma from which the cryoprecipitate was initially obtained. This step was performed to further demonstrate that the CIg has opsonic activity but that its isolation by ethanol extraction destroys biologic activity. Solutions of extracted and nonextracted cryoprecipitate were dialyzed against physiologic saline overnight, and 1 ml of each was used in the *in vitro* bioassay in place of serum; the results were compared to those of normal human serum.

Electroimmunoassay

Electroimmunoassay was used for quantification of the α_2-SB-glycoprotein in rat and human sera as initially developed by Laurell.[38] Monospecific antiserum to the rat protein was developed as previously described,[25] and immunospecificity of the rat and human antisera exists. However, cross-reactivity between species might occur with antisera with very high antibody titers.

Results

Figure 1A shows a typical pattern of changes in serum immunoreactive α_2-SB-glycoprotein concentrations after surgery (laparotomy plus jejunal enterotomy) in experimental rats. Consistent with previous data on bioassayable opsonic activity, a significant decline in immunoreactive opsonin levels occurs over a 1–3-hr interval after surgery, with subsequent restoration and a phase of elevated levels of this protein observed by 24 hr postoperation. A typical response in the rat after sublethal traumatic shock is also presented for comparison (Figure 1B). A similar pattern of acute deficiency in the early posttraumatic period, followed by recovery, was observed. The temporal pattern of opsonic activity that occurs after both surgical intervention and whole-body trauma in rats coincides closely with RES phagocytic activity, especially Kupffer cell phagocytosis of test colloids.[11,13,15] In humans, immunoreactive opsonin levels seen after various forms of elective surgery (Figure 1C) and after trauma (Figure 1D) reveal a similar pattern of deficiency in the early time intervals, with subsequent restoration dependent on the degree of injury. Changes in immunoreactive protein in surgical and trauma patients correlate with bioassay determinations.[18,19]

Presented in Figure 2 is a comparative analysis of human α_2-SB-glycoprotein and CIg in both the native and reduced states, as tested by polyacrylamide gel electrophoresis. A similar migration pattern for these two proteins was observed on SDS and gradient gels.

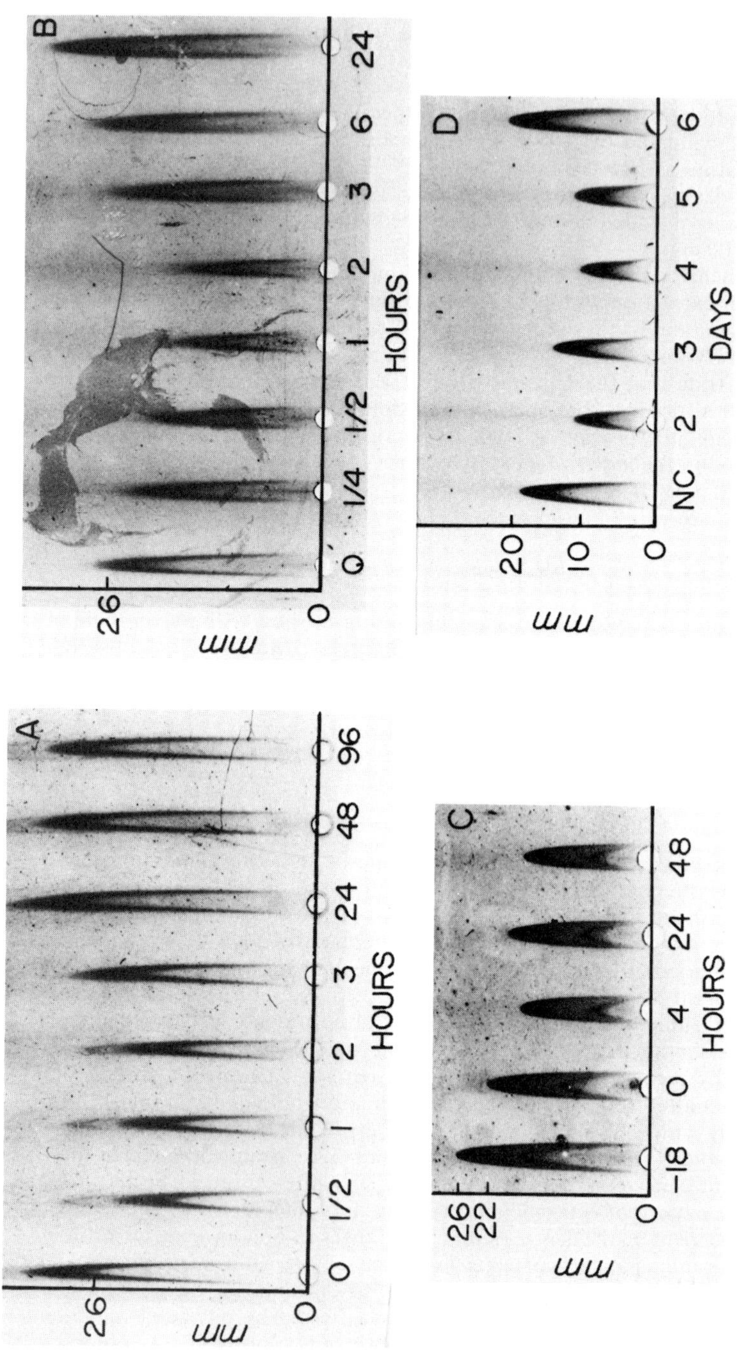

FIGURE 1. Representative changes in serum opsonic α_2-SB-glycoprotein levels in rats and humans after surgery and trauma. A: Immunoreactive protein levels prior to and at 30 min and 1, 2, 3, 24, 48, and 96 hr after surgery in rats (laparatomy plus jejunal enterotomy); B: immunoreactive protein levels prior to and at 15 and 30 min and 1, 2, 3, 6, and 24 hr after sublethal trauma in rats (NCD whole-body trauma); C: immunoreactive protein levels prior to (−18 & 0) and at 4, 24, 48 hr after elective abdominal surgery in patients; D: immunoreactive protein levels at 2, 3, 4, 5, and 6 days after severe traumatic injury as compared to serum levels in a comparably aged normal control (NC) volunteer.

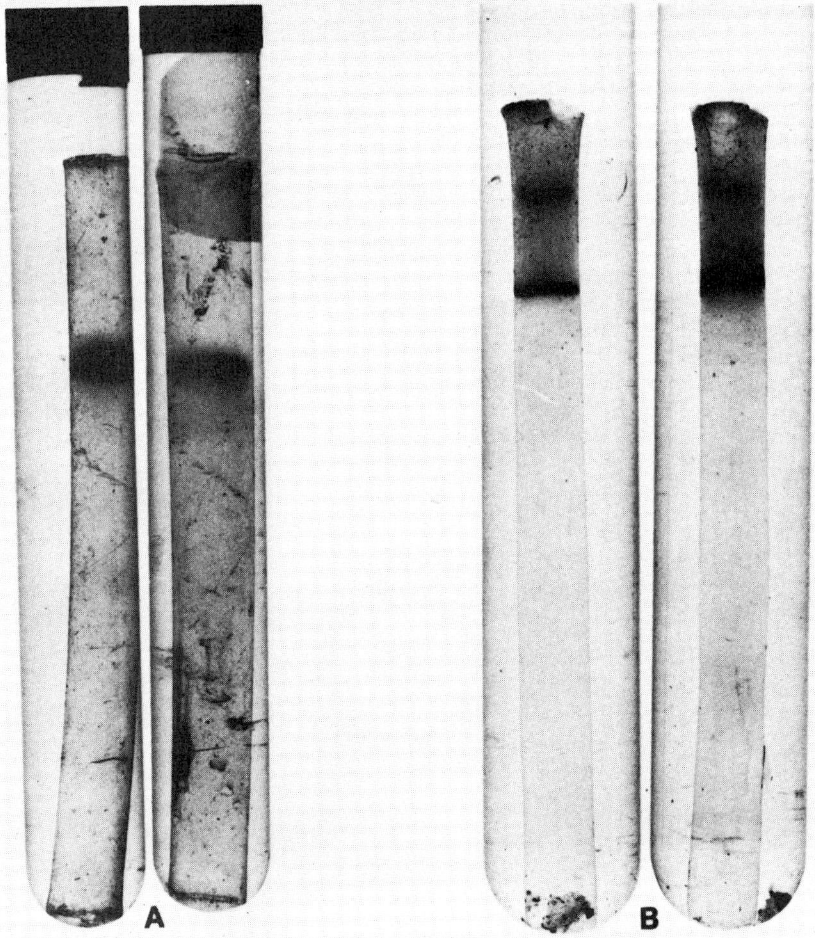

FIGURE 2. Polyacrylamide gel electrophoresis of purified human opsonic α_2-SB-glycoprotein (opsonin) and CIg. A: SDS polyacrylamide gel electrophoresis after 2-mercaptoethanol reduction; B: 2.7–27% gradient polyacrylamide gel electrophoresis of the native proteins.

FIGURE 3 gives the results of a double-diffusion experiment that demonstrates the common identity of these two proteins. In this study, monospecific antiserum to the human α_2-SB-glycoprotein and monospecific antiserum to plasma fibronectin (CIg) were diffused against normal human serum. Antibody to haptoglobin was also used as an example of nonidentity. Antiserum to both plasma CIg and α_2-SB-glycoprotein revealed a reaction of complete identity with human serum, indicating that these proteins are antigenically identical. Additional data in support of this antigenic identity were obtained by diffusion of monospecific antiserum to each of these two proteins against each purified antigen (FIGURE 4). Antiserum to the opsonic protein and antiserum to the fibronectin displayed a reaction of identity when diffused against the isolated opsonic protein, and a similar response was observed with respect to diffusion against the isolated fibronectin. Parallel studies presented elsewhere[39,40] demonstrate that *in vitro* colloid phagocytosis by Kupffer cells is inhibited by antisera to both proteins.

The amino acid analyses of both human proteins are presented in TABLE 1. The relative amino acid compositions of purified CIg and α_2-SB-glycoprotein are identical within the limits of the technique, and both values correspond closely to those published for CIg by Mosesson *et al.*[41] By our analysis, the CIg has a slightly lower content of sulfur-containing amino acids, which may relate to the lack of biologic activity in CIg-containing cryoprecipitate after ethanol extraction, as tested by the heparin-dependent liver slice bioassay and as compared to the α_2-SB-glycoprotein.[39] Inclusion of ME during purification was essential for retention of maximal activity during the isolation of the α_2-SB-glycoprotein, a finding that may relate to the above observations.

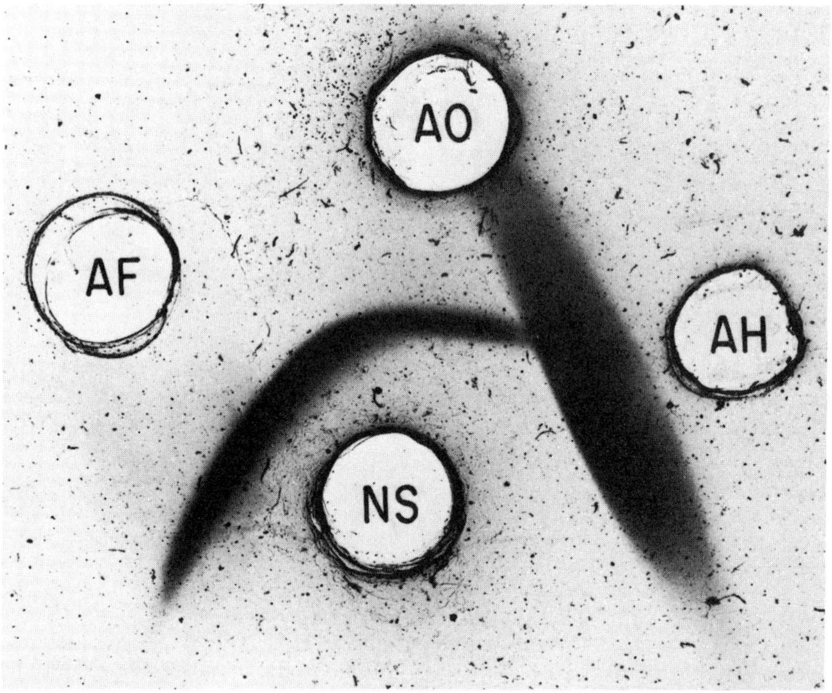

FIGURE 3. Ouchterlony double diffusion of normal human serum (NS) against antisera to the opsonic α_2-SB-glycoprotein (AO) and to CIg (AF). Antiserum to human haptoglobin (AH) was used to demonstrate nonidentity.

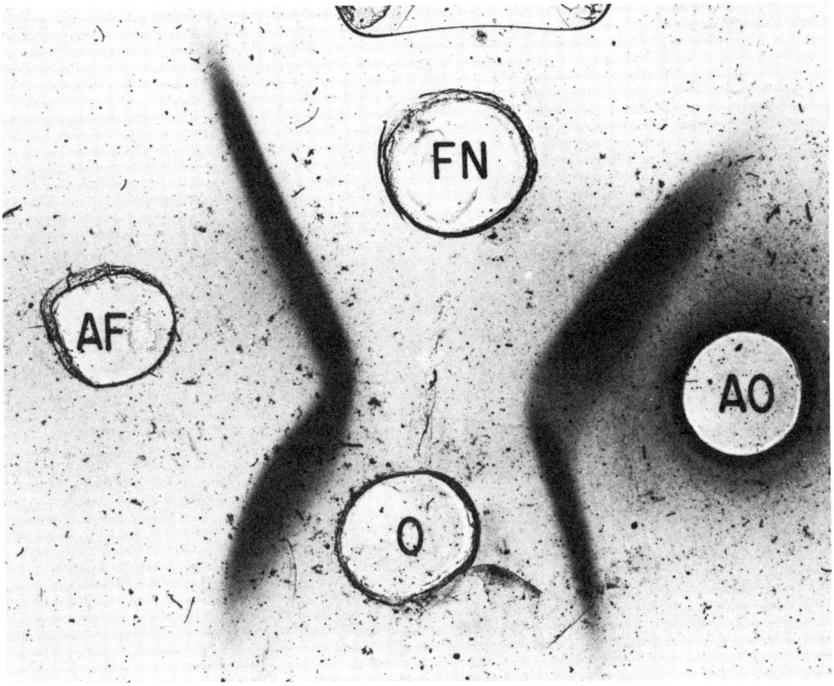

FIGURE 4. Ouchterlony double diffusion of human opsonic α_2-SB-glycoprotein (O) and CIg (FN) against antisera to the opsonic protein (AO) and to CIg (AF).

DISCUSSION

Previous studies of the opsonic α_2-SB-glycoprotein suggest an important function for this protein in the regulation of RES systemic host defense mechanisms.[28] In this regard, the ability of the RES to remove circulating foreign and abnormal particulate matter correlates with the serum levels of this protein.[5,24,25] A wealth of evidence demonstrates an important relationship between RES function and host defense against malignant disease, bacterial infection, and traumatic shock.[2,6,7,9,16,17] An equal abundance of evidence points to an important role for this protein as a determinant of Kupffer cell activity. Such data emphasize the need to understand the physiopathology of the RES and the manner in which this protein may be a determinant of its activity. The classic technique for measuring *in vivo* reticuloendothelial phagocytic activity involves the colloid clearance method.[1,2] However, the functional stability of the RES *in vivo* in humans has not been adequately studied due to the invasive nature of this technique. In addition, recent evidence demonstrates that intravenous colloid injection depletes the plasma level of this protein.[25] Thus, caution has to be exercised in utilizing the colloid clearance technique *in vivo* for routine clinical evaluation. These facts, coupled with the inadequacy of the tissue slice bioassay for routine clinical studies, stress the need to isolate this protein and develop an immunoassay that could potentially be used as a noninvasive measure of RE function.

Analysis of the isolated rat opsonic protein resulted in the realization that CIg or plasma fibronectin could be the α_2-SB-glycoprotein. The immunochemical and biochemical data obtained in the present study document for the first time their antigenic identity by double diffusion and their similar composition by amino acid analysis. Several other interesting observations have been made that deserve some comment. Fresh human plasma cryoprecipitate has both bioreactive and immunoreactive opsonic activity and is known to contain significant amounts of CIg. However, extraction of cryoprecipitate with 6.5% ethanol (a step in the purification of CIg) completely abolishes its biologic activity, as tested by bioassay.[39,40] Also, the human opsonic protein in its native state appears to be not very antigenic, and denaturation with SDS results in a more antigenic preparation.

TABLE 1

COMPARATIVE AMINO ACID ANALYSES OF HUMAN OPSONIC α_2-SB-GLYCOPROTEIN AND HUMAN COLD-INSOLUBLE GLOBULIN

	Moles % (amino acid content)	
Amino Acid*	Opsonic α_2-SB-Glycoprotein†	Cold-Insoluble Globulin†
Aspartic acid	9.14	9.44
Threonine	9.14	9.15
Serine	9.14	9.38
Glutamic acid	12.70	12.49
Glycine	10.69	9.61
Alanine	4.77	4.49
Valine	5.34	5.29
Methionine	2.13	1.21
Isoleucine	3.28	3.16
Leucine	5.00	5.93
Tyrosine	3.79	4.32
Phenylalanine	2.24	2.65
Lysine	3.62	3.97
Histidine	2.82	2.30
Arginine	5.29	5.75
Proline	8.79	9.09
Half-cystine	2.13	1.73

*Tryptophan analysis was not performed.

†Uncorrected for destruction of serine and threonine and for incomplete release of valine or isoleucine (24 hr, 110°C, 6 N HCl hydrolysis).

The present documentation of identity between these two proteins has important implications. Opsonic protein deficiency in animals and in humans after injury is presented in the present study and is well documented.[13-15,18,19] Moreover, there is an apparent depletion of this protein during advanced malignant disease.[6] Approaches designed to reverse this deficiency have been limited, due to the small quantity of pure biologically active protein that can be isolated with the technique developed by our laboratory. However, the similarities between these two proteins, coupled with the well-known content of CIg in plasma cryoprecipitate and, perhaps, in other fractions of fresh plasma, suggest that cryoprecipitate infusion may be an effective mechanism to reverse opsonic deficiency and RES systemic host defense depression. Preliminary studies have been initiated in surgical patients with severe septicemia or in trauma patients with coexistent sepsis and pulmonary insufficiency.[42] These studies demonstrate that intravenous infusion of cryoprecipitate results in an increase in both bioassayable and immunoreactive α_2-SB-glycoprotein levels, especially 30 min to 4 hr postinfusion, with a gradual decline over the next 24–48 hr at a rate somewhat related to the extent of injury. Significant, however, was the improvement in the clinical state of these patients, as reflected by a decrease in the febrile state, apparent attenuation of the severe septic condition, and improvement in pulmonary function. Persistent posttraumatic pulmonary insufficiency in humans exhibits a very high correlation with sepsis after major surgery and trauma[43] and is believed to be due in part to microembolization of the pulmonary bed by blood-borne particulate matter, for example, injured platelets, immune complexes, and fibrin microaggregates. Our clinical data, while preliminary in nature, suggest that restoration of RE function may augment the removal of such matter from the vascular compartment and circumvent their influence on pulmonary perfusion and vascular permeability, possibly resulting in pulmonary insufficiency.

Thus, CIg and α_2-SB-glycoprotein in human serum appear to be identical. Whether the isolated rat opsonic protein is identical to rat CIg remains to be determined. It is believed that these fundamental observations link two major areas of experimental biology and medicine that were previously separate and distinct. One area is related to the biochemical and immunochemical aspects of a plasma protein from a molecular standpoint and its relationship to membrane events. The second area deals with host defense against hematogenous dissemination of particulate matter, immune complexes, and tumor cells as potentially disturbed by surgery, trauma, burn injury, sepsis, and malignant disease. Further studies are required to enhance our somewhat primitive understanding of the significance of this relationship, but the relevance of these observations to surgical and posttraumatic care, in addition to cancer prognosis, appears predictable.

Acknowledgments

The authors thank Dr. Deane F. Mosher for providing purified plasma CIg and rabbit antiserum to plasma CIg (fibronectin). We also thank Ms. Maureen Kaiser for secretarial assistance.

References

1. Benacerraf, B., G. Biozzi, B. N. Halpern & C. Stiffel. 1975. Physiology of phagocytosis of particles by the RES. *In* Physiopathology of the Reticuloendothelial System. B. N. Halpern, Ed.: 52–79. Charles C. Thomas. Springfield, Ill.
2. Saba, T. M. 1970. Physiology and physiopathology of the reticuloendothelial system. Arch. Intern. Med. **126:** 1031–1052.
3. Metchnikoff, E. 1905. Immunity in Infective Diseases. F. G. Binnie, Transl. Cambridge University Press. Cambridge, England.
4. Unanue, E. R. 1972. The regulatory role of macrophages in antigenic stimulation. Adv. Immunol. **15:** 95–165.
5. Saba, T. M. & N. R. Di Luzio. 1969. Reticuloendothelial blockade and recovery as a function of opsonic activity. Amer. J. Physiol. **216:** 197–205.

6. Di Luzio, N. R. 1975. Macrophages, recognition factors and neoplasia. *In* The Reticuloendothelial System. Vol. 16: 49–64. IAP Monograph. The Williams & Wilkins Co. Baltimore, Md.
7. Pardy, B. J., R. C. Spencer & H. A. F. Dudley. 1977. Hepatic reticuloendothelial protection against bacteremia in hemorrhagic shock. Surgery **81:** 193–197.
8. Zweifach, B. W., B. Benacerraf & L. Thomas. Relationship between the vascular manifestations of shock produced by endotoxin, trauma, and hemorrhage. II. The possible role of the RES in resistance to each type of shock. J. Exp. Med. **106:** 403–414.
9. Rogers, D. E. 1960. Host mechanisms which act to remove bacteria from the bloodstream. Bacteriol. Rev. **24:** 50–66.
10. Alexander, J. W., M. A. McClellan, C. K. Ogle & J. P. Ogle. 1976. Consumptive opsoniopathy. Possible pathogenesis in lethal and opportunistic infections. Ann. Surg. **184:** 622–678.
11. Altura, B. M. & S. G. Hershey. 1972. Reticuloendothelial function in experimental injury and tolerance to shock. Adv. Exp. Med. Biol. **33:** 545–569.
12. Goldman, A. S., H. B. Rudolff, R. McNamee, L. D. Loose & N. R. Di Luzio. 1973. Deficiency of plasma humoral recognition factor activity following burn injury. J. Reticuloendothel. Soc. **15:** 192–197.
13. Kaplan, J. E. & T. M. Saba. 1975. Humoral deficiency and reticuloendothelial depression after traumatic shock. Amer. J. Physiol. **230:** 7–14.
14. Kaplan, J. E., T. M. Saba & E. Cho. 1976. Serological modifications of reticuloendothelial capacity and altered resistance to traumatic shock. Circ Shock **3:** 203–210.
15. Saba, T. M. 1972. Effect of surgical trauma on the clearance and localization of bloodborne particulate matter. Surgery **71:** 675–685.
16. Saba, T. M. 1975. Reticuloendothelial system host defense after surgery and traumatic shock. Circ. Shock **2:** 91–108.
17. Saba, T. M. & W. A. Scovill. 1975. Effect of surgical trauma on host defense. Surg. Annu. **7:** 71–102.
18. Scovill, W. A., T. M. Saba, J. E. Kaplan, H. Bernard & S. R. Powers. 1976. Deficits in reticuloendothelial humoral mechanisms after trauma. J. Trauma **16:** 898–904.
19. Scovill, W. A., T. M. Saba, J. E. Kaplan, H. Bernard & S. R. Powers. 1977. Disturbances in circulating opsonic activity in man after operative and blunt trauma. J. Surg. Res. **22:** 709–716.
20. Kaplan, J. E. & T. M. Saba. 1978. Low-grade intravascular coagulation and reticuloendothelial function. Amer. J. Physiol. **234.**
21. Schumacker, P. T. & T. M. Saba. 1977. Augmentation of pulmonary insufficiency during low-grade intravascular coagulation by prior impairment in hepatic RE clearance activity. Physiologist **20:** 85 (Abst.).
22. Esnouf, M. P. & R. Marshall. 1968. The effect of blockade of the reticuloendothelial system and of hypotension on the response of dogs to *Ancistrodon rhodostoma* venom. Clin. Sci. **35:** 261–272.
23. Pardy, B. J. & H. A. F. Padley. 1972. Post-traumatic pulmonary insufficiency. Surg. Gynecol. Obstet. **144:** 259–269.
24. Blumenstock, F. A., T. M. Saba, P. Weber & E. Cho. 1976. Purification and biochemical characterization of a macrophage stimulating alpha-2-globulin opsonic protein. J. Reticuloendothel. Soc. **19:** 157–172.
25. Blumenstock, F. A., P. Weber, T. M. Saba & R. Laffin. 1977. Electroimmunoassay of alpha-2-opsonic protein levels during reticuloendothelial blockade. Amer. J. Physiol. **232:** R80–R87.
26. Blumenstock, F. A., P. Weber & T. M. Saba. 1978. Purification of alpha-2-opsonic protein from human serum and its measurement by immunoassay. J. Reticuloendothel. Soc. **23:** 119–134.
27. Blumenstock, F. A., P. Weber & T. M. Saba. 1977. Isolation and biochemical characterization of an alpha-2-opsonic protein from rat serum. J. Biol. Chem. **252:** 7156–7162.

28. SABA, T. M. 1974. Aspecific opsonin. *In* The Immune System and Infectious Diseases. E. Neter & R. Milgrom, Eds.: 489–504. S. Karger. Basel, Switzerland.
29. MCLAIN, S., J. SIEGEL, J. MOLNAR, C. ALLEN & T. SABET. 1976. A phagocytosis promoting factor of rat serum independent of the complement system. J. Reticuloendothel. Soc. **19:** 127–138.
30. KAPLAN, J. E., J. MOLNAR, T. M. SABA & C. ALLEN. 1976. Comparative disappearance and localization of isotopically labelled opsonic protein and soluble albumin after surgical trauma. J. Reticuloendothel. Soc. **20:** 375–384.
31. SABA, T. M., J. P. FILKINS & N. R. DI LUZIO. 1966. Properties of the "opsonic system" regulating *in vitro* hepatic phagocytosis. J. Reticuloendothel. Soc. **3:** 398–414.
32. LOWRY, O., N. J. ROSEBROUGH, A. L. FARR & R. J. RANDALL. 1951. Protein measurement with the Folin phenol reagent. J. Biol. Chem. **193:** 265–275.
33. MOSHER, D. F. 1976. Changes in plasma cold-insoluble globulin concentration during experimental Rocky Mountain spotted fever infection in rhesus monkeys. Thromb. Res. **9:** 37–45.
34. OUCHTERLONY, O. 1958. Diffusion-in-gel methods for immunological analysis. *In* Progress in Allergy. D. Kallos, Ed. Vol. V: 1–78. S. Karger. Basel and New York.
35. WEBER, K. & M. OSBORN. 1975. Proteins and sodium dodecyl sulfate; molecular weight determination on polyacrylamide gels and related procedures. *In* The Proteins. N. H. Neurath & R. Hill, Eds.: 179–223. Academic Press, Inc. New York, N.Y.
36. FAIRBANKS, G. T., T. J. STECK & D. J. M. WALLACH. 1971. Electrophoretic analysis of the major polypeptides of the human erythrocyte membrane. Biochemistry **10:** 2606–2617.
37. SPACKMAN, D. H., W. H. STEIN & S. MOORE. 1958. Automatic recording apparatus for use in the chromatography of amino acids. Anal. Chem. **30:** 1190–1206.
38. LAURELL, C. B. 1972. Electroimmunoassay. Scand. J. Clin. Lab. Invest. **29**(Suppl. 124): 21–37.
39. BLUMENSTOCK, F. A., T. M. SABA, P. WEBER & R. LAFFIN. 1978. Biochemical and immunological characterization of human opsonic α_2SB glycoprotein: its identity with cold-insoluble globulin. J. Biol. Chem. In press.
40. BLUMENSTOCK, F. A., T. M. SABA, P. WEBER & R. LAFFIN. 1977. Physiological function for human cold-insoluble globulin: identity with human opsonic α_2SB glycoprotein. J. Reticuloendothel. Soc. **22:** 35 (Abst.).
41. MOSESSON, M. W., A. B. CHEN & R. M. HUSEBY. 1975. The cold-insoluble globulin of human: studies of its essential structural features. Biochim. Biophys. Acta **386:** 509–524.
42. SABA, T. M., F. A. BLUMENSTOCK, H. BERNARD & J. E. KAPLAN. 1977. Prevention of reticuloendothelial depression following surgery by intravenous opsonic glycoprotein therapy. J. Reticuloendothel. Soc. **22:** 16 (Abst.).
43. FULTON, R. L. & E. C. JONES. 1975. The etiology of post-traumatic pulmonary insufficiency in man. Rev. Surg. **32:** 84–85.
44. KAPLAN, J. E. & T. M. SABA. 1977. Unpublished results.

BOVINE PLASMA COLD-INSOLUBLE GLOBULIN: GROSS STRUCTURE AND FUNCTION*

Sadaaki Iwanaga

Institute for Protein Research
Osaka University
Yamada-Kami, Suita, Osaka 565, Japan

Koji Suzuki and Senichiro Hashimoto

Department of Clinical Pathology
Mie University School of Medicine
Tsu, Mie 514, Japan

Cold-insoluble globulin (CIg) is a major nonclottable plasma protein recognized since 1948 as a component of cryofibrinogen.[1] Cold-insoluble globulin has a tendency to coprecipitate with fibrinogen and its derivatives, factor VIII and factor XIII, in the cold.[2-6] This protein was first purified from human plasma in 1970 by Mosesson and Umfleet.[5] These workers characterized it as a glycoprotein with a molecular weight of 450,000 daltons.[6]

Although its biologic function in plasma is still unknown, it has recently been reported that CIg is incorporated into the fibrin clot in the presence of activated factor XIII.[7] Moreover, a protein very similar to CIg has been found on the surface of cultured normal fibroblasts, and several malignantly transformed fibroblasts have been reported to lack this protein on their surfaces.[8-11]

In the course of our studies on the purification of factor XIII from bovine plasma, we found a protein similar to human CIg. In this paper, we shall first describe the physicochemical properties and terminal structure of bovine plasma CIg. A specific interaction between CIg and fibrin monomer in the presence of activated factor XIII will then be discussed. Finally, stimulation of the urokinase-mediated transformation of plasminogen to plasmin by CIg will be described. Preliminary reports of some of these results have already appeared.[12-14]

Materials and Methods

Cold-insoluble globulin was purified from fresh bovine plasma by precipitation with ammonium sulfate at 20% saturation, followed by heat treatment at 56°C, ion-exchange chromatography on DEAE cellulose (Whatman DE-32), and gel filtration on Sepharose® 4B (Pharmacia Chemicals).[13] Analytic polyacrylamide gel electrophoresis was performed at pH 8.3 with 4 or 5% gels as described by Davis.[15] Sodium dodecyl sulfate (SDS) polyacrylamide gel electrophoresis was performed in 4 or 5% gels according to the method of Weber and Osborn.[16] The molecular weight of CIg was determined by SDS gel electrophoretic migration and by sedimentation equilibrium experiments according to the method of Yphantis.[17] Circular dichroic spectral analysis was achieved at 25°C with a JASCO J-20 spectral analyzer (Japan Spectroscopic Company). Electron micrographs of

*Supported in part by Grant 958020 from the Scientific Research Fund of the Ministry of Education, Science and Culture of Japan.

0077-8923/78/0312-0056 $0.175/0

CIg stained with 1% phosphotungstic acid (pH 5.6) on carbon-coated copper grids were obtained on Hitachi HY-12 electron microscopes. Antisera against bovine CIg and fibrinogen were prepared by immunizing rabbits with purified CIg and fibrinogen, respectively, in Freund's complete adjuvant. Amino acid analyses were performed with a JLC-5AH amino acid analyzer (Japan Electron Optics) according to the method of Spackman *et al.*[18] Samples were hydrolyzed in constant-boiling hydrochloric acid, at 110°C for 24, 48, or 72 hr in evacuated sealed tubes. Total hexose, hexosamine, and sialic acid values were determined by the orcinol-sulfuric acid reaction,[19] the Elson-Morgan reaction,[20] and Warren's method,[21] respectively. The amino terminal sequence was determined by use of the Edman direct degradation technique.[22] Phenylthiohydantoin (PTH) amino acids were identified by thin-layer chromatography on polyamide sheets with the solvents described by Kimura.[23] Digestion of CIg with bovine plasmin and bovine plasma kallikrein was performed at a substrate/enzyme ratio of 100:1 (w/w) at 37°C for various periods of time. Digestion was terminated by the addition of diisopropylphosphorofluoridate (DFP) to give a final concentration of 10 mM. Fibrinogen-agarose and fibrin monomer-agarose were prepared, respectively, according to the method of Heene and Matthias,[24] by use of bovine plasminogen-free fibrinogen (Daiichi Kagaku Company) and cyanogen bromide-activated Sepharose 4B. [^{125}I]Cold-insoluble globulin was prepared by a slight modification of the method of Hunter.[25] A cross-linked substance formed with CIg and fibrin in the presence of activated bovine factor XIII was prepared by the method described previously.[12] The activity of plasmin, transformed from purified bovine plasminogen[26] with urokinase (Mochida Pharmaceutical Company), was determined by esterolysis of synthetic substrate TAME[27] (α-*N*-toluenesulfonyl-L-arginine methyl ester), caseinolysis,[28] and fibrinolysis.

Results and Discussion

Physicochemical Properties

Figure 1 shows the elution pattern obtained by DEAE cellulose chromatography of the crude CIg fraction that remained after heat treatment. Cold-insoluble globulin was eluted in the fourth peak. At this stage, it was approximately 95% pure. This material was further purified on a Sepharose 4B column. By these procedures, approximately 300 mg of purified CIg were obtained from 10 liters of bovine plasma. The purified protein was not contaminated with fibrinogen, factor XIII, or immunoglobulin G, as indicated by immunodiffusion and immunoelectrophoresis. However, the preparation was contaminated by a small amount of an aggregating factor for human platelets.[29] Recently, we have succeeded in separating platelet-aggregating factor from our CIg preparations by use of human platelets fixed with paraformaldehyde.[29a]

Purified CIg gave a single band on polyacrylamide gel disc electrophoresis (Figure 2,1). However, on SDS gel electrophoresis, it gave two other faint bands, in addition to the single major CIg band, as shown in Figure 2 (2). This major band yielded two bands on reduction of CIg with 2-mercaptoethanol (Figure 2,3). The migration rates of these bands corresponded to those of the two faint bands observed in the unreduced preparation. Further, we found that the amino acid compositions of these protein bands isolated from SDS gels were almost identical with that of the major CIg protein band. Thus, these two proteins are apparently

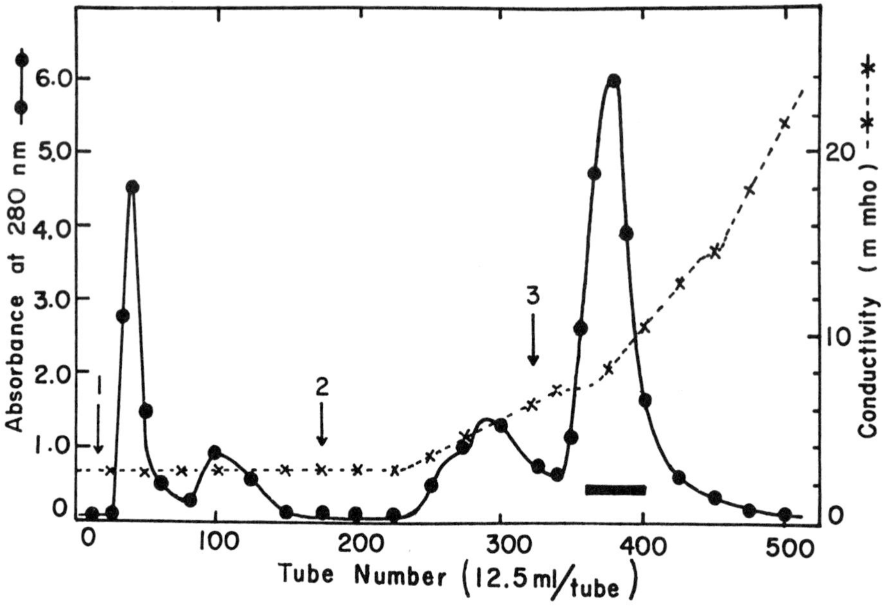

FIGURE 1. Chromatography of crude CIg on a DEAE cellulose (Whatman DE-32) column. The crude CIg sample (4.3 g), prepared by Loewy *et al.*,[33] was applied to the column (5 × 48 cm) at 4°C. Eluants: 1, 0.05 M Tris HCl (pH 7.5) that contained 1 mM EDTA; 2, a linear gradient with 0–0.1 M NaCl in eluant 1; 3, a linear gradient with 0.1–0.3 M NaCl in eluant 1. The flow rate was adjusted to 110 ml/hr.

not contaminants but, rather, degradation products of whole CIg. These results suggest that bovine CIg, like human CIg,[6] consists of two subunit polypeptides, the α and β chains.

Ultracentrifugal analysis of the purified material revealed a single component, and a sedimentation coefficient of 12.2S was obtained. The molecular weight of CIg, as estimated by the sedimentation equilibrium experiment in 0.1 M phosphate buffer (pH 7.0), is approximately 450,000 daltons. Moreover, the molecular weights of the α and β chains were estimated to be 220,000 and 212,000 daltons, respectively, based on a plot of log molecular weight of the nine reduced standard proteins versus relative mobility on 5% polyacrylamide gels in 0.1% SDS (FIGURE 3). Furthermore, circular dichroism indicated that large portions of the CIg molecule, like its human counterpart, contain a high proportion of β structure, as does silk fibroin and poly-L-lysine.[6]

TABLE 1 summarizes some of the known physicochemical properties of bovine CIg.

Electron microscopy of CIg after negative staining with phosphotungstic acid or shadow casting with platinum-carbon indicated a fibrous structure. FIGURE 4 is a photograph taken after negative staining.

TABLE 2 shows the amino acid and carbohydrate compositions of purified CIg. The overall composition was very similar to that of human CIg,[6] and the carbohydrate contents of these two proteins were almost identical. There was no

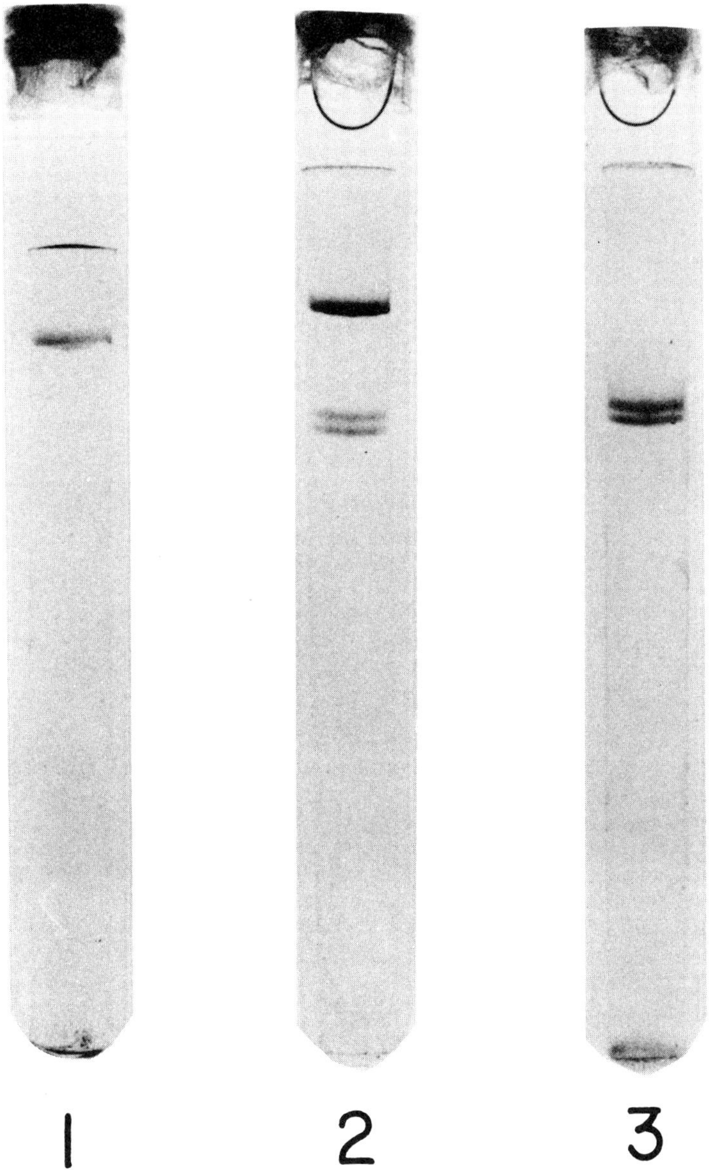

FIGURE 2. Polyacrylamide gel disc electrophoresis of purified CIg in the absence or presence of SDS. Electrophoresis in the absence of SDS (gel 1) was performed in 4% gels at pH 8.3 for 2 hr at 4 mA per tube. Electrophoresis in the presence of 0.1% SDS (gel 2, unreduced; gel 3, reduced with 2-mercaptoethanol) was performed in 4% gels at pH 7.3 for 4 hr at 8 mA per tube. The gels were stained with Coomassie brilliant blue. Migration from top (cathode) to bottom.

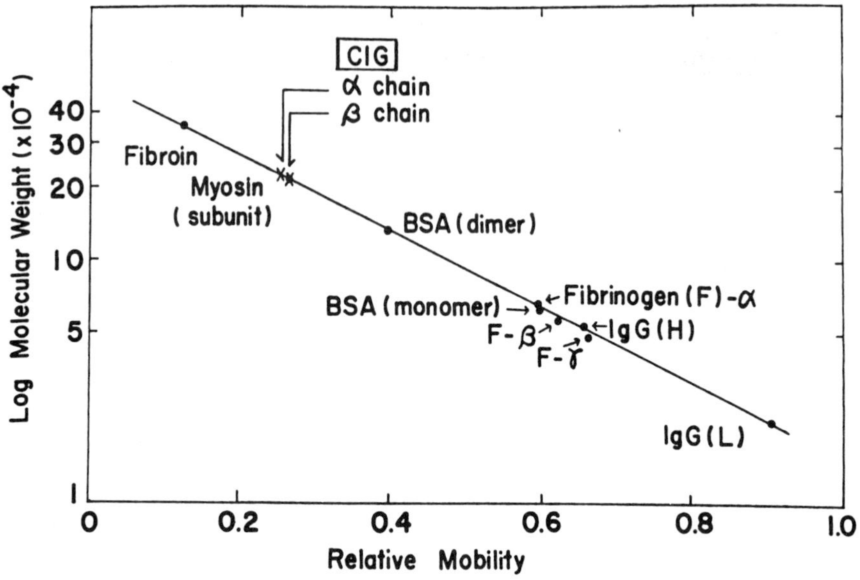

FIGURE 3. A plot of log molecular weight of reduced proteins versus relative mobility on 5% polyacrylamide gels in 0.1% SDS. The molecular weight of each reduced protein was calibrated as follows: silk fibroin, 350,000; muscle myosin subunit, 212,000; bovine serum albumin (BSA) dimer, 130,000; BSA monomer, 65,000; fibrinogen α chain, 63,500; fibrinogen β chain, 56,000; fibrinogen γ-chain, 47,000; immunoglobulin G heavy (H) chain [IgG(H)], 53,000; IgG light (L) chain [IgG(L)], 22,000.

unusual feature of the amino acid distribution compared with those of other proteins, except that threonine and valine residues seem to be relatively more numerous. The amino terminal residue of CIg was unreactive to Edman's reagent before treatment with pyrrolidone carboxylyl peptidase. However, after enzymatic treatment, a single PTH-alanine, followed by threonine, was identified, and the pyroglutamyl-alanyl sequence was the same as that of human CIg analyzed by Mosesson *et al.*[6]

TABLE 1
PHYSICOCHEMICAL PROPERTIES OF BOVINE PLASMA COLD-INSOLUBLE GLOBULIN

Molecular weight		
Sedimentation equilibrium		450,000
SDS gel electrophoresis		
unreduced		450,000
reduced	α chain	220,000
	β chain	212,000
Sedimentation coefficient		12.2S
Electrophoretic migration		fast β-globulin
Isoelectric point		pH 5.0
Partial specific volume		0.72 cm^3/g
$A^{1\%}_{280\,nm}$ in 8 M urea-0.2 N NaOH		12.9
Circular dichroism spectra		λ_{max} = 232 nm, λ_{min} = 218 nm

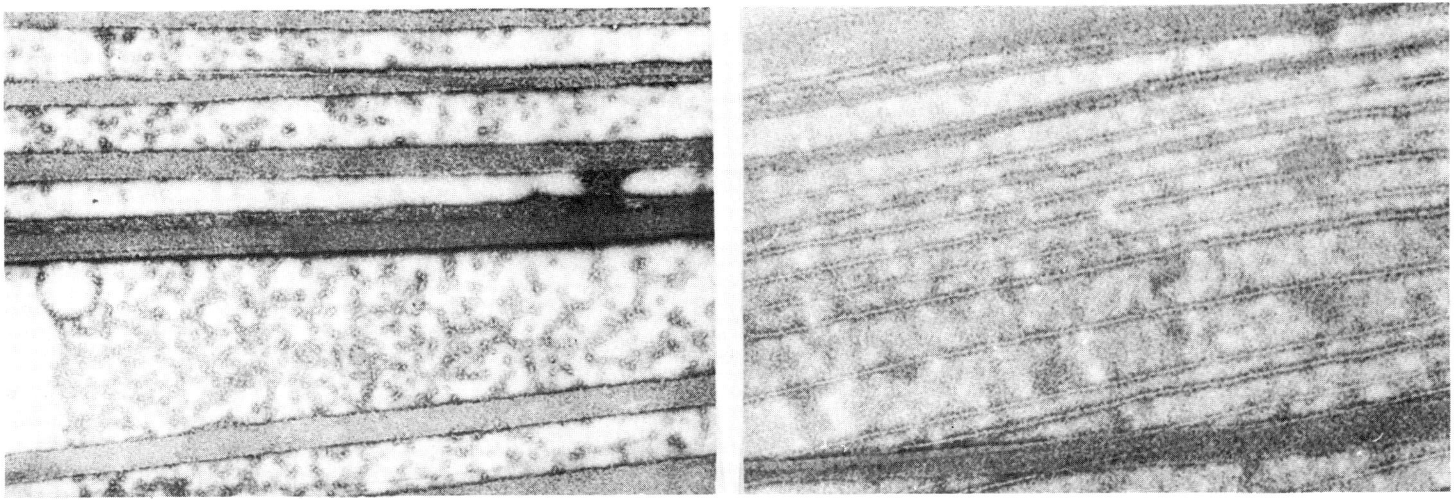

FIGURE 4. Electron micrographs of CIg after negative staining with phosphotungstic acid on carbon-coated copper grids. ×28,000.

TABLE 2

AMINO ACID AND CARBOHYDRATE COMPOSITIONS OF COLD-INSOLUBLE GLOBULIN

Amino Acid	Bovine	Human*
	(g of amino acid/100 g of protein)	
Aspartic acid	9.1	9.3
Threonine	9.8	9.7
Serine	8.0	6.8
Glutamic acid	11.4	11.6
Proline	7.9	7.6
Glycine	9.1	8.0
Alanine	4.6	4.3
Cystine	2.0	2.6
Valine	7.6	8.1
Methionine	1.0	1.1
Isoleucine	4.3	4.4
Leucine	5.6	5.7
Tyrosine	4.0	4.5
Phenylalanine	2.2	2.7
Lysine	3.9	3.6
Histidine	1.8	2.1
Arginine	5.2	5.2
Tryptophan	1.8	2.8
Carbohydrate		
Hexose (%)	1.8	1.8
Hexosamine (%)	2.2	2.1
Sialic acid (%)	1.1	1.2
Total (%)	5.1	5.1

*From Mosesson *et al.*[6]

Terminal Structure

To obtain some information regarding its gross structure, CIg was digested with bovine plasmin [substrate/enzyme ratio of 100:1 (w/w)], and the resultant digests were subjected to SDS gel electrophoresis without prior reduction. After incubation for 10 min, two new bands appeared on the gels, as illustrated in FIGURE 5 (*left*). These bands migrated at the positions of the α and β chains, respectively, suggesting that intermolecular disulfide bridging of CIg occurs near the end(s) of the molecule. On subsequent incubation, these bands were degraded gradually, accompanied by an increase of intensity of the fast-moving band, with a molecular weight of approximately 29,000 daltons. These digestion products were present for a long time and underwent only slow subsequent proteolysis. FIGURE 5 (*right*) shows the SDS gel patterns after digestion with highly purified bovine plasma kallikrein. High-molecular-weight products and a 29,000-dalton fragment were also formed. However, the rate of plasma kallikrein digestion of CIg was much slower than that of plasminolysis.

For the isolation of these fragments, a 1-hr plasmin digest of CIg was subjected to gel filtration on a column of Sephadex® G-100; the elution profile is shown in FIGURE 6. As judged from the SDS gel electrophoretic patterns of each of the five fractions, fractions 1–3 contained a high-molecular-weight product, in addition to undigested CIg. The amino terminal sequence of the major peak analyzed by the same procedures as those applied to whole CIg was found to be

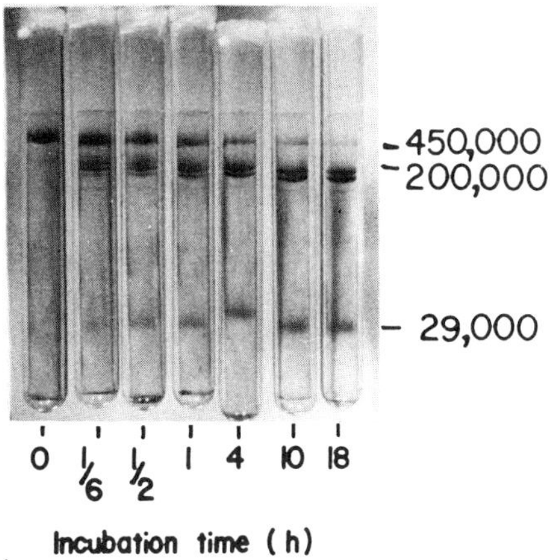

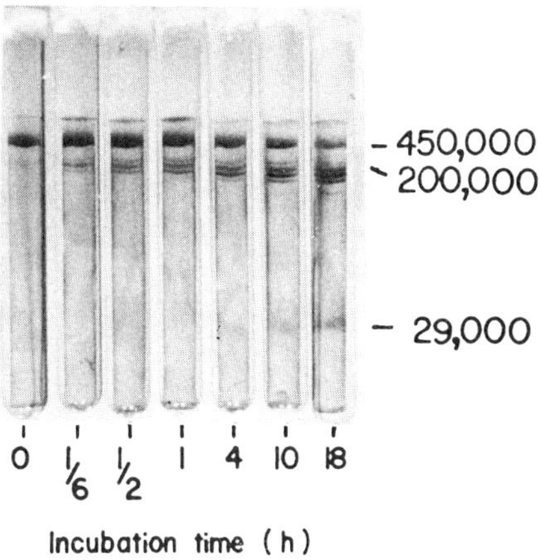

FIGURE 5. Sodium dodecyl sulfate gel electrophoretic patterns of digests of CIg with plasmin (*left*) and plasma kallikrein (*right*). Digestion was achieved at a substrate/enzyme ratio of 100:1 (w/w) at 37°C in 0.1 M NH_4HCO_3 (pH 8.0) for the various periods of time indicated on the Figure, and the reaction was terminated with DFP to give a final concentration of 10 mM. Sodium dodecyl sulfate gel electrophoresis was performed in 4% gels at pH 7.3 for 4 hr at 8 mA per tube. Migration from top (cathode) to bottom.

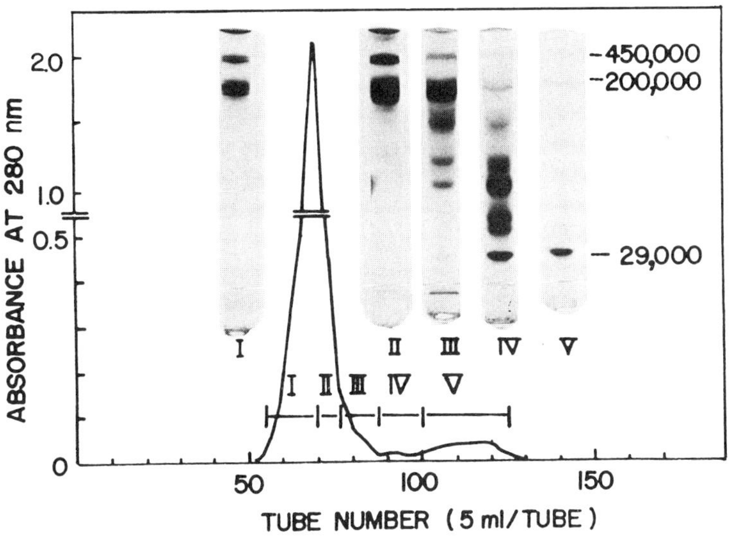

FIGURE 6. Gel filtration of a plasmin digest of CIg on a Sephadex G-100 column. Under the same conditions described in FIGURE 5, CIg was digested with plasmin for 1 hr, and the digest was applied to the column (3.5 × 95 cm) equilibrated with 0.2 M NH_4HCO_3 (pH 8.0) and eluted at room temperature. The flow rate was adjusted to 50 ml/hr. Fractions indicated with Roman numerals and solid bars were collected, lyophilized, and subjected to SDS gel electrophoresis on 5% gels. Migration was from top (cathode) to bottom.

pyroglutamyl-alanyl-threonine, identical to that of intact CIg. On the other hand, fraction 5 contained a fragment of 29,000 daltons and displayed a single band on SDS gel electrophoresis (FIGURE 6). The first three amino terminal residues of this fragment were found to be valyl-valyl-glutamine. These results suggest that the high-molecular-weight product and the 29,000-dalton fragment are localized, respectively, in the amino and carboxyl terminal portions of the parent molecule.

TABLE 3 lists the amino acid compositions of the amino terminal high-molecular-weight fragment and of the carboxyl terminal fragment thus isolated. The overall composition of the high-molecular-weight fragment was very similar to that of intact CIg, except that its cystine and glycine contents were significantly decreased. On the other hand, the carboxyl terminal fragment, which corresponded to the 29,000-dalton moiety, exhibited a different composition, as compared to that of the amino terminal high-molecular-weight fragment. It demonstrated a high concentration of proline, glycine, cystine, and valine residues. In particular, the integral number of cystine residues (shown in parentheses) was significantly higher, suggesting that the carboxyl terminal portion of CIg must contain a large number of disulfide bridges.

From these results, the linear polypeptide sequence of bovine CIg may be summarized as shown in FIGURE 7. The proposed point of insertion of the high-molecular-weight segment in the parent molecule is supported by analysis of the plasmic digest, because a masked amino terminal sequence found in the segment was identical to that of whole CIg. Furthermore, the 29,000-dalton fragment can be localized in the carboxyl terminal portion of whole CIg, and this portion appeared to contribute to the disulfide linkage between the α and β chains. Thus,

TABLE 3

AMINO ACID COMPOSITIONS* OF AMINO TERMINAL LARGE FRAGMENT AND CARBOXYL TERMINAL FRAGMENT OF COLD-INSOLUBLE GLOBULIN

Amino Acid	Amino Terminal Fragment	Carboxyl Terminal Fragment
	(g of amino acid/100 g of protein)	
Aspartic acid	9.0	10.0 (29)†
Threonine	9.7	8.9 (25)
Serine	7.6	6.6 (19)
Glutamic acid	12.4	12.8 (37)
Proline	8.4	4.4 (13)
Glycine	8.6	12.8 (39)
Alanine	4.8	4.2 (12)
Cystine	*1.7*	*5.0* (14)
Valine	8.1	4.1 (12)
Methionine	1.2	1.9 (5)
Isoleucine	4.5	4.4 (13)
Leucine	6.0	4.2 (12)
Tyrosine	4.2	5.0 (14)
Phenylalanine	2.3	2.1 (6)
Lysine	3.7	4.9 (14)
Histidine	2.2	2.0 (6)
Arginine	5.5	6.7 (19)

*Twenty-four-hour hydrolysate.

†The values in parentheses are nearest integers for residues per mole of protein (mol wt, 29,000).

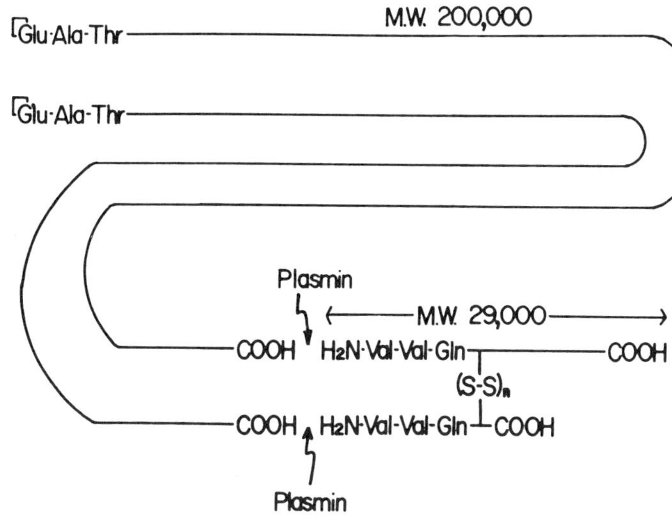

FIGURE 7. A possible linear polypeptide sequence of CIg.

of intact CIg. It was of interest that the large amino terminal domain of intact CIg strongly resists further plasminolysis.

Interaction of Cold-Insoluble Globulin and Fibrin

In the last International Congress on Hematology held in Kyoto in 1976,[12] we reported that if one increases the molar ratio of CIg to fibrin monomer, the degree of aggregation of fibrin monomer is decreased. This inhibitory effect was diminished when native CIg was replaced with its *S*-alkyl derivatives. These facts suggest that a specific interaction between CIg and fibrin monomer may be involved in fibrin aggregation. To test this possibility, we studied the affinity of CIg for fibrinogen or fibrin monomer covalently linked to agarose. As shown in FIGURE 8, CIg was strongly adsorbed by fibrin monomer-agarose, whereas no affinity for fibrinogen-agarose could be detected. Moreover, the affinity of CIg for fibrin monomer-agarose was enhanced in the cold. While this work was in progress, Mosher reported that in the presence of activated factor XIII, human CIg was covalently incorporated into the clot by forming cross-links with the α chain of fibrin.[7] In an attempt to obtain direct evidence for this conclusion, radioiodinated CIg and fibrin monomer-agarose were incubated with factor XIII and α-thrombin in the presence of calcium ions and dithiothreitol. After incubation for various periods of time at different temperatures, an aliquot of the suspension was taken and centrifuged, and the sediment was washed and counted. FIGURE 9 shows the effect of temperature on the incorporation of [^{125}I]CIg into fibrin monomer-agarose. Although there was no significant difference with respect to the initial rate of incorporation at different temperatures, CIg appeared to bind predominantly covalently to fibrin monomer-agarose at lower temperatures. FIGURE 10 demonstrates the effects of calcium ions and dithiothreitol (DTT) on the incorporation of CIg into fibrin monomer-agarose. Incorporation was in-

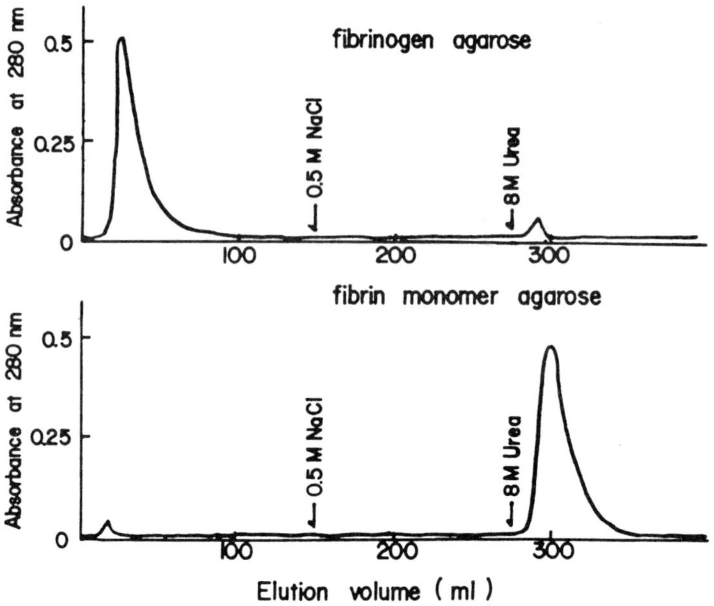

FIGURE 8. Affinity chromatography of CIg on fibrinogen and fibrin monomer-agarose columns. Cold-insoluble globulin was applied to an agarose column (1 × 5 cm) equilibrated with 0.05 M Tris HCl (pH 7.5) that contained 0.1 M NaCl and was eluted at room temperature, respectively, with 0.05 M Tris HCl (pH 7.5) that contained 0.5 M NaCl or 8 M urea.

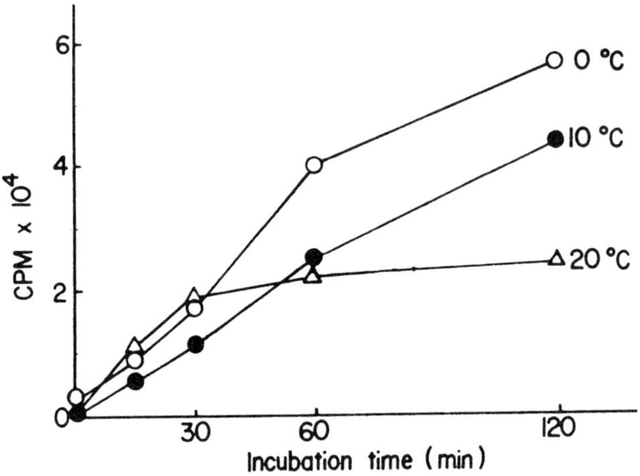

FIGURE 9. Effect of temperature on incorporation of [^{125}I]CIg into fibrin monomer-agarose. The reaction was performed as follows. A mixture that contained 200 μl of a fibrin monomer-agarose suspension [0.4 ml of agarose/ml of 0.05 M Tris HCl (pH 7.5) that contained 0.1 M NaCl], 10 μl of [^{125}I]CIg (2 μg), 40 μl of unlabeled CIg (20 μg), 50 μl of DTT (10 mM), and 50 μl of α-thrombin (20 U/ml) was incubated at 0, 10, and 20°C for various periods of time. After incubation, the reaction was terminated by addition of 100 μl of EDTA (100 mM). Then, 400 μl of the suspension were removed, washed three times with 0.05 M Tris HCl (pH 7.5) that contained 20 mM EDTA and 0.3 M NaCl, and counted.

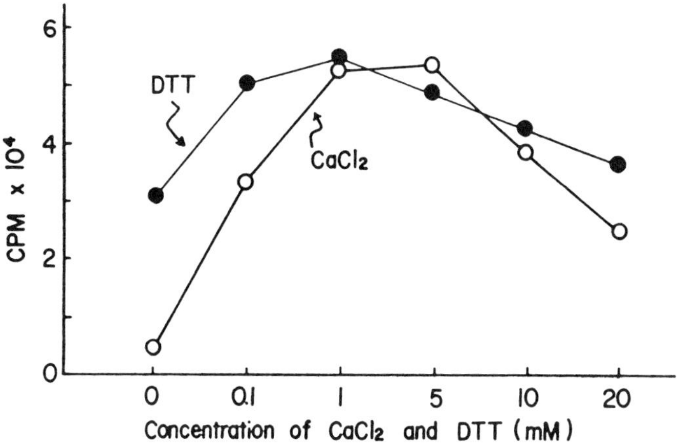

FIGURE 10. Effect of $CaCl_2$ and DTT on the incorporation of [^{125}I]CIg into fibrin monomer-agarose. For determination of the optimal concentration of $CaCl_2$ and of DTT, the final concentrations of DTT and $CaCl_2$ were kept at 1 and 5 mM, respectively. The reaction mixture was incubated at 0°C for 120 min; all other experimental conditions were the same as those described in FIGURE 9.

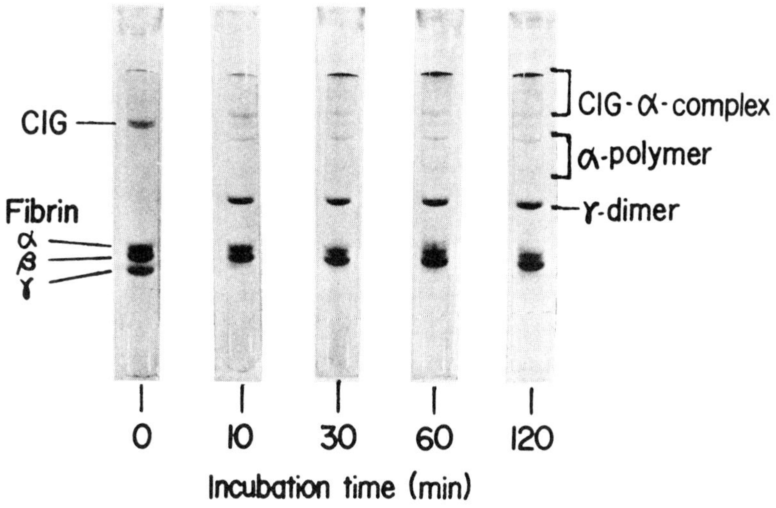

FIGURE 11. Evidence for incorporation of CIg into fibrin in the presence of activated factor XIII. The reaction was performed as follows. A mixture that contained 100 μl of fibrinogen (600 μg), 50 μl of CIg (120 μg), 25 μl of DTT (50 mM), 25 μl of $CaCl_2$ (50 mM), 25 μl of factor XIII (5 μg), and 25 μl of α-thrombin (20 U/ml) was incubated at 37°C for 10, 30, 60, and 120 min. The reaction was terminated by addition of 250 μl of a mixture that consisted of 4% SDS, 20 mM EDTA, and 4% 2-mercaptoethanol in 8 M urea. After the clot was completely dissolved, 10 μl of the reaction mixture were subjected to SDS gel electrophoresis on 5% gels.

TABLE 4

PREPARATION OF COLD-INSOLUBLE GLOBULIN-FIBRIN COMPLEX

Incubation mixture (37°C, 3 hr)	
CIg	94 mg
Fibrinogen	220 mg
DTT	1 mM
$CaCl_2$	5 mM
Factor XIII	500 μg
Thrombin	50 U

Reaction mixture was reduced and carboxymethylated, and the resultant material was applied to a Sepharose 4B column in 0.1 M Tris HCl (pH 7.5) that contained 0.1% SDS

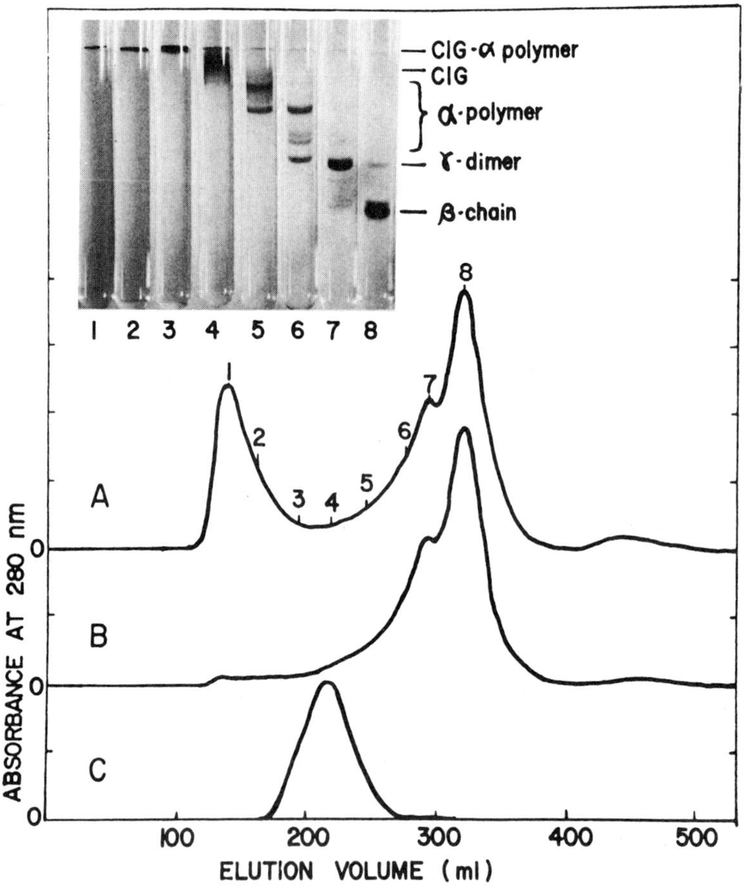

FIGURE 12. Separation of the *S*-alkyl derivatives of CIg-fibrin complex (A), cross-linked fibrin (B), and CIg treated with factor XIIIa (C) on a Sepharose 4B column. The experimental conditions are given in TABLE 4. Sodium dodecyl sulfate gel electrophoresis of samples of the fractions was performed on 5% gels for 3 hr. Migration was from top (cathode) to bottom.

creased by raising the concentrations of calcium ions and DTT, and the maximum incorporation observed occurred at concentrations of 5 and 1 mM, respectively, under the conditions employed.

FIGURE 11 illustrates SDS gel electrophoretic patterns of fibrin clots formed in the presence of CIg and factor XIII under optimal conditions. The CIg band disappeared after incubation for 10 min, and a highly polymerized substance, which appeared to represent covalently bound CIg and the α chain of fibrin, was detected on the tops of the gels. Further evidence for factor XIIIa-mediated cross-links between CIg and fibrin was obtained by use of factor XIII-deficient plasma. The α-thrombin-induced clots formed from normal and from deficient plasmas exhibited different electrophoretic patterns on SDS gels. A covalently bound complex of CIg and fibrin α chain could not be detected in clots formed from factor XIII-deficient plasma.

To obtain confirmation of the existence of such a complex, a mixture that contained bovine CIg, fibrinogen, α-thrombin, and factor XIII was incubated for 3 hr under the conditions described in TABLE 4. The resultant cross-linked CIg-fibrin complex was then *S*-carboxymethylated and applied to a Sepharose 4B column in SDS; the elution pattern is shown in FIGURE 12. The *S*-alkyl derivatives of cross-linked fibrin and of CIg that had been treated with factor XIII and

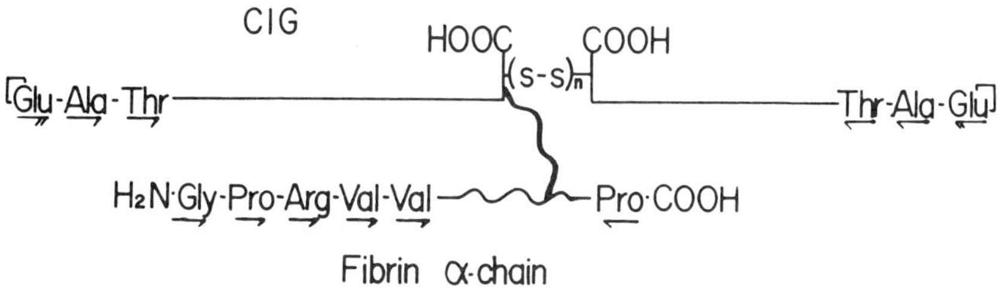

FIGURE 13. A gross molecular structure of cross-linked CIg and fibrin α chain.

α-thrombin in this system were prepared for use as controls. These preparations were applied, individually, to the same Sepharose 4B column. Under these conditions, which utilized a relatively low level of factor XIII, highly polymerized material was found only in the eluate of the *S*-alkylated CIg-fibrin complex. Other substances eluted in later fractions appeared to be CIg, fibrin α oligomers, fibrin γ dimer, and fibrin α chain, as indicated by their migrations on SDS gels. Amino terminal analysis of the highly polymerized substance (first peak in FIGURE 12) revealed the presence of only glycine, and subsequent residues were found to be Pro-Arg-Val-Val-. The carboxyl terminal prolyl residue, which is derived from the carboxyl terminal end of fibrin α chain,[30] was identified by hydrazinolysis. Moreover, the tripeptide Gly-Pro-Arg, which is located at the amino terminus of fibrin α chain,[31] was liberated from this substance by treatment with bovine α-thrombin. These results indicated clearly that the highly polymerized substance consists of CIg and fibrin α chain, as illustrated in FIGURE 13. The sites at which CIg is cross-linked to fibrin are not yet known; recent studies by Jilek and Hörmann,[32] however, have suggested that these sites may be located in the carboxyl terminal region, which corresponds to the 29,000-dalton segment.

Enhancement of Urokinase-Mediated Transformation of Plasminogen to Plasmin by Cold-Insoluble Globulin

As described previously, the carboxyl terminal fragment, which has a molecular weight of 29,000 daltons, had an amino terminal sequence of Val-Val-Gln-. This sequence is very similar to a unique sequence located in the amino terminal portion of various serine proteases, such as plasmin and α-thrombin. This similarity prompted us to test the possibility of whether CIg or its derivative has an effect on the urokinase-mediated activation of plasminogen to plasmin. The reaction mixture was composed of highly purified bovine plasminogen and urokinase, with or without CIg. After preincubation for various times, an aliquot of the reaction mixture was taken, and plasmin activity was measured by use of TAME, casein, and fibrin as substrates. FIGURE 14 shows the TAME esterolytic

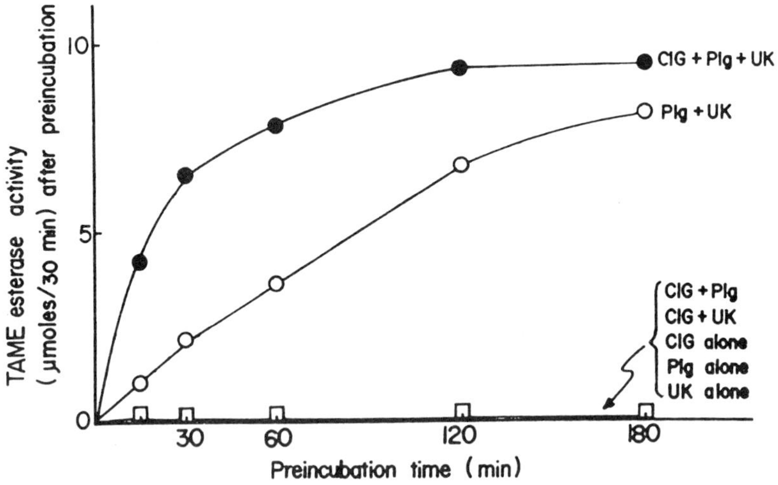

FIGURE 14. Effect of CIg on activation of plasminogen by urokinase as determined by esterolytic activity. A reaction mixture that contained 180 μg of CIg, 9 μg of bovine plasminogen, and 1 U of urokinase, in a total volume of 600 μl of 0.1 M NH_4HCO_3 (pH 8.3), was preincubated at 37°C for various periods of time. After preincubation, 100 μl of the reaction mixture were removed, and TAME esterase activity was determined.

activities of plasmin present at various stages of preincubation. In the presence of CIg, the transformation of plasminogen to plasmin by urokinase was significantly enhanced, as compared with that in the absence of CIg. The reaction mixtures that consisted of CIg plus urokinase and of CIg plus plasminogen did not display any plasmin activity. Under similar conditions, plasmin activity was measured with casein as the substrate. Activation of plasminogen by urokinase was also enhanced in the presence of CIg, as shown in FIGURE 15. FIGURE 16 shows the reduction of fibrin clot lysis time that was achieved by increasing the weight ratio of CIg to fibrinogen. These results suggest that intact CIg or its degradation product with plasmin may have a stimulatory effect on the initial rate of the urokinase-mediated transformation of plasminogen to plasmin. However, the mechanism of this phenomenon is still unknown.

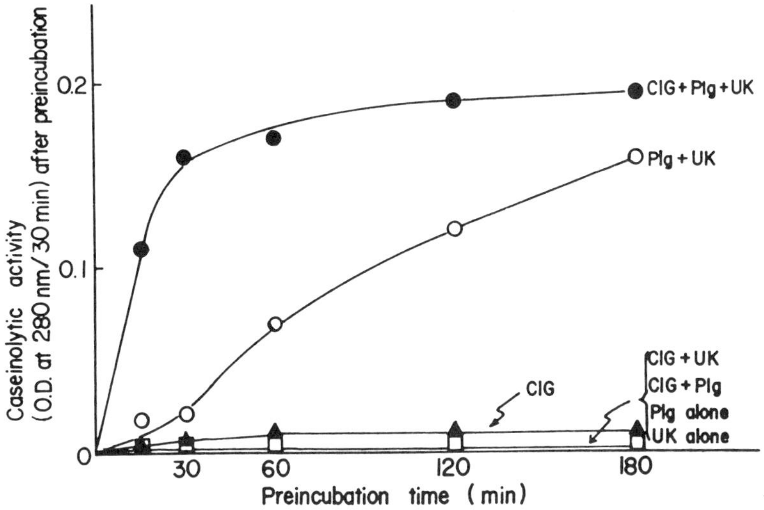

FIGURE 15. Effect of CIg on activation of plasminogen by urokinase as determined by caseinolytic activity. Preincubation of plasminogen and urokinase with or without CIg was performed under the same conditions as described in FIGURE 14. After preincubation, 100 μl of the reaction mixture were removed, and caseinolytic activity was determined.

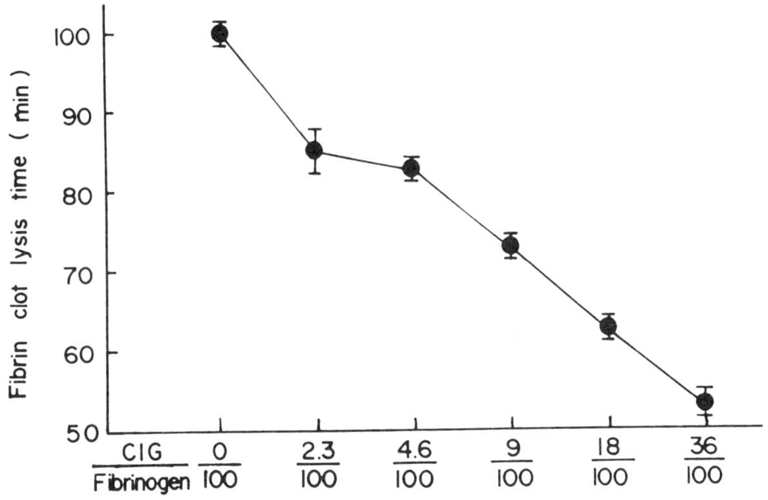

FIGURE 16. Effect of CIg on activation of plasminogen by urokinase as determined by fibrinolysis. A reaction mixture that consisted of 1.5 mg of fibrinogen, 40 μl of bovine plasminogen, 0.5 U of urokinase, and various amounts of CIg was incubated at 37°C in 500 μl of 0.1 M NH_4HCO_3 (pH 8.3) that contained 10 mM $CaCl_2$ and clotted with 5 U of α-thrombin. The clot formed was further incubated at 37°C, and lysis time was determined.

Summary

Bovine plasma CIg, like human CIg, is a glycoprotein with a molecular weight of approximately 450,000 daltons and consists of two homologous subunits, the α and β chains. These subunits are covalently linked through disulfide bridges in their carboxyl terminal domains. The carboxyl terminal regions are presumed to contain the fibrin-reactive transamidation site. The covalent incorporation of CIg into fibrin has been conclusively demonstrated by isolation of the *S*-carboxymethyl derivative of the CIg-fibrin-α chain complex and by determination of its terminal amino acid sequences. Cold-insoluble globulin has been shown to exert a stimulatory effect on the urokinase-mediated activation of bovine plasminogen to plasmin.

Acknowledgments

We express our thanks to Dr. S. Matsuura, Department of Physiology, Kansai Medical University, Osaka, for his kind assistance in the electron microscopic analysis. We also thank Dr. S. Nakamura, Department of Biochemistry, Primate Research Institute, Kyoto University, Inuyama, for his cooperation in the initial stage of this work. The assistance of Miss Kazue Fukunishi in amino acid analyses is appreciated.

References

1. Morrison, P. R., J. T. Edsall & S. G. Miller. 1948. Preparation and properties of serum and plasma proteins. XVIII. The separation of purified fibrinogen from fraction I of human plasma. J. Amer. Chem. Soc. **70:** 3103–3108.
2. Edsall, J. T., G. A. Gilbert & H. A. Scheraga. 1955. The non-clotting component of the human plasma fraction-1-1 ("cold insoluble globulin"). J. Amer. Chem. Soc. **77:** 157–161.
3. Hershgold, E. J., J. G. Pool & A. R. Pappernhagen. 1966. The potent antihemophilic globulin concentrate derived from a cold insoluble fraction of human plasma: characterization and further data on preparation. J. Lab. Clin. Med. **67:** 23–32.
4. Mosesson, M. W., R. W. Colman & S. Sherry. 1968. Chronic intravascular coagulation syndrome. Report of a case with special studies of an associated plasma cryoprecipitate, "cryofibrinogen." N. Engl. J. Med. **278:** 815–821.
5. Mosesson, M. W. & R. A. Umfleet. 1970. The cold-insoluble globulin of human plasma. I. Purification, primary characterization, and relationship to fibrinogen and other cold-insoluble fraction components. J. Biol. Chem. **245:** 5728–5736.
6. Mosesson, M. W., A. B. Chen & R. M. Huseby. 1975. The cold insoluble globulin of human plasma: studies of its essential structural features. Biochim. Biophys. Acta **386:** 509–524.
7. Mosher, D. F. 1975. Cross-linking of cold-insoluble globulin by fibrin stabilizing factor. J. Biol. Chem. **250:** 6614–6621.
8. Hynes, R. O. 1973. Alteration of cell-surface proteins by viral transformation and proteolysis. Proc. Nat. Acad. Sci. USA **70:** 3170–3174.
9. Gahmberg, C. G. & S. Hakomori. 1973. Altered growth behaviour of malignant cells associated with changes in externally labelled glycoprotein and glycolipid. Proc. Nat. Acad. Sci. USA **70:** 3329–3333.
10. Ruoslahti. E., A. Vaheri, P. Kuusela & E. Linder. 1973. Fibroblast surface antigen: a new serum protein. Biochim. Biophys. Acta **322:** 352–358.

11. VAHERI, A. & E. RUOSLAHTI. 1974. Disappearance of a major cell-type-specific surface glycoprotein antigen (SF) after transformation of fibroblasts by Rous sarcoma virus. Int. J. Cancer **13:** 579–586.
12. IWANAGA, S., S. NAKAMURA & K. SUZUKI. 1976. Bovine plasma factor XIII: its molecular structure and functions. Excerpta Med. Int. Congr. Ser. **415:** 393–396.
13. SUZUKI, K., S. NAKAMURA, S. IWANAGA, S. HASHIMOTO & S. MATSUURA. 1976. Bovine plasma cold insoluble globulin: purification, properties and biological function. Excerpta Med. Int. Congr. Ser. **415:** 311 (Abst.).
14. SUZUKI, K., S. IWANAGA & S. HASHIMOTO. 1977. Subunit structure of bovine cold insoluble globulin (CIG). Seikagaku **49:** 725 (in Japanese).
15. DAVIS, B. J. 1964. Disc electrophoresis, II. Method and application to human serum proteins. Ann. N. Y. Acad. Sci. **121:** 404–427.
16. WEBER, K. & M. OSBORN. 1969. The reliability of molecular weight determinations by dodecyl sulfate-polyacrylamide gel electrophoresis. J. Biol. Chem. **244:** 4406–4412.
17. YPHANTIS, D. A. 1964. Equilibrium ultracentrifugation of dilute solutions. Biochemistry **3:** 297–317.
18. SPACKMAN, D. H., W. H. STEIN & S. MOORE. 1958. Automatic recording apparatus for use in chromatography of amino acids. Anal. Chem. **30:** 1190–1206.
19. FRANCOIS, C., R. D. MARSHALL & A. NEUBERGER. 1962. Carbohydrates in protein. The determination of mannose in hen's egg albumin by radioisotope dilution. Biochem. J. **83:** 335–341.
20. GARDELL, S. 1953. Separation on Dowex-50 ion-exchange resin of glucosamine and galactosamine, and their quantitative determination. Acta Chem. Scand. **7:** 207–215.
21. WARREN, L. 1960. Unbound sialic acids in fish eggs. Biochim. Biophys. Acta **44:** 347–351.
22. EDMAN, P. & A. HENSCHEN. 1975. Sequence determination. *In* Molecular Biology, Biochemistry and Biophysics. Protein Sequence Determination. S. B. Needleman, Ed. Vol. **8:** 232–279. Springer-Verlag. Berlin, Federal Republic of Germany.
23. KIMURA, S. 1974. Japan Analyst **23:** 563–575 (in Japanese).
24. HEENE, D. L. & F. R. MATTHIAS. 1973. Adsorption of fibrinogen derivatives on insolubilized fibrinogen and fibrinomonomer. Thromb. Res. **2:** 137–154.
25. HUNTER, W. M. 1973. Radioimmunoassay. *In* Handbook of Experimental Immunology. D. M. Weir, Ed. Vol. **17:** 1–35. Academic Press, Inc. New York, N.Y.
26. ROBBINS, K. C. & L. SUMMARIA. 1976. Plasminogen and plasmin. Methods Enzymol. **45** (part B): 257–273.
27. WALSH, K. A. 1970. Trypsinogens and trypsin of various species. Methods Enzymol. **19:** 41–63.
28. REIMERDES, E. H. & H. KLOSTERMYER. 1976. Determination of proteolytic activities on casein substrates. Methods Enzymol. **45** (part B): 26–28.
29. ALLAIN, J. P., H. A. COOPER, R. H. WAGNER & K. M. BRINKHOUS. 1975. Platelets fixed with paraformaldehyde: a new reagent for assay of von Willebrand factor and platelet aggregating factor. J. Lab. Clin. Med. **85:** 318–328.
29a. SUZUKI, K., S. IWANAGA & S. HASHIMOTO. 1977. Unpublished results.
30. IWANAGA, S. & M. OKUDE. 1972. Carboxyl-terminal structures of mammalian fibrinogen and fibrin. Ann. N.Y. Acad. Sci. **202:** 308–317.
31. IWANAGA, S., P. WALLÉN, N. J. GRÖNDAHL, A. HENSCHEN & B. BLOMBÄCK. 1969. On the primary structure of human fibrinogen: isolation and characterization of N-terminal fragments from plasmic digests. Eur. J. Biochem. **8:** 189–199.
32. JILEK, F. & H. HÖRMANN. 1977. Cold-insoluble globulin, III. Cyanogen bromide and plasminolysis fragments containing a label introduced by transamidation. Hoppe-Seyler's Z. Physiol. Chem. **358:** 1165–1168.
33. LOEWY, A. G., K. DUNATHAN, R. KRIEL & H. L. WOLFINGER. 1961. Fibrinase. I. Purification of substrate and enzyme. J. Biol. Chem. **236:** 2625–2633.

DISTRIBUTION OF COLD-INSOLUBLE GLOBULIN IN PLASMA AND TISSUES*

Michio Matsuda,† Nobuhiko Yoshida,† Nobuo Aoki,†
and Kunio Wakabayashi‡

†Institute of Hematology
and ‡Department of Surgery
Jichi Medical School
Tochigi-Ken 329-04, Japan

Cold-insoluble globulin (CIg) in plasma is an α_2-glycoprotein with a molecular weight of 440,000 that consists of two nearly identical 220,000-dalton subunit polypeptides, held together by disulfide linkages.[1,2] Cold-insoluble-globulin has been shown to be cross-linked to the α-chain of fibrin by the catalytic action of activated factor XIII of blood coagulation (fibrin-stabilizing factor, FSF[3,4]), and the FSF-mediated cross-linking was also demonstrated on the cultured fibroblasts.[5] Cold-insoluble globulin on the plasma membrane of fibroblasts seems to be important in the regulation of growth and behavior of cultured fibroblasts, particularly in regard to virus-mediated malignant transformation.[6-11] Immunologically cross-reactive substances were shown to be widely distributed in tissues of vertebrates.[12,13] Cold-insoluble globulin is also present in cryofibrinogen obtained from the plasma of patients under a certain pathologic conditions.[14,15] Taking these data into consideration, we have been concerned about the clinical aspects of this protein under physiologic and pathologic conditions.

Materials and Methods

Blood samples were collected by venipuncture. To 10 mg of dipotassium ethylenediaminetetraacetate (EDTA) were added 2.0 mg of blood, and plasma was separated by centrifugation at 2500g for 30 min. Nine volumes of blood were added to trisodium citrate (3.8%), and plasma was separated as mentioned previously. Serum was separated after 2 hr of incubation of blood at 37°C in a regular glass tube.

Antihuman Plasma CIg Rabbit Antiserum. Rabbits were immunized by toepad injection intradermally with 500 μg of chromatographically purified human CIg fraction,[16] emulsified by completely mixing with Freund's complete adjuvant. After two sets of additional injections of 200 μg given 2 weeks apart, the animals were bled, and immune sera were harvested. The sera were then made monospecific by absorption with insolubilized plasma proteins[17] from which CIg had been depleted.

Quantitation of Parameter Proteins. Six parameter proteins (CIg, antithrombin III, plasminogen, α_1-antitrypsin, α_2-plasmin inhibitor, α_2-macroglobulin) were quantitated by either the single radial immunodiffusion technique[18] or Laurell's method.[19] Antisera against α_1-antitrypsin and α_2-macroglobulin were obtained from Behringwerke (Marburg-Lahn, Federal Republic of Germany), and antisera

*Supported by research grants from the Ministry of Education and the Intractable Diseases Division, Public Health Bureau, the Ministry of Health and Welfare of Japan.

0077-8923/78/0312-0074 $01.75/0© 1978, The New York Academy of Sciences against the remaining four proteins were prepared in a similar manner by immunizing rabbits with purified materials, as described above. Fibrinogen was determined by the method of Ratnoff and Menzie,[20] and fibrin/fibrinogen degradation products (FDP) were assessed by the method of Merskey *et al.*[21]

Distribution of CIg in tissues was studied by an indirect fluorescein antibody technique[22] on 3–4 μm thick frozen sections. Tissues obtained by surgical resection were embedded in cryomolds with O.C.T. compound (Lab-Tek Products, Naperville, Ill.) and quickly frozen in acetone-dry ice. Fluorescein isothiocyanate (FITC) labeled antirabbit IgG goat γ-globulin (Behringwerke) was used for staining. Control studies were always run with nonimmunized normal rabbit serum instead of antihuman CIg rabbit antiserum.

Preparation of cryofibrinogen was performed according to the method of Mosesson *et al.*[14] with minor modifications of their technique (FIGURE 1). Since CIg is susceptible to proteolysis by plasmin, care was taken to protect CIg by removing plasminogen in the initial step and by using protease inhibitors throughout the procedure. The final product was dissolved in 0.3 M NaCl to give absorbance at 280 nm of 14 ± 2.

Immunoelectrophoresis was performed in 0.8% agarose gel and barbital buffer [ionic strength 0.05 (pH 8.6)] at 3 mA/cm for 60–90 min. The gel was then washed extensively with physiologic saline and distilled water, dried, and stained with amido black. Antihuman fibrinogen rabbit antiserum was prepared in a similar manner in rabbits as described previously.

Ultracentrifugation was performed at 22–24°C in a Spinčo model E analytic ultracentrifuge equipped with Schlieren optics.

Amino acid composition was determined with a Beckman model 121 amino acid analyzer after hydrolyzation of samples in 6 N HCl for 24 hr at 110°C.[23]

Sodium dodecyl sulfate (SDS) polyacrylamide gel electrophoresis was performed according to the method of Weber and Osborn[24] with 5% gels. The samples were pretreated for 30 min at 37°C with an equal volume of 2% SDS in 0.02 M phosphate buffer (pH 7.2) with or without 1.5% dithiothreitol (DTT).

Analysis of NH_2 termini of the fibrinogen moiety was accomplished according to the method of Kulbe.[25]

The quasielastic light scattering technique was performed according to the principles and method of Pecora.[26]

Determination of secondary structures by circular dichroism was performed according to the method of Chen *et al.*[27]

Distribution of CIg in Normal Plasma and Serum

Cold-insoluble globulin is normally present in plasma and serum at concentrations of 27.52 ± 4.60 and 23.46 ± 5.18 mg/dl, respectively (TABLE 1). These values were determined on samples obtained from 75 healthy adults and were expressed as means ± SD. The approximate 15% decrease in serum is explained by incorporation of CIg into fibrin via activated FSF-mediated cross-linking and not by simple adsorption onto fibrin clots. This notion was confirmed by experiments that demonstrated the requirement of calcium ions to decrease CIg levels in the clot liquor of EDTA plasma clotted with thrombin (TABLE 1). This value of 15% is somewhat less than that reported by Mosesson and Umfleet[16] and may be due to the higher temperature (37°C) we employed for incubation of blood samples to separate sera, because incorporation of CIg into fibrin was shown to be increased when temperature is reduced.[4]

ACD-plasma

Lysine-Sepharose

+ Heparin, 10 u/ml

Kept at 0-2°C for 48 hours

Ppt

Repeated extractions at 0-2°C with
0.005 M phosphate buffer, pH 6.0*

0.055 M citrate buffer, pH 6.0 containing
1 M glycine and 6.5 % ethanol**

Dissolved at 30°C in 0.055 M citrate buffer,
pH 6.35# (original plasma volume)

Kept at 0-2°C for 24 hours

Ppt

Dissolved in the same buffer at 30°C
(1/2 of original plasma volume)

Kept at 0-2°C for 24 hours

Ppt (CRYOFIBRINOGEN)

*, ** and # contained 0.01 M EACA and
10 u/ml of Trasylol

FIGURE 1. Preparation procedure of cryofibrinogen according to Mosesson *et al.*[14] with minor modifications.

TABLE 1

CONCENTRATION OF COLD-INSOLUBLE GLOBULIN (MEANS ± SD, mg/dl)

Plasma	27.52 ± 4.60*
Serum	23.46 ± 5.18*
Liquid phase of EDTA plasma to which was added†	
Buffer	26.86 ± 4.04‡
Thrombin plus EDTA	26.04 ± 4.25‡
Thrombin plus $CaCl_2$	17.05 ± 5.19‡

*Determined on samples obtained from 75 healthy adults.

†The reaction mixture consisted of 0.8 ml of plasma and 0.2 ml of additives: 5.0 U/ml of thrombin, 0.01 M $CaCl_2$, or 0.1% EDTA. The reaction was allowed to proceed at 37°C for 2 hr.

‡Corrected for the volume of additives.

Distribution of CIg in Blood under Pathologic Conditions

Concentration of CIg in Plasma in the Disseminated Intravascular Coagulation Syndrome

Cold-insoluble globulin was shown to undergo limited proteolysis with a variety of proteases, including plasmin. This reaction would account for the heterogeneity of circulatory CIg reported by Chen *et al.*[2] By means of incorporation into fibrin clots, digestion by plasmin, or both mechanisms, the concentration of CIg would be decreased under certain pathologic conditions, such as disseminated intravascular coagulation (DIC) syndromes, in which thrombi are rapidly formed and/or fibrinolysis is induced.

Indeed, CIg levels in plasma of patients with DIC syndromes were found to be significantly reduced. The diagnosis of a DIC syndrome was based on clinical signs and symptoms and on laboratory results, including decreased platelet counts, plasma antithrombin III (AT-III), plasminogen (PIg), and α_2-plasmin inhibitor (α_2-PI),[28,29] increased FDP, and a prolonged prothrombin time. The concentration of CIg was determined in blood samples obtained before treatment had been

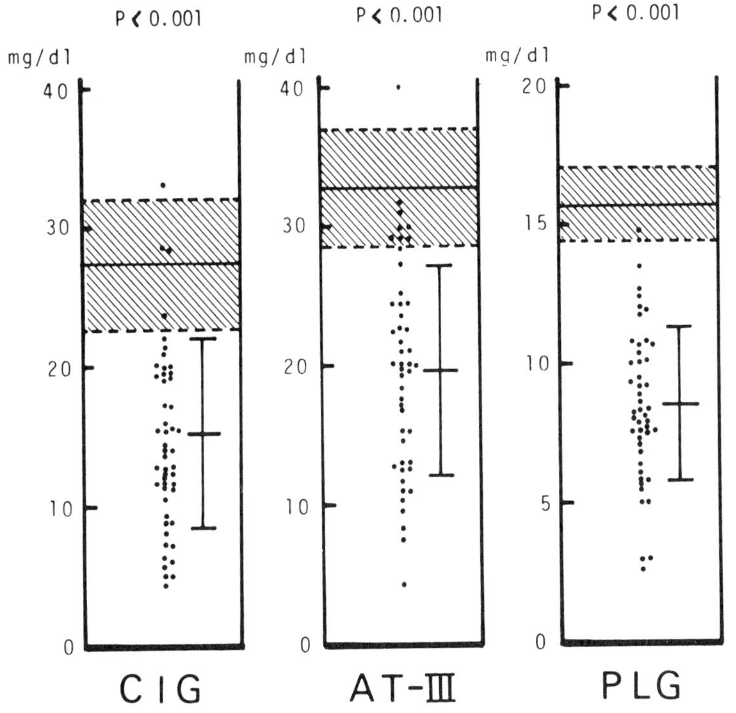

FIGURE 2. Distribution of CIg, antithrombin III (AT-III), and plasminogen (PLg) in plasma of patients with DIC syndromes. Hatched areas represent means ± SD of normal values, and longitudinal lines with crossed bars indicate those of levels in the plasma of affected patients. Probability values refer to the difference in concentration of each parameter between normal plasma and affected plasma.

initiated, and the values observed were compared with those of other parameters, including plasma AT-III, Plg, α_2-PI, α_1-antitrypsin (α_1-AT), and α_2-macroglobulin (α_2-M). The results are shown in FIGURES 2 and 3. The hatched areas indicate means ± SD of normal values, and longitudinal lines with crossed bars represent those of levels in patients' plasmas. As shown in FIGURES 2 and 3, the level of CIg in DIC syndromes is significantly reduced ($p < 0.001$). Antithrombin-III, Plg, and α_2-PI are also significantly decreased ($p < 0.001$), but α_2-M was to be not significantly reduced ($p > 0.05$). Conversely, α_1-AT was increased ($p < 0.05$). TABLE 2 shows correlation coefficients between CIg levels and those of

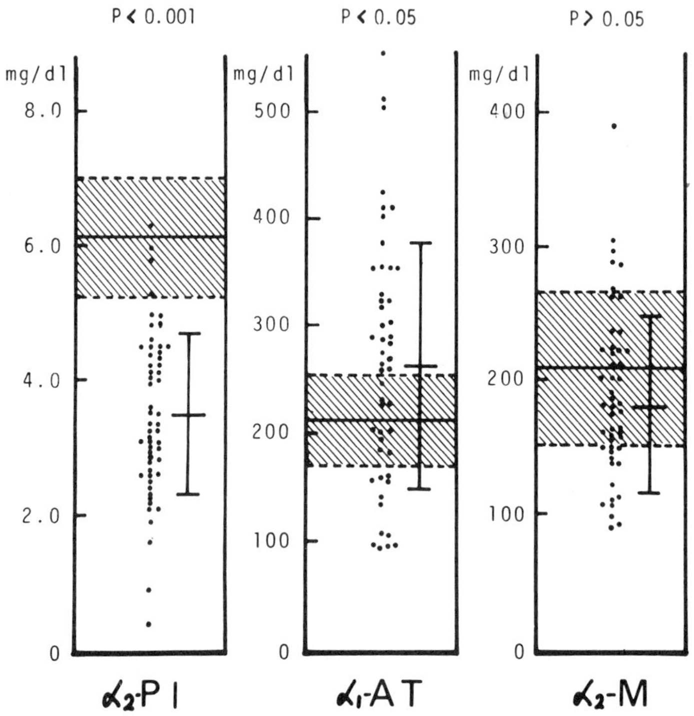

FIGURE 3. Distribution of α_2-plasmin inhibitor (α_2-PI), α_1-antitrypsin (α_1-AT), and α_2-macroglobulin (α_2-M) in the plasma of patients with DIC syndromes. For details, see the legend to FIGURE 1.

other parameters, including fibrinogen (Fbg) and FDP. Values of CIg were found to be strongly positively correlated with those of AT-III and less strikingly with those of Plg and FDP. However, a weak inverse correlation between CIg and α_2-PI levels was found. The reason for the latter finding is not clear, but we hypothesize that there exist two different mechanisms in the early phase of DIC syndromes, namely, consumption of CIg through intravascular coagulation and a decrease in α_2-PI levels as a consequence of secondary fibrinolysis. In support of this theory, we noted in individual cases that plasma CIg concentrations rapidly decrease at the very beginning of the DIC episode, whereas α_2-PI levels continue

TABLE 2

CORRELATIONS BETWEEN CONCENTRATION OF CIg AND THOSE OF OTHER PARAMETERS IN PLASMA OF PATIENTS WITH DIC SYNDROME

	Mean ± SD (mg/dl)	Correlation Coefficient (r)	Level of Significance (%)
AT-III	19.17 ± 7.37	0.68	1
Plg	8.61 ± 2.81	0.39	1
α_2-PI	3.51 ± 1.18	−0.29	5
α_1-AT	264.54 ± 110.96	0.22	ns*
α_2-M	184.37 ± 63.78	0.18	ns*
Fbg	285.75 ± 219.84	−0.24	ns*
FDP	11.08 ± 15.32	0.31	5

*Not significant.

to decrease over the next few days. These different plasma concentration patterns may account for the inverse correlation seen between CIg and α_2-PI levels in the early stage of the DIC syndromes. α_2-Plasmin inhibitor has been shown to bind to plasmin *in vitro* almost instantaneously,[29] and the rapid formation of an α_2-PI-plasmin complex has also been demonstrated *in vivo* in the plasma of patients treated with urokinase.[30] When patients with acute cerebral infarction were treated with an intravenous infusion of 240,000 U of urokinase within 2 hr, α_2-PI values decreased continuously, at least for 20 hr after the start of infusion.[30] In contrast, concentrations of CIg in the same samples did not change, as shown in TABLE 3. Thus, it would appear that α_2-PI, after a brief urokinase infusion, effectively blocks plasmin and protects CIg. However, we cannot exclude the possibility that part of the immunoreactive CIg in the plasma sample is degraded through limited proteolysis by plasmin.

Concentration of CIg in Plasma After Major Surgery

Plasma levels of CIg exhibit a characteristic pattern after major surgery. TABLE 4 shows the changes of CIg levels (means ± SD) in 43 patients who underwent major surgery with uneventful postoperative courses. The surgical operations included total or partial gastrectomy with lymph node dissection for gastric cancer, radical mastectomy for breast cancer, and cholecystectomy for bile duct and gallbladder stones. Patients who had been transfused with more than 600 ml of blood during and after their operation were excluded from the study. As shown

TABLE 3

CONCENTRATIONS OF CIg IN PLASMA BEFORE AND AFTER 2-HR INFUSION OF 24,000 U OF UROKINASE

Concentration of CIg (MEANS ± SD, mg/dl)	
Before	After
26.90 ± 5.51	26.33 ± 4.35
(N = 10, not significant)	

TABLE 4

CHANGES IN CONCENTRATION OF CIg* IN PLASMA AFTER MAJOR SURGERY

	Mean ± SD	Range	†
Preoperative day	27.19 ± 8.38	17.5–49.0	
Postoperative			<0.001
Day 1	19.61 ± 7.66	7.8–33.0	
Day 3	22.88 ± 7.69	9.6–42.0	<0.001
Day 7	27.15 ± 7.63	9.2–47.0	<0.001
Day 14	27.13 ± 7.14	13.3–44.0	>0.5

*Milligrams per decaliter.

†Values refer to the differences in concentration of CIg between paired samples within groups.

in TABLE 4, CIg values fell initially on postoperative Day 1 ($p < 0.001$), then slowly returned to preoperative levels by Day 14. FIGURE 4 shows a typical CIg patterns observed in a 55-year-old female patient after gastrectomy. This profile resembles that of Plg but not those of Fbg and α_1-AT, which are "acute-phase reactants" (data not shown).

DISTRIBUTION OF CIg IN TISSUES

The distribution of CIg or immunologically cross-reactive substances was studied in tissues obtained by surgical resection. Fluorescence was observed to

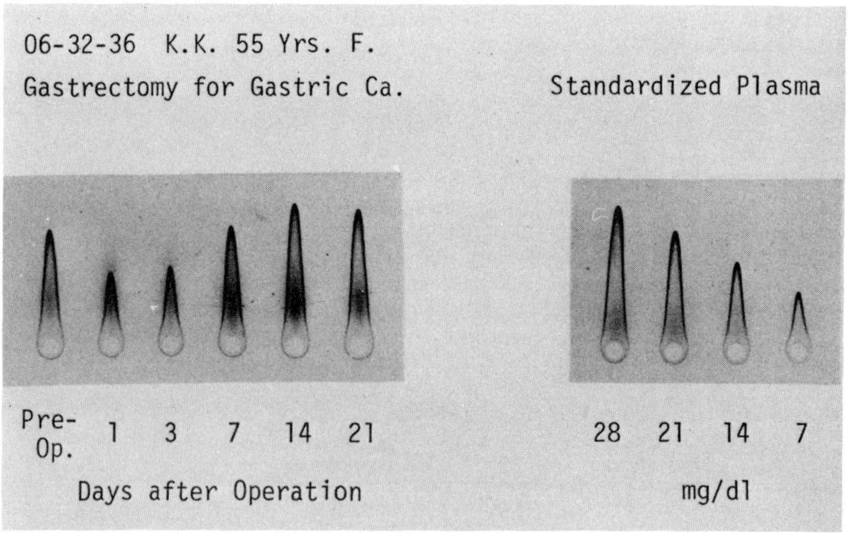

FIGURE 4. A typical pattern of CIg levels in plasma from a 55-year-old female after gastrectomy; Laurell's method, in 1.0% agarose that contained 1.5% antiserum with a constant current of 1.0 mA/cm for 10 hr at 4°C. The dilution series of standard plasma was subjected to electrophoresis simultaneously as a reference.

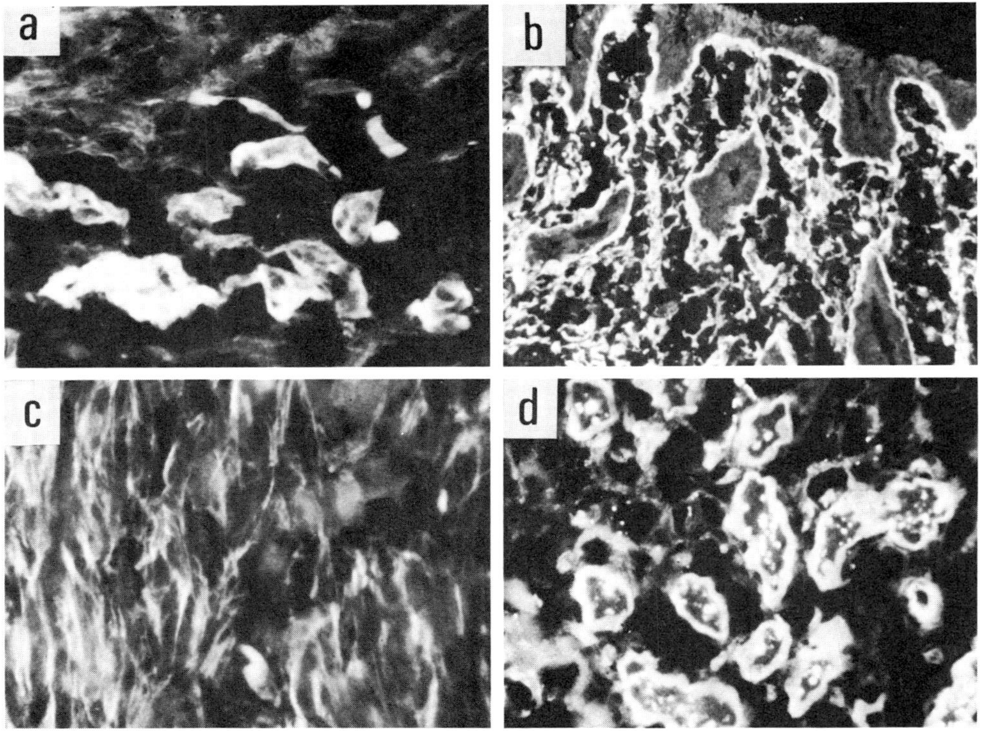

FIGURE 5. Immunofluorescence study of frozen sections from surgically removed tissues. a: FITC-labeled antibody stains fibroblasts located in the subcutaneous tissue. b: Appendix mucosa. Note associations of fluorescence with basal lamina and fibroblasts in the submucosal layer. c: Eighteen-day-old skin scar tissue. Fluorescence occurs around new fibroblasts but not in mature collagen fibers. d: Sweat glands. Fluorescence is associated with basement membranes. a, × 200; b, × 80; c & d, × 200.

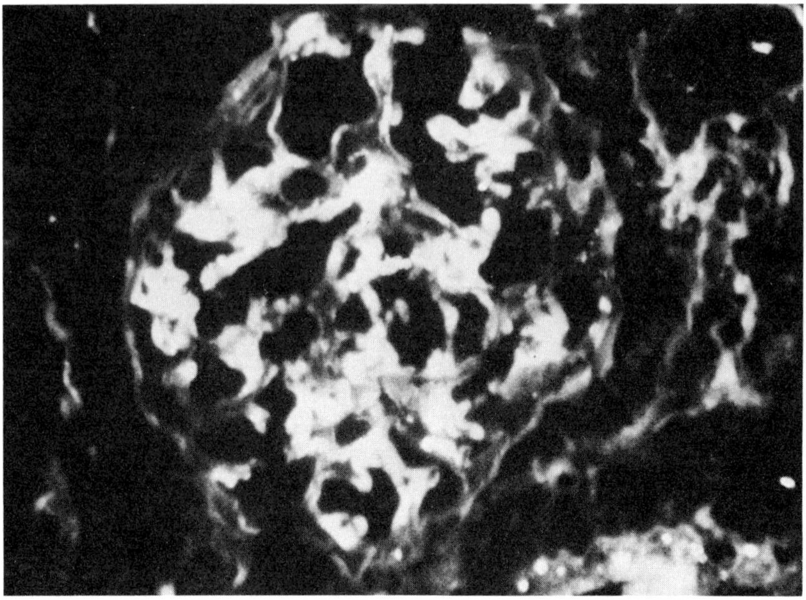

FIGURE 6. Immunofluorescence study of section from renal cortex. Fluorescence is associated with the mesangium but not with the glomerular basement membrane. × 175.

occur widely in association with cells and tissues of mesenchymal origin, including fibroblasts, adipose cells, and smooth muscle cells. In addition, we found CIg fluorescent antibody associated with basement membranes from a wide variety of tissues and exocrine glands, including skin, mucous membranes, sweat, and sebaceous, mammary, and parotid glands. FIGURE 5 shows examples of these CIg distribution patterns. An exception to this finding was the glomerular basement membrane, which is currently believed to have a different origin. In a section of the renal cortex, fluorescence was found associated with the mesangium but not with the glomerular basement membrane (FIGURE 6).

CRYOFIBRINOGEN IN PLASMA OF PATIENTS WITH ULCERS ON THE LEGS

Cold-insoluble globulin is also present in plasma as a component of cryofibrinogen obtained from patients under certain pathologic conditions. Cryofibrinogen was prepared from the plasma of a 49-year-old male patient with leg dermal ulcers. The ulcers were located on both malleolus regions and were shallow and irregular. Peripheral arterial occlusive diseases, including Buerger's disease and atherosclerosis, malignant tumors, and paraproteinemia were not seen on repeated examinations. This cryofibrinogen, which corresponds to Mosesson's cryofibrinogen A, forms a solid gel in 10–15 sec in melting ice, and the gel readily redissolves at 37°C. The material tolerates repeated manipulations.

Immunoelectrophoresis revealed a single precipitin line against antihuman fibrinogen rabbit antiserum and also a single line at the α_2-region against antihuman CIg rabbit antiserum. When a mixture (1:1, v/v) of these antisera was

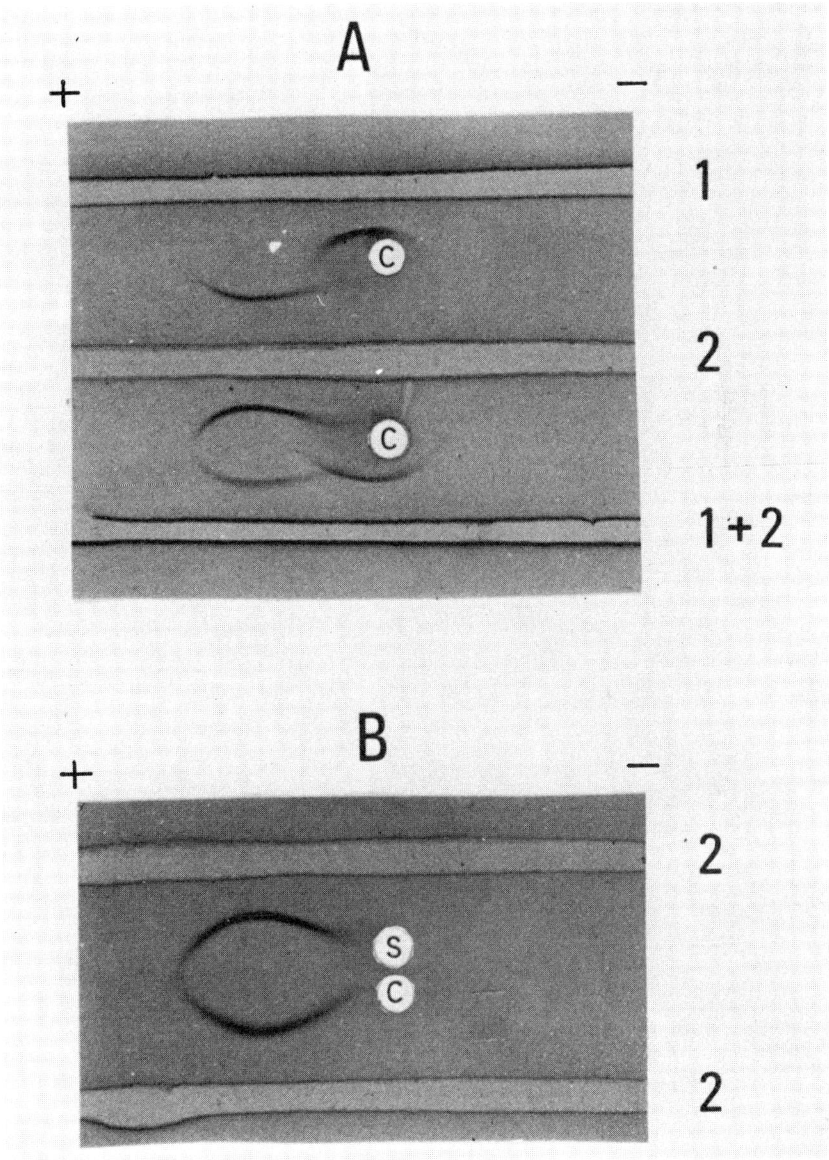

FIGURE 7. Immunoelectrophoresis of cryofibrinogen. A: Dissociated coexistence of CIg and fibrinogen. c, Cryofibrinogen; 1, antihuman fibrinogen rabbit antiserum; 2, antihuman CIg rabbit antiserum. B: Immunologic identity of CIg in cyrofibrinogen with that in normal serum. c, Cryofibrinogen; s, normal serum; 2, antihuman CIg rabbit antiserum.

placed in the trough, two independent arcs were observed (FIGURE 7, A). This finding suggests that these two components were present in the solution as independent entities at ambient temperatures. Immunoelectrophoretically, CIg in the cryofibrinogen was identical with that in normal plasma or serum (FIGURE 7, B).

Ultracentrifugation also revealed two separate peaks, further evidence of the dissociated coexistence of CIg and fibrinogen, as previously stated. Sedimentation coefficients ($S^0_{20,w}$) determined on the CIg component of two samples were 11.3S and 12.8S.

The amino acid compositions of the heat-stable (56°C, 3 min) fractions are shown in TABLE 5 and were almost identical to those reported on CIg in plasma by Mosher.[3]

Sodium dodecyl sulfate polyacrylamide gel electrophoresis of CIg isolated from cryofibrinogen was performed under both unreducing and reducing conditions (FIGURE 8). A protein molecular weight of 440,000 daltons and a protein subunit molecular weight of 215,000 daltons were found, in good agreement with molecular weight estimates for plasma CIg reported by others.[2,3]

Sodium dodecyl sulfate polyacrylamide gel electrophoresis of cryofibrinogen clearly revealed that the major components in cryofibrinogen are CIg and fibrinogen (FIGURE 9). The possibility of microheterogeneity of CIg in this fraction cannot be determined from this Figure; however, when a small amount of sample was applied, closely migrating doublets in the 440,000-dalton region were observed. No other fragments of CIg with single-chain structures were present in the cryofibrinogen fraction. Thus, intact dimeric CIg appears to be the form of the molecule that complexes with fibrinogen.

It would be interesting to determine whether the fibrinogen moiety retains fibrinopeptide A, because Shainoff and Page[30] showed that their cryoprofibrin, which was a complex of fibrinogen and monomeric fibrin devoid of fibrinopeptide A, was cryoprecipitable. Stemberger and Hörmann[31] also reported a high-

TABLE 5

AMINO ACID COMPOSITION OF CIg IN CRYOFIBRINOGEN*

Amino Acid	CIg in Cryofibrinogen[15]	CIg in Plasma[3]
Aspartic acid	9.88	9.8
Threonine	9.50	10.8
Serine	7.19	8.0
Glutamic acid	11.30	11.5
Proline	8.34	8.2
Glycine	9.63	8.6
Alanine	5.65	4.2
Half-cystine	2.31	2.5
Valine	8.73	8.2
Isoleucine	4.36	4.5
Leucine	6.03	5.4
Tyrosine	4.36	4.3
Phenylalanine	2.82	2.4
Lysine	3.72	3.6
Histidine	1.93	2.3
Arginine	4.24	5.4
Methionine	0	0.8

*Molar amounts of all amino acids were summed, and the amount of each amino acid was expressed as a percentage of the total.

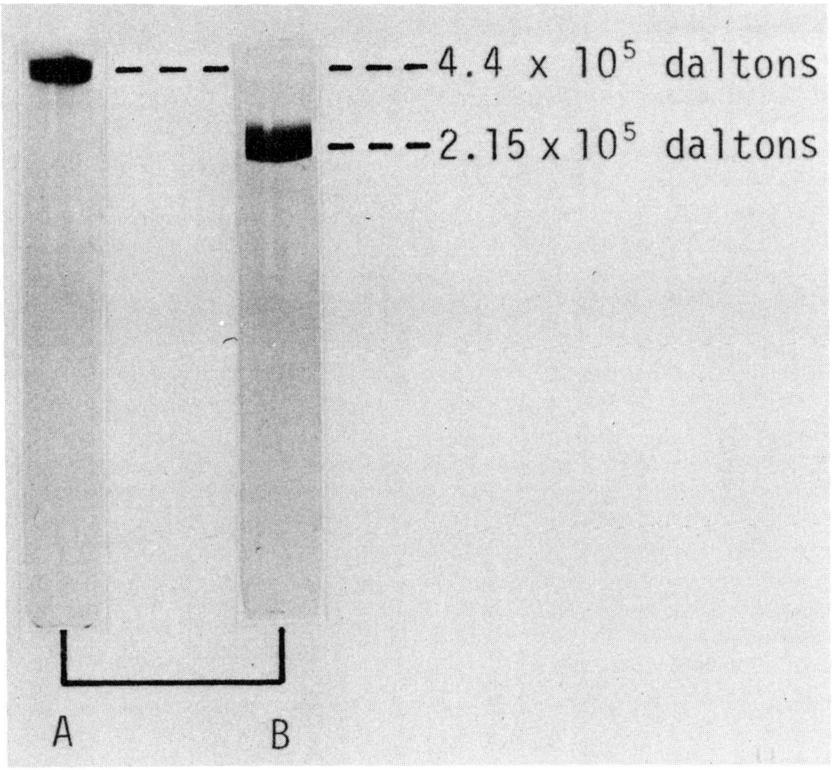

FIGURE 8. Molecular weight estimates by SDS polyacrylamide gel electrophoresis of CIg isolated from cryofibrinogen. A, Unreduced; B, reduced.

affinity complex of soluble fibrin monomer and CIg. As shown in FIGURE 10, the α-chain of fibrinogen moiety is homogeneous and appears to preserve fibrinopeptide A (FIGURE 10, A) better than does the thrombin-treated fraction, which migrated a little faster (FIGURE 10, T). This result was confirmed by analysis of the NH_2 termini of the fibrinogen moiety isolated from cryofibrinogen, which contained only alanine, tyrosine, and a small quantiy of aspartic acid (FIGURE 11). These NH_2 termini were similar to those of normal fibrinogen but not those of soluble fibrin monomer complex. These data suggest that fibrinopeptide A is preserved in the α-chain of the fibrinogen moiety.

The data discussed above indicate that cryofibrinogen apparently represents a simple complex between normal plasma CIg and normal plasma fibrinogen. However, the following experiments on the physical properties of cryofibrinogen and mixtures of normal fibrinogen and normal plasma CIg or cryofibrinogen-derived CIg suggest that cryofibrinogen from pathologic plasmas may have unique properties distinct and different from those of comparable mixtures.

The properties of cryofibrinogen and comparable mixtures were studied at different temperatures, utilizing the technique of quasielastic light scattering.[27] This technique provides more accurate estimates of changes in particle size and of

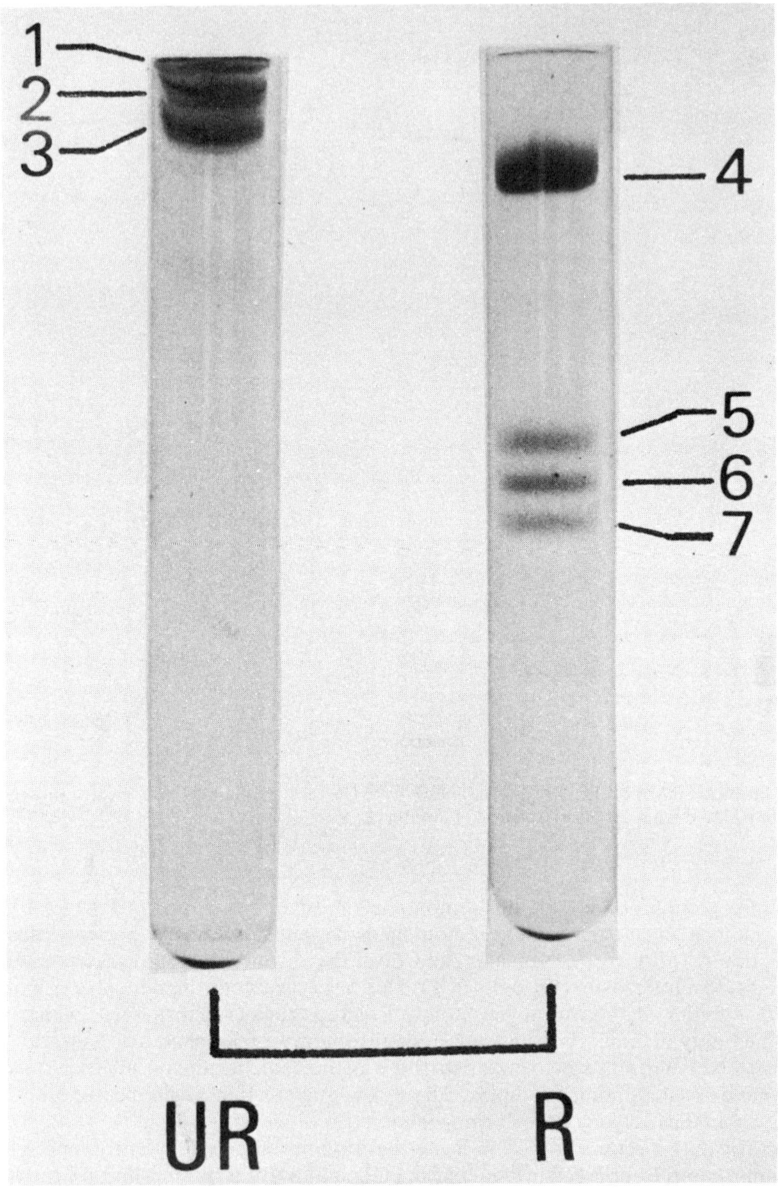

FIGURE 9. Sodium dodecyl sulfate polyacrylamide gel electrophoresis of cryofibrinogen. UR, Unreduced; R, reduced. 1, Unidentified but could be aggregated CIg and/or fibrinogen; 2, CIg in dimeric form with a molecular weight of 440,000 ± 20,000 daltons; 3, fibrinogen; 4, CIg in monomeric form with a molecular weight of 215,000 ± 5000 daltons; 5–7, subunits of fibrinogen.

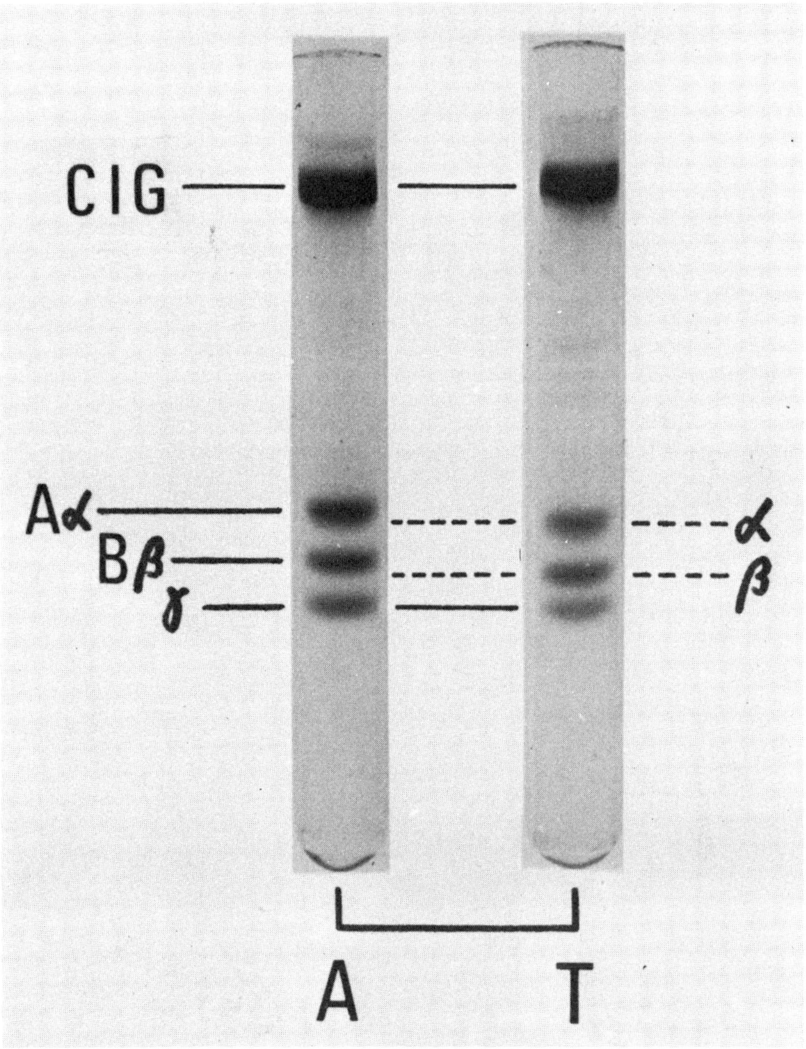

FIGURE 10. Comparison of subunits before (A) and after (T) treatment with thrombin; SDS polyacrylamide gel electrophoresis.

the number of molecules per particle than are obtainable through the conventional light-scattering technique. Samples of identical absorbance at 280 nm of 0.5 were examined. Scattering intensity did not differ appreciably between the normal control mixture of chromatographically purified CIg and fibrinogen, even at 8.5°C, as shown in FIGURE 12 (○--○). However, the scattering intensity of cryofibrinogen increased at 15.5°C, as indicated by the wavy arrow at A in the same Figure. At

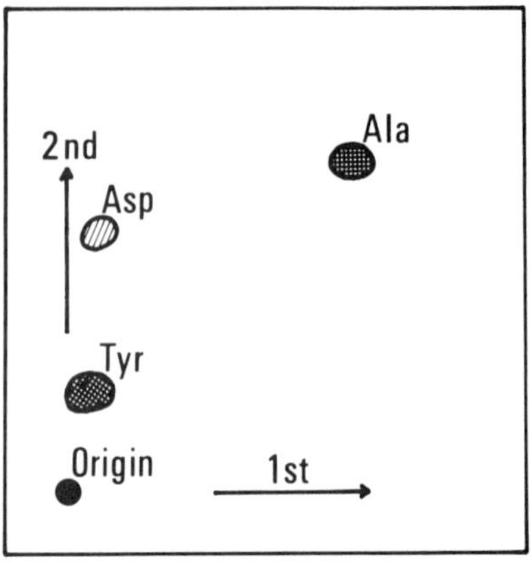

FIGURE 11. Amino terminal amino acid analysis of the fibrinogen moiety of cryofibrinogen by micropolyamide thin-layer chromatography. Phenylthiohydantoin (PTH) sample was obtained by Edman degradations. The PTH sample was applied to the spot "origin" and separated two-dimensionally as indicated; 3.75 × 3.75-cm micropolyamide sheet. First solvent: toluene/*n*-pentaneacetic acid (60:30:16, v/v) that contained 250 mg of the fluorescent indicator 2-(4′-*tert*-butylphenyl)-5-(4″-biphenyl)-1,3,4-oxadiazole; second solvent: 25% aqueous acetic acid.

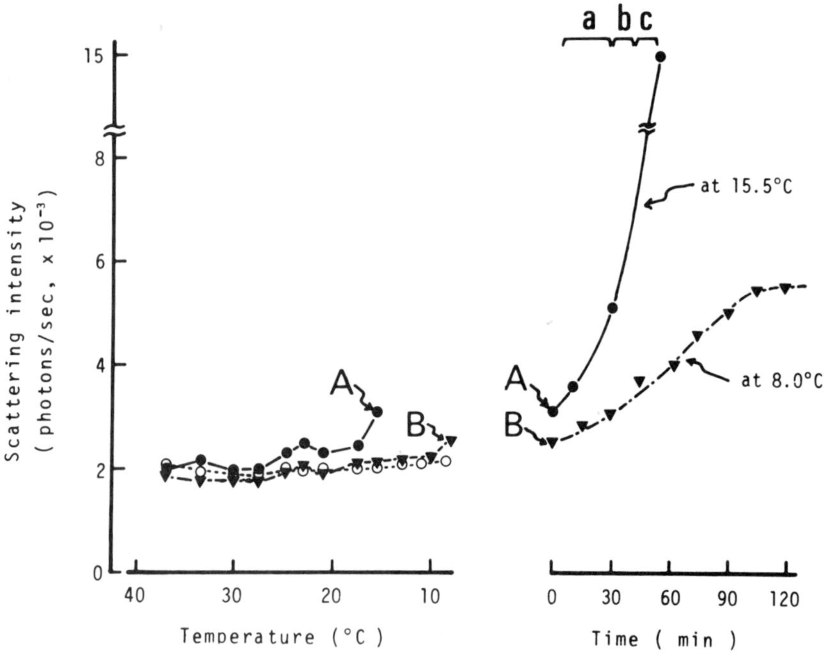

FIGURE 12. Changes of light-scattering intensity at various temperatures (*left*) and those obtained as a function of time at 15.5°C for cryofibrinogen and at 8°C for a mixture of pathologic CIg and normal fibrinogen (*right*). For details, see the text. Concentrations of samples were adjusted to give absorbance at 280 nm of 0.5. ●, Cryofibrinogen; ▾, cryo-CIg plus normal fibrinogen; ○, normal CIg plus normal fibrinogen.

this temperature, the fraction was still in the sol phase, without any visible filamentous structures. A comparable mixture of CIg from cryofibrinogen and normal fibrinogen exhibited an intermediate profile (FIGURE 12, ▼), and an increase of scattering intensity was noted at 8°C, as indicated by the wavy arrow at B. Scattering intensity was then determined as a function of time at 15.5°C for cryofibrinogen and at 8°C for the mixture of cryofibrinogen-derived CIg and normal fibrinogen (FIGURE 12, *right*). The scattering intensity for cryofibrinogen increased by a geometric progression, whereas that for the mixture of fibrinogen and cryofibrinogen-derived CIg reached a plateau after 2 hr, suggesting that the complexes were in a state of equilibrium, with no further polymerization occurring.

Circular dichroism analysis was performed to determine whether there is any difference in secondary structure between normal and pathologic CIg. The results are shown in TABLE 6. We observed a relatively high proportion of β-structure but little or almost nonexistent α-helices in both forms of the CIg fractions. However, there was no significant difference between normal and pathologic CIg, that is, CIg from plasma and cryofibrinogen.

TABLE 6

DETERMINATION OF SECONDARY STRUCTURES BY CIRCULAR DICHROISM

	Normal Fibrinogen	CIg		Cryofibrinogen
		Normal	Pathologic	
α-Helix	49*	0	0	13*
β-Structure	9	$\gtrsim 30$	$\gtrsim 10$	26
Unordered form	42	—	—	61

*All values are expressed as percentages of the total.

The circular dichroism patterns obtained from cryofibrinogen at various temperatures, from 37° down to 12°C, showed no appreciable change. Below 12°C, determination of circular dichroism was technically impractical and was not attempted. However, the information obtained provides evidence that the secondary structure of cryofibrinogen is not significantly affected when complex formation between CIg and fibrinogen first begins to occur. Because the fibrinogen moiety had not been separated from cryofibrinogen without denaturation, data on the mixture of normal CIg and fibrinogen from cryofibrinogen are not available. Nevertheless, from the data so far obtained, we suggest that both CIg and fibrinogen have been altered to produce binding sites likely to be exposed at low temperatures. Further studies will be required to confirm this hypothesis.

SUMMARY

Cold-insoluble globulin is normally present in plasma and serum at concentrations of 27.52 ± 4.60 and 23.46 ± 5.18 mg/dl, respectively (means ± SD). The concentration of CIg in blood samples was significantly decreased in DIC syndromes (14.69 ± 6.55 mg/dl; $p < 0.001$). A strong, positive correlation was found with AT-III ($r = 0.68$) and a less striking one with Plg. Although α_2-PI was shown to be significantly decreased in DIC syndromes ($p < 0.001$), a weak, inverse correlation was found between CIg and α_2-PI ($r = -0.29$). Immunologically cross-reactive substances were found to be widely distributed in association with the cells and tissues of mesenchymal origin, including fibroblasts, adipose cells, smooth muscle cells, and basement membranes. The glomerular basement membrane was an exception and is currently believed to be of different origin. In the kidney, fluorescence was found in the mesangium.

Cold-insoluble globulin is also present as a component of cryofibrinogen that forms a solid gel at low temperatures. Sodium dodecyl sulfate polyacrylamide gel electrophoresis revealed that CIg in this fraction was rather homogeneous. Although closely migrating doublets were occasionally seen in the 440,000-dalton region on gels of unreduced samples, monomeric derivatives with a molecular weight of 220,000 or less, which have been claimed to occur in circulating plasma, were not observed. Thus, intact dimeric CIg appears to be the form of the molecule that complexes with fibrinogen. Cold-insoluble globulin is the fraction that was shown to exist as an independent entity from fibrinogen at an ambient temperature by immunoelectrophoresis and ultracentrifugation. However, very rapid formation of highly polymerized complexes in the sol phase at low temperatures was manifested by the finding of a sharp increase in light-scattering intensity using the technique of quasielastic light scattering. A control study on a mixture of normal CIg and fibrinogen disclosed no appreciable change in the temperature range between 37 and 8.5°C. A comparative study on a mixture of cryofibrinogen-derived CIg and normal fibrinogen revealed an intermediate light-scattering pattern. After 2 hr at 8°C, this mixture reached a state of equilibrium, where no further polymerization occurred. The secondary structures of normal and cryofibrinogen-derived CIg, determined by circular dichroism, showed no appreciable difference. A noteworthy finding was the almost complete absence of α-helices and a relatively high proportion of β-structure in both forms of CIg.

Amino termini of the fibrinogen moiety of cryofibrinogen were found to consist of alanine, tyrosine, and a small quantity of aspartic acid, consistent with the NH_2 terminal moiety composition of normal fibrinogen but not of soluble fibrin monomer complex.

Acknowledgments

We are indebted to Prof. A Wada and Drs. I. Nishio and N. Suzuki, Department of Biophysics, University of Tokyo, for their laborious work of quasielastic light scattering and to Dr. H. Kihara, Department of Physics, Jichi Medical School, for his cooperation in the experiment of circular dichroism. Skillful technical assistance was provided by Misses M. Yamaguchi, M. Chikugo, and C. Nakamikawa.

References

1. Suzuki, K., S. Nakamura, S. Iwanaga, S. Hashimoto & S. Matsumura. 1976. Bovine plasma cold insoluble globulin: purification, properties and biological function (Abst.). 16th International Congress of Haematology, Kyoto, Japan.
2. Chen, A. B., D. L. Amrani & M. W. Mosesson. 1977. Heterogeneity of the cold-insoluble globulin of human plasma (Cig), a circulating cell surface protein. Biochim. Biophys. Acta **493:** 310–322.
3. Mosher, D. F. 1975. Cross-linking of cold-insoluble globulin by fibrin-stabilizing factor. J. Biol. Chem. **250:** 6614–6621.

4. MOSHER, D. F. 1976. Action of fibrin stabilizing factor on cold-insoluble globulin and α_2-macroglobulin in clotting plasma. J. Biol. Chem. **251:** 1639–1645.
5. MOSHER, D. F. 1977. Labeling of a major fibroblast surface protein (fibronectin) catalyzed by blood coagulation factor XIII. Biochim. Biophys. Acta **491:** 205–210.
6. WARTIOVAARA, J., E. LINDER, E. RUOSLAHTI & A. VAHERI. 1974. Distribution of fibroblast surface antigen. Association with fibrillar structures of normal cells and loss upon viral transformation. J. Exp. Med. **140:** 1522–1533.
7. STONE, K. R., R. E. SMITH & W. K. JOKLIK. 1974. Changes in membrane polypeptides that occur when chick embryo fibroblasts and NRK cells are transformed with avian sarcoma virus. Virology **58:** 86–100.
8. CHAMBERG, C. C., D. KIEHN & S. HAKOMORI. 1974. Changes in a surface-labeled galactoprotein and in glycolipid concentrations in cells transformed by a temperature-sensitive polyoma virus mutant. Nature (London) **248:** 413–415.
9. VAHERI, A. & E. RUOSLAHTI. 1974. Disappearance of a major cell-type specific surface glycoprotein antigen (SF) after transformation of fibroblasts by Rous sarcoma virus. Int. J. Cancer **13:** 579–586.
10. BURRIDGE, K. 1976. Changes in cellular glycoproteins after transformation: identification of specific glycoproteins and antigens in sodium dodecyl sulfate gels. Proc. Nat. Acad. Sci. USA **73:** 4457–4461.
11. YAMADA, K. M., S. S. YAMADA & I. PASTAN. 1976. Cell surface protein partially restores morphology, adhesiveness, and contact inhibition of movement to transformed fibroblasts. Proc. Nat. Acad. Sci. USA **73:** 1217–1221.
12. RUOSLAHTI, E., A. VAHERI, P. KUUSELA & E. LINDER. 1973. Fibroblast surface antigen: a new serum protein. Biochim. Biophys. Acta **322:** 352–358.
13. RUOSLAHTI, E. & A. VAHERI. 1974. Novel human serum protein from fibroblast plasma membranes. Nature (London) **248:** 789–791.
14. MOSESSON, M. W., R. W. COLMAN & S. SHERRY. 1968. Chronic intravascular coagulation syndrome. Report of a case with special studies of an associated plasma cryoprecipitate ("cryofibrinogen"). N. Engl. J. Med. **278:** 815–821.
15. MATSUDA, M., T. SAIDA & R. HASEGAWA. 1976. Cryofibrinogen in the plasma of patients with skin ulcerative lesions on the legs: a complex of fibrinogen and cold insoluble globulin. Thromb. Res. **9:** 541–552.
16. MOSESSON, M. W. & R. A. UMFLEET. 1970. The cold-insoluble globulin of human plasma. I. Purification, primary characterization, and relationship to fibrinogen and other cold insoluble fraction components. J. Biol. Chem. **245:** 5728–5736.
17. AVRAMEAS, S. & T. TERNYNCK. 1969. The cross-linking of proteins with glutaraldehyde and its use for the preparation of immunoadsorbents. Immunochemistry **6:** 53–66.
18. MANCINI, G., A. O. CARBONARA & J. F. HEREMANS. 1965. Immunochemical quantitation of antigens by single radial immunodiffusion. Immunochemistry **2:** 235–254.
19. LAURELL, C. B. 1966. Quantitative estimation of proteins by electrophoresis in agarose gel containing antibodies. Anal. Biochem. **15:** 45–52.
20. RATNOFF, O. D. & C. MENZIE. 1951. A new method for the determination of fibrinogen in small samples of plasma. J. Lab. Clin. Med. **37:** 316–320.
21. MERSKEY, C., G. J. KLEINER & A. J. JOHNSON. 1966. Quantitative estimation of split products of fibrinogen in human serum, relation to diagnosis and treatment. Blood **28:** 1–18.
22. COONS, A. H. 1956. Histochemistry with labeled antibody. Int. Rev. Cytol. **5:** 1–23.
23. SPACKMAN, D. H., W. H. STEIN & S. MOORE. 1958. Automatic recording apparatus for use in the chromatography of amino acids. Anal. Chem. **30:** 1190–1206.
24. WEBER, K. & M. OSBORN. 1969. The reliability of molecular weight determinations by dodecyl sulfate-polyacrylamide gel electrophoresis. J. Biol. Chem. **244:** 4406–4412.
25. KULBE, K. D. 1974. Micropolyamide thin-layer chromatography of phenylthiohydantoin amino acids (PTH) at subnanomolar level. A rapid microtechnique for simultaneous multisample identification after automated Edman degradations. Anal. Biochem. **59:** 564–573.

26. Pecora, R. 1964. Doppler shifts in light scattering from pure liquids and polymer solutions. J. Chem. Phys. **40:** 1604–1614.
27. Chen, Y. H., J. T. Yang & H. M. Martinez. 1972. Determination of the secondary structures of proteins by circular dichroism and optical rotatory dispersion. Biochemistry **11:** 4120–4131.
28. Moroi, M. & N. Aoki, 1976. Isolation and characterization of α_2-plasmin inhibitor from human plasma. J. Biol. Chem. **251:** 5956–5965.
29. Aoki, N., M. Moroi, M. Matsuda & K. Tachiya. 1977. The behavior of α_2-plasmin inhibitor in fibrinolytic states. J. Clin. Invest. **60:** 361–369.
30. Shainoff, J. R. & I. H. Page. 1962. Significance of cryoprofibrin in fibrinogen-fibrin conversion. J. Exp. Med. **116:** 687–707.
31. Stemberger, A. & H. Hörmann. 1976. Affinity chromatography on insolubilized fibrinogen and fibrin monomer, II. The behavior of cold-insoluble globulin. Hoppe-Seyler's Z. Physiol. Chem. **357:** 1003–1005.

PART II. CHARACTERISTICS OF A HIGH-MOLECULAR-WEIGHT GLYCOPROTEIN IN CONNECTIVE TISSUE AND BASEMENT MEMBRANES

ORGANIZATION OF EXTRACELLULAR PROTEINS ON THE CONNECTIVE TISSUE CELL SURFACE: RELEVANCE TO CELL-MATRIX INTERACTIONS *IN VITRO* AND *IN VIVO**

Paul Bornstein, Dan Duksin, Gary Balian,†
Jeffrey M. Davidson, and Ed Crouch

Department of Biochemistry
University of Washington
Seattle, Washington 98195

The fluid mosaic model for the structure of the plasma membrane[1] postulates the existence of integral proteins that, within limits, are free to diffuse laterally in the plane of the membrane formed by a fluid lipid bilayer, and the occurrence of interactions of such proteins with peripheral proteins either on the cytoplasmic or on the external surface of the cell membrane.[2-4] It is thought that cytoplasmic microfilaments are linked directly or indirectly to integral proteins via peripheral proteins subjacent to the cell membrane and that many cellular properties, including adhesion, shape, and motility, are controlled by dynamic interactions between these protein components.[3-6]

As a consequence of these concepts and of the studies they engendered, it has been possible to achieve a better understanding of the binding of drugs and hormones to membrane receptors,[7] the clustering of receptors by antibodies and lectins,[3,4,8] and cell-cell interactions.[9] Considerable light has also been shed on alterations in the properties of neoplastic and transformed cells, many of which are reflected in changes in cell surface functions.[10-12]

In contrast, the interactions of the connective tissue cell surface with the extracellular matrix and the consequences of this relationship for cellular function have received less attention. These interactions are likely to account, in part, for such diverse phenomena as directed cellular migration, maintenance of a characteristic shape, the ability of the cell to organize a complex matrix, adherence to surfaces, and the role of the cell surface in tissue differentiation.

We have developed a model for the organization of the external cell surface (FIGURE 1) designed to account for the nature of cell-matrix interactions and for the cytoplasmic control of these interactions. We propose that in a connective tissue cell attached to a substratum (FIGURE 1,A), fibronectin (see below) serves as an external peripheral protein, one of whose functions is to bind collagen or collagen precursors, or both. The length of the collagen molecule (~3000 Å) and its tendency to aggregate promote clustering of fibronectin-binding integral proteins, resulting in a meshwork on the external cell surface. This protein meshwork may be linked, in turn, to microfilaments via transmembrane integral proteins and cytoplasmic peripheral proteins. Collagen, in its role as a "secondary"

*Supported by National Institutes of Health Grants AM 11248, DE 02600, and HL 18645 and Training Grant GM 07266 (to E.C.).

†Established Investigator of the American Heart Association.

0077-8923/78/0312-0093 $01.75/0

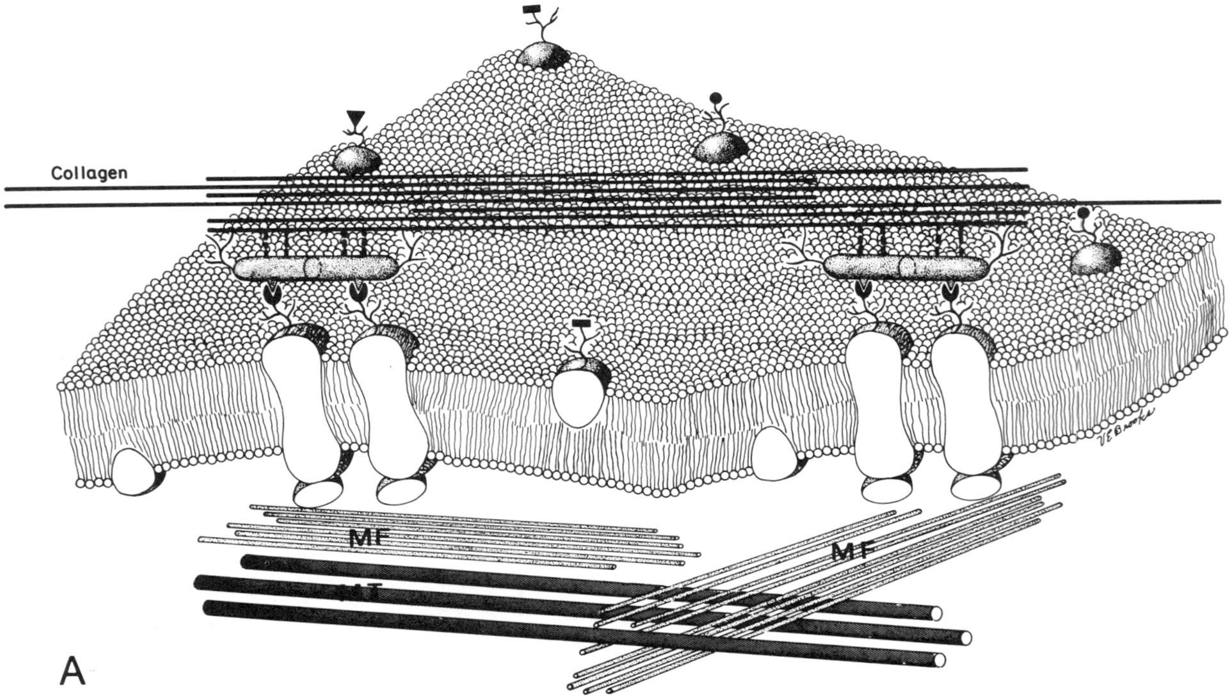

FIGURE 1. The molecular organization of the cell surface in a connective tissue cell attached to a substratum (A) and in a similar cell detached from the surface by the action of proteases, in mitosis, or after viral transformation (B). The model is based, in part, on the fluid mosaic structure of the cell membrane proposed by Singer and Nicolson[1] and on the relation of the internal cytoskeleton, which consists of microfilaments and microtubules, to the cell surface as discussed by Edelman[3] and Nicolson.[4] The relative dimensions of cell membrane constituents and submembranous structures have been distorted to illustrate processes that involve membranous proteins; the implications of depicting these various elements to scale were presented by Loor.[13] The quantity of membrane proteins, relative to lipid, has also been reduced for the purpose of clarity.

In a cell attached to a substratum (A), integral membrane proteins are clustered by a cell surface protein complex, which consists of fibronectin (shown as a capsule-shaped molecule) and collagen. The dimeric nature of fibronectin permits clustering of integral proteins; interactions are depicted to involve the oligosaccharide side chains of integral proteins, but carbohydrate side chains of fibronectin may also participate, or amino acid side chains in both proteins may be primarily involved. Binding of fibronectin to the cell membrane may involve divalent cations. The tendency of collagen to aggregate serves to form a collagen-fibronectin meshwork on the surface of the cell. This meshwork is linked via membrane proteins to microfilaments (MF) and

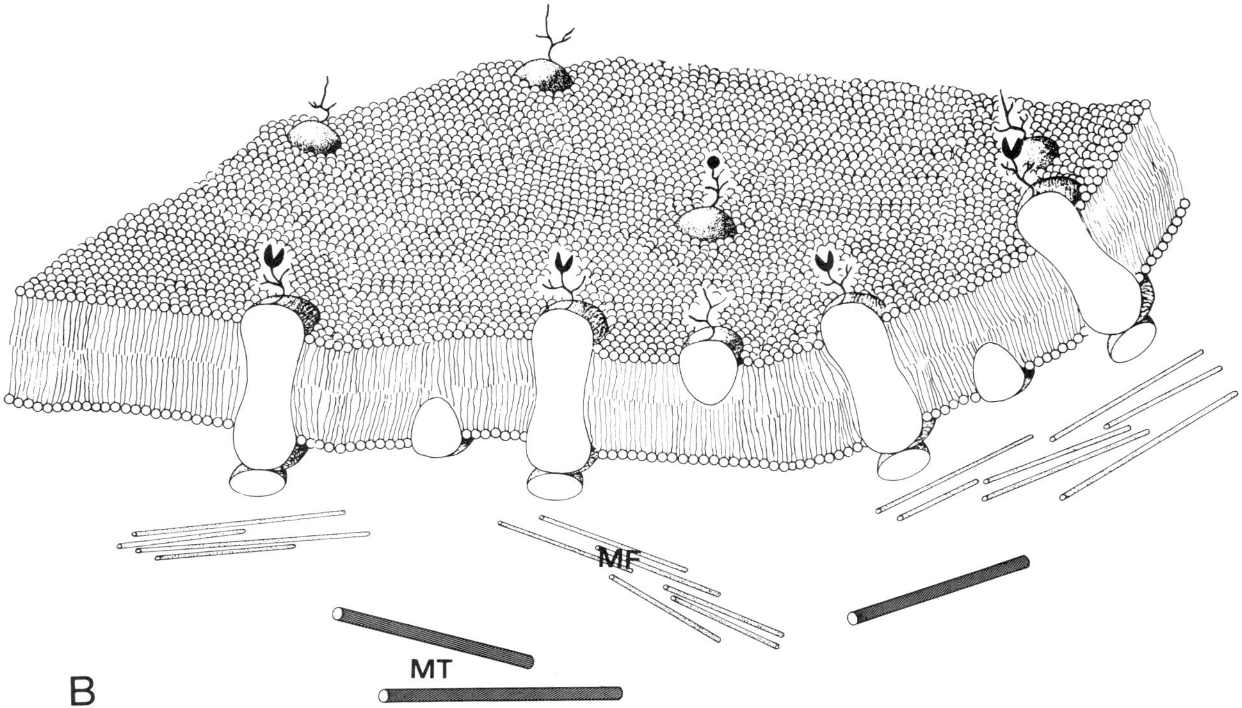

FIGURE 1, *continued*
microtubules (MT), thus enveloping the bilayer membrane in a two-ply system. This linkage is shown to be direct but could function indirectly through ionic fluxes or changes in cyclic nucleotides.

In cells detached from a surface by proteases or in mitotic or transformed cells (B), fibronectin and collagen are missing from the cell surface, or are markedly reduced, and the system of microtubules and microfilaments is disorganized. Fibronectin-binding integral proteins are more dispersed. Some oligosaccharide side chains are shown to lack normal determinants. Interactions between the external protein meshwork and the internal cytoskeleton are thought to be cooperative, so that disturbances in either component could lead to disruption of the other. This model depicts protease-treated, mitotic, and transformed cells identically to emphasize common surface features of these cells. Clearly, important differences among these three types of cells must exist.

peripheral protein (secondary in the sense that it does not interact directly with an integral membrane protein), adds cohesiveness to the external meshwork and mediates some of the interactions of the cell with its extracellular environment. Additional stabilization may be provided by formation of disulfide bonds between fibronectin monomers and between fibronectin and other cell surface proteins[14] (not shown in Figure).

Alterations in any of the components of the system by genetic or environmental changes would be expected to disrupt the integrity of the entire meshwork (FIGURE 1,B). Thus, changes in components of the internal cytoskeleton caused by drugs, mitosis, viral transformation, failure to synthesize or secrete collagen or fibronectin normally, or failure to assemble these proteins on the cell surface should all have major predictable consequences for cell adhesion, shape, and motility and matrix organization. In the following discussion, we summarize the data that led to the development of this model and examine some of its consequences.

FIBRONECTIN IS INVOLVED IN CELL-SUBSTRATE INTERACTIONS

Recently, a glycoprotein, or a group of glycoproteins, variously known as cell surface protein,[15] fibroblast surface antigen,[16] LETS (large external transformation-sensitive) protein,[17] galactoprotein a,[18] Z protein,[19] and fibronectin,[20,21] has been found on the surface of fibroblasts,[17,22] smooth muscle cells,[23] glial cells,[24] and both epithelial[25] and endothelial[26] cells. The protein has been identified primarily by cell surface labeling[27] and exists as a dimer with a subunit molecular weight of approximately 230,000 daltons[10]; its linkage to the cell surface is known to be sensitive to trypsin and other proteases. Fibronectin (the term used here) is reduced or absent from the surfaces of many transformed and malignant cells of mesenchymal origin[10-12] and is thought to be involved in the attachment of a cell to a matrix (see below). Of considerable interest is the recent finding that fibronectin also exists as an important structural component of the extracellular matrix.[28,29]

While intercellular interactions are likely to involve complex specificities that function in cellular recognition, the molecular specificities that govern cell-matrix interactions may be more limited in number, since motile connective tissue cells, such as fibroblasts, interface primarily with collagen fibrils and proteoglycan aggregates and, to lesser extents, with elastic fibers, extracellular connective tissue microfibrils, and other cells. Several studies implicate fibronectin in cell-substrate interactions: (1) During attachment and spreading, cells in culture deposit a "microexudate" on the artificial surface that consists of hyaluronic acid, collagen, and several glycoproteins, one of which is similar in size to fibronectin.[30-32] (2) Mutagenized Balb/c 3T3 cells, selected for reduced adhesiveness to a plastic surface, were examined for cell surface proteins by lactoperoxidase-catalyzed iodination; some of these cells were shown to lack several proteins, including fibronectin.[33] The defect appeared to be primarily in adhesion, because addition of dibutyryl cyclic adenosine monophosphate to the culture medium caused rounded mutant cells to flatten and to revert partially to the morphology of the wild-type cells but did not increase the adhesion of these cells to the substratum. (3) Fibronectin isolated from embryonic chick or NIL hamster fibroblasts restored a fibroblast-like morphology and increased adhesiveness and contact inhibition of movement in several transformed cell lines.[34-36] Fibronectin can also agglutinate formalinized sheep erythrocytes, and this reaction was inhibited by agents that chelated calcium.[37] The latter observation is consistent with the ability of chelating agents to detach cells from a substratum. (4) Tunicamycin, an inhibitor of lipid carrier-dependent protein glycosylation,[38,39] inhibited the secretion of fibronectin by embryonic chick fibroblasts[40] and 3T3 cells[41] in culture. Mouse 3T3 and virally transformed 3T3 cells showed dramatic changes in surface morphology and reduced adhesiveness in the presence of tunicamycin.[42] Transformed cells were far more sensitive to the drug, while effects with normal 3T3 cells were seen primarily when cells were in the logarithmic growth phase. Cell surface labeling demonstrated that tunicamycin interfered with insertion of fibronectin on the cell surface. The increased tendency of transformed 3T3 cells to detach from a surface in the presence of tunicamycin is consistent with a marked reduction of cell surface fibronectin in these cells in comparison with most normal cells (see below).

A Reduction in Fibronectin Is Associated with Decreased Cell-Substrate Adhesion

Numerous studies have documented that fibronectin is either absent or reduced on the surfaces of virally transformed[10-12] and neoplastic cells.[24,43] A good correlation was observed between lack of fibronectin, as detected by surface immunofluorescence, and oncogenicity in several virally transformed cells and in cells derived from spontaneously occurring neoplasms.[43]

Virally transformed and neoplastic cells differ from normal cells in cell-cell and cell-substrate adhesive properties.[44,45] Indeed, reduction in adhesion to an extracellular matrix may play a role in metastatic spread of malignant tumors. Although the factors involved in these cell surface modifications are complex,[11,12] there are increasing indications that loss of fibronectin from the surface of transformed cells contributes to reduced cell-substrate adhesion. Perhaps the most direct evidence for this hypothesis is the restoration of adhesiveness in transformed cells induced by addition of fibronectin.[34,36]

Cells in mitosis possess certain of the altered cell surface properties displayed by transformed cells, including reduced adhesion to a surface.[46] In mitosis, a marked reduction of lactoperoxidase-catalyzed labeling of fibronectin was observed in hamster NIL and mouse L-929 cells.[47,48] Mitotic cells also demonstrate reduced cell surface, but not intracellular, fibronectin concentrations when examined by immunofluorescence microscopy.[49] Conversely, cells arrested in the G1 phase of the cycle, as a result of density-dependent inhibition of growth or serum deprivation, show high levels of cell surface fibronectin.[47,48]

Fibronectin Binds Both to the Cell Surface and to Collagen

The binding of fibronectin to the cell surface is apparent from its localization by immunofluorescence[21-23] and from its ability to restore some normal surface properties to transformed cells.[34-36] A surface location is supported by the finding that antifibronectin serum is cytotoxic to fibroblasts in the presence of complement.[52]

We believe that there is also sufficient evidence to implicate binding of cell surface fibronectin to collagen. A high-molecular-weight protein in serum, thought to be related to fibronectin, mediates attachment of SV40-transformed 3T3 and CHO cells to collagen-coated surfaces and requires divalent cations and cellular metabolic energy for its function.[53,54] This factor binds to collagen, dissociated collagen α-chains, and to a cyanogen bromide-produced collagen fragment, $\alpha 1$(I)-CB7.[55] An interaction between collagen or gelatin and cold-insoluble globulin (CIg), a serum protein similar in molecular weight and immunologically related to fibronectin, has been shown both immunochemically[23] and by direct binding experiments.[52,87,88]

It is of interest that collagen chains and the $\alpha 1$(I)-CB7 fragment enhance attachment of myoblasts to a plastic surface.[56] Fusion of myoblasts and myotube formation *in vitro* are promoted by interaction of the cells with collagen or collagen-related proteins[57,58] and depend on the presence of calcium ions.[59] Fibronectin appears to play a central role in this process, because there are alterations in the distribution of surface fibronectin at the time of fusion,[60] and virally transformed myoblasts, which had markedly reduced levels of surface fibronectin, failed to fuse.[61]

Collagen Is a Cell Surface Component of Attached Connective Tissue Cells

Studies dealing with a variety of cells attached to a substratum[62,63] and in suspension[64] indicate that antibodies to collagen, or to a collagen-related synthetic polytripeptide, induce complement-mediated cytotoxicity. In some of these and other studies,[65] collagen was located on the cell surface by indirect immunofluorescent staining. Recently, Bornstein and Ash[23] demonstrated collagen or procollagen, or both, on the surface of monolayer normal rat kidney (NRK) fibroblasts by use of affinity-purified antibodies to collagen. As judged by indirect immunofluorescence, the protein was distributed in a reticular fashion on the cell surface, and antibody-induced translational movement of this network in the plane of the membrane was severely restricted. Similar distribution and behavior were observed for fibronectin in fibroblasts[22,23] and in smooth muscle cells.[23] When NRK fibroblasts were dissociated with trypsin and examined in suspension, neither collagen nor fibronectin was observed on the cell surface; mitotic figures observed in monolayer cultures of these cells also showed markedly reduced or absent immunofluorescent staining with antibodies to these proteins.[23] These observations suggest that the external protein meshwork is released or shielded when cells are detached from a substratum.

Model for the Organization of Extracellular Proteins on the Connective Tissue Cell Surface

We propose that connective tissue cells *in vitro* and *in vivo* synthesize and organize a complex extracellular cell surface meshwork that contains both fibronectin and collagen and that can interact with other cell surface proteins, with the internal cytoskeleton, and with the extracellular matrix. It should be emphasized that fibronectin and collagen are fundamentally viewed as extracellular matrix glycoproteins. Under the constraint of two-dimensional growth in the fluid environment of cell culture, they are the major secreted proteins of connective tissue cells. However, the considerations discussed below have led us to reconsider the conventional concept that a distinct dividing line separates the cell surface and the matrix.

The interaction of fibronectin with collagen (FIGURE 1,A) probably involves discrete regions of the collagen molecule[55,56,87] and may be dependent on divalent cations[54]; cations may also mediate binding of fibronectin to the cell surface. We are currently unable to exclude the possibility that procollagen rather than collagen (or perhaps a combination of the two proteins) binds to fibronectin, because antisera to rat procollagen also stain NRK fibroblasts.[23] If so, the nontriple-helical domains at both the NH_2 and COOH termini of procollagen may interact separately with other cell surface binding sites. We have also not distinguished among the several known collagen types[66] in assigning a role for collagen as a cell surface protein. Very possibly, different connective tissue cells have different collagen types on their surfaces, and this difference may contribute to the characteristic properties of these cells. In cells known to possess a structurally distinct basal lamina, the cell surface-associated collagen may serve as a scaffold for this extracellular structure.

The similar, nonrandom immunofluorescent patterns observed for cell surface-associated fibronectin[22,67] and collagen,[23] the interaction between these macromolecules, and the relatively restricted movement, at the light microscope level, induced by divalent antibody[22,23] all suggest that fibronectin and collagen form a discrete meshwork on the surface of normal cells attached to a substratum. Preliminary experiments performed with separate fluorescein and rhodamine labels for fibronectin and collagen indicate that these proteins are partially coextensive on the surfaces of chicken and rat embryo fibroblasts.[68] Additional support for the nonrandom distribution of fibronectin on the cell surface is provided by the segregation of the major part of the externally labeled protein into a high-density subclass of plasma membrane vesicles after homogenization by nitrogen cavitation.[69] Direct evidence for the immobility of cell surface fibronectin has also come from fluorescence photobleaching recovery experiments.[70]

We further propose that this external protein meshwork interacts, via transmembrane and cytoplasmic peripheral proteins, with microfilaments subjacent to the plasma membrane and that, together, these assemblies envelop the plasma membrane in a two-ply system. It is of interest that the addition of fibronectin to transformed cells was associated with the appearance of well-defined actin cables that were not evident in the untreated transformed cells.[36] We envision that the integrated function of the internal cytoskeleton and external surface meshwork dictates several properties of normal cells, including adhesion to a substratum, and that the interactions with neighboring cells are manifest in the phenomenon of density-dependent inhibition of movement.[71]

While we have focused on two protein constituents of this putative external protein meshwork, it is likely that other proteins contribute to the complex. Fibrinogen and CIg interact,[72,73] and plasma transglutaminase (blood coagulation factor XIII), which can cross-link fibronectin on the surface of human fibroblasts,[20] is also capable of forming hybrid polymers of fibrin and CIg.[73] Other components of the extracellular matrix, including proteoglycans, may also interact with the cell surface meshwork.

Virally transformed cells, cells in mitosis, and cells dissociated by proteolytic enzymes have certain surface characteristics in common (FIGURE 1,B). The internal cytoskeleton in these cells is partially disassembled or organized differently.[74,75] These alterations could lead to a reduced level of surface fibronectin if, as we suggest, the organization of integral fibronectin-binding proteins is dependent on the organization of the internal cytoskeleton. On the other hand, the external protein meshwork could also exert a stabilizing influence on the internal cytoskeleton. However, measurement of synthetic rates for fibronectin[76] and the finding of reduced levels of mRNA for both fibronectin and procollagen in transformed cells[77,78] suggest that reduced synthesis of these proteins contributes to the altered cell surface characteristics that are observed in these cells. Cells dissociated with proteases may lack fibronectin as a consequence of the protease sensitivity of the protein, and the disassembly of actin-containing microfilament bundles may, in turn, result from this surface proteolysis.[75] The interrelation of these events is underscored by the findings that cell surface proteolytic activity is increased after transformation[12] and that transformed cells demonstrate altered membrane-associated cytoskeletal elements and transmembrane control.[4,11]

These observations suggest that the external protein meshwork and elements of the internal cytoskeleton interact as suggested in FIGURE 1(A). The very close apposition of cytoplasmic cortical microfilaments with the cell membrane[79] is consistent with linked protein interactions. Alternatively, alterations in surface proteins may lead to ionic fluxes or changes in cyclic nucleotides that secondarily affect the assembly of microfilaments and microtubules.

Enzymatically dissociated cells and cells in mitosis lack cell surface-associated collagen,[23] a finding in accord with the model proposed in FIGURE 1(B). Our model predicts that virally transformed cells will also have reduced cell surface-associated collagen. Since transformed cells generally synthesize far less collagen than do their normal counterparts,[80] a distinction between reduced synthesis and binding may be difficult to make.

POSSIBLE FUNCTIONS OF A FIBRONECTIN-COLLAGEN CELL SURFACE MESHWORK

There are many functions that might be well served by a collagen-fibronectin complex on the connective tissue cell surface, in addition to adhesion to a substratum, density-dependent inhibition of movement, and effects related to cell shape and motility. In specialized tissues, such as the cornea, a precise arrangement of collagen fibrils is observed; intricate spatial orientation of extracellular structures of this type could be effected, in part, by the ability of the cell to use its surface as a nucleation site for a growing fibril. Directed migration of connective tissue cells over considerable distances occurs in the developing embryo.[81] Such movement could take advantage of cell surface collagen-collagen fiber interactions. Indeed, the inductive properties of collagen[82] may be mediated by such interactions. Mobility would require transient interactions. This property could be provided by subjecting the extracellular protein meshwork to transmembrane control by cellular metabolic changes that act via the internal cytoskeleton. Finally, the ability of the external protein meshwork to interact with other proteins, such as fibrin, could serve an important role in the localization of fibroblasts in a healing wound.

Collagen chains and fragments have been reported to be chemotactic for both fibroblasts[83] and monocytes.[84] The ability to bind collagen may therefore not be limited to connective tissue cells. Collagen is known to be capable of binding to platelets, resulting in platelet aggregation and release of granular contents. The effectiveness of collagen in this process is enhanced markedly by prior aggregation of collagen molecules.[85] The requirement for polymeric collagen may reflect the need to redistribute fibronectin-like platelet membrane proteins to elicit the physiologic response. Platelets, and possibly other cells, bind the Clq component of complement, a protein known to contain a collagen-like domain.[86] It would be of interest to determine whether fibronectin or a related protein is involved in these interactions and whether a process analogous to the formation of a cell surface meshwork might occur in these cells.

Summary

A model has been developed that proposes a cell surface-associated protein meshwork, composed in part of fibronectin and collagen, for a connective tissue cell attached to a substratum. In support of this model are the observations that collagen and fibronectin interact and that these proteins are similarly distributed on the fibroblast cell surface. We suggest that this external meshwork interacts directly or indirectly with the internal cytoskeleton and with the extracellular matrix and thereby mediates several cellular properties, including adhesion, shape, and motility. Loss of cell surface fibronectin as a result of viral transformation, or due to treatment of normal cells with tunicamycin, an inhibitor of protein glycosylation, may contribute to the reduced adhesion and altered morphology observed in these circumstances. We therefore predict that the changes in these properties observed with virally transformed cells, mitotic cells, and cells treated with proteolytic enzymes are related to alterations in the external protein meshwork.

Acknowledgment

We thank Virginia Brooks for assistance with the illustrations.

References

1. Singer, S. J. & G. L. Nicolson. 1972. The fluid mosaic model of the structure of cell membranes. Science **175:** 720–731.
2. Singer, S. J. 1974. The molecular organization of membranes. Annu. Rev. Biochem. **43:** 805–833.
3. Edelman, G. M. 1976. Surface modulation in cell recognition and cell growth. Science **192:** 218–226.
4. Nicolson, G. L. 1976. Transmembrane control of the receptors on normal and tumor cells. I. Cytoplasmic influence over cell surface components. Biochim. Biophys. Acta **457:** 57–108
5. Poste, G., D. Papahadjopoulos & G. L. Nicolson. 1975. Local anesthetics affect transmembrane cytoskeletal control of mobility and distribution of cell surface receptors. Proc. Nat. Acad. Sci. USA **72:** 4430–4434.
6. Rees, D. A., C. W. Lloyd & D. Thom. 1977. Control of grip and stick in cell adhesion through lateral relationships of membrane glycoproteins. Nature (London) **267:** 124–128.
7. Cuatrecasas, P. & M. D. Hollenberg. 1976. Membrane receptors and hormone action. Adv. Protein Chem. **30:** 251–451.
8. Singer, S. J. 1974. Molecular biology of cellular membranes with applications to immunology. Adv. Immunol. **19:** 1–66.
9. Singer, S. J. 1976. The fluid mosaic model of membrane structure. Some applications to ligand-receptor and cell-cell interactions. *In* Surface Membrane Receptors: Interface between Cells and Their Environment. R. Bradshaw, Ed.: 1–24. Plenum Publishing Corporation. New York, N.Y.
10. Hynes, R. O. 1976. Cell surface proteins and malignant transformation. Biochim. Biophys. Acta **458:** 73–107.
11. Nicolson, G. L. 1976. Trans-membrane control of the receptors on normal and tumor cells. II. Surface changes associated with transformation and malignancy. Biochim. Biophys. Acta **458:** 1–72.
12. Robbins, J. C. & G. L. Nicolson. 1975. Surfaces of normal and transformed cells. *In* Cancer: A Comprehensive Treatise. Volume 4, Biology of Tumors: Surfaces, Im-

munology and Comparative Pathology. F. F. Becker, Ed.: 3–54. Plenum Publishing Corporation. New York, N.Y.
13. LOOR, F. 1976. Cell surface design. Nature (London) **264:** 272–273.
14. HYNES, R. O. & A. DESTREE. 1977. Extensive disulfide bonding at the mammalian cell surface. Proc. Nat. Acad. Sci. USA **74:** 2855–2859.
15. YAMADA, K. M. & J. A. WESTON. 1974. Isolation of a major cell surface glycoprotein from fibroblasts. Proc. Nat. Acad. Sci. USA **71:** 3492–3496.
16. RUOSLAHTI, E., A. VAHERI, P. KUUSELA & E. LINDER. 1973. Fibroblast surface antigen: a new serum protein. Biochim. Biophys. Acta **322:** 352–358.
17. HYNES, R. O. 1973. Alteration of cell-surface proteins by viral transformation and by proteolysis. Proc. Nat. Acad. Sci. USA **70:** 3170–3174.
18. GAHMBERG, C. G. & S.-I. HAKOMORI. 1975. Surface carbohydrates of hamster fibroblasts. II. Interaction of hamster NIL cell surfaces with *ricinus communis* lectin and concanavalin A as revealed by surface galactosyl label. J. Biol. Chem. **250:** 2447–2451.
19. BLUMBERG, P. M. & P. W. ROBBINS. 1975. Effect of proteases on activation of resting chick embryo fibroblasts and on cell surface proteins. Cell **6:** 137–147.
20. KUUSELA, P., E. RUOSLAHTI, E. ENGVALL & A. VAHERI. 1976. Immunological interspecies cross-reactions of fibroblast surface antigen (fibronectin). Immunochemistry **13:** 639–642.
21. KESKI-OJA, J., D. F. MOSHER & A. VAHERI. 1976. Cross-linking of a major fibroblast surface-associated glycoprotein (fibronectin) catalyzed by blood coagulation factor XIII. Cell **9:** 29–35.
22. WARTIOVAARA, J., E. LINDER, E. RUOSLAHTI & A. VAHERI. 1974. Distribution of fibroblast surface antigen. Association with fibrillar structures of normal cells and loss upon viral transformation. J. Exp. Med. **140:** 1522–1533.
23. BORNSTEIN, P. & J. F. ASH. 1977. Cell surface-associated structural proteins in connective tissue cells. Proc. Nat. Acad. Sci. USA **74:** 2480–2484.
24. VAHERI, A., E. RUOSLAHTI, B. WESTERMARK & J. PONTEN. 1976. A common cell-type specific surface antigen in cultured human glial cells and fibroblasts: loss in malignant cells. J. Exp. Med. **143:** 64–72.
25. CHEN, L. B., N. MAITLAND, P. H. GALLIMORE & J. K. MCDOUGALL. 1977. Detection of the large external transformation-sensitive protein on some epithelial cells. Exp. Cell Res. **106:** 39–46.
26. BORNSTEIN, P. & S. SCHWARTZ. 1977. Unpublished observations.
27. CRITCHLEY, D. R., J. A. WYKE & R. O. HYNES. 1976. Cell surface and metabolic labelling of the proteins of normal and transformed chicken cells. Biochim. Biophys. Acta **436:** 335–352.
28. LINDER, E., A. VAHERI, E. RUOSLAHTI & J. WARTIOVAARA. 1975. Distribution of fibroblast surface antigen in the developing chick embryo. J. Exp. Med. **142:** 41–49.
29. STENMAN, S. & A. VAHERI. 1977. Distribution of fibronectin—a new connective tissue protein—in normal human tissues. Uppsala J. Med. Sci. **82:** 154.
30. TERRY, A. H. & L. CULP. 1974. Substrate-attached glycoproteins from normal and virus-transformed cells. Biochemistry **13:** 414–425.
31. LEMBACH, K. J., R. E. BRANSON, P. B. HEWGLEY & L. W. CUNNINGHAM. 1977. The synthesis of macromolecular 3-hydroxyproline by attaching and confluent cultures of human fibroblasts. Eur. J. Biochem. **72:** 379–383.
32. CULP, L. A. 1976. Electrophoretic analysis of substrate-attached proteins from normal and virus-transformed cells. Biochemistry **15:** 4094–4104.
33. POUYSSEGUR, J. M. & I. PASTAN. 1976. Mutants of Balb/c 3T3 fibroblasts defective in adhesiveness to substratum: evidence for alteration in cell surface proteins. Proc. Nat. Acad. Sci. USA **73:** 544–548.
34. YAMADA, K. M., S. S. YAMADA & I. PASTAN. 1976. Cell surface protein partially restores morphology, adhesiveness, and contact inhibition of movement to transformed fibroblasts. Proc. Nat. Acad. Sci. USA **73:** 1217–1221.
35. YAMADA, K. M., S. H. OHANIAN & I. PASTAN. 1976. Cell surface protein decreases microvilli and ruffles on transformed mouse and chick cells. Cell **9:** 241–245.

36. Ali, I. U., V. Mautner, R. Lanza & R. O. Hynes. 1977. Restoration of normal morphology, adhesion and cytoskeleton in transformed cells by addition of a transformation-sensitive surface protein. Cell **11:** 115–120.
37. Yamada, K. M., S. S. Yamada & I. Pastan. 1975. The major cell surface glycoprotein of chick embryo fibroblasts is an agglutinin. Proc. Nat. Acad. Sci. USA **72:** 3158–3162.
38. Takatsuki, A. & G. Tamura. 1971. Effect of tunicamycin on the synthesis of macromolecules in cultures of chick embryo fibroblasts infected with Newcastle disease virus. J. Antibiot. **24:** 785–794.
39. Tkacz, J. S. & J. O. Lampen. 1975. Tunicamycin inhibition of polyisoprenyl N-acetylglucosaminyl pyrophosphate formation in calf-liver microsomes. Biochem. Biophys. Res. Commun. **65:** 248–257
40. Duksin, D. & P. Bornstein. 1977. Impaired conversion of procollagen to collagen by fibroblasts and bone treated with tunicamycin, an inhibitor of protein glycosylation. J. Biol. Chem. **252:** 955–962.
41. Duksin, D. & P. Bornstein. 1977. Changes in surface properties of normal and transformed cells caused by tunicamycin, an inhibitor of protein glycosylation. Proc. Nat. Acad. Sci. USA **74:** 3433–3437.
42. Duksin, D., K. Holbrook & P. Bornstein. 1978. Cell surface morphology and adhesive properties of normal and virally transformed cells treated with tunicamycin, an inhibitor of protein glycosylation. Exp. Cell Res. In press.
43. Chen, L. B., P. H. Gallimore & J. K. McDougall. 1976. Correlation between tumor induction and the large external transformation sensitive protein on the cell surface. Proc. Nat. Acad. Sci. USA **73:** 3570–3574.
44. Shields, R. & K. Pollock. 1974. The adhesion of BHK and PyBHK cells to the substratum. Cell **3:** 31–38.
45. Sachs, L. 1974. Regulation of membrane changes, differentiation, and malignancy in carcinogenesis. Harvey Lect. **68:** 1–35.
46. Petersen, D. F., E. C. Anderson & R. A. Tobey. 1968. Mitotic cells as a source of synchronized cultures. *In* Methods in Cell Physiology. D. M. Prescott, Ed. Vol. 3: 347–370. Academic Press, Inc. New York, N.Y.
47. Hynes, R. O. & J. M. Bye. 1974. Density and cell cycle dependence of cell surface proteins in hamster fibroblasts. Cell **3:** 113–120.
48. Hunt, R. C., E. Gold & J. C. Brown. 1975. Cell cycle dependent exposure of a high molecular weight protein on the surface of mouse L cells. Biochim. Biophys. Acta **413:** 453–458.
49. Stenman, S., J. Wartiovaara & A. Vaheri. 1977. Changes in the distribution of a major fibroblast protein, fibronectin, during mitosis and interphase. J. Cell Biol. **74:** 453–467.
50. Pearlstein, E. & M. D. Waterfield. 1974. Metabolic studies on ^{125}I-labeled baby hamster kidney cell plasma membranes. Biochim. Biophys. Acta **362:** 1–12.
51. Gahmberg, C. G. & S. I. Hakomori. 1974. Organization of glycolipids and glycoproteins in surface membranes: dependency on cell cycle and on transformation. Biochem. Biophys. Res. Commun. **59:** 283–291.
52. Ruoslahti, E. & H. O. Sjögren. 1976. Specific anti-fibroblast cytotoxicity of antibodies to fibroblast surface antigen. Int. J. Cancer **18:** 375–378.
53. Klebe, R. J. 1974. Isolation of a collagen-dependent cell attachment factor. Nature (London) **250:** 248–251.
54. Klebe, R. J. 1975. Cell attachment to collagen: the requirement for energy. J. Cell. Physiol. **86:** 231–236.
55. Kleinman, H. K., E. B. McGoodwin & R. J. Klebe. 1976. Localization of the cell attachment region in types I and II collagens. Biochem. Biophys. Res. Commun. **72:** 426–432.
56. Hauschka, S. D. & N. K. White. 1972. Studies of myogenesis in vitro. I. Temporal changes in the proportion of muscle-colony-forming cells during the early stages of limb development. II. Myoblast-collagen interaction: molecular specificity required. Excerpta Med. Int. Congr. Ser. **240:** 53–71.
57. Hauschka, S. D. & I. R. Konigsberg. 1966. The influence of collagen on the development of muscle clones. Proc. Nat. Acad. Sci. USA **55:** 119–126.

58. KETLEY, J. M., R. W. ORKIN & G. R. MARTIN. 1976. Collagen in developing chick muscle in vivo and in vitro. Exp. Cell Res. **99:** 261–268.
59. PATTERSON, B. & R. C. STROHMAN. 1972. Myosin synthesis in cultures of differentiating chicken embryo skeletal muscle. Dev. Biol. **29:** 113–138.
60. CHEN, L. B. 1977. Alteration in cell surface LETS protein during myogenesis. Cell **10:** 393–400.
61. HYNES, R. O., G. S. MARTIN, M. SHEARER, D. R. CRITCHLEY & C. J. EPSTEIN. 1976. Viral transformation of rat myoblasts: effects on fusion and surface properties. Dev. Biol. **48:** 35–46.
62. LUSTIG, L. 1970. Cytotoxic action of an antiserum to soluble collagen on tissue culture of fibroblasts. Proc. Soc. Exp. Biol. Med. **133:** 207–211.
63. DUKSIN, D., A. MAOZ & S. FUCHS. 1975. Differential cytotoxic activity of anti-collagen serum on rat osteoblasts and fibroblasts in tissue culture. Cell **5:** 83–86.
64. LICHTENSTEIN, J. R., E. A. BAUER, R. HOYT & H. J. WEDNER. 1976. Immunologic characterization of the membrane-bound collagen in normal human fibroblasts: identification of a distinct membrane collagen. J. Exp. Med. **144:** 145–154.
65. PAGE FAULK, N., L. B. CONOCHIE & A. TEMPLE. 1975. Immunobiology of membrane-bound collagen on mouse fibroblasts. Nature (London) **256:** 123–125.
66. MILLER, E. J. 1976. Biochemical characteristics and biological significance of the genetically-distinct collagens. Mol. Cell. Biochem. **2:** 165–192.
67. HYNES, R. O., A. T. DESTREE & V. MAUTNER. 1976. Spatial organization at the cell surface. *In* Membranes and Neoplasia: New Approaches and Strategies. V. T. Marchesi, Ed.: 189–201. Alan R. Liss, Inc. New York, N.Y.
68. VON DER MARK, K. 1977. Personal communication.
68a. BORNSTEIN, P., J. F. ASH & G. BALIAN. 1978. In preparation.
69. GRAHAM, J. M., R. O. HYNES, E. A. DAVIDSON & D. F. BAINTON. 1975. The location of proteins labeled by the ^{125}I-lactoperoxidase system in the NIL 8 hamster fibroblast. Cell **4:** 353–365.
70. SCHLESSINGER, J., L. S. BARAK, G. G. HAMMES, K. M. YAMADA, I. PASTAN, W. W. WEBB & E. L. ELSON. 1977. Mobility and distribution of a cell surface glycoprotein and its interaction with other membrane components. Proc. Nat. Acad. Sci. USA **74:** 2909–2913.
71. ABERCROMBIE, M. 1970. Contact inhibition in tissue culture. In Vitro **6:** 128–142.
72. EDSALL, J. T., G. A. GILBERT & H. A. SCHERAGA. 1955. The non-clotting component of the human plasma fraction I-l ("cold insoluble globulin"). J. Amer. Chem. Soc. **77:** 157–161.
73. MOSHER, D. F. 1975. Cross-linking of cold-insoluble globulin by fibrin-stabilizing factor. J. Biol. Chem. **250:** 6614–6621.
74. POLLACK, R., M. OSBORN & K. WEBER. 1975. Patterns of organization of actin and myosin in normal and transformed cultured cells. Proc. Nat. Acad. Sci. USA **72:** 994–998.
75. POLLACK, R. & D. RIFKIN. 1975. Actin-containing cables within anchorage-dependent rat embryo cells are dissociated by plasmin and trypsin. Cell **6:** 495–506.
76. OLDEN, K. & K. M. YAMADA. 1977. Mechanism of the decrease in the major cell surface protein of chick embryo fibroblasts after transformation. Cell **11:** 957–969.
77. ADAMS, S. L., M. E. SOBEL, B. H. HOWARD, K. OLDEN, K. M. YAMADA, B. DE CROMBRUGGHE & I. PASTAN. 1977. Levels of translatable mRNAs for cell surface protein, collagen precursors, and two membrane proteins are altered in Rous sarcoma virus transformed chick embryo fibroblasts. Proc. Nat. Acad. Sci. USA **74:** 3399–3403.
78. ROWE, D. W., R. C. MOEN, J. M. DAVIDSON, P. H. BYERS, P. BORNSTEIN & R. D. PALMITER. 1978. Correlation of procollagen mRNA levels in normal and transformed chick embryo fibroblasts with different rates of procollagen synthesis. Biochemistry. In press.
79. POLLARD, T. D. & R. R. WEIHING. 1974. Actin and myosin and cell movement. CRC Crit. Rev. Biochem. **2:** 1–65.
80. PETERKOFSKY, B. & W. B. PRATHER. 1974. Increased collagen synthesis in Kirsten

sarcoma virus-transformed BALB 3T3 cells grown in the presence of dibutyryl cyclic AMP. Cell **3:** 291–299.

81. TRELSTAD, R. L., E. D. HAY & J. P. REVEL. 1967. Cell contact during early morphogenesis in the chick embryo. Dev. Biol. **16:** 78–106.
82. HAY, E. D. & S. MEIER. 1976. Stimulation of corneal differentiation by interaction between cell surface and extracellular matrix. II. Further studies on the nature and site of transfilter "induction." Dev. Biol. **52:** 141–157.
83. POSTLETHWAITE, A. E., R. SYNDERMAN & A. H. KANG. 1976. The chemotactic attraction of human fibroblasts to a lymphocyte-derived factor. J. Exp. Med. **144:** 1188–1203.
84. POSTLETHWAITE, A. E. & A. H. KANG, 1976. Collagen and collagen-peptide induced chemotaxis of human blood monocytes. J. Exp. Med. **143:** 1299–1307.
85. BRASS, L. F. & H. B. BENSUSAN. 1974. Role of collagen quaternary structure in platelet-collagen interaction. J. Clin. Invest. **54:** 1480–1487.
86. REID, K. B. M. & R. R. PORTER. 1976. Subunit composition and structure of subcomponent of Clq of the first component of human complement. Biochem J. **155:** 19–23.
87. DESSAU, W. & F. JILEK. 1977. Interaction of the serum-derived AGF and denatured collagen. Uppsala J. Med. Sci. **82:** 128.
88. ENGVALL, E. & E. RUOSLAHTI. 1977. Binding of soluble form of fibroblast surface protein, fibronectin, to collagen. Int. J. Cancer **20:** 1–5.

NATURE OF THE GLYCOPROTEIN COMPONENTS OF BASEMENT MEMBRANES*

Robert G. Spiro

Departments of Biological Chemistry and Medicine
Harvard Medical School
and Elliott P. Joslin Research Laboratory
Boston, Massachusetts 02215

It has become apparent from recent reports[1-4] that there is an intriguing relationship between the cell surface protein, fibronectin, and basement membranes that may have vast biologic significance in terms of such phenomena as morphogenesis and tissue regeneration.

Basement membranes appear very early in the developing embryo[5] and form a microskeleton on which the various epithelial cells of the body rest. These cells that cover the body and line its various tubes and passages are attached to basement membranes that separate them from the underlying connective tissue. A critical interaction must occur between the periphery of these cells and their basement membranes, which leads to their orderly orientation, and it seems possible that a surface component of the cells may be involved in the attachment process. The properties of fibronectin make it a suitable component to perform this adhesive function, because, in addition to being a cell surface molecule, it has been detected immunochemically in several basement membranes in both embryonic and adult tissues[1-4] and, furthermore appears to have a high affinity for collagens.[6-8]

In any evaluation of these cell-basement membrane interactions, which are so important in the organization of tissues, information about the chemical nature of macromolecular components of basement membranes is relevant; accordingly, in this paper, a brief review of our studies on basement membrane biochemistry will be presented.

Distribution and Isolation of Basement Membranes

Although basement membranes are widely distributed, as can be readily appreciated by an examination of tissue sections treated with the periodic acid-Schiff (PAS) stain, by which they are sharply outlined,[9] only a few of these membranes can be isolated in sufficient purity to be suitable for structural investigations. The glomerular and tubular basement membranes of the kidney and the lens capsule of the eye can readily be obtained free of cellular material and other insoluble proteins and have therefore been the primary subjects of our investigations. Lens capsules can be obtained by simple dissection, followed by ultrasonic treatment to remove any adherent cells.[10] Glomeruli and tubules can be isolated from renal cortex by sieving techniques, and, on ultrasonic disruption, filtration, and centrifugation, they yield basement membranes in purified form.[11] The ultrasonic treatment serves to detach the closely apposed cells, and the basement membranes are recovered morphologically free of recognizable cellular elements.

*Supported by National Institutes of Health Grants 17325 and 17477.

0077-8923/78/0312-0106 $01.75/0

The isolated basement membrane, like the membrane *in situ,* shows little organization under the electron microscope and lacks the periodicity of fibrillar collagens (FIGURE 1).

COMPOSITION OF BASEMENT MEMBRANES

Basement membranes are essentially devoid of lipid and nucleic acids and consist of collagen-like glycoprotein material. Although basement membranes contain a substantial amount of hydroxyproline and hydroxylysine, they differ from the fibrillar collagens in several important respects (TABLE 1), including a high carbohydrate content, a large number of half-cystine residues, and the

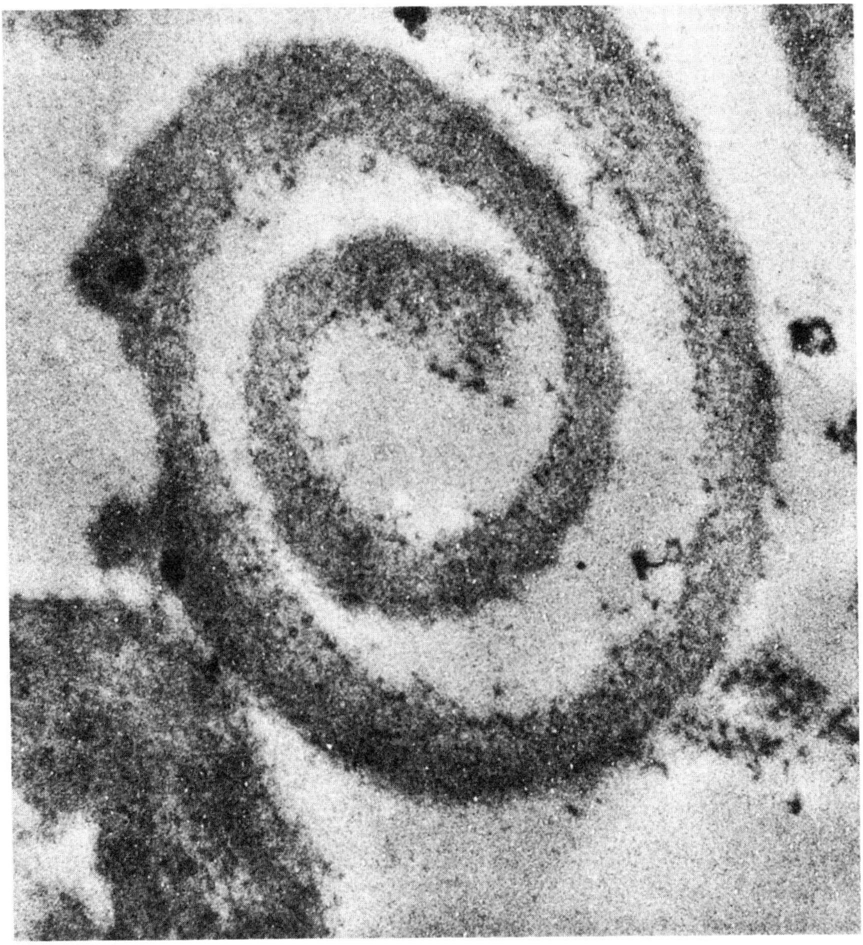

FIGURE 1. Electron micrograph of isolated bovine glomerular basement ×33,000. (From Spiro.[11] By permission of *Journal of Biological Chemistry.*)

presence of less glycine but a greater amount of polar amino acids. The composition of basement membrane is strikingly like that of the Clq component of complement, a rather unusual protein of plasma (TABLE 1). A comparison of the analyses of several basement membranes (TABLE 2) indicates that they are generally similar in composition. Carbohydrate usually comprises about 10% of the

TABLE 1

COMPARISON OF COMPOSITION OF GLOMERULAR BASEMENT MEMBRANE, SKIN TROPOCOLLAGEN, AND Clq PROTEIN OF COMPLEMENT

Component	Glomerular Basement Membrane, Bovine[11]	Skin Tropocollagen, Calf[12]	Clq Protein of Complement, Human[13]
	Residues per 1000 Total Amino Acid Residues		
Glycine	208	320	173
Alanine	61	112	44
Valine	38	20	55
Isoleucine	29	11	38
Leucine	59	25	60
Proline	69	138	60
Phenylalanine	28	13	42
Tyrosine	18	3	30
Tryptophan	6	—*	9
Serine	55	36	55
Threonine	37	18	53
Half-cystine	31	—	24
Methionine	14	4	25
Lysine	27	27	37
Histidine	16	5	14
Arginine	49	50	43
Aspartic acid	68	45	85
Glutamic acid	96	72	92
Hydroxyproline†	68	94	39
Hydroxylysine	22	7	19
Glucose	16	1	16
Galactose	20	2	17
Mannose	5	—	6
Fucose	2	—	1
Glucosamine	10	—	6
Galactosamine	1	—	3
Sialic acids	4	—	1

*Dashes indicate that less than one residue per 1000 amino acids is present.

†Refers to 4-hydroxyproline; bovine glomerular basement membrane also contains eight residues of 3-hydroxyproline/1000 amino acid residues.

membrane's weight[12] and is distributed between two major types of carbohydrate moieties (TABLE 3). One is a disaccharide (2-*O*-α-D-glucosyl-D-galactose) that exists in a *O*-β-glycosidic linkage to about 80% of the hydroxylysine residues (FIGURE 2); the other carbohydrate is a branched heteropolysaccharide composed of sialic acid, fucose, galactose, mannose, and *N*-acetylglucosamine that is at-

TABLE 2

COMPARISON OF COMPOSITION OF SEVERAL BASEMENT MEMBRANES*

Component	Glomerular Basement Membrane, Human[14]	Tubular Basement Membrane, Human[15]	Anterior Lens Capsule, Calf[10]	Posterior Lens Capsule, Calf[10]
	Residues per 1000 Total Amino Acid Residues			
Glycine	214	237	236	222
4-Hydroxyproline	81	77	94	84
Hydroxylysine	25	31	37	32
Lysine	25	18	14	17
Aspartic acid	66	63	57	61
Tyrosine	16	15	14	14
Half-cystine	20	17	24	24
Glucose	18	26	29	24
Galactose	20	32	33	28
Mannose	5	4	5	5
Fucose	1	1	2	2
Hexosamines	10	8	11	17
Sialic acids	3	3	1	3

*Abbreviated composition; only values for key components are given.

tached to asparagine residues on the peptide chains.[16,17] The former unit, in view of the linkage amino acid, is unique to the collagen family of proteins,[18] while the latter has been observed in a large variety of glycoproteins.[19] The presence of small amounts of galactosamine in most basement membranes[12] and of glucuronic acid in lens capsule[10] suggests that some other, less abundant saccharide units may also occur.

POLYDISPERSITY OF THE GLYCOPROTEIN COMPONENTS

Although basement membranes are essentially insoluble in salt solutions at physiologic pH, they can usually be substantially solubilized by reduction of their

TABLE 3

CARBOHYDRATE UNITS OF BOVINE GLOMERULAR BASEMENT MEMBRANE[16]

	Disaccharide	Heteropolysaccharide
Monosaccharides (residues per unit)		
Glucose	1	
Galactose	1	4
Mannose		3
N-acetylglucosamine		5
Sialic acid		3
Fucose		1
Molecular weight	324	3170
Percentage of total carbohydrate of membrane	50	50
Relative number of units in membrane	10	1
Amino acid involved in glycopeptide bond	hydroxylysine	asparagine

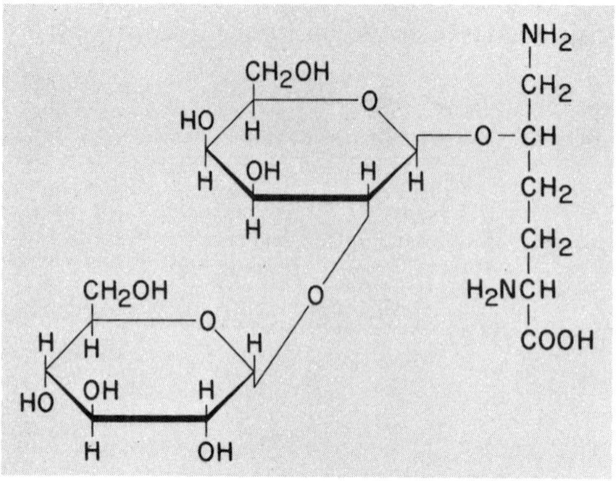

FIGURE 2. Structure and peptide attachment of the disaccharide unit of basement membranes. (From Spiro.[17] By permission of *Journal of Biological Chemistry*.)

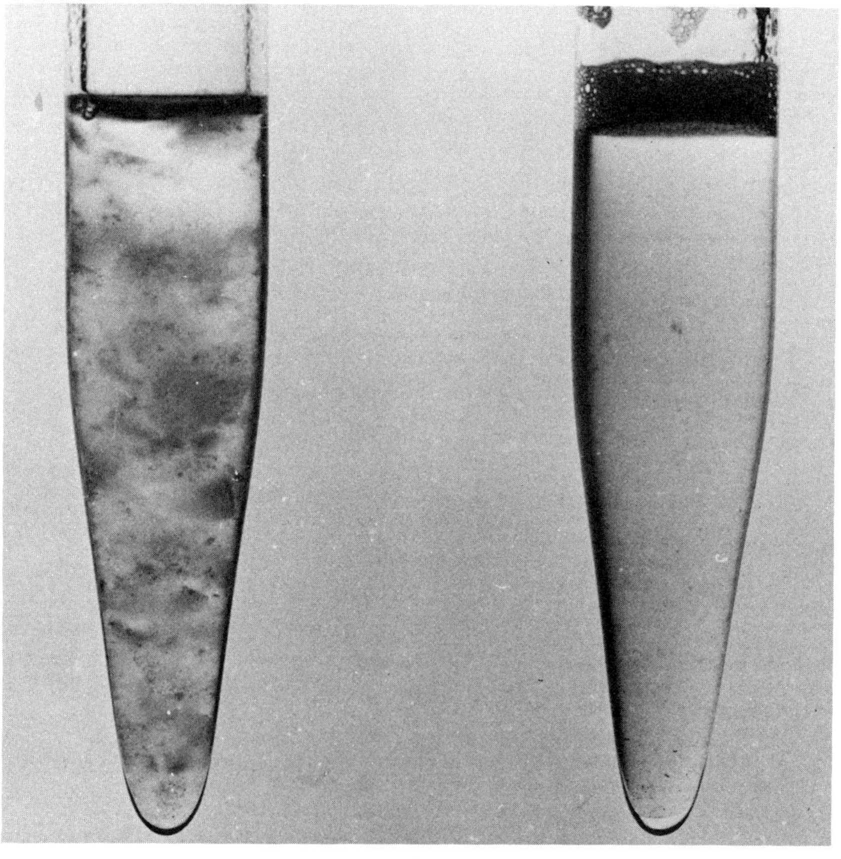

FIGURE 3. Effect of reduction of disulfide bonds on the solubility of glomerular basement membrane. Basement membrane suspended in 8 M urea buffered at pH 8.5 before (*left*) and 15 min after (*right*) the addition of 2-mercaptoethanol. (From Hudson & Spiro.[20] By permission of *Journal of Biological Chemistry*.)

disulfide bonds in the presence of urea or sodium dodecyl sulfate (SDS) (FIGURE 3). When the reduced bovine glomerular basement membrane is examined by polyacrylamide gel electrophoresis, a large number of polypeptide components are observed, which range in molecular weight from about 25,000 to 200,000 daltons[21,22] (FIGURE 4). This multiplicity of bands is evident even when preparation and solubilization of the bovine membrane are performed in the presence of protease inhibitors,[15,23] indicating that it represents the *in vivo* state of the membrane. The reduced tubular basement membrane also exhibits several peptide components when resolved by polyacrylamide gel electrophoresis (FIGURE 5).

Since after alkylation the reduced glomerular basement membrane is more than 80% soluble in SDS, it has been feasible to fractionate its numerous polypeptide components according to size and charge by gel filtration, ion-exchange chromatography, and polyacrylamide gel electrophoresis (FIGURE 6). The peptide subunits of the reduced and alkylated glomerular basement membrane demonstrate pronounced compositional diversity, even when those of apparently identical molecular weight are compared (TABLE 4). Although all have been found to be glycoprotein in nature, they differ in the proportion of disaccharide versus polysaccharide units. The former carbohydrate units were found to be more abundant in polypeptides with a more collagen-like composition, as judged by glycine, hydroxyproline, and hydroxylysine content, whereas the latter saccharides were present in greater numbers in more polar subunits (TABLE 4). The amino acids

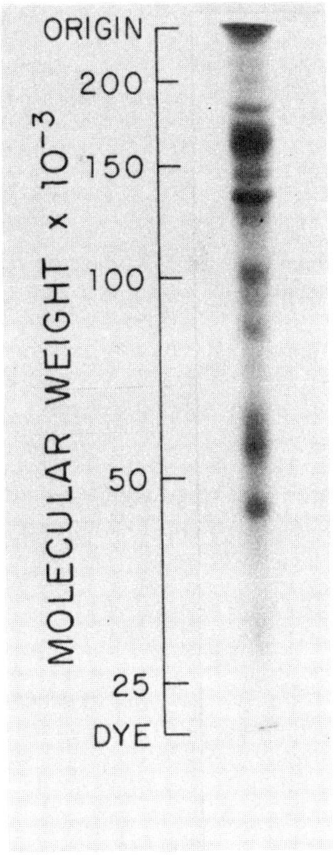

FIGURE 4. Polyacrylamide gel electrophoresis in SDS of reduced bovine renal glomerular basement membrane. The membrane was solubilized by extraction with SDS in the presence of 2-mercaptoethanol and was electrophoresed in a 5% gel in the presence of the reducing agent. The polypeptide components were detected with Coomassie blue. The molecular weight scale is based on the migration of reduced standard proteins. (From Sawada & Spiro.[15])

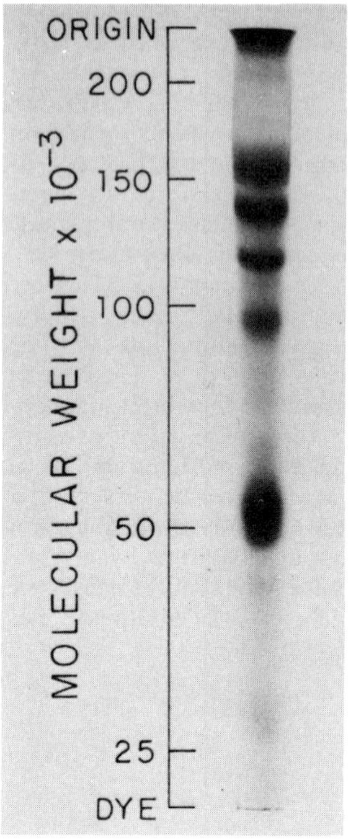

FIGURE 5. Polyacrylamide gel electrophoresis in SDS of reduced human renal tubular basement membrane. Conditions were the same as those described in the legend to FIGURE 4. (From Sawada & Spiro.[15])

characteristic of collagen varied in a parallel manner with each other and in a reciprocal fashion with such constituents as lysine, aspartic acid, tyrosine, and half-cystine. These analyses suggested the presence of collagen-like and polar domains on the same peptide chain, with the disaccharide being attached primarily to the collagen-like regions and the polysaccharide being linked to more polar sequences of amino acids.[24] We have obtained further evidence that these two major types of saccharide are indeed attached to the same peptide sequence and that they may sometimes be located quite close together, by the isolation from a proteolytic digest of glomerular basement membrane of a glycopeptide, 15 amino acids in length, that contained both a disaccharide and a heteropolysaccaride unit.[25]

STRUCTURAL MODEL OF BASEMENT MEMBRANES

These studies on basement membrane structure can best be represented by a structural model in which cross-linked peptide chains that vary in length and in the proportion of collagen-like segments that they contain are layered over each other (FIGURE 7). Disulfide bonds constitute the major interchain links, although there are indications that as yet undefined cross-links may also occur, because very high-molecular-weight material, which does not enter 5% polyacrylamide gels, is always evident after reduction of the disulfide bonds.[21] A small number of peptide chains, quite polar in nature but not devoid of hydroxyproline and hydroxylysine, are attached to the membrane by noncovalent bonds.[20] These chains represent about 20% of the peptide and carbohydrate of glomerular basement membrane and can be solubilized by urea or SDS without the presence of reducing reagents. The numerous bulky, hydrophilic carbohydrate substituents may play an impor-

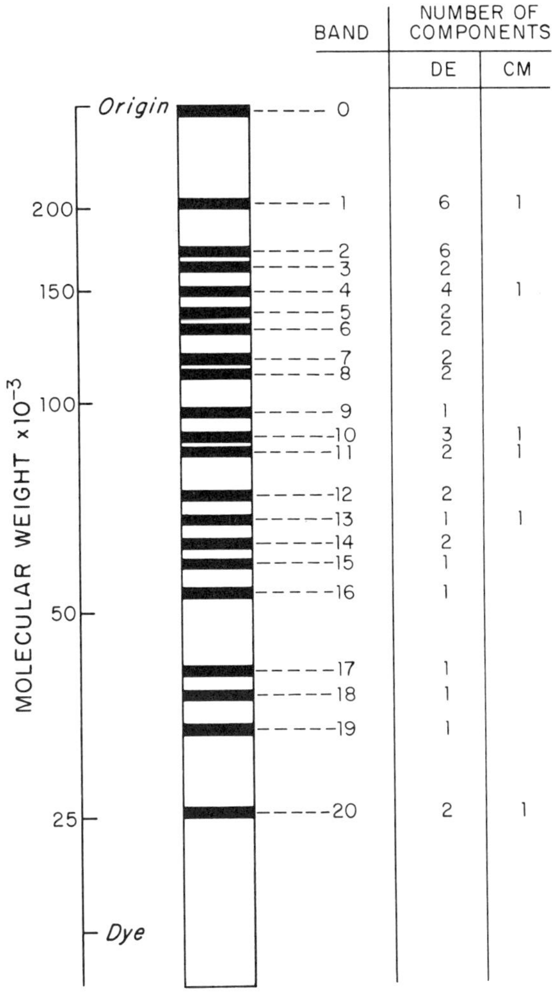

FIGURE 6. Molecular weight distribution by polyacrylamide gel electrophoresis of reduced alkylated glomerular basement membrane polypeptide components. The migration of 20 bands that represent 58 distinct polypeptides resolved by DEAE cellulose (DE) or carboxymethyl cellulose (CM) chromatography, followed by preparative electrophoresis, are shown. The molecular weights of polypeptide components appear somewhat higher after reduction and alkylation than after reduction alone. (From Sato & Spiro.[22] By permission of *Journal of Biological Chemistry*.)

TABLE 4
COMPOSITION OF SEVERAL PURIFIED GLYCOPROTEIN COMPONENTS OF BOVINE GLOMERULAR BASEMENT MEMBRANE*

	B-4‡	G-4	D-5	A-8	B-10	D-10	B-14	C-16
	(158,000)§		(140,000)	(111,000)	(93,000)		(63,000)	(53,000)
Constituent†	Residues per 1000 Total Amino Acid Residues							
Glycine	288	178	169	304	259	104	218	95
4-Hydroxyproline	109	57	34	123	96	15	70	20
Hydroxylysine	42	20	15	39	31	4	22	5
Lysine	14	27	30	14	19	39	23	48
Aspartic acid	48	69	73	50	53	92	61	83
Tyrosine	9	12	17	8	27	18	19	22
Half-cystine	11	25	29	9	15	35	23	21
Disaccharide¶	34	16	12	31	24	3	18	4
Polysaccharide¶	1	2	3	—‖	1	5	1	6

*Data from Sato and Spiro.[22]

†Abbreviated composition; only values for key components are given.

‡Glycoproteins are designated by letter and number; letters refer to DEAE cellulose fractions from which they were obtained, while numbers refer to the band position to which they migrated on polyacrylamide gel electrophoresis (FIGURE 6).

§Values in parentheses give the apparent molecular weights (daltons) of the glycoprotein component.

¶Values calculated from saccharide analyses.

‖ Analysis not performed.

tant role in determining the packing of the peptide chains and could, in the capacity of spacers, help to determine the porosity of basement membranes. The amorphous nature of basement membranes could at least partly be due to interference with fibril formation by the closely spaced saccharide units.[18]

It has become evident in recent years that the occurrence of substantial polar regions, such as those present in the polypeptide components of the basement membrane, is not uncommon in members of the collagen family of proteins. In their procollagen stage, the interstitial collagens contain extensive polar domains that may involve several hundred amino acid residues.[26] Likewise, the C1q component of complement has been shown to contain substantial polar sequences that comprise more than half of the amino acid residues in the peptide subunits[27] and that account for the relatively low glycine and hydroxyproline content of this protein (TABLE 1). Acetylcholinesterase appears to be another functionally important protein that has a collagen-like portion in a molecule that consists primarily of noncollagenous sequences.[28]

Perhaps the most striking aspect of basement membrane structure is the polydispersity of its peptide components, which we have observed in glomerular and tubular basement membranes and in lens capsule.[15] This phenomenon has been reported from several other laboratories[29-31] and indeed appears to extend to basement membranes from invertebrate sources. In the latter instance, a study of intestinal basement membrane of the round worm (*Ascaris suum*) has revealed the presence of at least 17 polypeptide components that, as in the glomerular basement membrane, vary greatly in size and in extent of collagenous domains.[31]

The physiologic basis of the multiple peptide components of the basement membrane presents an intriguing problem, because it appears unlikely that each could be a biosynthetic subunit. Although postribosomal modifications, such as hydroxylation and glycosylation, could contribute to the heterogeneity, the remarkable differences in size and composition noted among the polypeptide components cannot be entirely a result of the biosynthetic steps. It would appear more likely that the polydispersity is the result of processing of the basement membrane by limited proteolysis (FIGURE 8). Cleavage of the peptide chain of a small number, or even a single, biosynthetic subunit could yield variously sized segments. Some of these proteolytic fragments might diffuse away, but most would remain connected to the parent membrane by the numerous disulfide bonds and would be liberated only after isolation of the basement membrane and reductive scission of these cross-links. Glomerular basement membrane antigens are normally found in urine,[32] and it is possible that these fragments might represent diffusable products of proteolytic modification of the membrane. Neutral proteases capable of digesting basement membrane have been detected in the granules of polymorphonuclear leukocytes[33] and may also be found in such tissues as the kidney in which the membranes are located.

In contrast to the rather complex polypeptide pattern that has been observed in basement membranes by many investigators.[15,20,22,29-31] Kefalides has reported that basement membranes (from the glomerulus, lens capsule, and Descemet's membrane) contain a collagen, which he has termed "type IV," that consists of three identical $\alpha 1$ chains 108,000 daltons in molecular weight and that contains as carbohydrate only the hydroxylysine-linked disaccharide units.[34,35] Although this material was isolated after pepsin digestion of the basement membranes, the report of a collagen comparable in many ways to the discrete interstitial collagens (types I–III) appeared to simplify an otherwise very complicated picture. To evaluate precisely how this product of pepsin digestion fits into the overall basement membrane structure, we have prepared this fraction from glomerular basement membrane according to the published procedure.[34]

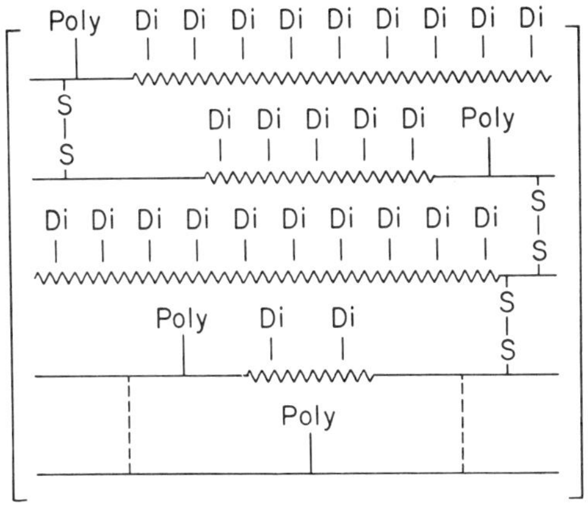

FIGURE 7. Diagrammatic representation of a hypothetical basement membrane structure. The jagged lines represent the collagen-like portion of peptide chains; the straight lines represent more polar segments. The position of the disaccharide (Di) and heteropolysaccharide (Poly) units on the chains are indicated. Interchain disulfide bonds are shown. A few peptide chains are attached to the membrane by noncovalent bonds (dashed lines).

We found that the protein that has been termed "type-IV collagen" represents only 1.4% of the total basement membrane and yields, on carboxymethyl cellulose chromatography with the suggested gradient,[6] an even smaller amount (0.26%) of the total basement membrane protein.[25] Polyacrylamide gel electrophoresis in the presence of 2-mercaptoethanol moreover indicated that this protein is quite heterogeneous and contains some material greater than 200,000

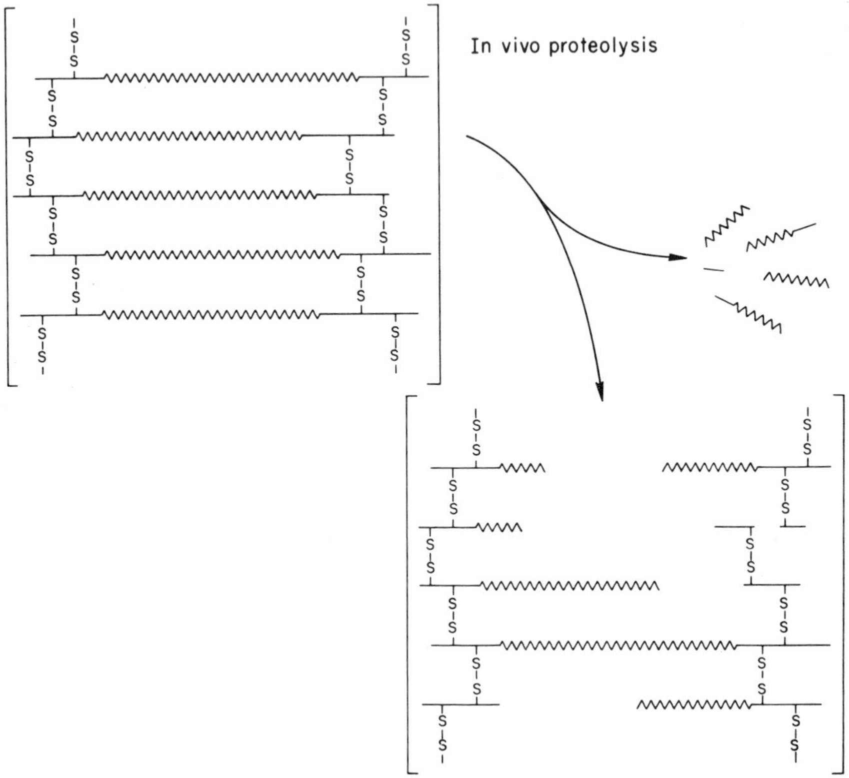

FIGURE 8. Representation of a hypothetical processing of basement membrane by limited *in vivo* proteolysis to account for the observed multiplicity of polypeptide components. The action of proteases on a biosynthetic subunit held together by interchain disulfide bonds is depicted, yielding variously sized polypeptide segments still held together by these cross-links and some diffusable peptide fragments. Collagen-like portions of the peptide chains are depicted by jagged lines; the straight lines represent more polar segments.

daltons in molecular weight, several polypeptides in the 100,000–200,000 dalton range in addition to some lower-molecular-weight components (FIGURE 9), a pattern that is not consistent with the reported 108,000 dalton molecular weight. Furthermore, analysis of this "collagen" fraction indicated, contrary to the previous report,[34] that sufficient mannose and glucosamine were present to account for at least one heteropolysaccharide unit per 1000 amino acid residues.[25]

It would appear, then, that a distinct collagen that merits the designation of "basement membrane collagen" has not yet been isolated. Pepsin treatment of

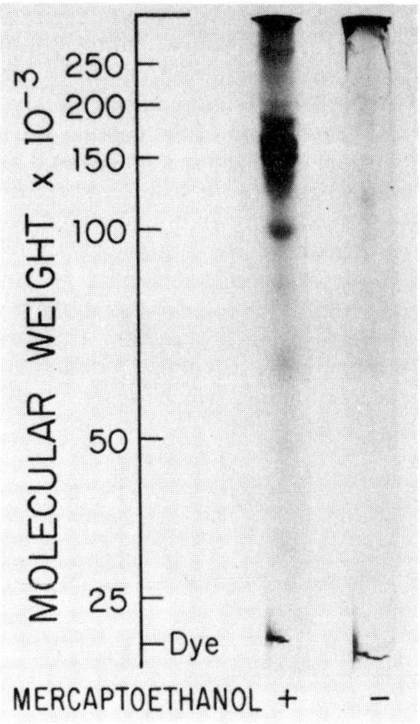

FIGURE 9. Polyacrylamide gel electrophoresis in SDS of a bovine glomerular basement membrane fraction prepared by pepsin digestion according to the procedure of Kefalides.[34] The electrophoretic patterns of this fraction in 5% gels in the presence (+) and absence (−) of 2-mercaptoethanol are shown. The polypeptide components were detected with Coomassie blue. The molecular weight is based on the migration of reduced standard protein. (From Levine & Spiro.[25])

basement membrane is essentially a degradative procedure that tends to remove the extensive polar regions that are characteristic of the basement membrane protein. The products of pepsin digestion of basement membranes apparently represent some peptide sequences of the native membrane, extensively cross-linked by disulfide bonds, that resist the proteolytic treatment. In view of the very limited yield in which this fraction is obtained after pepsin treatment, it is difficult to evaluate how representative this fraction is of basement membrane structure.

METABOLISM OF BASEMENT MEMBRANES

An investigation of glomerular basement membrane metabolism in the rat with the aid of injected tracer doses of tritiated amino acids has revealed that whereas turnover of this structure as a whole is very slow, portions are degraded more rapidly.[36] The slow turnover is apparent when decay of radioactivity from

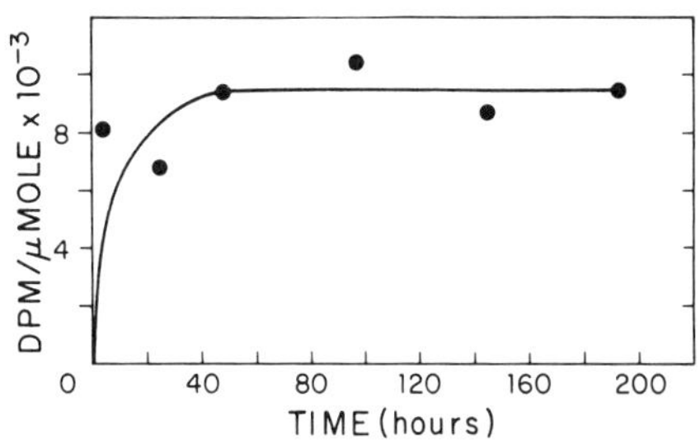

FIGURE 10. Specific activity of glycine in glomerular basement membrane at various times after injection of [2-^{3}H] glycine into rats. The calculated turnover time for this constituent was greater than 100 days. (From Price & Spiro.[36] By permission of *Journal of Biological Chemistry*.)

the glycine (or hydroxyproline) residues of the basement membrane was determined (FIGURE 10), while more rapid metabolism was evident on measurement of the specific radioactivity of the phenylalanine (or lysine) residues (FIGURE 11). This apparently nonuniform turnover could be due to different rates of synthesis and degradation of the peptide subunits, with the less collagen-like components being metabolized at a more rapid rate, or to the physiologic action of proteases, which could result in the preferential elimination of peptide sequences outside of the collagen-like domains. Indeed, subtle morphologic and functional heterogeneities have been observed by Walker in electron microscopic studies employing the silver labeling technique.[37]

Despite the relative metabolic inertness of basement membranes, certain dynamic aspects are evident from compositional changes observed in various physiologic and pathologic states.[14,38-40] Representative of such changes are the effects of aging, which we have observed in the rat glomerular basement mem-

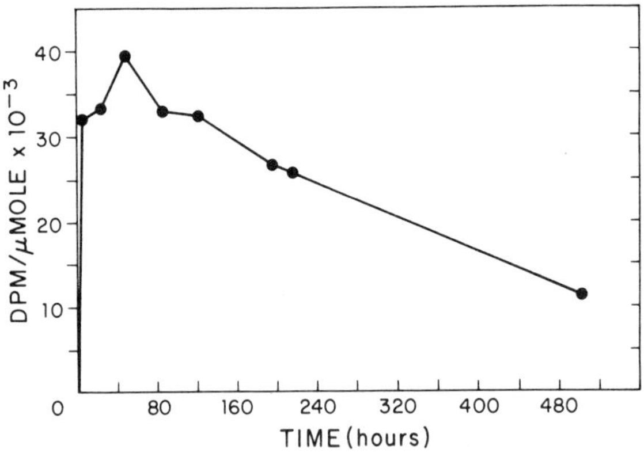

FIGURE 11. Specific activity of phenylalanine in glomerular basement membrane at various times after injection of L-[3-^{3}H]phenylalanine into rats. The calculated turnover time for this component was 23 days. (From Price & Spiro.[36] By permission of *Journal of Biological Chemistry*.)

brane[41] and the bovine lens capsule.[10] As the basement membrane thickens, it acquires a more collagen-like composition, as reflected by an increase in hydroxyproline and hydroxylysine (Fig. 12) content and by an augmentation in the number of disaccharide units linked to the latter amino acid; concomitantly, there is a significant decrease in other amino acid constituents, including half-cystine, serine, threonine, alanine, valine, and lysine.[41] These age-related changes could be due to an increased synthesis or decreased degradation of the more collagen-like polypeptide components of the basement membrane.

CONCLUSIONS

This survey of basement membrane biochemistry indicates that these extracellular structures are composed of collagenous glycoprotein components relatively rich in carbohydrate and cross-linked primarily by disulfide bonds. The poly-

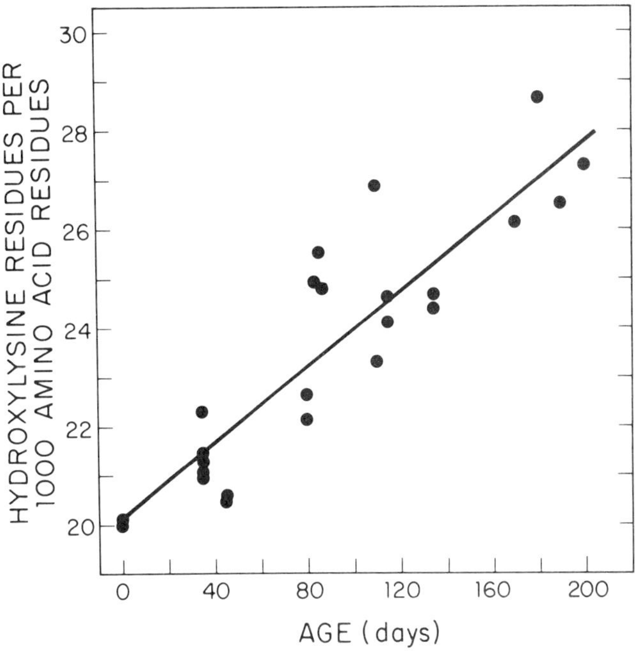

FIGURE 12. Effect of age on the hydroxylysine content of rat glomerular basement membrane. Each point represents a membrane preparation from the glomeruli of at least six animals. The correlation coefficient that relates hydroxylysine content to age was 0.868 ($p < 0.001$). (From Hoyer & Spiro.[41] By permission of *Archives of Biochemistry and Biophysics.*)

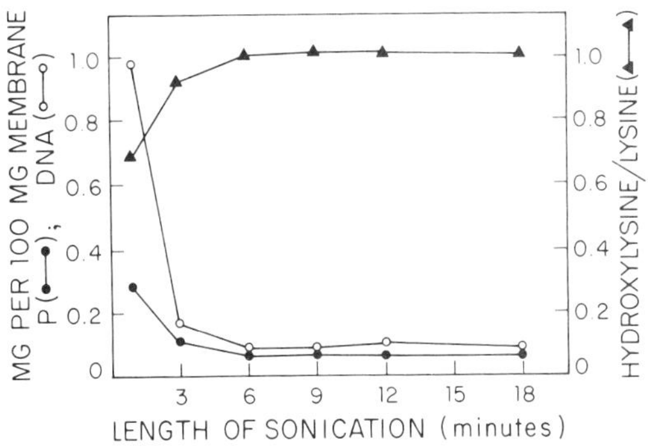

FIGURE 13. Effect of varying length of ultrasonic treatment of bovine renal glomeruli on the purity of the subsequently isolated basement membrane preparation. After 6 min of sonication, there was no detectable change in the composition of these membranes nor was there any alteration in the electrophoretic pattern of the polypeptide components. (From Levine & Spiro.[25])

dispersity of the basement membrane polypeptides and the compositional changes observed in the membrane in various situations may be the result of an interplay between synthetic and degradative processes.

If one views the basement membrane as a substratum for cellular attachment and orientation, there appear to be several possible ways by which a cell surface protein, such as fibronectin, could interact with it and thereby serve as the cementing substance. The high affinity that fibronectin has been shown to possess for collagen, particularly in its denatured form,[8] suggests that the collagen-like domains of the basement membrane may serve as the site of interaction with the cell surface. It is not known to what extent the collagenous regions of basement membrane are in the helical form; if the numerous bulky saccharide groups limit helix formation, a greater affinity for fibronectin would be expected because of the less highly organized state that would prevail. Indeed, because of their location, the basement membranes would, much more than would fibrillar collagens, appear to be the logical site for fibronectin attachment under physiologic conditions. While the initial interaction between the cell surface glycoprotein and basement membrane may involve hydrophobic or hydrogen bonds, covalent cross-links might ultimately be formed, and the latter might include ϵ-(γ-glutamyl)lysine[42] or disulfide bonds. It will be of interest to determine if the carbohydrate units on either fibronectin or the basement membrane are in any way involved in the binding phenomenon. This is particularly relevant in view of the important biologic role that has been attributed in recent years to saccharides of glycoproteins of the cell surface.[43]

A search for fibronectin in isolated basement membranes may present some problems, because it is difficult to determine whether this glycoprotein is a *bona fide* component of the cell membrane or of the basement membrane. If fibronectin is indeed involved in the attachment of cells to their basement membranes, it would be sandwiched in between these two structures and might properly be considered to belong to either. In the purification of basement membranes, an attempt is made to remove all cellular material, as judged by chemical analysis and morphologic examination. The most commonly employed procedures use ultrasonic treatment in 1 M saline to dislodge cellular components from the basement membranes (FIGURE 13); negligible phosphorus and DNA values and consistent hydroxylysine/lysine ratios have been found to be reliable guides for monitoring the length of sonication needed. Basement membranes prepared in this manner might be divested of cell surface components, such as fibronectin, if they are noncovalently attached. On the other hand, basement membranes that contain significant phosphorus (particularly lipid-phosphorus) and nucleic acids must be considered to be grossly contaminated with cellular material, and, consequently, the presence of a protein like fibronectin might merely represent an impurity. It is likely, however, that these difficulties can be overcome by carefully controlled studies and that in combination with immunofluorescent investigations on tissue sections and various binding studies, the role of fibronectin in cell-basement membrane interactions can be defined.

REFERENCES

1. VAHERI, A., D. MOSHER, J. WARTIOVAARA, J. KESKI-OJA, M. KURKINEN & S. STENMAN. 1977. *In* Cell Interactions in Differentiation. M. Karkinen-Jääskeläinin, L. Saxén & L. Weiss, eds.: 311–323. Academic Press, Ltd. London, England.

2. Linder, E., A. Vaheri, E. Ruoslahti & J. Wartiovaara. 1975. J. Exp. Med. **142:** 41–49.
3. Wartiovaara, J., S. Stenman & A. Vaheri. 1976. Differentiation **5:** 85–89.
4. Bray, B. A. This monograph.
5. Pierce, G. B. 1966. Devel. Biol. **13:** 231–249.
6. Klebe, R. J. 1974. Nature (London) **250:** 248–251.
7. Pearlstein, E. 1976. Nature (London) **262:** 497–500.
8. Engvall, E. & E. Ruoslahti. 1977. Int. J. Cancer **20:** 1–5.
9. Leblond, C. P., R. E. Glegg & D. Eidinger. 1957. J. Histochem. Cytochem. **5:** 445–458.
10. Fukushi, S & R. G. Spiro. 1969. J. Biol. Chem. **244:** 2041–2048
11. Spiro, R. G. 1967. J. Biol. Chem. **242:** 1915–1922.
12. Spiro, R. G. 1972. *In* Glycoproteins. A. Gottschalk, Ed. 2nd edit. Part B: 964–999. Elsevier. Amsterdam, The Netherlands.
13. Yonemasu, K., R. M. Stroud, W. Niedermeier & W. T. Butler. 1971. Biochem. Biophys. Res. Commun. **43:** 1388–1394.
14. Beisswenger, P. J. & R. G. Spiro. 1973. Diabetes **22:** 180–193.
15. Sawada, T. & R. G. Spiro. 1977. Unpublished results.
16. Spiro, R. G. 1967. J. Biol. Chem. **242:** 1923–1932.
17. Spiro, R. G. 1967. J. Biol. Chem. **242:** 4813–4823.
18. Spiro, R. G. 1969. J. Biol. Chem. **244:** 602–612.
19. Spiro, R. G. 1973. Adv. Protein Chem. **27:** 349–467.
20. Hudson, B. G. & R. G. Spiro. 1972. J. Biol. Chem. **247:** 4229–4238.
21. Hudson, B. G. & R. G. Spiro. 1972. J. Biol. Chem. **247:** 4239–4247.
22. Sato, T. & R. G. Spiro. 1976. J. Biol. Chem. **251:** 4062–4070.
23. Freytag, J. W., M. Ohno & B. G. Hudson. 1976. Biochem. Biophys. Res. Commun. **72:** 796–802.
24. Spiro, R. G. 1973. N. Engl. J. Med. **288:** 1337–1342.
25. Levine, M. L. & R. G. Spiro. 1975. Unpublished results.
26. Bornstein, P. 1974. Annu. Rev. Biochem. **43:** 567–603.
27. Reid, K. B. M. 1974. Biochem. J. **141:** 189–203.
28. Lwebuga-Mukasa, J. S., S. Lappi & P. Taylor. 1976. Biochemistry **15:** 1425–1434.
29. Myers, C. & P. Bartlett. 1972. Biochim. Biophys. Acta **290:** 150–157.
30. Marquardt, H., C. B. Wilson & F. J. Dixon. 1973. Biochemistry **12:** 3260–3266.
31. Hung, C. H., M. Ohno, J. W. Freytag & B. G. Hudson. 1977. J. Biol. Chem. **252:** 3995–4001.
32. Lerner, R. A. & F. J. Dixon. 1968. J. Immunol. **100:** 1277–1287.
33. Janoff, A. & J. D. Zeligs. 1968. Science **161:** 702–704.
34. Kefalides, N. A. 1971. Biochem. Biophys. Res. Commun. **45:** 226–234.
35. Kefalides, N. A. 1973. Annu. Rev. Connect. Tissue Res. **6:** 63–105.
36. Price, R. G. & R. G. Spiro. 1977. J. Biol. Chem. **252:** 8597–8602.
37. Walker, F. 1973. J. Pathol. **110:** 233–244.
38. Mahieu, P. 1972. Kidney Int. **1:** 115–123.
39. Westberg, N. G. & A. F. Michael. 1973. Acta Med. Scand. **194:** 39–47, 49–57.
40. Spiro, R. G. 1976. Diabetologia **12:** 1–14.
41. Hoyer, J. R. & R. G. Spiro. 1978. Arch. Biochem. Biophys. **185:** 496–503.
42. Mosher, D. F. 1975. J. Biol. Chem. **250:** 6614–6621.
43. Hughes, R. C. 1976. Membrane Glycoproteins. Butterworths. London, England.

SYNTHESIS OF FIBRONECTIN BY CULTURED HUMAN ENDOTHELIAL CELLS

Eric A. Jaffe* and Deane F. Mosher

Division of Hematology-Oncology
Department of Medicine
Cornell University Medical College
New York, New York 10021
and Division of Hematology
Department of Medicine
University of Wisconsin Medical School
Madison, Wisconsin 53706

Fibronectin (FN),[1] a glycoprotein composed of high-molecular-weight polypeptide subunits, is present in substantial amounts in the plasma of vertebrates.[1,2] It was first identified in 1948 and was previously referred to as cold-insoluble globulin (CIg).[3] In this paper, the FN species found in plasma that contain a 200,000-dalton polypeptide subunit will be designated "plasma FN." In the last several years, an immunologically similar protein with a 220,000-dalton polypeptide subunit has been found in the postculture medium of fibroblasts; the same antigen has been identified by immunofluorescence microscopy in vertebrate basement membrane and loose connective tissue.[4-6] This FN species, which we shall call "cellular FN," is also known as surface fibroblast antigen (SFA),[4] large, external, transformation-sensitive (LETS) protein,[7] and cell surface protein (CSP).[8] Cellular FN is one of the major proteins synthesized by cultured fibroblasts[9-12] and astroglial cells[13] and is also found in myoblasts and some epitheloid cells.[14,15] Fibronectin in fibroblasts is found both in a soluble form in conditioned medium and in insoluble forms in the extracellular connective tissue matrix and on the cell surface. The insoluble forms may be spatially related to collagen: when fibroblast cultures are studied by immunofluorescence microscopy, the same filamentous extracellular structures are labeled by both FN and collagen antibodies.[16] Plasma FN has recently been shown to bind to collagen and gelatin.[17] Plasma FN is probably the protein in serum-containing tissue culture medium that mediates initial cell attachment and spreading on the surfaces of culture vessels.[18,19] The cellular FN synthesized and deposited underneath cells may mediate adhesion of cells to culture dishes and other, more biologically relevant surfaces.[20] Transformed cultured fibroblasts generally lack cell surface FN[7,8,11-13] reviewed in Reference 21). When purified cellular FN is added to transformed cells, the transformed cells assume a more normal morphology and state of adhesiveness.[22,23] The relationship between plasma and cellular FN is not clear. Both the insoluble and the soluble FN synthesized by cultured fibroblasts have the same-sized polypeptide subunit which has a molecular weight of 220,000 daltons.[12] While it is hypothesized that plasma FN is a breakdown product of cellular FN, fibroblast cell surface FN appears to have a slightly larger polypeptide subunit than does plasma FN.[6,24] Thus, the site of synthesis of plasma FN may not be any of those cells previously shown to synthesize cellular FN.

*Recipient of National Institutes of Health Research Career Development Award 1 K04 HL 00237 and a Career Scientist Award from the Irma T. Hirschl Trust.

0077-8923/78/0312-0122 $01.75/0 © 1978, The New York Academy of Sciences

We and others have previously shown that cultured human endothelial cells synthesize extracellular material that is morphologically and immunologically like amorphous basement membrane and biochemically like basement membrane collagen.[25,26] Cultured endothelial cells also appear to synthesize extracellular material morphologically similar to microfibrils and elastic fibers.[25] Recent immunofluorescence microscopic studies have demonstrated FN in the subendothelium of vessels in developing chicks[4] and in human tissues.[27] In this paper, we report that cultured human endothelial cells synthesize FN with a polypeptide subunit molecular weight of 200,000. Cultured endothelial cells secrete FN into culture medium and incorporate FN into the extracellular matrix. We suggest that endothelial cells may be a major site of synthesis of plasma FN.

Materials and Methods

Cell Culture Techniques and Culture Media

Human endothelial cells were derived from umbilical cords and cultured by use of methods and materials described previously.[28] Endothelial cells were cultured in plastic T-25 or T-75 flasks (Corning). The flasks were pretreated by incubation overnight at 37°C with 0.2% gelatin (Difco) in water. The gelatin was removed just before usage. The culture medium consisted of medium 199 (Flow Laboratories) that contained either 20% fetal calf serum (Reheis Chemical Company) or 20% normal rabbit serum (Pel-Freez Biologicals). The culture media also contained penicillin (100 U/ml), streptomycin (100 μg/ml), L-glutamine (2 mM), and Hepes buffer (15 mM, pH 7.4, Sigma Chemical Company).

Antibodies and Antigens

Plasma FN and goat and rabbit antibodies to human plasma FN (designed anti-FN) were prepared as described previously.[2,29] The unabsorbed rabbit anti-FN had no detectable antifactor VIII procoagulant activity or antivon Willebrand factor activity when assayed by methods described previously.[30,31] However, due to the possibility of small amounts of contaminating factor-VIII antigen in the material used to elicit the anti-FN sera, all of the anti-FN sera used in this study were absorbed with purified plasma factor VIII[32] before use. Anti-FN sera were absorbed by reacting 4 vol of anti-FN with 1 vol of purified factor VIII (0.98 mg/ml) at 37°C for 1 hr and incubating overnight at 4°C. The anti-FN sera were then centrifuged at 8000*g* for 20 min at 4°C. When the absorbed anti-FN sera and a well-characterized antifactor-VIII serum[32] were reacted by immunodiffusion against a mixture that contained FN and factor VIII, lines of nonidentity were seen, and antifactor VIII also detected free factor VIII antigen in the anti-FN sera.

Radioactively Labeled Fibronectin Synthesized by Endothelial Cells

Radioactively labeled FN was prepared by culturing monolayers of human endothelial cells in T-75 flasks in leucine-free MEM that contained 10% heat-inactivated rabbit serum, L-[4,5-^{3}H]leucine (20 μCi/ml; sp act, 50 Ci/mmol;

Amersham/Searle Corporation), and Trasylol® (50 U/ml, FBA Pharmaceuticals) and incubating the cells for 24 hr at 37°C. The postculture medium was removed, and the following inhibitors were added to it to yield the listed final concentration: 0.4 mM phenylmethylsulfonyl fluoride (Sigma Chemical Company), 5 mM *N*-ethylmaleimide (Sigma), 10 mM EDTA, and 1 μM pepstatin (Protein Research Foundation, Osaka, Japan). The postculture medium was precipitated with 50% saturated ammonium sulfate; the precipitate was then redissolved in phosphate-buffered saline (PBS, 1/10 original volume) that contained the four inhibitors listed above, dialyzed at 4°C against two changes of 2 liters each of PBS, and stored frozen at –40°C. This procedure yielded approximately 2 ml of a protein solution that contained 18.6 mg/ml of protein and 4.54 × 10^7 dpm/ml (sp act, 2.44 × 10^6 dpm/mg).

Immunoisolation of [^{3}H]Fibronectin

Fibronectin synthesized by cultured endothelial cells was isolated from fractions derived from the postculture medium by a method that employes anti-FN coupled to a protein A-Sepharose column. The method is a modification of that of Kessler.[33] To inhibit proteolytic enzymes, phenylmethylsulfonyl fluoride (0.4 mM), EDTA (10 mM), *N*-ethylmaleimide (5 mM), and pepstatin (1 μM) were added to all buffer and protein solutions used in these procedures.

Protein A bound to Sepharose 4B (Pharmacia Fine Chemicals) was washed twice with 0.15 M NaCl, 0.02 M Tris HCl (pH 7.4). Two-tenths of a milliliter of the washed beads, 0.2 ml of Tris-buffered saline (TBS), and 0.15 ml of anti-FN were incubated with end-over-end rotation for 2 hr at 20°C. To saturate the remaining protein A-binding sites, 2 ml of heat-inactivated normal rabbit serum were added, the mixture was incubated for 1 hr at 4°C, and the beads were then washed six times with PBS. The washed beads were suspended in 0.75 ml of PBS, 0.25 ml of the ^{3}H-labeled ammonium sulfate-precipitated postculture medium was added, and the mixture was incubated overnight at 4°C. The suspension was transferred to a 0.6 × 15 cm chromatographic column (Glenco Scientific Inc.), and the excess fluid was drained off. The beads (0.2 ml packed volume) were washed sequentially with PBS (6 ml), 5 M KI (8 ml), 1% Triton® X-100 (Sigma) in PBS (8 ml), and PBS (5 ml). The beads were then eluted with 1 M acetic acid, and the eluate was analyzed by sodium dodecyl sulfate (SDS) polyacrylamide gel electrophoresis. The washes with 5 M KI and 1% Triton X-100 were added because in preliminary experiments, FN was noted to adhere nonspecifically to protein A-Sepharose. Kessler[33] has shown that 5 M KI and the nonionic detergent NP-40 do not release appreciable amounts of antigen-antibody complexes bound to protein A. We found that washes with 5 M KI and 1% Triton X-100 in PBS release nonspecifically bound FN but do not release FN bound to anti-FN. Control experiments were performed with antiovalbumin and antihuman serum albumin.

Sodium Dodecyl Sulfate Polyacrylamide Gel Electrophoresis

This procedure was performed in 3% polyacrylamide-0.5% agarose gels by the method of Weinstein *et al.*[34] Samples in 1 M acetic acid were prepared for electrophoresis by heating in sample buffer for 1 hr at 37°C. The samples were then dialyzed against running buffer that contained 1% SDS, 0.4 mM phenylmethylsulfonyl fluoride, 5 mM *N*-ethylmaleimide, 10 mM EDTA, and 1 μM pepstatin for 2 hr at 20°C. Samples to be reduced were treated with 3 μl/100 μl of a 1.4 M dithiothreitol solution by boiling for 5 min. Gels were sectioned into 2-mm thick slices, processed for liquid scintillation counting as described previously,[32] and counted in a Searle Mark III liquid scintillation counter. Purified human plasma factor VIII (subunit mol wt, 202,000) used as a marker was prepared as described previously.[32] Purified plasma FN used as a marker (subunit mol wt, 200,000) was purified as described previously.[2]

The 3% polyacrylamide-0.5% agarose system was chosen because when samples are run unreduced in this system, factor VIII and plasma FN are completely separated from each other, whereas with reduction, they have the same mobility.[34]

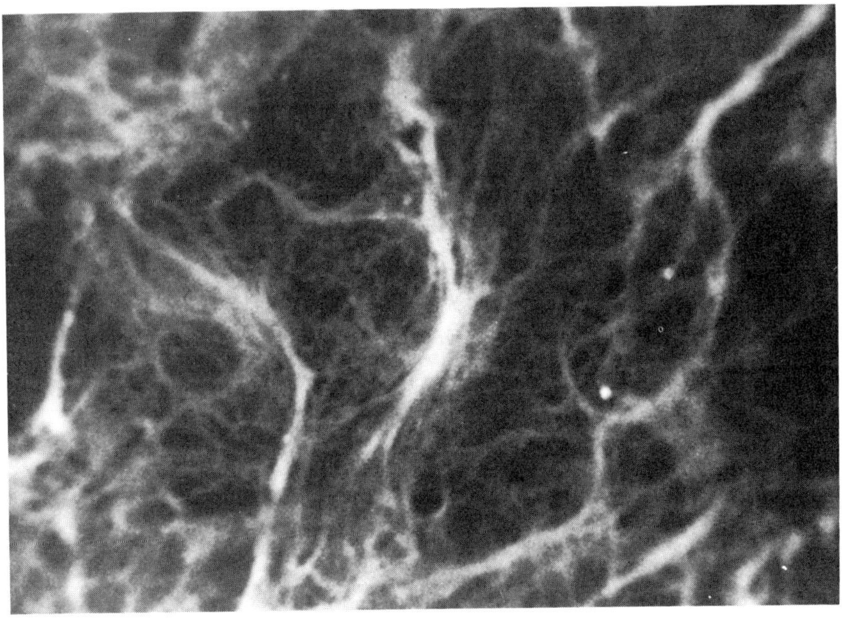

FIGURE 1. Immunofluorescence study of cultured human endothelial cells. Cells were treated with anti-FN and then with fluorescein-conjugated goat antirabbit IgG. A meshwork of extracellular fibrils is brightly stained. x960.

RESULTS

Immunofluorescence Studies

When coverslips of confluent cultures of human endothelial cells, cultured in normal rabbit serum and fixed in acetone, were incubated sequentially with rabbit anti-FN and fluorescein-conjugated goat antirabbit IgG, meshworks of extracellular fibrils were brightly stained (FIGURE 1). When subconfluent endothelial cells were similarly stained, only rare immunofluorescent extracellular fibrils were seen, although large numbers of brightly fluorescent intracellular granules were present. As the cells became more confluent, the numbers of intracellular immunofluorescent granules decreased, and the extent of the extracellular immunofluorescent fibrillar meshwork increased. By the time the cells were well confluent, only rare immunofluorescent intracellular granules were seen. The fibrils were not related to areas of cell-cell junctions; instead, they appeared to form a continuous meshwork underlying the cells. No staining was evident when the cultured endothelial cells were stained with normal rabbit serum, antiprothrombin, or antisera to antithrombin III instead of anti-FN.

Immunofluorescence staining of endothelial cells was completely inhibited by prior absorption at equivalence of anti-FN serum by purified plasma FN. Supernatants from the zone of antigen excess did not stain cultured endothelial cells, whereas supernatants in the zone of antibody excess did.

Endothelial Cell Fibronectin

For this study, endothelial cells were cultured in media that contained 20% fetal calf serum, because the goat anti-FN was unable to detect FN in fetal calf serum by immunodiffusion analysis. Pre- and postculture media were concentrated fivefold and assayed for FN by radial immunodiffusion by use of goat anti-FN. Concentrated preculture media contained no detectable FN (TABLE 1). Concentrated postculture media, however, did contain significant amounts of FN.

Immunodiffusion Studies

On immunodiffusion analysis, goat anti-FN reacted with a line of identity when tested against endothelial cell postculture medium and human plasma. Fetal calf serum, which was present in the culture medium utilized for these studies, did not form a precipitin line when tested against the goat anti-FN serum.

Immunoisolation of Endothelial Cell Fibronectin

Fibronectin synthesized by endothelial cells was isolated from fractions of postculture media derived from endothelial cells cultured in 20% normal rabbit

TABLE 1

FIBRONECTIN CONTENT OF PRE- AND POSTCULTURE MEDIA*

Material Tested†	Fibronectin‡ (μg/culture flask)
Preculture media	<8
Postculture media	
Culture 1	50
2	25
3	55
4	77

*From Jaffe & Mosher.[42]

†Four separate strains of endothelial cells were cultured for 3 days, and the media were concentrated fivefold and analyzed by radial immunodiffusion.

‡The minimal amount of FN detectable by this technique was 8 μg per culture flask. The values represent averages of two sets of analyses on multiple dilutions of each sample.

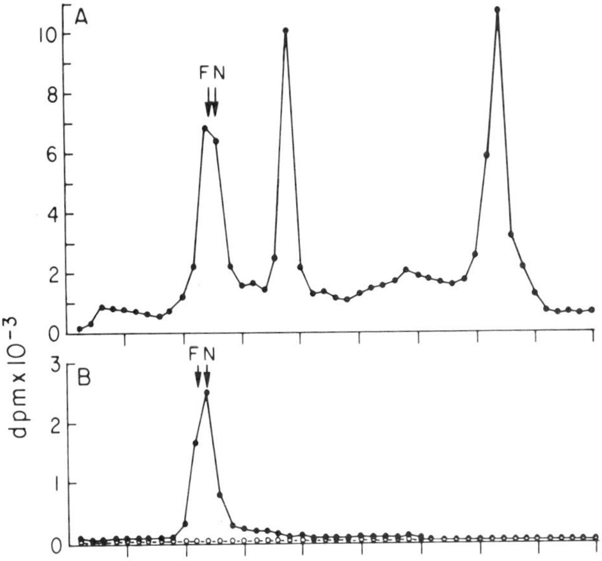

FIGURE 2. Protein A-Sepharose isolation of FN synthesized by cultured endothelial cells. Distribution of radioactivity in SDS polyacrylamide gels of (A) the fraction derived from [^{3}H]leucine-labeled endothelial cell postculture medium by precipitation with 50% saturated ammonium sulfate (reduced with dithiothreitol) and of (B) the fraction derived by acetic acid elution of an anti-FN protein A-Sepharose column that had been reacted with the [^{3}H]leucine-labeled, ammonium sulfate-precipitated endothelial cell postculture medium shown in (A) and then washed as detailed in MATERIALS AND METHODS (●). Control experiments were performed with antihuman serum albumin and antiovalbumin (○) instead of anti-FN (all reduced with dithiothreitol). The arrows labeled FN represent the locations of the plasma FN marker run on companion gels. When reduced, plasma FN electrophoresis reveals a closely spaced doublet with an average molecular weight of 200,000 daltons, hence the double arrow. (From Jaffe & Mosher.[42] By permission of *Journal of Experimental Medicine.*)

or fetal calf serum with [^{3}H]leucine in the presence of Trasylol (50 U/ml). The [^{3}H]FN was isolated by two different methods: a technique that employs protein A-Sepharose and double-antibody immunoprecipitation. The protein A-Sepharose technique utilized a rabbit anti-FN serum to avoid interspecies cross-reactivity. The isolated [^{3}H]FN was characterized by SDS polyacrylamide gel electrophoresis with and without reduction by dithiothreitol.

A fraction obtained from [^{3}H]leucine-labeled endothelial cell postculture medium by precipitation with ammonium sulfate at 50% saturation was electrophoresed, sliced, and counted for radioactivity (FIGURE 2, A). This gel contained three major radioactive peaks, one of which comigrated with the human plasma FN used as a marker. The radioactively labeled ammonium sulfate-precipitated postculture medium was subjected to immunoisolation by use of anti-FN coupled to protein A-Sepharose. After this procedure, analysis of the depleted labeled postculture media by SDS polyacrylamide gel electrophoresis revealed that the radioactive peak that comigrated with the plasma FN marker (FIGURE 2, A) had almost completely disappeared (data not shown). Sodium dodecyl sulfate polyacrylamide gel electrophoresis (after reduction) of the eluate from the anti-FN protein A-Sepharose column (FIGURE 2, B), revealed one radioactive band with a molecular weight of 200,000 that comigrated with the marker plasma FN. When the acetic acid eluate was electrophoresed without reduction (data not shown), one radioactive band was seen, and it comigrated with the unreduced plasma FN marker. Control studies performed with antihuman serum albumin and antiovalbumin instead of anti-FN did not isolate any labeled material (FIGURE 2, B). In this gel system (3% polyacrylamide-0.5% agarose), after reduction, the plasma factor-VIII marker comigrated with the plasma FN marker. However, when factor VIII was electrophoresed without reduction, it did not enter the gel, whereas the plasma FN marker did. Thus, this system could be used to completely separate FN and factor VIII.[34]

Labeling patterns identical to those shown in FIGURE 2 were seen when the radioactively labeled ammonium sulfate-precipitated postculture medium was subjected to double-antibody immunoprecipitation with anti-FN and goat anti-rabbit IgG. Control double-antibody immunoprecipitation experiments with antihuman serum albumin and antiovalbumin instead of anti-FN precipitated no labeled material.

DISCUSSION

The studies reported here demonstrate that cultured human endothelial cells synthesize and release FN into culture media and incorporate FN into the extracellular matrix. The radioactive FN secreted by endothelial cells into the culture medium comigrated with the plasma FN marker on SDS polyacrylamide gel electrophoresis both with and without reduction and had a single polypeptide subunit with a molecular weight of 200,000 daltons, similar to the plasma FN marker. Unlike endothelial cell FN, the FN synthesized by some fibroblasts has a subunit molecular weight of 220,000 in both its soluble and insoluble (cell surface-bound and extracellular matrix-embedded) forms.[4-6,24] Since the fibronectin synthesized by fibroblasts consists of 220,000-dalton subunit molecular weight polypeptides, which are apparently not transformed into the polypeptides of 200,000 molecular weight found in plasma,[24] endothelial cells may well synthesize the bulk of plasma fibronectin that consists of polypeptides of 200,000 molecular weight.

The validity of these studies is dependent on the specificity of the anti-FN sera used. These antibodies have previously been shown to form a single line of identity when tested against human plasma by immunodiffusion analysis and immunoelectrophoresis.[2,29] The quantitative immunoprecipitin curve and the loss of immunofluorescence at equivalence (FIGURE 2) also argue strongly for monospecificity. We have previously shown that cultured endothelial cells contain, synthesize, and release factor-VIII antigen.[30,32,35] To avoid the possibility of contamination of the anti-FN sera with trace amounts of antifactor VIII, all anti-FN sera were absorbed with purified plasma factor VIII before use. These absorptions and the pattern of mobility of the [^{3}H]FN peak seen in FIGURE 2 and that obtained on electrophoresis without reduction rule out the possibility that the isolated radioactive peak seen in FIGURE 2 is factor-VIII antigen.

Fibronectin is one of the major proteins synthesized and secreted by cultured endothelial cells. It appears to constitute approximately 15% of the protein released by the cells into the culture media. By comparison, factor-VIII antigen represented only 5.5% of a high-molecular-weight fraction (separated by Sepharose 6B) prepared from the same starting material[35] and thus probably represented less than 1% of the total protein released by the cells into the culture medium. These estimates are consistent with the concentrations of these proteins in plasma, namely, factor VIII antigen, approximately 5–10 μg/ml, and plasma FN, about 330 μg/ml.[1] These estimates are also consistent with the results obtained when anti-FN bound to protein A-Sepharose was reacted with the radioactively labeled, ammonium sulfate-precipitated postculture medium. Practically the entire 200,000-dalton peak seen in FIGURE 2 (A) was removed; the remaining radioactivity in the peak area (~5% of the initial amount) probably represents factor-VIII antigen. Cultures of human endothelial cells secreted an average of 51.8 μg of FN/25 cm^2 of dish during 3 days in culture (TABLE 1). By comparison, a variety of first-passage human embryonic cell strains secreted an average of 106 μg of FN/25 cm^2 of dish (equivalent) during 3 days in culture.[12,36]

Fibronectin has been shown to be present in the subendothelium of a variety of blood vessels. Linder *et al.*,[4] in studies on the developing chick embryo, showed that in addition to loose connective tissue, the subendothelium of arteries and veins and also of brain capillaries contained FN when tissue sections were studied by immunofluorescence microscopy. Fibronectin was absent in parenchymal cells in a variety of organs, such as liver, striated and smooth muscle, brain (except neural sheaths), cartilage, and bone. Similar studies on sections of human thyroid have shown that FN is present in the basement membrane that surrounds the colloid follicles and in the subendothelium of blood vessels.[27]

The physiologic role of FN is not clear, although several of its effects are known. Fibronectin is necessary for the attachment of tissue culture cells to culture vessels.[19,20] Fibronectin increases the strength of attachment of transformed cells to culture vessels and thus changes their morphology (but not their growth characteristics and malignant potential).[22,23] The mechanism(s) responsible for these two effects is unknown. Fibronectin also seems to be an important constituent of basement membranes and loose connective tissue, where it forms extensive meshworks of extracellular fibrils. The same fibrils have also been shown to contain collagen when examined by immunofluorescence microscopy.[16] The basis for this congruence of location may be the recently described noncovalent interactions of FN with other FN molecules[37] and with collagen and gelatin.[17] In addition, FN cross-links with itself and other proteins through disulfide bonds.[6,38] Fibronectin on cell surfaces is known to be immobile and thus may serve as an anchorage point for cell attachment.[37] Fibronectin can be covalently cross-linked by factor XIII both to other FN molecules and to the α-chain of fibrin.[2,29] This interaction presumably allows fibrin in wounds to attach to fibroblasts. Fibroblasts are known to be able to contract fibrin clots *in vitro*[39]; *in vivo,* these reactions are probably important in wound healing. Muir *et al.* have suggested that FN may be similar to a microfibrillar protein isolated from cultured monkey smooth muscle cell extracellular matrix.[40] We suggest an additional possibility: that FN in the extracellular matrix is the high-molecular-weight glycoprotein (subunit polypeptide molecular weight after reduction, ~200,000 daltons) isolated by Kefalides from collagenase-digested glomerular basement membranes.[41] Thus, endothelial cell fibronectin may be an important component of the subendothelium that influences normal blood vessel physiology.

REFERENCES

1. MOSESSON, M. W. & R. A. UMFLEET. 1970. The cold-insoluble globulin of human plasma. I. Purification, primary characterization, and relationship to fibrinogen and other cold-insoluble fraction components. J. Biol. Chem. **245:** 5728–5736.

2. Mosher, D. F. 1975. Cross-linking of cold-insoluble globulin by fibrin-stabilizing factor. J. Biol. Chem. **250:** 6614–6621.
3. Morrison, P. R., J. T. Edsall & S. G. Miller. 1948. Preparation and properties of serum and plasma proteins. XVIII. The separation of purified fibrinogen from fraction I of human plasma. J. Amer. Chem. Soc. **70:** 3103–3108.
4. Linder, E., A. Vaheri, E. Ruoslahti & J. Wartiovaara. 1975. Distribution of fibroblast surface antigen in the developing chick embryo. J. Exp. Med. **142:** 41–49.
5. Ruoslahti, E. & A. Vaheri. 1975. Interaction of soluble fibroblast surface antigen with fibrinogen and fibrin. Identity with cold-insoluble globulin of human plasma. J. Exp. Med. **141:** 497–501.
6. Keski-Oja, J., D. F. Mosher & A. Vaheri. 1977. Dimeric character of fibronectin, a major cell surface associated glycoprotein. Biochem. Biophys. Res. Commun. **74:** 699–706.
7. Hynes, R. O. & J. M. Bye. 1974. Density and cell cycle dependence of cell surface proteins in hamster fibroblasts. Cell **3:** 113–120.
8. Yamada, K. M. & J. A. Weston. 1974. Isolation of a major cell surface glycoprotein from fibroblasts. Proc. Nat. Acad. Sci. USA **71:** 3492–3496.
9. Ruoslahti, E., A. Vaheri, P. Kuusela & E. Linder. 1973. Fibroblast surface antigen: a new serum protein. Biochim. Biophys. Acta **322:** 352–358.
10. Kuusela, P., E. Ruoslahti & A. Vaheri. 1975. Polypeptides of a glycoprotein antigen (SF) present in serum and surface of normal but not of transformed chick fibroblasts. Biochim. Biophys. Acta **379:** 295–303.
11. Vaheri, A. & E. Ruoslahti. 1975. Fibroblast surface antigen produced but not retained by virus-transformed human cells. J. Exp. Med. **142:** 530–535.
12. Mosher, D. F., O. Saksela, J. Keski-Oja & A. Vaheri. 1977. Distribution of a major surface-associated glycoprotein, fibronectin, in cultures of adherent cells. J. Supramol. Struct. **6:** 551.
13. Vaheri, A., E. Ruoslahti, B. Westermark & J. Pontén. 1976. A common cell-type specific surface antigen (SF) in cultured human glial cells and fibroblasts: loss of malignant cells. J. Exp. Med. **143:** 64–72.
14. Hynes, R. O., G. S. Martin, M. Shearer, D. R. Critchley & C. J. Epstein. 1976. Viral transformation of rat myoblasts: effects on fusion and surface properties. Devel. Biol. **48:** 35–46.
15. Chen, L. B., N. Maitland, P. H. Gallimore & J. K. McDougall. 1977. Detection of the large external transformation-sensitive protein on some epithelial cells. Exp. Cell Res. **106:** 39–46.
16. Bornstein, P. & J. F. Ash. 1977. Cell surface-associated structural proteins in connective tissue cells. Proc. Nat. Acad. Sci. USA **74:** 2480–2484.
17. Engvall, E. & E. Ruoslahti. 1977. Binding of soluble form of fibroblast surface protein, fibronectin, to collagen. Int. J. Cancer **20:** 1–5.
18. Jilek, F. & H. Hörmann. 1977. Cold insoluble globulin. II. Plasminolysis of cold insoluble globulin. Hoppe-Seyler's Z. Physiol. Chem. **358:** 133–136.
19. Grinnell, F. & D. G. Hays. 1978. Cell adhesion and spreading factor: identity with cold insoluble globulin in human serum. Exp. Cell Res. In press.
20. Pearlstein, E. 1976. Plasma membrane glycoprotein which mediates adhesion of fibroblasts to collagen. Nature (London) **262:** 497–500.
21. Hynes, R. O. 1976. Cell surface proteins and malignant transformation. Biochim. Biophys. Acta **458:** 73–107.
22. Yamada, K. M., S. S. Yamada & I. Pastan. 1976. Cell surface protein partially restores morphology, adhesiveness, and contact inhibition of movement to transformed fibroblasts. Proc. Nat. Acad. Sci. USA **73:** 1217–1221.
23. Ali, I. U., V. Mautner, R. Lanza & R. O. Hynes. 1977. Restoration of normal morphology, adhesion, and cytoskeleton in transformed cells by addition of a transformation-sensitive surface protein. Cell **11:** 115–126.
24. Mosher, D. F. 1977. Labeling of a major fibroblast surface protein (fibronectin) catalyzed by blood coagulation factor XIIIa. Biochim. Biophys. Acta **491:** 205–210.
25. Jaffe, E. A., C. R. Minick, B. Adelman, C. G. Becker & R. Nachman. 1976. Syn-

thesis of basement membrane collagen by cultured human endothelial cells. J. Exp. Med. **144:** 209–225.

26. Howard, B. V., E. J. Macarak, D. Gunson & N. A. Kefalides. 1976. Characterization of the collagen synthesized by endothelial cells in culture. Proc. Nat. Acad. Sci. USA **73:** 2361–2364.
27. Vaheri, A., D. F. Mosher, J. Wartiovaara, J. Keski-Oja, M. Kurkinen & S. Stenman. 1977. Interactions of fibronectin, a cell-type specific surface-associated glycoprotein. *In* Cell Interactions in Differentiation. M. Karkinen-Jääskeläinen, L. Saxén & L. Weiss, Eds.: 311. Academic Press, Inc. New York, N.Y.
28. Jaffe, E. A., R. L. Nachman, C. G. Becker & C. R. Minick. 1973. Culture of human endothelial cells derived from umbilical veins. Identification by morphologic and immunologic criteria. J. Clin. Invest. **52:** 2745–2756.
29. Mosher, D. F. 1976. Action of fibrin-stabilizing factor on cold-insoluble globulin and α_2-macroglobulin in clotting plasma. J. Biol. Chem. **251:** 1639–1645.
30. Jaffe, E. A., L. W. Hoyer & R. L. Nachman. 1974. Synthesis of von Willebrand factor by cultured human endothelial cells. Proc. Nat. Acad. Sci. USA **71:** 1906–1909.
31. Weksler, B. B., T. Chang & C. Ley. 1977. Convenient microtiter assay for von Willebrand's factor. Thromb. Haemostasis **38:** 350.
32. Jaffe, E. A. & R. L. Nachman. 1975. Subunit structure of factor VIII antigen synthesized by cultured human endothelial cells. J. Clin. Invest. **56:** 698–702.
33. Kessler, S. W. 1975. Rapid isolation of antigens from cells with a staphylococcal protein A-antibody absorbent: parameters of the interaction of antibody-antigen complexes with protein A. J. Immunol. **115:** 1617–1624.
34. Weinstein, M. J., D. Deykin & E. W. Davie. 1976. Quantitative determination of factor VIII protein by two-stage gel electrophoresis. Brit. J. Haematol. **33:** 343–355.
35. Jaffe, E. A., L. W. Hoyer & R. L. Nachman. 1973. Synthesis of antihemophilic factor antigen by cultured human endothelial cells. J. Clin. Invest. **52:** 2757–2764.
36. Mosher, D. F., O. Saksela & A. Vaheri. 1977. Synthesis and secretion of alpha 2 macroglobulin by cultured adherent lung cells. Comparison with cell strains derived from other tissues. J. Clin. Invest. **60:** 1036–1045.
37. Schlesinger, J., L. S. Barak, G. G. Hammes, K. M. Yamada, I. Pastan, W. W. Webb & E. L. Elson. 1977. Mobility and distribution of a cell surface glycoprotein and its interactions with other membrane components. Proc. Nat. Acad. Sci. USA **74:** 2909–2913.
38. Hynes, R. O. & A. Destree. 1977. Extensive disulfide bonding at the mammalian cell surface. Proc. Nat. Acad. Sci. USA **74:** 2855–2859.
39. Niewiarowski, S. & S. Goldstein. 1973. Interaction of cultured human fibroblasts with fibrin: modification by drugs and aging in vitro. J. Lab. Clin. Med. **82:** 605–610.
40. Muir, L. W., P. Bornstein & R. Ross. 1976. A presumptive subunit of elastic fiber microfibrils secreted by arterial smooth-muscle cells in culture. Eur. J. Biochem. **64:** 105–114.
41. Kefalides, N. A. 1973. Structure and biosynthesis of basement membrane. Int. Rev. Connect. Tissue Res. **6:** 63–104.
42. Jaffe, E. A. & D. F. Mosher. 1978. Synthesis of fibronectin by cultured human endothelial cells. J. Exp. Med. In press.

CELL SURFACE AND EXTRACELLULAR MATRIX GLYCOPROTEIN FIBRONECTIN: EXPRESSION IN EMBRYOGENESIS AND IN TERATOCARCINOMA DIFFERENTIATION*

Jorma Wartiovaara, Ilmo Leivo, Ismo Virtanen, and Antti Vaheri†

Departments of Pathology and †Virology
University of Helsinki
SF-00290 Helsinki 29, Finland

Christopher F. Graham

Department of Zoology
Oxford University
Oxford, England

The cell surface glycoprotein fibronectin is characteristically present in fibroblasts and sera of various vertebrate species. Similar proteins have been given various names (SF antigen, LETS protein, CIG, CSP, Z protein). Although definite information on their chemical and immunologic similarities is still lacking, the proteins appear to be largely alike.[1] In adult[17] and embryonic tissue,[2] fibronectin is largely confined to basal laminae and extracellular matrix material.

It is our purpose here to describe the formation of fibronectin as seen by indirect immunofluorescence techniques in different stages of ontogenesis in the mouse and during mouse teratocarcinoma differentiation into structures called embryoid bodies,[3,4] a process that to a certain degree mimics early development.[5] At present, fibronectin is believed to play a role in the positioning of cells in morphogenetic phenomena and in the maintenance of orderly tissue organization.

Early Embryos

The presence of fibronectin in early embryos was studied in inbred mice (CBA × Balb/C) by indirect immunofluorescence techniques with antifibronectin serum.[6]

The early development of the mouse embryo consists of following stages[7]: preimplantation development, including cleavage, compaction, and blastocyst stages; and postimplantation stages, which consist of egg cylinder, primitive streak, and further stages of development.

During preimplantation development, the cellular relationships in the embryo are characterized by the formation of inside cells and outside cells. This stage is considered to be of decisive importance in the determination and differentiation of the cells.[8] Beginning with implantation, another major change in cellular relations occurs with the formation of an inner cell mass within the blastocyst. The development of the inner cell mass into embryonic ectoderm surrounded by a

*Supported by grants from the Sigrid Jusélius Foundation, the British Medical Research Council, and the National Cancer Institute.

0077-8923/78/0312-0132 $01.75/0 © 1978, The New York Academy of Sciences

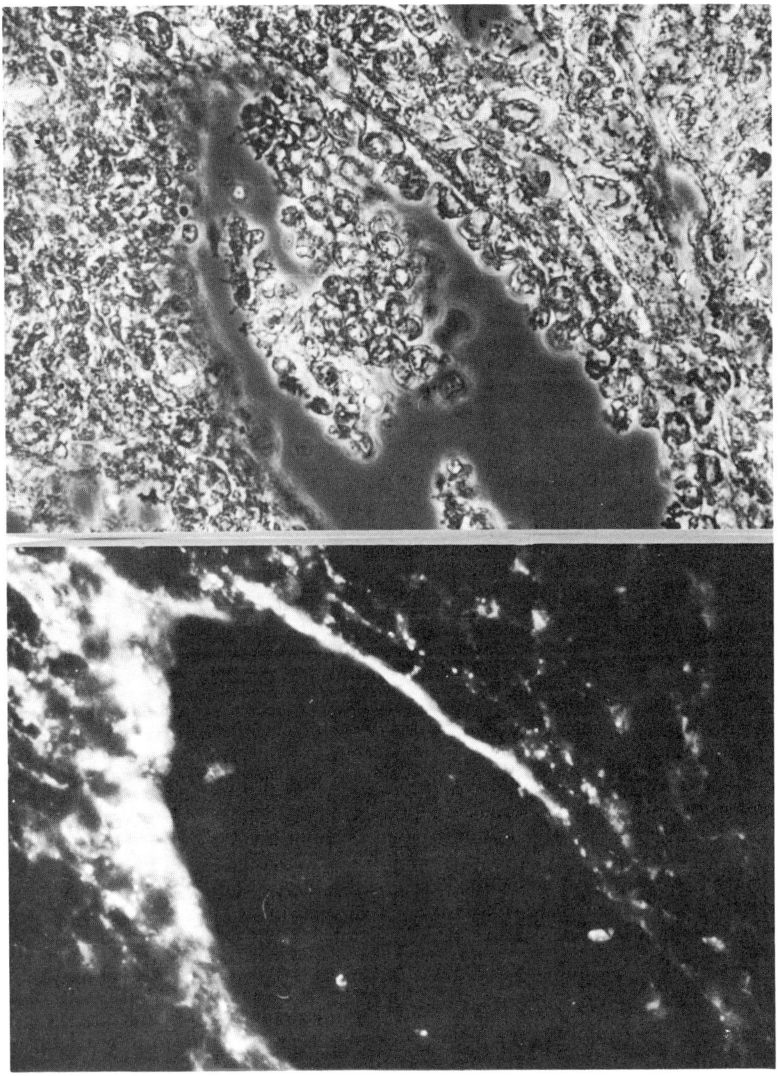

FIGURE 1. A 6-day (CBA × BALB/C) mouse embryo at late blastocyst stage seen by phase-contrast microscopy (*top*) and by indirect immunofluorescence (*bottom*) after staining with antifibronectin antiserum.[6] Tissue was fixed with ethanol-glacial acetic acid (99:1, v/v) for 24 hr and processed according to the method of Sainte-Marie.[9] Sections were stained with rabbit antihuman fibronectin serum and treated with fluorescein isothiocyanate-conjugated sheep antirabbit IgG. No staining is detectable in the embryo, but some fluorescence is visible in the uterine tissue. ×550.

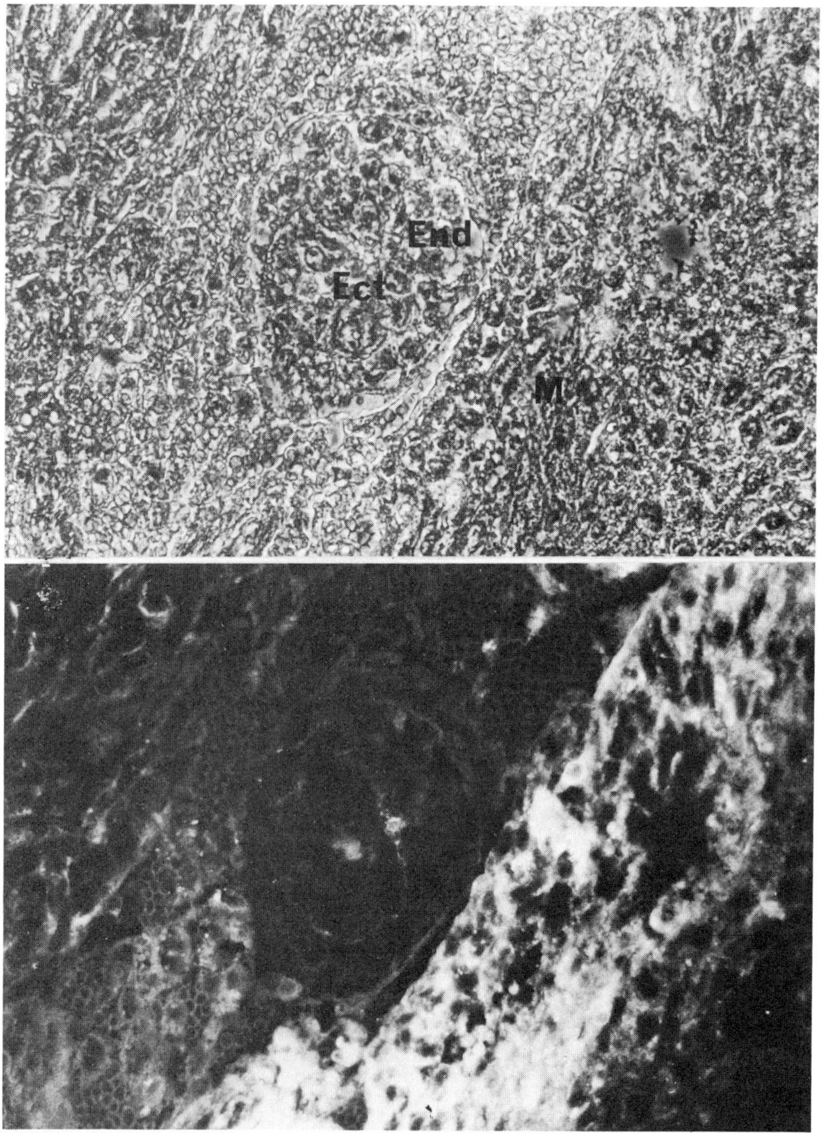

FIGURE 2. Early 7-day mouse embryo. Fibronectin-associated fluorescence is visualized as a band between the proximal endoderm (End) and ectoderm (Ect). The uterine mucosa (M) shows strong fluorescence. × 330.

layer of proximal endoderm cells is of importance for the discussion section in this paper.

When preimplantation embryos from the two-cell stage onward were studied for expression of fibronectin, no detectable fluorescence was seen. In postimplantation embryos at the sixth day of development (day of plug equals first day of development), the late blastocyst stage, again, no visible fibronectin fluorescence was seen (FIGURE 1). Some staining was observed in the uterine tissue at the site of implantation.

At the seventh and eighth days of development, distinct fibronectin fluorescence was observed in the uterine wall and also in the embryo. In the latter, fluorescence was visualized largely as a band underneath the proximal (visceral) endoderm in most parts of the embryo (FIGURES 2 & 3). The ectoderm appeared negative.

In the 8-day embryo, Reichert's membrane was strongly fluorescent (FIGURE 3). Reichert's membrane is an extracellular basal lamina-like structure located between the distal endoderm and trophectoderm cell layers, which form a saclike structure around the embryo proper that separates it from the uterus.

At later stages in more complex embryos, increased fluorescence was seen between cell layers in different parts of the embryo.

ADVANCED STAGES OF DEVELOPMENT

It is characteristic to the structure of organs that epithelial and mesenchymal components are separated by basal lamina-like structures of extracellular material. In adult tissues, fibronectin is localized mainly within these structures.[19] Previously, it has been established that the same holds true for chick embryo tissues.[2] In the mouse embryo, fibronectin is similarly localized in developing organs in the basal laminae. During the formation of epithelial secretory tubules in mouse kidney development, basal lamina-type structures appear around aggregating mesenchymal cells.[10] The expression of fibronectin has been studied in this morphogenetic phenomenon.[11] The mesenchymal cells stain strongly positive for fibronectin. As the cells aggregate and differentiate into epithelial cells of the secretory tubules, they lose their fibronectin fluorescence. On the other hand, the basal lamina that is formed around the tubules is strongly fluorescent for fibronectin (FIGURE 4).

EMBRYOID BODIES FORMED BY EMBRYONAL CARCINOMA CELLS *in Vitro*

Teratocarcinomas are tumors that contain cells that have differentiated along various morphogenetic pathways and also cells with many developmental capacities, embryonal carcinoma (EC) cells. Mouse teratocarcinomas can be obtained in various ways: Tumors can arise spontaneously both in the testis and in the ovary; in certain mouse strains, the incidence of teratocarcinomas is very high.[12] They can also be obtained experimentally by transplanting the fetal genital ridge to the adult mouse or by implanting a normal embryo up to $7\frac{1}{2}$ days old into an extrauterine site.[13]

The embryonal carcinoma line (OC15S1) used in the studies described herein was derived from a tumor obtained in 7-day embryos.[14] When EC cells are grown for several days in tissue culture, the cells form loosely attached aggregates. If such

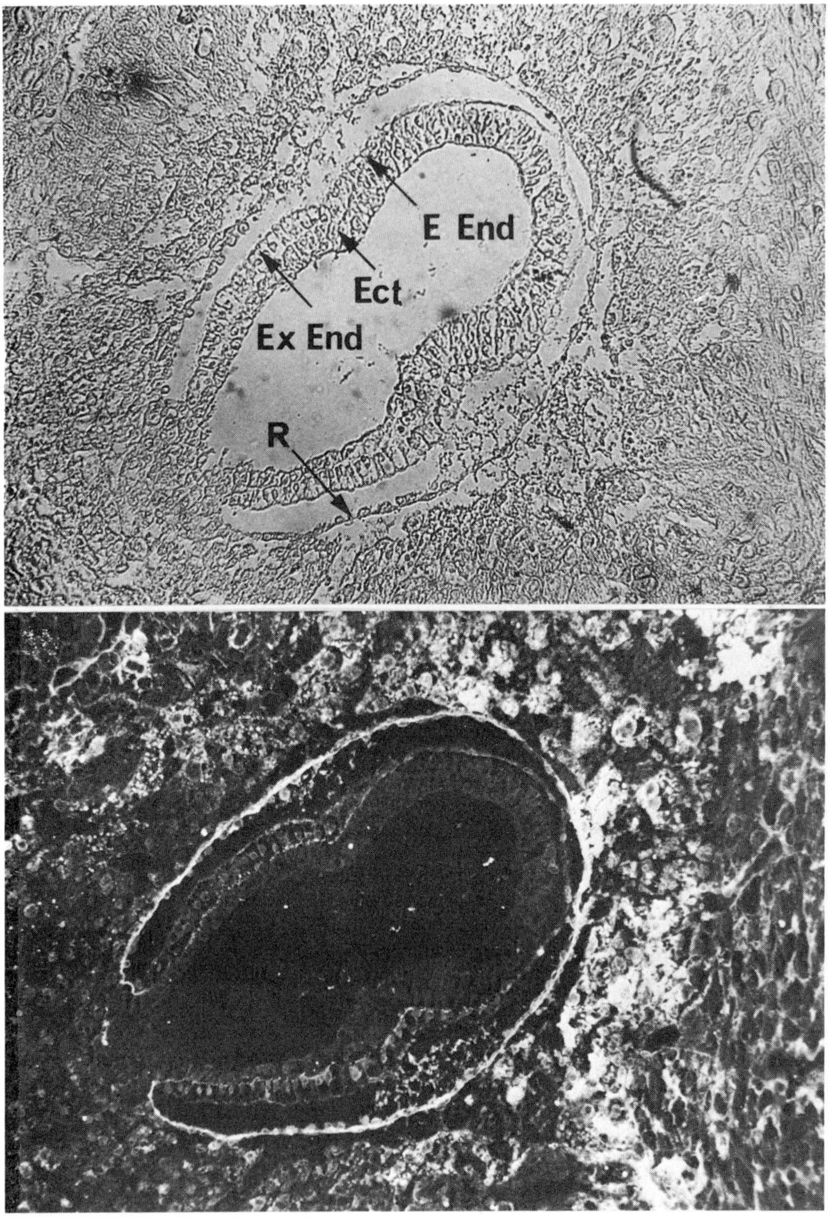

FIGURE 3. An 8-day mouse embryo. Fibronectin-associated fluorescence is seen in the embryonic endoderm (E End), extraembryonic endoderm (Ex End), and in a layer between them and the corresponding ectodermal (Ect) cells. The layer of fluorescence continues into the strongly stained Reichert's membrane (R). Also, many cells in the uterine wall stain brightly. × 130.

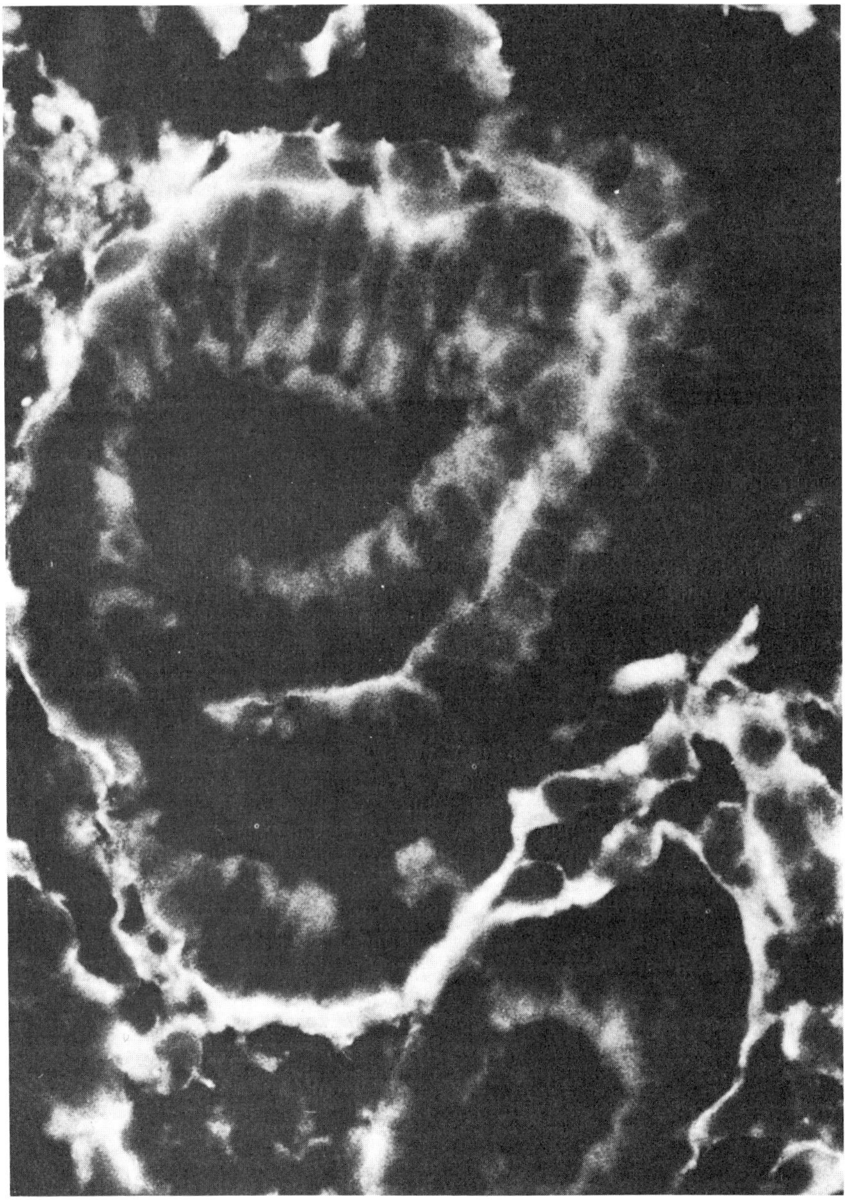

FIGURE 4. Section through the kidney of a 13-day mouse embryo shows secretory tubules and accompanying mesenchymal cells. The tubular epithelium is relatively negative for fibronectin-associated fluorescence, whereas the basal laminae surrounding the tubules and the mesenchymal cells exhibit strong fluorescence. × 715.

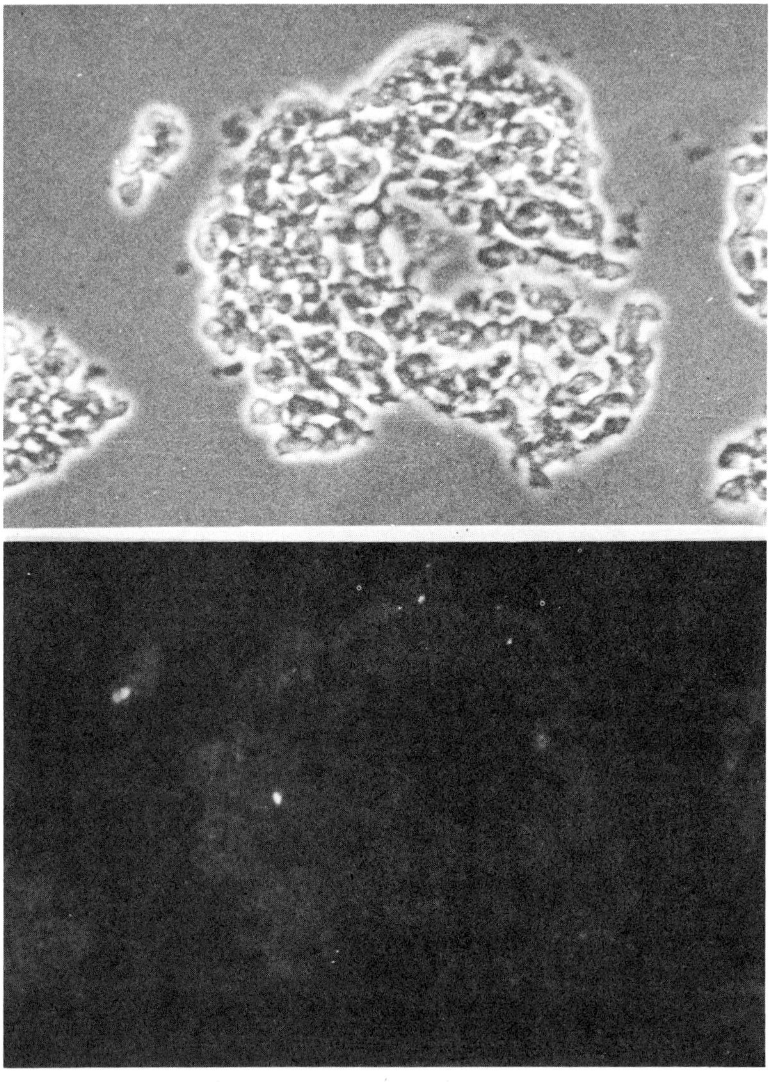

FIGURE 5. A simple embryoid body (embryonal carcinoma cell line OC15S1) on the first day after transfer to a bacteriologic petri dish. No fibronectin-associated fluorescence is visible. × 550.

aggregates are transferred to nonadhesive bacteriologic petri dishes, embryoid bodies are formed during further cultivation.[15] The development of the embryoid bodies somewhat resembles normal embryogenesis.[5]

When studied by indirect immunofluorescence techniques with antifibronectin antibodies, EC cells were found to show little or no detectable fibronectin fluorescence.[18] At the onset of cultivation on bacteriologic petri dishes, embryoid bodies were also negative for fibronectin (FIGURE 5). During further cultivation, the peripheral EC cells of the embryoid bodies differentiate into endodermal cells.[15] At that stage, a basal lamina is formed beneath the endodermal cells. In our experiments, fibronectin-associated fluorescence appeared around the differentiating endodermal cells and became localized in the basal lamina. Further differentiation of the embryoid bodies is characterized by the formation of cysts within the inner cell mass of the embryoid bodies. The fibronectin-associated

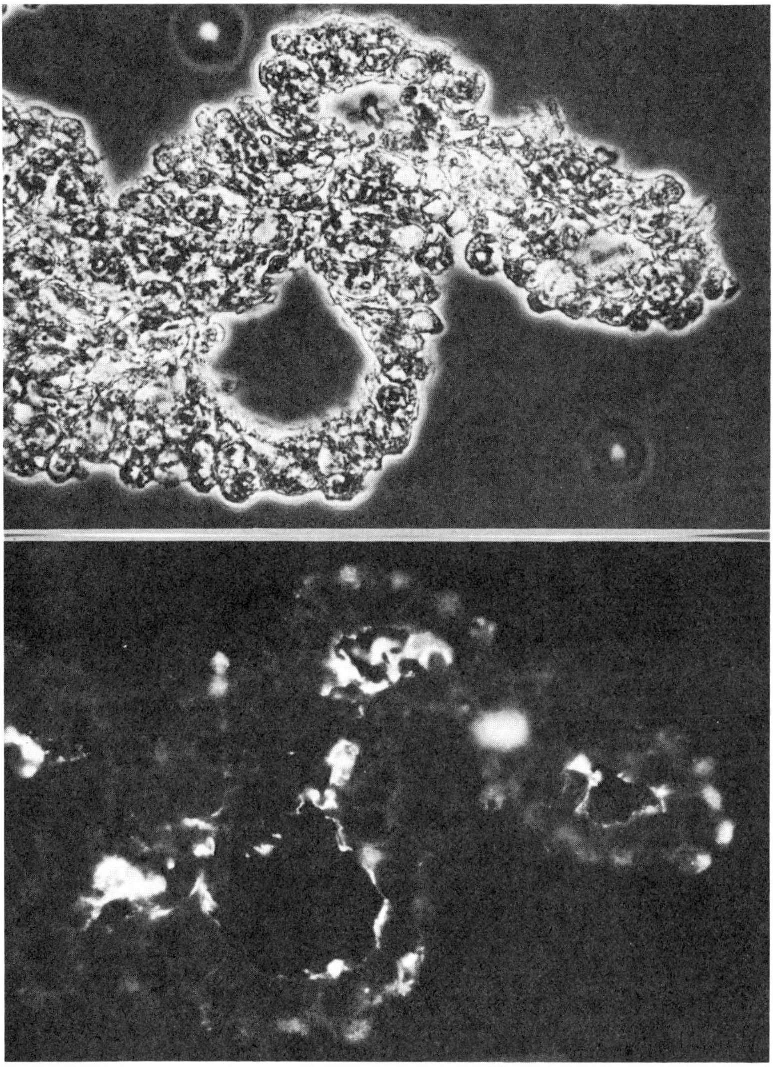

FIGURE 6. A cystic embryoid body on the eighth day of culture in a bacteriologic petri dish. Several cysts of various sizes are observed where fibronectin-associated fluorescence lines the inner surface of the cyst. × 550.

fluorescence was at that stage very distinct. Still later, the cysts enlarged into cavities, often with walls only one cell thick. In such cases, fibronectin-associated fluorescence was found to line pericellularly the internal surfaces of the cysts (FIGURE 6). In the indirect immunofluorescence experiments, when antifibronectin serum was substituted for normal rabbit serum or when antifibronectin serum was blocked with purified fibronectin, no fluorescence was seen.

MOUSE EMBRYOS AND EMBRYOID BODIES

The results obtained with early mouse embryos and mouse embryoid bodies indicate that fibronectin is localized in both cases at the same sites where extracellular material is deposited in basal lamina form. The basal lamina underneath the layer of endodermal cells of the embryoid body is analogous to the extracellular material seen in the embryo under the proximal endodermal cells, both staining positive for fibronectin. The basal lamina between the distal endodermal cells and the trophectodermal cells, namely, Reichert's membrane, was also observed to stain positively. This observation parallels the results of Pierce,[16] who obtained strong fluorescence in Reichert's membrane with an antibasement membrane antibody, even before interphase under the proximal endoderm became positive.

The results presented here and elsewhere[20] on the localization of fibronectin in various basal laminae and in the connective tissue matrix both *in vivo* and *in vitro* suggest that fibronectin may play a structural role in morphogenesis.

ACKNOWLEDGMENT

We thank Miss Elina Waris for technical assistance.

REFERENCES

1. VAHERI, A., D. MOSHER, J. WARTIOVAARA, J. KESKI-OJA, M. KURKINEN & S. STENMAN. 1977. Interactions of fibronectin, a cell-type specific surface-associated glycoprotein. *In* Cell Interactions in Differentiation. M. Karkinen-Jääskeläinen, L. Saxén & L. Weiss, Eds.: 311–323. Academic Press. London, England.
2. LINDER, E., A. VAHERI, E. RUOSLAHTI & J. WARTIOVAARA. 1975. Distribution of fibroblast surface antigen in the developing chick embryo. J. Exp. Med. **142:** 41–49.
3. STEVENS, L. C. 1959. Embryology of testicular teratomas in strain 129 mice. J. Nat. Cancer Inst. **23:** 1249–1295.
4. PIERCE, G. B., JR., F. J. DIXON & E. L. VERNEY. 1960. Teratocarcinogenic and tissue-forming potentialities of the cell types comprising neoplastic embryoid bodies. Lab. Invest. **9:** 583–602.
5. HSU, Y.-C. & J. BASKAR. 1974. Differentiation in vitro of normal mouse embryos and mouse embryonal carcinoma. J. Nat. Cancer Inst. **53:** 177–185.
6. STENMAN, S., J. WARTIOVAARA & A. VAHERI. 1977. Changes in the distribution of a major fibroblast protein, fibronectin, during mitosis and interphase. J. Cell Biol. **74:** 453–467.
7. SNELL, G. D. & L. C. STEVENS. 1966. Early embryology. *In* Biology of the Laboratory Mouse. E. L. Green, Ed.: 205–245. McGraw Hill Book Co. New York, N.Y.
8. GRAHAM, C. F. 1976. The formation of different cell types in animal embryos. *In* The

Developmental Biology of Plants and Animals. C. F. Graham & P. F. Wareing, Eds.: 14–28. W. B. Saunders Co. Philadelphia and Toronto.
9. SAINTE-MARIE, G. 1962. A paraffin embedding technique for studies employing immunofluorescence. J. Histochem. Cytochem. **10:** 250–256.
10. SAXÉN, L., O. KOSKIMIES, A. LAHTI, H. MIETTINEN, J. RAPOLA & J. WARTIOVAARA. 1968. Differentiation of kidney mesenchyme in an experimental model system. Adv. Morphog. **7:** 251–293.
11. WARTIOVAARA, J., S. STENMAN & A. VAHERI. 1976. Changes in expression of fibroblast surface antigen (SFA) during cytodifferentiation and heterokaryon formation. Differentiation **5:** 85–89.
12. STEVENS, L. C. 1973. A new inbred subline of mice (129/terSv) with a high incidence of spontaneous congenital testicular teratomas. J. Nat. Cancer Inst. **50:** 235–242.
13. DAMJANOV, I. & D. SOLTER. 1974. Experimental teratoma. Curr. Top. Pathol. **59:** 69–130.
14. MCBURNEY, M. W. 1976. Clonal lines of teratocarcinoma cells in vitro: differentiation and cytogenetic characteristics. J. Cell. Physiol. **89:** 441–455.
15. MARTIN, G. R. & M. J. EVANS. 1975. Differentiation of clonal lines of teratocarcinoma cells: formation of embryoid bodies in vitro. Proc. Nat. Acad. Sci. USA **72:** 1441–1445.
16. PIERCE, G. B. 1966. The development of basement membranes of the mouse embryo. Dev. Biol. **13:** 231–249.
17. STENMAN, S. & A. VAHERI. 1978. Distribution of a major connective tissue protein, fibronectin, in normal human tissues. J. Exp. Med. **147:** 1054–1064.
18. WARTIOVAARA, J., I. LEIVO, I. VIRTANEN, A. VAHERI & C. F. GRAHAM. 1978. Appearance of fibronectin during differentiation of mouse teratocarcinoma in vitro. Nature (London) **272:** 355–356.
19. LINDER, E., S. STENMAN, V.-P. LEHTO & A. VAHERI. This monograph.
20. HEDMAN, K., A. VAHERI & J. WARTIOVAARA. 1978. External fibronectin of cultured human fibroblasts is predominantly a matrix protein. J. Cell Biol. **76:** 748–760.

PRESENCE OF FIBRONECTIN IN BASEMENT MEMBRANES AND ACIDIC STRUCTURAL GLYCOPROTEINS FROM HUMAN PLACENTA AND LUNG*

Bonnie Anderson Bray

Department of Medicine
College of Physicians & Surgeons
Columbia University
New York, New York 10032

Attempts by two different groups[1-4] to isolate trophoblast basement membrane (TBM) from term placenta have afforded membranous products, antisera to which localized in all basement membranes of placenta, lung, and kidney. However, the placental membranes were primarily glycoprotein in nature and thus quite different from glomerular basement membrane (GBM). The antiserum against membranes prepared in this laboratory[1,2] detected at least three antigens. One of the antigens carried fibrinogen determinants and appeared to be complexed with a second noncollagenous antigen. Absorption of the fluorescein-conjugated antiserum with certain preparations of fibrin or fibrinogen reduced the linear staining of the placental basement membranes markedly and of lung basement membranes to a lesser extent. In lung and kidney, the linear staining of basement membranes was sharpened with the absorbed antiserum due to a loss of nonspecific background (not basement membrane) staining in these organs. Since antifibrinogen could neither block the tissue-staining properties of the antiserum nor duplicate them, it was suspected that the fibrinogen determinants were part of a much larger antigen as it existed in the tissues or that the antigen responsible for the linear staining in placenta was a substance that associated with fibrinogen.

Many of the properties of cold-insoluble globulin (CIg),[5,6] which is antigenically identical to LETS protein and to fibroblast surface antigen[7] and is now called fibronectin,[8] suggested that CIg was the second placental antigen. For instance, CIg is known to form complexes with fibrinogen and fibrin[7,9,10] and to combine covalently with fibrin in the presence of fibrin-stabilizing factor (blood coagulation factor XIII).[11] Since both the insoluble form of fibronectin[8] and the soluble plasma form[11] are cross-linked by factor XIII into high-molecular-weight multimers and since both forms associate with collagen,[12-14] it is reasonable to assume that fibronectin could contribute to basement membrane structures. Indeed, Linder *et al.*[15] have detected the antigen in basement membranes of the developing chick embryo by immunofluorescence techniques. In this study, isolated human basement membranes from placenta, lung, and kidney were solubilized by enzymatic digestion, and the digests were tested for the presence of CIg.

It was also observed that antigenically reactive material could be solubilized from TBM with 0.3 M acetic acid, 0.01 N NaOH and by reduction and alkylation of the disulfide bonds.[2] The fractions solubilized by these chemical procedures appeared to be predominantly acidic structural glycoproteins[16-19] and also contained CIg.

*Supported by National Heart and Lung Institute Grant HL-15832 and by a grant from the New York Lung Association.

0077-8923/78/0312-0142 $01.75/0

Materials and Methods

Preparation of the basement membranes, which involved sieving and sonicating placental villi (TBM), lung parenchyma (alveolar basement membrane), and renal cortex (GBM) have been described previously.[2,20] To obtain soluble, immunologically reactive fragments from the membranes, they were digested with elastase, plasmin, or collagenase as previously described.[2,20] Porcine pancreatic elastase, chromatographically purified, was obtained from Worthington Biochemical Corporation. Plasmin activated with matrix-bound activator was obtained from A. B. Kabi, Stockholm, Sweden. The collagenase (form III, Advance Biofactures Corporation) was found to be free of contaminating proteases when assayed by the method of Peterkofsky and Diegelmann.[21]

Acidic structural glycoproteins (ASG) were extracted from TBM and from intact terminal villi of placenta[2] and intact lung parenchyma[19] with 0.3 M acetic acid at 4° C and were recovered by adjusting the pH of the extracts to 4.7. An ASG fraction from placental villi was partially purified by chromatography on DEAE cellulose.[2] The fraction that remained on the column during washing and during elution with 0.1 M NaCl–0.017 M phosphate (pH 8) but that eluted with 0.1 M acetic acid was used in double-diffusion and immunoelectrophoresis experiments.

The trophoblast basement membranes were also extracted with 0.01 N NaOH. To 182 mg of TBM were added 18.2 ml of 0.01 N NaOH, and the suspension was stirred at 19°C for 16 hr and centrifuged. The supernatant (extract 1) was neutralized with 1 N HCl and lyophilized. Two additional extractions (2 and 3) were performed on the residue at 37°C for 24 hr each.

For reduction of the disulfide bonds, 200 mg of a TBM preparation were suspended in 20 ml of an 8 M urea—0.05 M Tris buffer (pH 7.5). After the vessel was flushed with nitrogen, 0.2 ml of mercaptoethanol was added. The vessel was again flushed with nitrogen, and the suspension was stirred at room temperature for 16 hr. The suspension was centrifuged and the clear supernatant was diluted with water to reduce the urea concentration to 2 M for the alkylation step. One half of the supernatant was alkylated with iodoacetic acid at pH 8.5. An amount of iodoacetic acid equivalent to the mercaptoethanol used was added just before the pH was adjusted to 8.5. Small increments of iodacetic acid were added until the nitroprusside test for sulfhydryl groups was negative. The pH was adjusted to 3.4 with glacial acetic acid, and the mixture was chromatographed on Sephadex® G-25 to remove salts before it was lyophilized. For alkylation with 2-vinylquinoline,[22] half of the supernatant from the reduction step was reacted for 16 hr at room temperature with 0.2 ml of 2-vinylquinoline (Eastman Kodak). The reaction mixture was dialyzed and lyophilized.

An estimate of the percentage of membrane or tissue solubilized by the chemical procedures was arrived at indirectly by washing the insoluble residue extensively with water to remove salts, lyophilizing it, and weighing it.

Crude CIg was prepared by clotting Cohn fraction I with bovine thrombin (Parke Davis). After 5 min, the clot was withdrawn, and the liquid was then centrifuged to remove clot fragments. The clear supernatant was shown to be free of fibrinogen by its failure to react with antifibrinogen in double-diffusion and immunoelectrophoresis experiments.

An antiserum to the placental membranes was prepared in rabbits.[1,2] This antiserum reacted with fibrinogen, ASG from placental villi, and with fragments from placental membranes, alveolar basement membrane (ABM), and GBM. After removal of antifibrinogen antibodies by absorption with chromatographically purified fibrinogen, the antiserum still gave a precipitin line with GBM and an in-

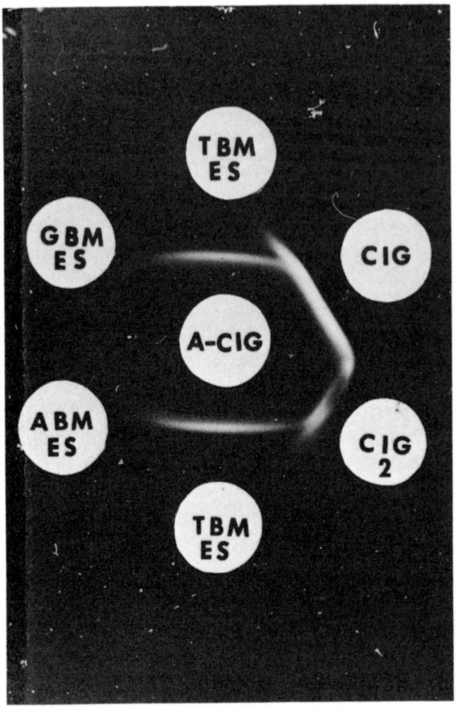

FIGURE 1. Double-diffusion reactions in agarose between anti-CIg antiserum (A-CIG) and elastase-solubilized fragments of basement membranes. TBM ES, trophoblast basement membrane, elastase soluble; CIG, cold-insoluble globulin; CIG 2, crude cold-insoluble globulin (cryofibrinogen); ABM ES, alveolar basement membrane, elastase soluble; GBM ES, glomerular basement membrane, elastase soluble.

dependent line with placental ASG.[2] The specific rabbit antiserum to human CIg was a generous gift of Dr. M. W. Mosesson, Downstate Medical Center. Rabbit antiserum to human fibrinogen, which did not react with CIg, was provided by Dr. Konrad C. Hsu of this institution.

RESULTS

Immunologic determinants of CIg (fibronectin) were detected in basement membrane preparations from placenta (TBM ES, FIGURE 1) by use of a specific antiserum to CIg. In the experiment presented here, in which the basement membranes were solubilized by digestion with elastase, the determinants were not detected in alveolar basement membrane (ABM ES) or in glomerular basement membrane (GBM ES). The reactions of CIg with anti-CIg spurred over the TBM-anti-CIg reactions. This finding indicated that only a portion of the CIg molecule was present in the digests, perhaps as a result of proteolysis by the elastase.

Since a negative result was obtained with ABM and GBM, these two basement membranes were also digested with collagenase, and the soluble material was reacted with the anti-CIg antiserum. In this experiment (FIGURE 2), a faint reaction was seen for ABM, which formed a reaction of identity with known CIg. Again, GBM did not react. Plasmin digestion also produced fragments from both TBM and ABM that gave precipitin lines with anti-CIg, but GBM was negative. These impressions have been upheld in preliminary quantitative experiments with the Laurell technique.[23] No CIg has been detected in GBM, although as little as 0.1% could easily have been detected.

Under the conditions used here, dilute alkali (0.01 N NaOH) solubilized 53% of TBM, and reduction with 2-mercaptoethanol in 8 M urea solubilized 36% of TBM. Extraction with 0.3 M acetic acid solubilized only 10–15% of the membranes but with complete disruption of their morphologic characteristics. As can be seen in FIGURE 3, all of these methods of solubilization afforded fragments that gave precipitin reactions with anti-CIg.

The material solubilized by these procedures was negatively charged and migrated to the region of serum α_1-globulins (shown in FIGURE 4 for the alkali extracts). The bands also stained pink in the periodic acid-Schiff reaction (not shown) and therefore represent glycoproteins. In addition, the material extracted with 0.3 M acetic acid from intact terminal villi of placenta (see below) could be recovered by isoelectric precipitation at pH 4.7, redissolved in 0.3 M acetic acid, and reprecipitated at pH 4.7 (reversible isoelectric precipitation). On the basis of these properties, and on analytic values, these materials are designated acidic structural glycoproteins.[16–18]

Acidic structural glycoprotein fractions were also extracted from lung parenchyma[19] and placental villi[2] with 0.3 M acetic acid and were recovered by adjusting the pH to 4.7. These fractions gave strong precipitin lines with anti-CIg (not shown).

The ASG fractions reacted with antifibrinogen (not shown), and this reactivity was retained by the fraction partially purified on DEAE cellulose. In FIGURE 5, this fraction (A2) is shown reacting with an antiserum to placental membranes in

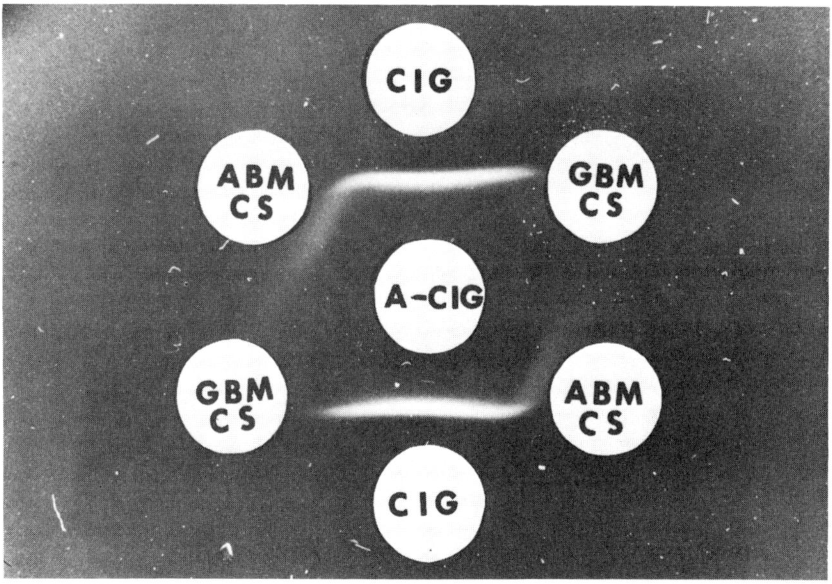

FIGURE 2. Double-diffusion reactions in agarose between anti-CIg antiserum (A-CIG) and collagenase-solubilized fragments of basement membranes. CIG, cold-insoluble globulin; GBM CS, glomerular basement membrane, collagenase soluble; ABM CS, alveolar basement membrane, collagenase soluble.

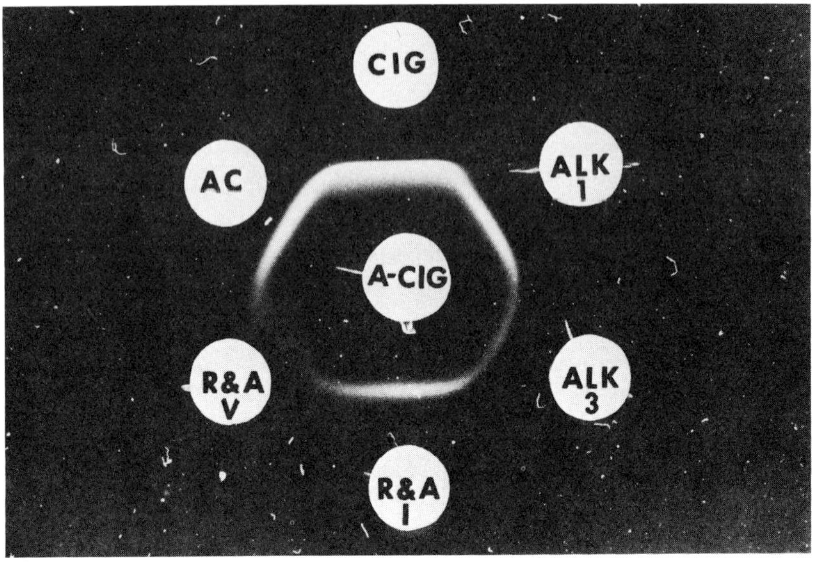

FIGURE 3. Double-diffusion reactions in agarose between anti-CIg antiserum (A-CIG) and chemically solubilized fragments of trophoblast basement membrane. CIG, cold-insoluble globulin; ALK 1 and ALK 3, alkali extracts 1 and 3; R&A I, reduced followed by alkylation with iodoacetic acid; R&A V, reduced followed by alkylation with 2-vinylquinoline; AC, acetic acid (0.3 M) extract. The extract was adjusted to pH 4.7, and the precipitate was dissolved in 0.1 M barbital buffer (pH 8.6).

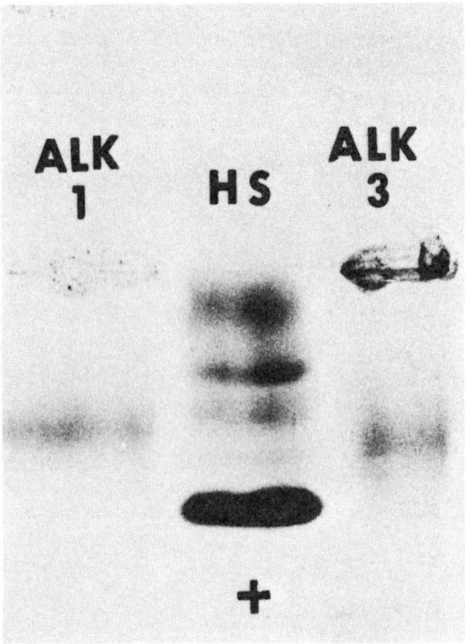

FIGURE 4. Electrophoresis on Sepraphore® III at pH 8.6 of alkali extracts 1 and 3 (ALK 1 and ALK 3) from a TBM preparation (see MATERIALS AND METHODS for details of the extraction procedure). For comparison, human serum (HS) was electrophoresed in the center. The anode (+) was at the bottom. This strip was fixed in absolute methanol and stained with 0.1% amido black 10B in methanol-glacial acetic acid-water (5:1:5).

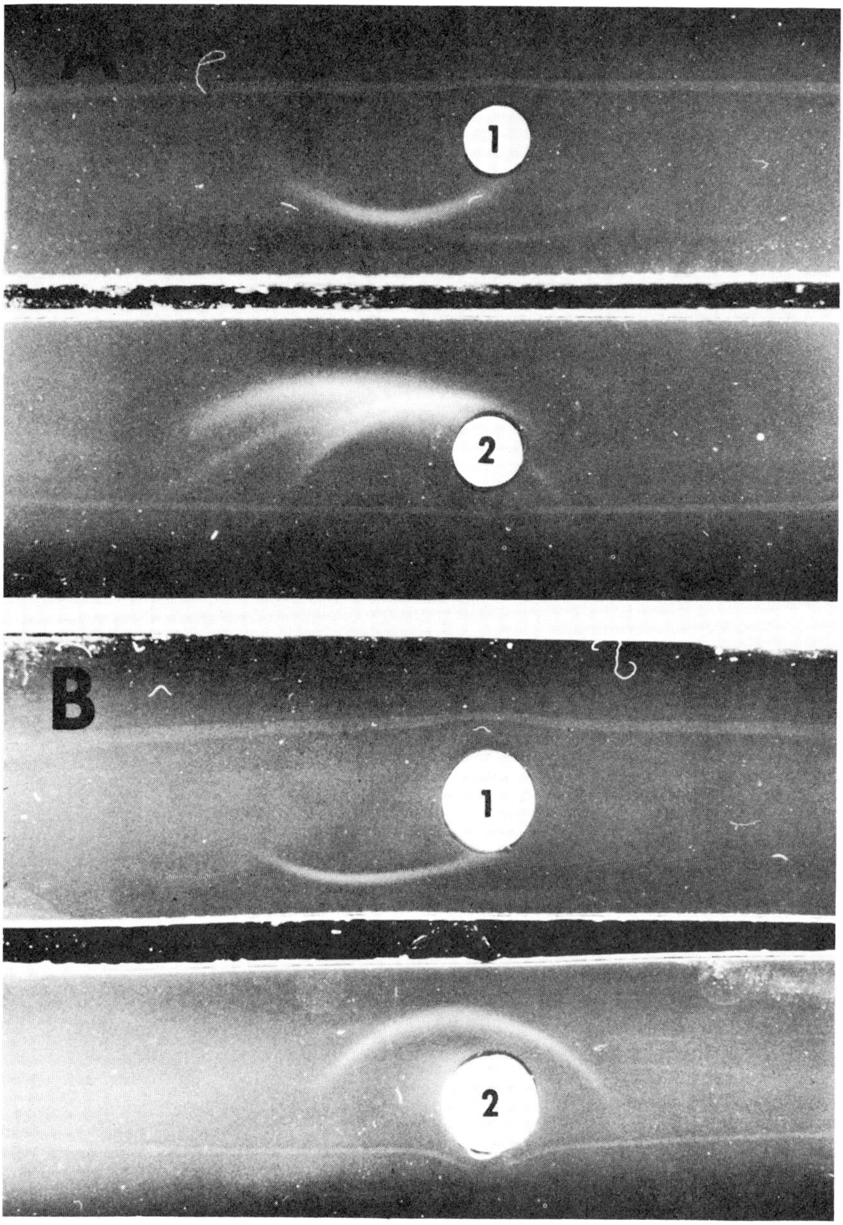

FIGURE 5. Immunoelectrophoresis of CIg, placental ASG, and fibrinogen (FI) with unabsorbed antitrophoblast basement membrane antiserum. The anode was to the left. A1, cold-insoluble globulin; A2, acidic structural glycoproteins from terminal villi of placenta. This fraction was partially purified on DEAE cellulose. B1, cold-insoluble globulin; B2, fibrinogen (FI). This was the same batch of FI from which the CIg (A1 and B1) was made.

an immunoelectrophoresis experiment that compares it to reference CIg (A1). Although four or five components with the mobility of CIg were detected, no component with the mobility of free fibrinogen (B2) was present. It is clear that both types of determinants (fibrinogen and CIg) occur on molecular species that are negatively charged. In double-diffusion experiments, this fraction reacted with both anti-CIg and antifibrinogen, and the two precipitin lines appeared to fuse, suggesting that both types of determinant were present on the same molecular species. However, a single component must be isolated before this point can be resolved.

Although extraction with 0.3 M acetic acid completely disrupted the organization of the tissues, it dissolved only 10% of lung parenchyma and only 30% of placental villi. The tissues that remained insoluble were digested with collagenase, and the collagenase-solubilized fractions were also found to contain CIg.

DISCUSSION

The finding that CIg (fibronectin) is present in trophoblast and alveolar basement membranes was anticipated (see page 142). However, the failure to detect it in isolated GBM is puzzling, especially in view of the immunofluorescence data of Linder *et al.*[15] and of Tan and Kaplan.[24] One possible explanation is that CIg is present but is bound in such a way that it is either not released by collagenase and plasmin or is degraded by them. Another possibility is that fibronectin is present in structures, such as the connective tissue microfibrils, that associate with basement membranes and that, in the case of GBM, sonication caused the removal of the microfibrils. As seen ultrastructurally, both ABM[25] and all basement membranes of placenta[2,26] have numerous microfibrils in and around them. These microfibrils are easily demonstrated by uranyl acetate and lead citrate staining. Although numerous studies have been performed on the ultrastructure of GBM using the same techniques, microfibrils have seldom been observed in association with this basement membrane. Therefore, it is possible that CIg may be present in the tissue as a chemical constituent of connective tissue microfibrils. The recovery of the antigen in ASG fractions supports this idea, because Kadar *et al.*[27] have suggested that the microfibrils of connective tissues are composed of ASG. Kewley *et al.*[28] have shown that the microfibrillar protein associates with collagen fibrils and with elastic fibers. The release of CIg from tissues by collagenase digestion is consistent with this association. Finally, a presumptive subunit of elastic fiber microfibrils[29] has been shown to have a molecular weight of 220,000 (on gels cross-linked with methylene bisacrylamide), which is within the molecular weight estimates of CIg.[6] However, immunologic studies at the ultrastructural level with appropriate antisera will be required before morphologic assignments can be made.

The prominence of the CIg antigen in these isolated placental fractions, and the observation of Chen *et al.*[30] that CIg is concentrated in amniotic fluid relative to maternal and fetal plasmas, suggests that it may contribute to a specialized function of this organ. One possible function concerns the ability of the fetal trophoblast to create for itself a unique immunologic status, that of a homograft that does not elicit transplantation immunity, even in a presensitized mother. Two obvious mechanisms for achieving this status are the secretion by the trophoblast of a substance(s) that suppresses the cellular immune response of the mother and the elaboration by the trophoblast of a high-molecular-weight substance that in effect surrounds the cells with a physical barrier, such as was postulated by Medawar.[31]

In reality, a combination of mechanisms may be operative, because it would be difficult to make and maintain an intact physical barrier, and it would be dangerous to the mother if cellular immunity were suppressed too far. The facility of the association between CIg and fibrinogen suggests an attractive mechanism by which the trophoblast can produce high-molecular-weight complexes (compounds) that would provide a barrier between it and the maternal decidua.

Summary

Cold-insoluble globulin was detected in both trophoblast and alveolar basement membrane preparations, whereas it was not detected in GBM preparations. Acidic structural glycoproteins from both placental villi and lung parenchyma, which were extracted with 0.3 M acetic acid and recovered by adjusting the pH to 4.7, also contained CIg. Fractions of TBM, solubilized by either dilute alkali (0.01 N NaOH), by reduction and alkylation of the disulfide bonds, or by 0.3 M acetic acid extraction, all contained the antigen and possessed properties similar to those of ASG.

The ASG fractions also reacted with antifibrinogen, but proof that the two types of determinants occur on a given molecular species is lacking at present.

Purified collagenase solubilized CIg from ABM, from lung parenchyma, and from the stroma of placental villi, and this finding is strong evidence for an association of CIg with collagen in these connective tissues.

Acknowledgments

It is a pleasure to acknowledge the encouragement and support of Drs. G. M. Turino and Ines Mandl. I also thank Drs. George Wilner and M. W. Mosesson for many helpful suggestions and Dr. Mosesson for his generous gift of specific antiserum to CIg.

References

1. Bray, B. A., E. C. Leroy & K. C. Hsu. 1972. Human basement membrane from placenta and lung. Fed. Proc. **31:** 761 (Abst.).
2. Bray, B. A., K. C. Hsu, H. J. Wigger & E. C. Leroy. 1975. Association of fibrinogen and microfibrils with trophoblast basement membrane. Connect. Tissue Res. **3:** 55–71.
3. Gang, N. F. & M. M. Gelfand. 1972. Isolation of trophoblast basement membrane from full-term placenta. Proc. Soc. Exp. Biol. Med. **140:** 188–193.
4. Schwartz, E. S., N. F. Gang & M. M. Gelfand. 1974. Studies on human placenta. I. Isolation and partial characterization of a glycoprotein from the chorionic villus. Amer. J. Obstet. Gynecol. **118:** 857–863.
5. Mosesson, M. W. & R. A. Umfleet. 1970. The cold-insoluble globulin of human plasma. I. Purification, primary characterization, and relationship to fibrinogen and other cold-insoluble fraction components. J. Biol. Chem. **245:** 5728–5736.
6. Mosesson, M. W., A. B. Chen & R. M. Huseby. 1975. The cold-insoluble globulin of human plasma: studies of its essential structural features. Biochim. Biophys. Acta **386:** 509–524.
7. Ruoslahti, E. & A. Vaheri. 1975. Interaction of soluble fibroblast surface antigen with fibrinogen and fibrin. Identity with cold insoluble globulin of human plasma. J. Exp. Med. **141:** 497–501.
8. Keski-Oja, J., D. F. Mosher & A. Vaheri. 1976. Cross-linking of a major fibroblast

surface-associated glycoprotein (fibronectin) catalyzed by blood coagulation factor XIII. Cell **9:** 29–35.
9. Edsall, J. T., G. A. Gilbert & H. A. Scheraga. 1955. The non-clotting component of the human plasma fraction I-1 ("cold insoluble globulin"). J. Amer. Chem. Soc. **77:** 157–161.
10. Mosesson, M. W., R. W. Colman & S. Sherry. 1968. Chronic intravascular coagulation syndrome. N. Engl. J. Med. **278:** 815–821.
11. Mosher, D. F. 1975. Cross-linking of cold-insoluble globulin by fibrin-stabilizing factor. J. Biol. Chem. **250:** 6614–6621.
12. Pearlstein, E. 1976. Plasma membrane glycoprotein which mediates adhesion of fibroblasts to collagen. Nature (London) **262:** 497–500.
13. Bornstein, P. & J. F. Ash. 1977. Cell surface-associated structural proteins in connective tissue cells. Proc. Nat. Acad. Sci. USA **74:** 2480–2484.
14. Engvall, E. & E. Ruoslahti. 1977. Binding of soluble form of fibroblast surface protein, fibronectin, to collagen. Int. J. Cancer **20:** 1–5.
15. Linder, E., A. Vaheri, E. Ruoslahti & J. Wartiovaara. 1975. Distribution of fibroblast surface antigen in the developing chick embryo. J. Exp. Med. **142:** 41–49.
16. Timpl, R., I. Wolff & M. Weiser. 1969. Acidic structural proteins of connective tissue. I. Solubilization and preliminary chemical characterization. Biochim. Biophys. Acta **194:** 112–120.
17. Furthmayr, H. & R. Timpl. 1970. Acidic structural proteins of connective tissue. II. Evidence for disulfide cross-links. Biochim. Biophys. Acta **220:** 396–398.
18. Francis, G. & J. Thomas. 1975. Isolation and characterization of acidic structural glycoproteins in pulmonary tissues. Biochem. J. **145:** 299–304.
19. Bray, B. A. & G. M. Turino. 1976. Isolation of both acidic structural glycoproteins (ASG) and intact collagen from human lung parenchyma. Fed. Proc. **35:** 1741 (Abst.).
20. Bray, B. A. & E. C. Leroy. 1976. Human alveolar basement membrane. Chemical and immunologic comparisons with glomerular basement membrane and trophoblast basement membrane. Microvasc. Res. **12:** 77–89.
21. Peterkofsky, B. & R. Diegelmann. 1971. Use of a mixture of proteinase-free collagenases for the specific assay of radioactive collagen in the presence of other proteins. Biochemistry **10:** 988–994.
22. Krull, L. H., D. E. Gibbs & M. Friedman. 1971. 2-Vinylquinoline, a reagent to determine protein sulfhydryl groups spectrophotometrically. Anal. Biochem. **40:** 80–85.
23. Laurell, C. B. 1966. Quantitative estimation of proteins by electrophoresis in agarose gel containing antibodies. Anal. Biochem. **15:** 45–52.
24. Tan, E. M. & M. H. Kaplan. 1963. Immunological relation of basement membrane and a serum beta globulin in the mouse. Demonstration of renal basement membrane alteration in mice injected with streptolysin S. Immunology **6:** 331–344.
25. Low, F. N. 1962. Microfibrils: fine filamentous components of the tissue space. Anat. Rec. **142:** 131–138.
26. Verbeek, J. H., E. M. Robertson & M. D. Haust. 1967. Basement membranes (amniotic, trophoblastic, capillary) and adjacent tissue in term placenta. Amer. J. Obstet. Gynecol. **99:** 1136–1146.
27. Kadar, A., B. Robert & L. Robert. 1973. Etude biochimique et électromicroscopique des microfibrilles du tissue conjonctif. Pathol.-Biol. **21**(Suppl.); 80–88.
28. Kewley, M. A., F. S. Steven & G. Williams. 1977. Preparation of a specific antiserum towards the microfibrillar protein of elastic tissues. Immunology **32:** 483–489.
29. Muir, L. W., P. Bornstein & R. Ross. 1976. A presumptive subunit of elastic fiber microfibrils secreted by arterial smooth-muscle cells in culture. Eur. J. Biochem. **64:** 105–114.
30. Chen, A. B., M. W. Mosesson & G. I. Solish. 1976. Identification of the cold-insoluble globulin of plasma in amniotic fluid. Amer. J. Obstet. Gynecol. **125:** 958–961.
31. Medawar, P. B. 1953. Some immunological and endocrinological problems raised by the evolution of viviparity in vertebrates. Soc. Exp. Biol. Symp. **VII:** 320–338.

DISTRIBUTION OF FIBRONECTIN IN HUMAN TISSUES AND RELATIONSHIP TO OTHER CONNECTIVE TISSUE COMPONENTS*

E. Linder,† S. Stenman,‡ V.-P. Lehto,‡ and A. Vaheri§

Departments of † Immunology and Bacteriology of ‡ Pathology and of § Virology University of Helsinki 00290 Helsinki 29, Finland

The cell surface-associated and intercellular fibrillar distributions of fibronectin [fibroblast surface (SF) antigen] have been well substantiated by results from numerous studies on cultured fibroblasts.[1-4]

The *in vitro* distribution of fibronectin (FN) seems to correspond to its *in vivo* localization in loose undifferentiated connective tissue of the developing fetus.[5] In addition, FN *in vivo* is associated with limiting membranes.[5,6] The *in vivo* localization suggests that FN is an extracellular connective tissue (CT) component.

This study was undertaken to focus on the tissue distribution of FN in relation to other connective tissue components (TABLE 1) under normal and in some pathologic conditions.

MATERIALS AND METHODS

Antibodies against FN were raised in sheep by immunization with purified human plasma fibronectin.[7]

Antibodies against collagen were produced by immunization of rabbits with collagen purified from human fetal skin.[8] The antihuman skin collagen antiserum reacted with both types-I and -III collagen.[9] This finding was established by use of specific rabbit antibodies against interspecies cross-reacting calf types-I and -III collagen as reference sera.

Antibodies against connective tissue microfibrils were spontaneous human autoantibodies.[10]

Human tissues were obtained from autopsies, surgical specimens, and legal abortions. Cryostat sections fixed in cold acetone were used for immunofluorescence microscopic studies.

The immunofluorescence (IF) techniques employed have been described previously.[5,10] Double-staining experiments were performed with fluorescein- and rhodamine-conjugated antibodies. For direct and indirect IF staining, a Zeiss standard microscope equipped with epiilluminator IV FL was used.

Histologic staining was performed as follows. Cryostat sections of formalin-fixed tissues processed for routine light microscopic examination were stained for various connective tissue components: Van Gieson's picric acid fuchsin and Masson's trichrome for collagen, Gomori's method for reticulin, and Unna-

*Supported by the Finnish Culture Foundation, the Finnish Medical Research Council, and the Sigrid Jusélius Foundation.

0077-8923/78/0312-0151 $01.75/0

TABLE 1

METHODS THAT DIFFERENTIATE AMONG FIBRILLAR CONNECTIVE TISSUE COMPONENTS

Histology	Electron Microscopy	Immunofluorescence
Collagen fibers	elastic fibers	collagen fibers
Elastic fibers	elastin	"reticulin fibers"
Reticulin fibers	microfibrils	"reticulin"
		"fibronectin"
		collagen type III
		microfibrils

Taenzer's orcein, Verhoeff's iron hematoxylin, Weigert's elastin, the iron gallein elastic method, and Gomori's aldehyde fuchsin for elastic fibers.

RESULTS

Localization of Fibronectin in Normal Tissues

Fibronectin was seen in loose "reticular" connective tissue of various organs. Fibronectin was also seen in various limiting membranes, both in epithelio-mesenchymal basement membranes and vascular walls and in capillary walls (FIGURES 1–4). In some organs, for example, the kidney, there seemed to be a reduced amount of FN in adult as compared to fetal tissue. This difference may be related to a decrease in the amount of loose connective tissue during differentiation.

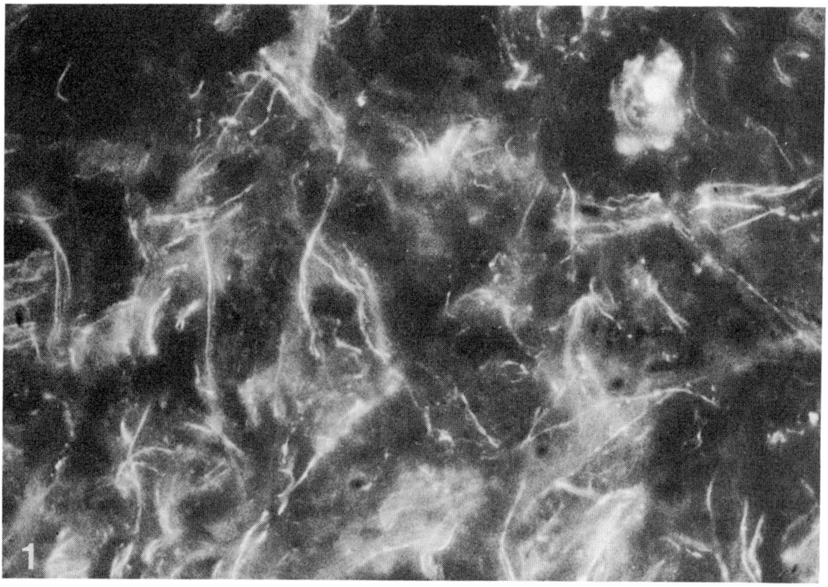

FIGURE 1. Fibronectin in fetal dermis is distributed as thin fibrils. ×450.

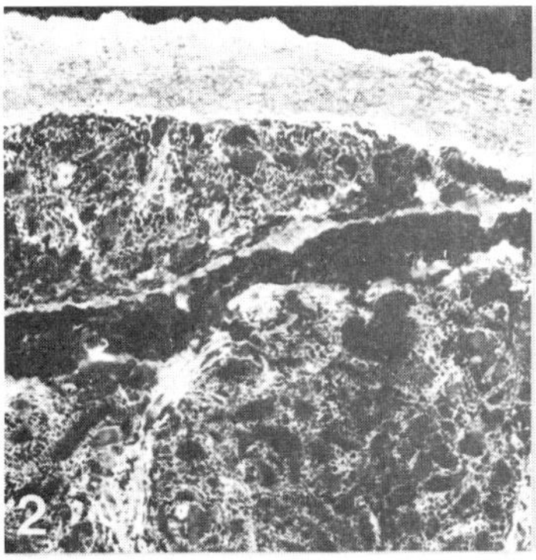

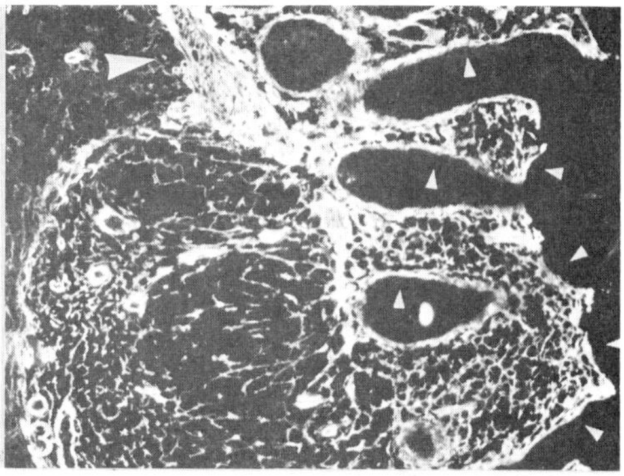

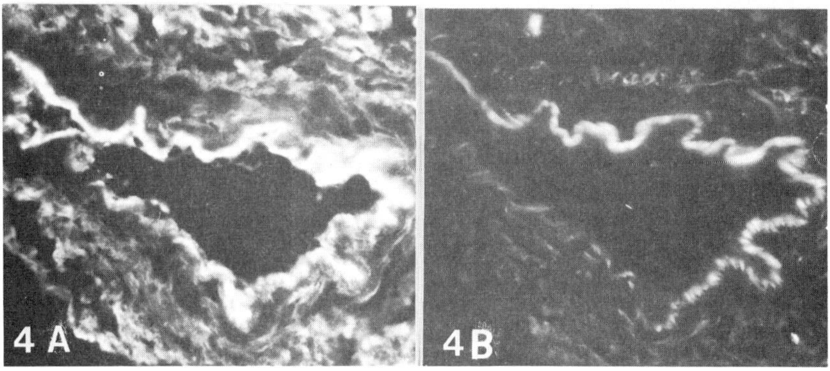

FIGURE 2. Fibronectin in human spleen. The capsule contains large amounts of fibronectin, and in the stroma fibronectin is detected as a network that corresponds to the reticulin stroma. ×58.

FIGURE 3. Fibronectin in normal mucosa of the small intestine is detected in epithelial basement membranes (small arrowheads), around smooth muscle cells of the muscularis mucosae (large arrowhead), in the loose connective tissue of the lamina propria, and as a loose network within a Peyer's patch (lower left corner). × 58.

FIGURE 4. Fibronectin and elastic fibers in a nonmuscular artery. Fibronectin is visualized by indirect IF and the elastic fibers of the same section by autofluorescence by use of excitation with green light from a high-pressure mercury lamp. Fibronectin is distributed in the basal lamina region of the endothelium (FIGURE 4A). The autofluorescence of the internal elastic lamina is outside this region (FIGURE 4B). Fibronectin occurs also in the vessel wall. × 144.

Codistribution with Fibrillar Connective Tissue Components

Based on brief examination, the similar distributions of FN and collagen were evident (FIGURE 5). Double staining of the same tissue sections revealed partial codistribution between FN fibrils and collagen bundles. In fetal connective tissue, an association of FN and microfibrils was occasionally observed (FIGURE 6). Usually, there was no association between microfibrils and fibronectin (FIGURE 7). Localizations of FN not shared by other fibrillar CT components were capillary walls (FIGURE 7) and glomerular mesangial cells (FIGURE 8).

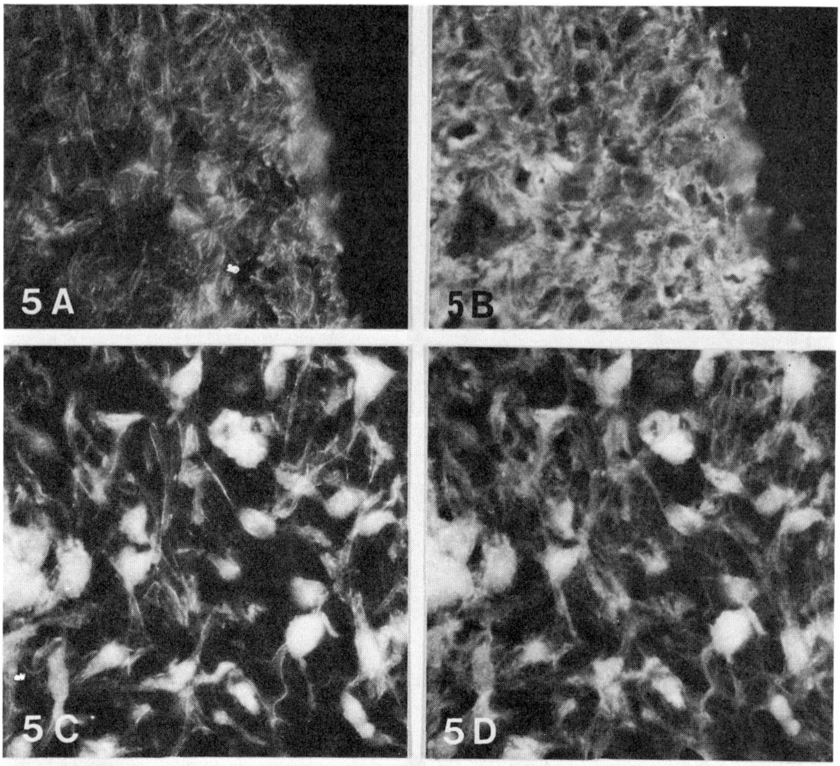

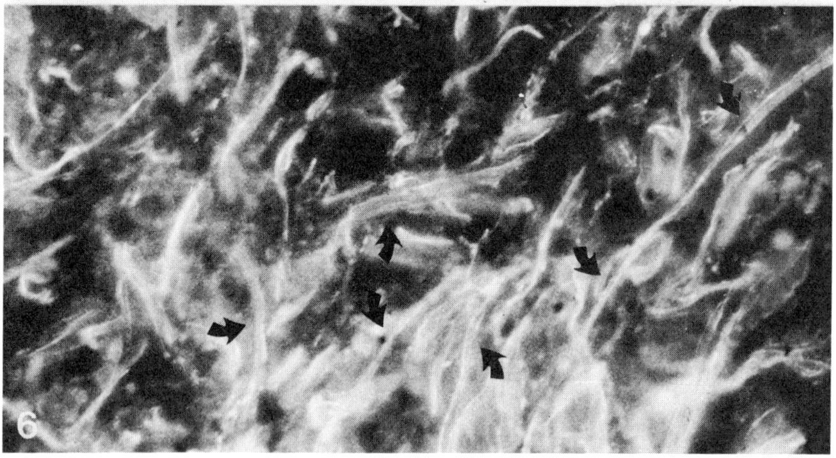

FIGURE 5. Fetal dermis stained with antibodies against fibronectin (FIGURES 5A & 5C), a crude preparation of skin collagen (FIGURE 5B), and type-III collagen (FIGURE 5D). The codistribution between fibronectin and type-III collagen is extensive. FIGURES 5A & 5B, ×450; FIGURES 5C & 5D, ×720.

FIGURE 6. Double staining of fibronectin (rhodamine fluorescence) and microfibrils (fluorescein isothiocyanate fluorescence) in fetal dermis. Occasionally, fine fibers of fibronectin (arrows) are associated with thicker microfibrils. ×900.

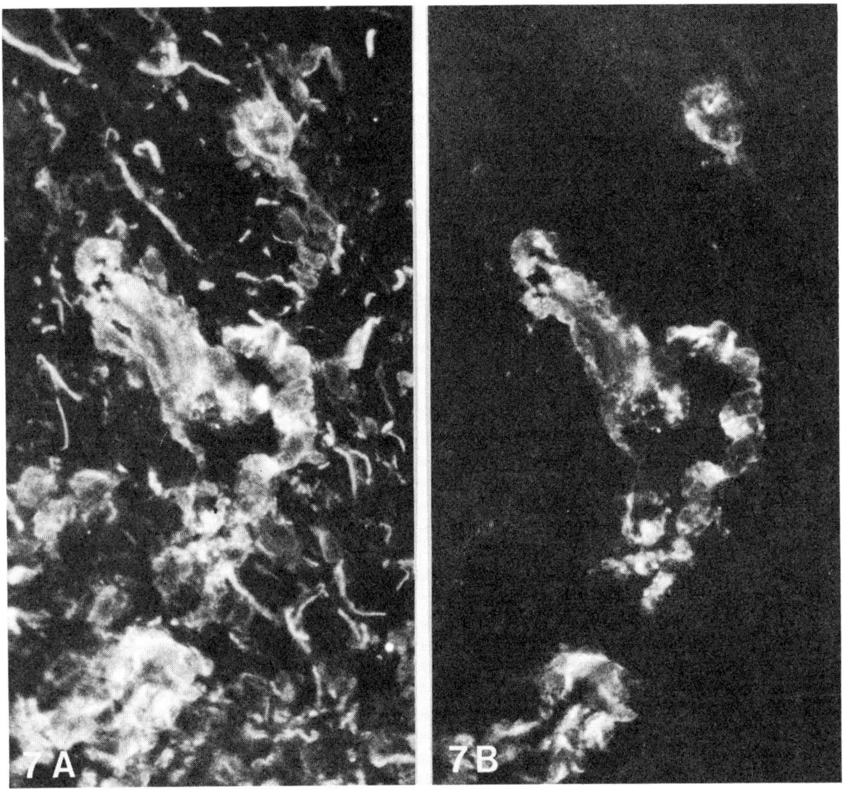

FIGURE 7. Double staining (see FIGURE 6) of microfibrils and fibronectin (FIGURE 7A) in fetal dermis reveals no association (see text). Fibronectin is detected only in the capillary walls (FIGURE 7B). ×900.

TABLE 2
DIFFERENTIATION OF FIBRILLAR CONNECTIVE TISSUE COMPONENTS BY IMMUNOFLUORESCENCE

	Reference
Collagen type	
I	9
II	
III	9
IV	
Glycoproteins	
"reticulin"	13
microfibrils of elastic fibers	10
fibronectin	5

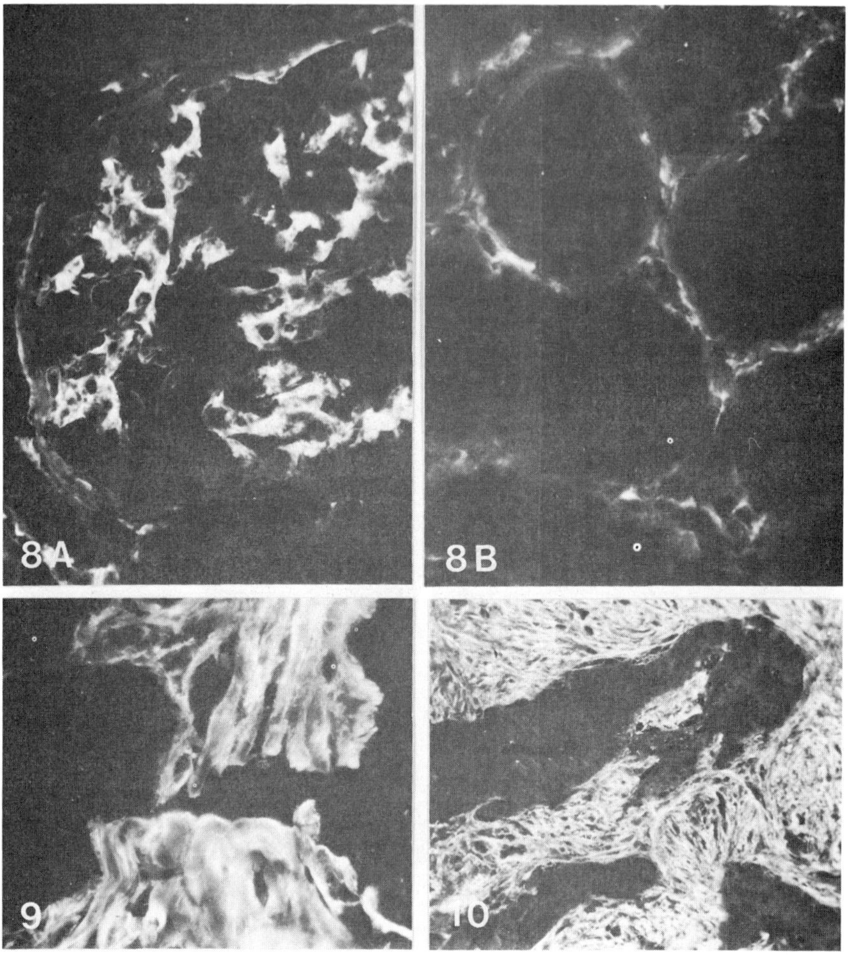

FIGURE 8. Fibronectin in normal kidney glomerulus (FIGURE 8A) and medulla (FIGURE 8B). Mesangial cells contain fibronectin; basement membranes are negative. The fibronectin detected around tubules (FIGURE 8B) is probably located in capillary walls. ×495.

FIGURE 9. Fibronectin in the connective tissue stroma of an adenocarcinoma of the rectum. ×144.

FIGURE 10. Fibronectin in the stroma of a breast carcinoma. ×360.

Distribution of Fibronectin in Pathologic Conditions

Increased amounts of FN were observed in connective tissues surrounding epithelial tumors and in inflammatory reactions. The connective tissue stroma of solid tumors stained intensely (FIGURES 9 & 10). In the kidney, increased mesangial staining was seen in some cases of glomerulonephritis, and increased interstitial and basement membrane staining was noted in a case of acute rejection of a kidney graft (FIGURE 11). In the latter case, the enlarged intertubular connective tissue spaces were filled with fibrin. In the glomerulonephritic kidneys, deposits of complement were found in the mesangium.

Correlation between Histologic Staining and Fibronectin Distribution

The IF microscopic localization of FN was very similar to that of reticulin fibers (FIGURE 12), and reticulin fibers were shown to contain FN. However,

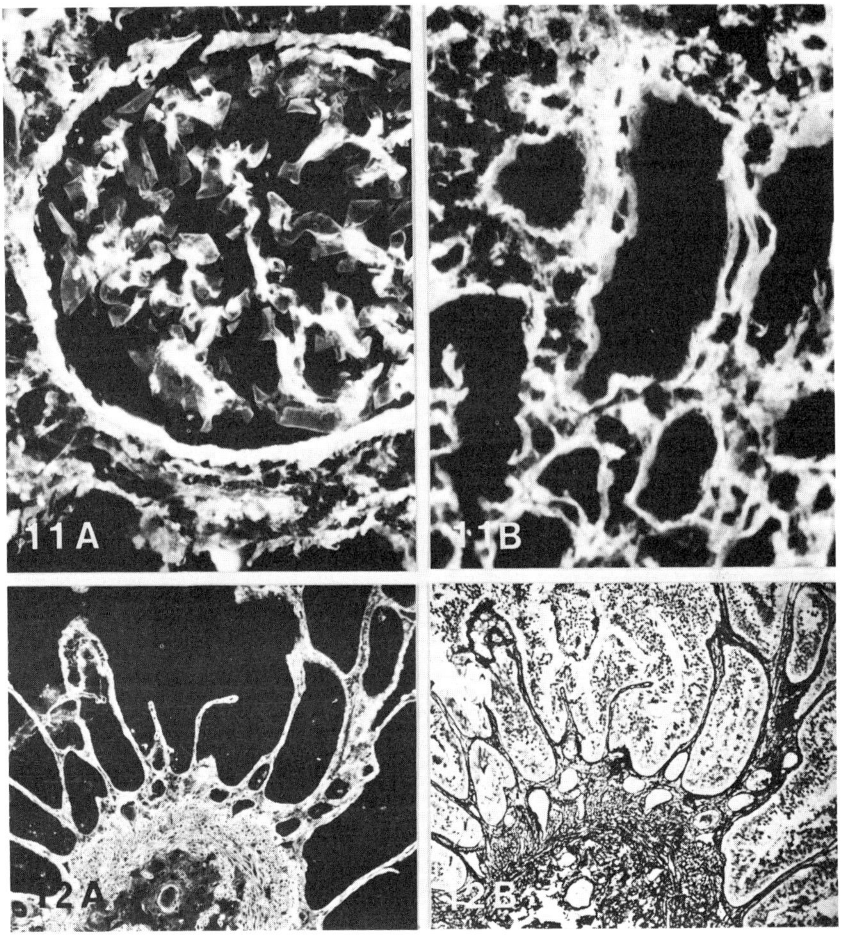

FIGURE 11. Fibronectin in an allotransplanted human kidney rejected hyperacutely 12 days after transplantation. Tubular basement membranes and Bowman's capsule contain large amounts of fibronectin. In normal kidney, these structures contain very little, if any, fibronectin (see FIGURE 8).

FIGURE 12. Fibronectin and reticulin in a villous adenoma of the rectum. Fibronectin was visualized with indirect IF (FIGURE 12A), and the same section was subsequently stained for reticulin by use of alkaline silver impregnation (FIGURE 12B). Fibronectin and reticulin codistribute. ×58.

reticulin fibers that did not contain FN could also be demonstrated, for example, in the kidney. The distributions of collagen and elastin were different from that of FN.

Discussion

The location of CT components can be studied by histological and electron microscopic techniques. To confine the location of specific protein components to structures so detected, immunohistologic methods are, however, necessary. In this work, we have used antibodies against collagen and microfibrils in combination with antibodies against fibronectin to study the distribution of fibronectin in relationship to other connective tissue components.

The results confirm earlier localization findings[5,11] and suggest that FN is an important noncollagenous glycoprotein constituent of the extracellular connective tissue.[12] Histologic studies demonstrated that FN is a component of reticulin fibers. This result is compatible with the fact that reticulin stains react with glycoproteins.[13,14] This result is also consistent with the codistribution of FN and type-III collagen, which has been shown to be a constituent of reticulin.[9] The observation that FN is associated partly with microfibrils of fetal connective tissue is compatible with earlier observations[15] but suggests that FN is a minor component of microfibrils.

Several observations suggest that actively proliferating connective tissue cells contain fibronectin: increased amounts of FN are present in fetal connective tissues, around tumor cells, and in inflammation. The observed presence of both FN and fibrin in the rejected kidney graft is compatible with the observed affinity of FN for fibrin[7] and suggests an alternative mechanism for tissue distribution of FN under pathologic conditions: it is possible that the tissue repair mechanism that follows fibrin deposition is dependent on the transglutaminase (factor XIII) -induced polymerization of circulating plasma FN.[7]

References

1. Ruoslahti, E., A. Vaheri, P. Kuusela & E. Linder, 1973. Fibroblast surface antigen: a new serum protein. Biochim. Biophys. Acta **322:** 352.
2. Wartiovaara, J., E. Linder, E. Ruoslahti & A. Vaheri. 1974. Distribution of fibroblast surface antigen. Association with fibrillar structures of normal cells and loss upon viral transformation. J. Exp. Med. **140:** 1522.
3. Chen, L. B. 1977. Alteration in cell surface LETS protein during myogenesis. Cell **10:** 393.
4. Furcht, L. T. , D. F. Mosher & G. Wendelschafer-Crabb. 1978. Submitted for publication.
5. Linder, E., A. Vaheri, E. Ruoslahti & J. Wartiovaara. 1975. Distribution of fibroblast surface antigen in the developing chick embryo. J. Exp. Med. **142:** 41.
6. Wartiovaara, J., S. Stenman & A. Vaheri. 1976. Changes in expression of fibroblast surface antigen (SFA) in induced cytodifferentiation and in heterokaryon formation. Differentiation **5:** 85.
7. Mosher, D. F. 1975. Cross-linking of cold-insoluble globulin by fibrin-stabilizing factor. J. Biol. Chem. **250:** 6614.
8. Rothbard, S. & R. F. Watson. 1972. Demonstration of collagen in human tissues by immunofluorescence. Lab. Invest. **27:** 76.
9. Nowack, H., S. Gay, G. Wick, U. Becker & R. Timpl. 1976. Preparation and use in

immunohistology of antibodies specific for type I and type III collagen and procollagen. J. Immunol. Methods **12:** 117.
10. LINDER, E. & V.-L. LEHTO. 1977. Antibodies against microfibrils of developing connective tissue in patients with inflammatory conditions. Scand. J. Immunol. In Press.
11. STENMAN, S. & A. VAHERI. 1978. Distribution of a new major connective tissue protein, fibronectin, in normal human tissues. J. Exp. Med. In press.
12. ANDERSON, J. C. 1976. Glycoproteins of the connective tissue matrix. Int. Rev. Connect. Tissue Res. **7:** 251.
13. PRAS, M. & L. E. GLYNN. 1973. Isolation of a non-collagenous reticulin component and its primary characterization. Brit. J. Exp. Pathol. **54:** 449.
14. LILLIE, R. D. & H. M. FULLMER. 1976. Histopathologic Technic and Practical Histochemistry. 4th edit. McGraw-Hill Book Company. New York, N.Y.
15. MUIR, L., P. BORNSTEIN & R. ROSS. 1976. A presumptive subunit of elastic fiber microfibrils secreted by arterial smooth-muscle cells in culture. Eur. J. Biochem. **64:** 105.

PART III. CHARACTERISTICS OF THE HIGH-MOLECULAR-WEIGHT FIBROBLAST SURFACE-ASSOCIATED GLYCOPROTEIN

CHEMICAL COMPOSITION, GROSS STRUCTURE, AND ORGANIZATION OF TRANSFORMATION-SENSITIVE GLYCOPROTEINS*

William G. Carter,† Minoru Fukuda,
Clifford Lingwood,‡ and Sen-itiroh Hakomori

Biochemical Oncology
Fred Hutchinson Cancer Research Center
and Departments of Pathobiology and Microbiology
University of Washington
Seattle, Washington 98104

In our laboratory, we have been studying the surface-labeled galactoproteins of hamster NIL and NILpy cells since we introduced the surface labeling procedure with galactose oxidase and NaB^3H_4 in 1973.[1] Three classes of transformation-sensitive, or transformation-dependent, surface-labeled glycoproteins have been detected. Some of them have been isolated, and their chemical and organizational properties have been studied.

A glycoprotein with a subunit molecular weight of 230,000 was termed as galactoprotein a (Gap a), which was deleted in transformed cells and showed high sensitivity to exogenous treatment of cells with trypsin.[2-4] In contrast, increased amount of label was found in a glycoprotein with a molecular weight of 130,000 daltons in NILpy cells and was termed galactoprotein b (Gap b).[2-4] The Gap b fraction of transformed NIL cells contained a glycoprotein whose mobility on gel electrophoresis is faster than normal Gap b and was called "Gap bT."[4]

Recently, a new glycoprotein with a subunit molecular weight of 170,000 daltons was found as a transformation-sensitive but protease-resistant glycoprotein. This glycoprotein was labeled only faintly by surface labeling procedures but labeled strongly by metabolic labeling. The glycoprotein termed as 170K was lost in transformed cells, similar to Gap a.

A similar or identical glycoprotein as Gap a has been described in various laboratories during 1973–4 and was variously termed LETS (large external transformation-sensitive protein),[5] zeta,[6] SFA,[7] CSP,[8] fibronectin,[9] and so on (for review, see Hynes[10]). The Gap a increased at cell contact, at the G_1 phase of the cell cycle,[11-13] and interacted with concanavalin A, *Ricinis communis* lectin,[3,4] collagen,[14,15] and possibly with fibrinogen or fibrin.[16] Both Gap a and b have been isolated recently by affinity chromatography on *R. communis* (RCA) columns after sequential solubilization of cells.[4] A serum glycoprotein that is immunologically identical to Gap a has been detected in conditioned culture media and presumably is shed from the cells as a soluble form of Gap a. This soluble form of Gap a shall be referred to as "released-form Gap a."

Both Gap a and 170K seem to be associated with particular cytoskeletal proteins. A possible association of 170K with an "intermediate filament protein" (IFP) is now suggested by affinity copurification and coprecipitation (or copoly-

*Supported by National Cancer Institute Contract CB43920.
†Recipient of a Cystic Fibrosis Foundation Fellowship.
‡Present address: Hospital for Sick Children, Toronto, Ontario, Canada.

merization). Some data are available for a possible association of Gap a and actin,[4] although this possiblity needs further extensive study.

Galactoprotein a (Gap a, or LETS)

Gap a in hamster embryonic fibroblasts was not readily extractable by non-ionic detergents or by hypotonic urea, in contrast to a previous report[17]; it was extracted by high concentrations of urea (5–8 M) at elevated temperature from the insoluble cell residue that remained after sequential exhaustive extraction with sucrose-ATP and with zwitterionic detergent Empigen® BB (see Figure 1). The glycoprotein was isolated from the diluted urea extract by affinity chromatography on an RCA-polyacrylhydrazido agarose (PAHA) column, as shown in Figures 1 and 2.

On gel electrophoresis of isolated Gap a, in the absence of 2-mercaptoethanol, a large quantity of material either did not enter the 5% stacking gel or migrated

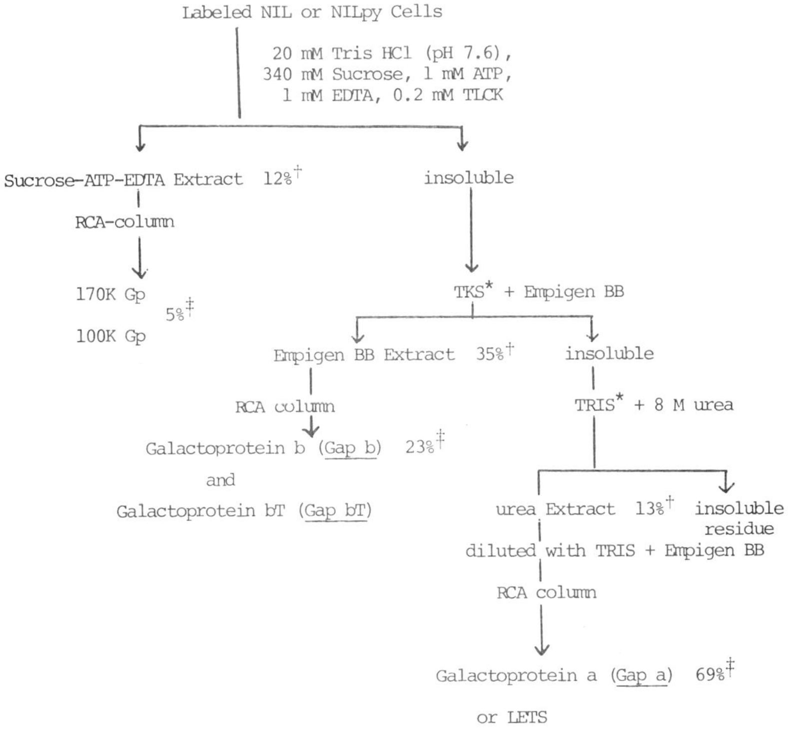

Figure 1. Sequential extraction of surface-labeled galactoproteins of NIL and NILpy cells. Abbreviations: *TKS, 50 mM Tris HCl buffer (pH 7.6) that contains 25 mM KCl, 5 mM $CaCl_2$, and 5 mM $MgCl_2$; TLCK, *N*-α-*p*-tosyl-L-lysine chloromethyl ketone hydrochloride; Tris, tris(hydroxymethyl)aminomethane. †Percentage of activity of each extract to the total label on cells. ‡Percentage of activity of glycoprotein isolated by the RCA-affinity column to the total activity of each extract. §For details of the procedure, see Carter and Hakomori.[4]

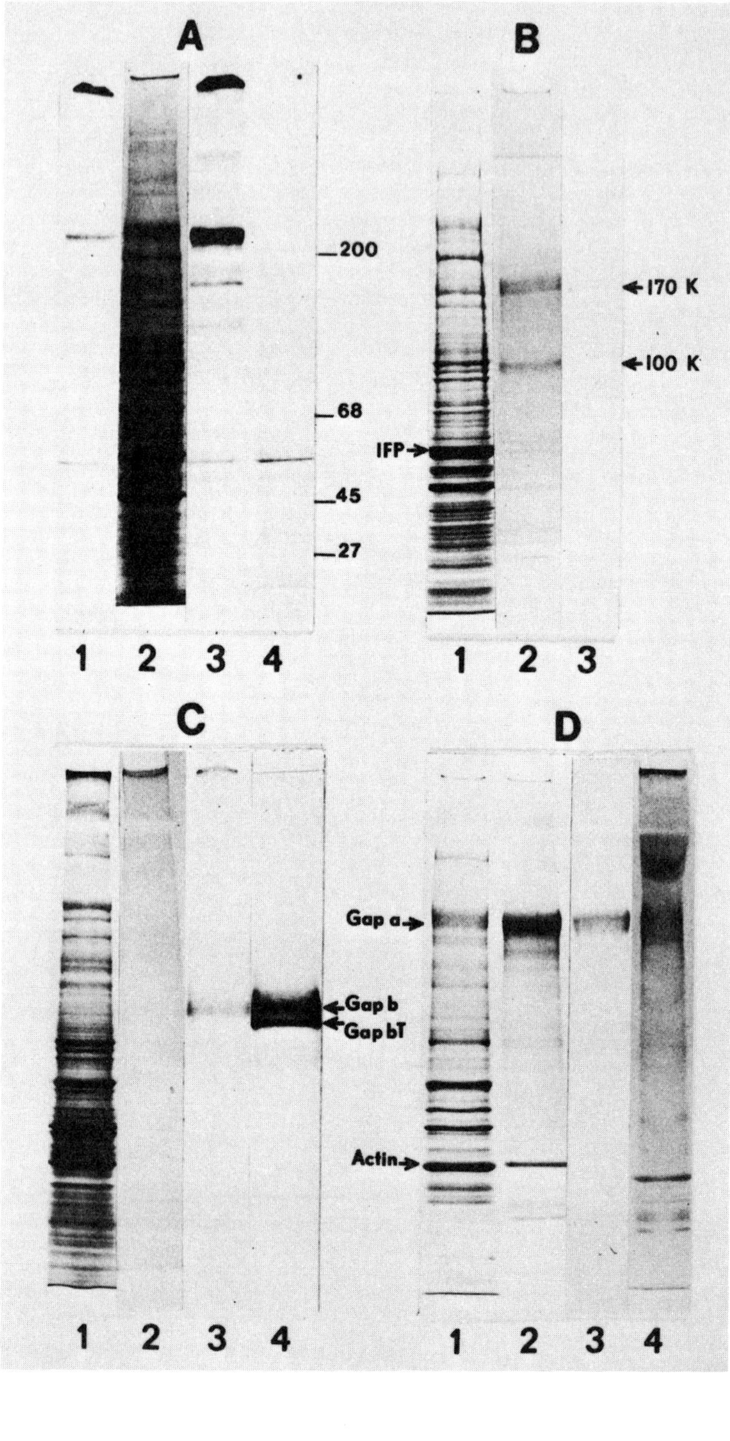
A
B
200
68
45
27
170 K
100 K
IFP
1 2 3 4
1 2 3
C
D
Gap a
Gap b
Gap bT
Actin
1 2 3 4
1 2 3 4

with an apparent molecular weight greater than 300,000 daltons (see FIGURE 2,D, gel 4). This finding indicates that Gap a is present as a polymeric form, with subunits connected through disulfide bonds, in agreement with recent reports.[18,19]

For carbohydrate studies, hamster embryo fibroblasts were labeled metabolically with [^{3}H]glucose to constant specific activity[20] prior to isolation of Gap a by affinity chromatography on RCA-PAHA columns. Monitoring of radioactivity in detected sugar residues of isolated Gap a was utilized to ensure that the sugar was not from a noncellular contaminant. Pronase digestion of the isolate, metabolically labeled Gap a, followed by gel filtration chromatography on Biogel® P-10, resolved two major radioactive peaks (FIGURE 3, *top*). Peak II contained approximately 50% of the applied radioactivity and 100% of the detectable free amino groups. Carbohydrate analysis of peak II detected only trace quantities of mannose, galactose, glucose, and no sialic acid, and this peak was therefore not subject to further analysis. Peak I was subject to a second pronase digestion and rechromatography on Biogel P-6 (FIGURE 3, *bottom*). A single major radioactive peak was obtained and was subject to amino acid and carbohydrate analyses (TABLE 1). Our results suggest that Gap a, with a subunit molecular weight of 230,000, contains 5.5% carbohydrate by weight, essentially all of which is contained in asparagine-linked glycopeptides of molecular weight 2000. The glycopeptide possesses a carbohydrate composition (normalized to 1 mol of asparagine) of fucose$_{0\ \mathrm{or}\ 1}$, mannose$_{2\ \mathrm{or}\ 3}$, galactose$_2$, glucosamine$_{3\ \mathrm{or}\ 4}$, sialic acid$_{0\ \mathrm{or}\ 1}$. We can calculate that each Gap a subunit contains five or six asparagine-linked carbohydrate chains. The carbohydrate composition and an observed insensitivity of the glycopeptide to digestion by *endo-β-N*-acetylglucosaminidase H[22-24] suggest that the glycopeptide is of the complex type.

The amino acid analysis of the isolated Gap a glycoprotein is presented in TABLE 2.

RELEASED-FORM GALACTOPROTEIN a

To investigate the chemical properties of Gap a, we isolated released-form Gap a, which was produced by cells and released into culture medium, because

---

FIGURE 2. Electrophoretic patterns of isolated RCA receptors from galactose oxidase-NaB^3H_4-labeled NIL and NILpy cells on pH 7.6 polyacrylamide gradient (5–14%) slab gels in the presence of 0.1% sodium dodecyl sulfate. **A:** Pattern of total cells: 1, Fluorograph of NIL cells reduced with NaB^3H_4 (no galactose oxidase); 2, protein stain of NIL cells; 3, fluorograph of gel 2; 4, fluorograph of NIL cells labeled with galactose oxidase-NaB^3H_4, followed by trypsin digestion (10 μg of trypsin/ml) for 10 min at 37°C; 200, 68, 45, and 27 indicate mobilities of proteins with known molecular weights (rabbit muscle myosin, 200,000 daltons; bovine serum albumin, 68,000; rabbit muscle actin, 45,000; *Dolichos biflorus* lectin subunit, 27,000). **B:** Pattern of sucrose-ATP-EDTA extract (see FIGURE 1): 1, Protein stain of sucrose-ATP-EDTA extract of labeled NIL cells; 2, protein stain of RCA receptors isolated from 1; 3, fluorograph of gel 2. Arrows indicate light bands visible on original fluorograph. **C:** Pattern of Empigen BB extract (see FIGURE 1): 1, Protein stain of Empigen BB extract of labeled NIL cells; 2, protein stain of RCA receptor isolated from 1 (Gap b); 3, fluorograph of gel 2; 4, fluorograph of RCA receptor isolated from Empigen BB extract of labeled NILpy cells. Arrows indicate extra band present in NILpy cells. **D:** Pattern of urea extract (see FIGURE 1): 1, Protein stain of urea extract of labeled NIL cells; 2, protein stain of RCA receptor isolated from 1 (Gap a); 3, fluorograph of gel 2; 4, protein stain of RCA receptor isolated from 1 (sample not reduced with 2-mercaptoethanol). Arrow indicates mobility of isolated hamster muscle actin.

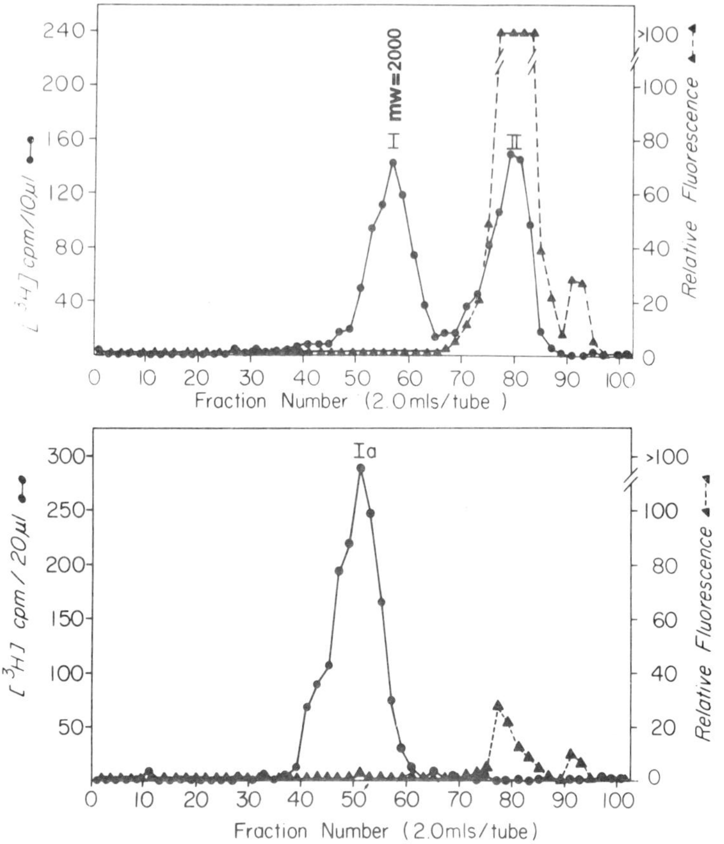

FIGURE 3. Fractionation of glycopeptides that result from pronase digestion of metabolically labeled, isolated Gap a by gel filtration chromatography on Biogels P-10 and P-6. Galactoprotein a was isolated from NIL cells after metabolic labeling with [^{3}H]Glc for 3 days. The labeled Gap a was pronase digested (96 hr) and subjected to gel filtration chromatography on Biogel P-10 (*top*). Carbohydrate analysis of peak II detected only trace quantities of carbohydrate, and peak II was therefore not subject to further analysis.

Peak I (relative molecular weight 2000) was subjected to a second pronase digestion and chromatographed on Biogel P-6 (*bottom*). Peak Ia (relative molecular weight 2000) was collected and utilized for further studies.

the released form of Gap a is more abundant than is cellular Gap a. Cells were metabolically labeled with either [^{3}H]Glc, [^{3}H]GlcN, [^{3}H]Fuc, [^{14}C]Man, or [^{14}C]Gal, and the culture medium was fractionated by $(NH_4)_2SO_4$ precipitation. Released-form Gap a was purified by affinity chromatography by use of insolubilized gelatin columns, as described previously.[25] Isolated glycoprotein appears as a single band on sodium dodecyl sulfate polyacrylamide gel electrophoresis, with a molecular weight approximately the same as that of cellular Gap a. The amino acid composition of this purified glycoprotein is shown in TABLE 2 and is compared to that of cellular Gap a. No significant difference was observed.

Among the various radioactive sugars utilized, [^{3}H]GlcNac is incorporated most efficiently and specifically. [^{3}H]Fucose and [^{14}C]Man are incorporated rather specifically, whereas [^{3}H]Glc and [^{14}C]Gal are metabolized to other components during labeling.

Biogel P-6 column chromatography of the pronase digest of isolated metabolically labeled released-form Gap a is shown in FIGURE 4. It is apparent that the carbohydrate moiety is labeled more specifically by [^{3}H]GlcNac than by [^{3}H]Glc. FIGURE 4 (*bottom*) also shows that most of the carbohydrate chains are resistant to alkaline sodium borohydride treatment, thus indicating that more than 80% of the carbohydrate units seem to be an asparagine-linked type. The carbohydrate composition of this glycopeptide (TABLE 3) supports this conclusion. Based on the molecular weight and carbohydrate composition of the glycopeptide(s), the carbohydrate unit of released-form Gap a seems to have the composition of sialic $acid_{1\ or\ 0}$, $Gal_{2\ or\ 3}$, $GlcN_4$, Man_3, $Fuc_{1\ or\ 0}$.

Preliminary structural studies were performed employing glycosidase digestion and permethylation of the isolated glycopeptide(s). Most of the galactose is released by β-galactosidase without pretreatment with neuraminidase; however,

TABLE 1

CARBOHYDRATE COMPOSITION OF GALACTOPROTEIN a AND A MAJOR GLYCOPEPTIDE ISOLATED FROM GALACTOPROTEIN a*

	Glycopeptide		Glycoprotein
Residue	Moles of Sugar/ Mole of Asparagine	Moles of Sugar/ Mole of Gap a†	Moles of Sugar/ Mole of Gap a‡
Fucose	0.6	4.0	5.9
Mannose	2.3	14.8	16.8
Galactose	1.9	12.0	11.5
Glucosamine	3.5	22.3	22.4
Galactosamine	trace	trace	5.2
Sialic acid§	0.3	2.1	2.0
Glucose	0	0	contaminant¶
Asparagine‖	1.00		
Proline	0.85		
Threonine	0.64		
Alanine	0.47		
Glycine	0.39		
Serine	0.32		
Other amino acids	trace		

*The carbohydrate content was determined by gas-liquid chromatography and mass spectrometry of the alditol acetate derivatives of sugars released from known dry weights of glycoprotein after acetolysis. A subunit molecular weight of 230,000 was utilized for calculations. Glycoproteins and glycopeptides were isolated from cells labeled metabolically with [^{3}H]glucose, and the cellular origin of sugars was confirmed by radioactivity detection.

†Values based on recovery of radioactive glycopeptide from Gap a of known specific activity.

‡Values based on the average of three determinations.

§Sialic acid was determined by fluorometric assay.[21]

¶Variable trace quantities of glucose was detected; however, the glucose was not radioactive and therefore not of cellular origin.

‖Amino acid analysis performed by AAA Laboratory, Seattle, after 24-hr 6 N HCl hydrolysis at 110°C; serine increased by 10%, and theronine increased by 5% to compensate for destruction by acid.

TABLE 2

AMINO ACID COMPOSITION OF GALACTOPROTEIN a AND GALACTOPROTEIN b ISOLATED FROM NIL CELLS* (EXPRESSED IN MOLES/100 MOLES)

	Gap a	Released-Form Gap a	Gap b
Asparagine	9.5	10.3	11.1
Threonine	10.4	10.0	6.0
Serine	9.6	7.3	7.7
Glutamine	12.4	11.6	11.1
Proline	7.5	10.0	5.2
Glycine	10.2	10.5	8.5
Alanine	7.1	5.2	7.1
Valine	7.0	7.7	5.9
Isoleucine	4.7	5.5	4.3
Leucine	5.8	6.7	8.3
Tyrosine	3.8	3.7	3.0
Phenylalanine	2.4	2.1	4.4
Histidine	2.2	1.6	2.3
Lysine	3.1	3.2	4.5
Arginine	4.6	4.5	5.3
Methionine†	ND	ND	1.9
Half-cystine‡	ND	ND	3.5
Totals	100.0	99.9	100.1

*Twenty-four-hour 6 N HCl hydrolysis at 110°C; serine increased by 10%, and threonine increased by 5% to compensate for destruction by acid. Analysis performed by AAA Laboratory, Seattle.

†Methionine plus methionine sulfoxide.

‡Performic acid oxidized prior to acid hydrolysis. Calculated from cystic acid/alanine ratio.

release of galactose was slightly increased with prior removal of sialic acid (FIGURE 5). *N*-Acetylglucosamine was then released by β-*N*-acetylglucosaminidase. The glycopeptide(s), which was treated by neuraminidase, β-galactosidase, and β-*N*-acetylglucosaminidase sequentially, was further hydrolyzed by *endo*-β-*N*-acetylglucosaminidase D.[22,24] In contrast, the glycopeptide(s) was not hydrolyzed by either *endo*-β-*N*-acetylglucosaminidase H[23] or *endo*-β-galactosidase.[26]

Permethylation studies[27] support the above results and gave us further information (TABLE 4). By comparing the permethylated sugars from intact and neuraminidase and galactosidase-treated glycopeptides, it was concluded that sialic acid is linked to galactose by a 2 → 6 linkage and that galactose is linked to *N*-acetylglucosamine by a 1 → 4 linkage. One residue of mannose serves as a branching point in the core portion.

Based on the above results, the carbohydrate unit of released-form Gap a is composed of two or three side chains that contain Gal-(β1 → 4)-GlcNAc or sialyl-(α2 → 6)-Gal-(β1 → 4)-GlcNAc and a core portion that contains $Man_3GlcNAc_2Fuc$. Further studies are necessary to elucidate the complete structure.

GALACTOPROTEINS b AND bT (GAP b AND GAP bT)

As shown in FIGURE 1, Gap b and bT were isolated from the Empigen BB extract of cells. The isolated glycoprotein was stained very faintly with protein stain-

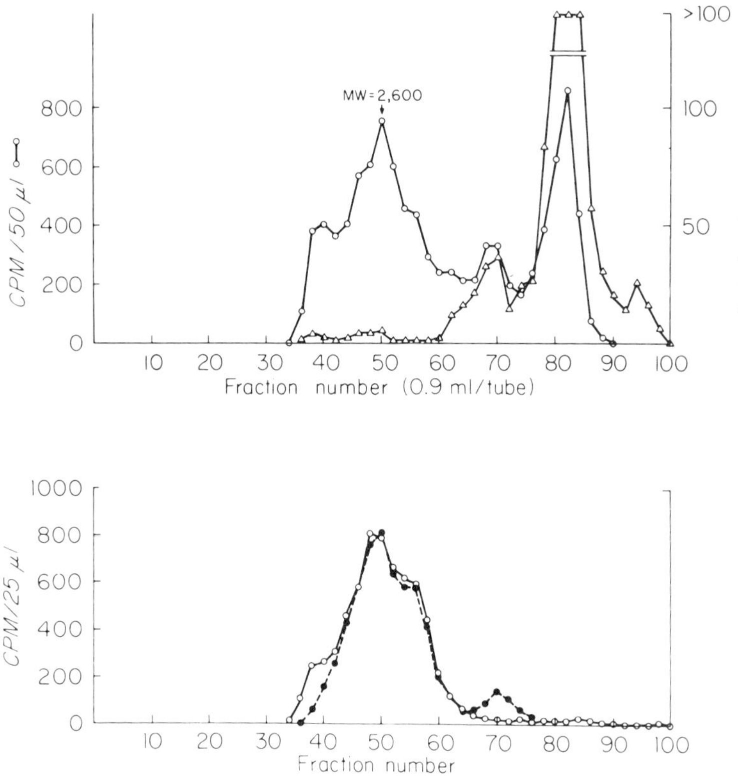

FIGURE 4. Biogel P-6 column chromatography of pronase digest of released-form Gap a. *Top:* Elution profiles derived from the [^{3}H]Glc (*top*) or [^{3}H]GlcN (*bottom*) metabolically labeled glycoprotein. *Bottom:* Dashed line indicates the elution profile after alkaline (sodium borohydride) treatment.

TABLE 3

CARBOHYDRATE COMPOSITION OF RELEASED-FORM GAP a*

Sugar	Molar Ratio†
Fucose	0.6
Galactose	2.8
Mannose	3.0
N-acetylglucosamine	4.0
N-acetylgalactosamine	0.2
Sialic acid	0.6

*Carbohydrate analysis was performed as TMS derivatives by gas-liquid chromatography after methanolysis.

†Molar ratio was expressed as mannose to 3.0.

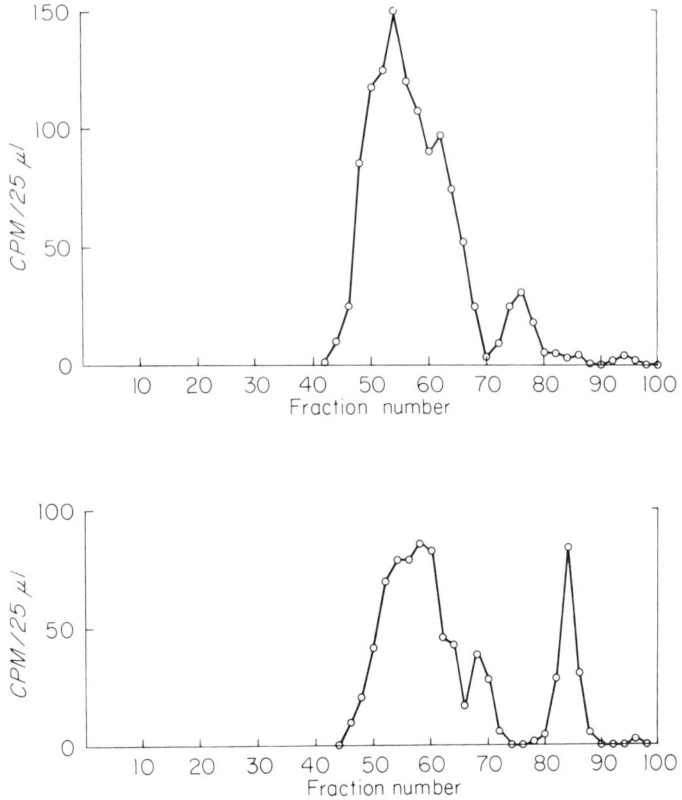

FIGURE 5. Biogel P-6 column chromatography of glycopeptide treated with neuraminidase (*top*) and with neuraminidase, followed by β-galactosidase (*bottom*). The glycopeptide was labeled metabolically with [^{3}H]Glc.

ing but was readily detectable by fluorography of surface-labeled material with apparent molecular weight of 125–130 × 10^5. The glycoprotein was present in nontransformed NIL cells, but the label was greatly enhanced in NILpy cells. The sample with or without reduction in 2-mercaptoethanol had the same mobility; therefore, the glycoprotein may not interact through disulfide bonds. The amino acid and carbohydrate compositions of Gap b are shown in TABLES 2 and 5. The surface-labeled glycopeptide from Gap b, after exhaustive digestion with pronase, was separated on a Biogel P-10 column (see FIGURE 6).

A new glycoprotein characteristic for NILpy cells and absent in NIL cells has been detected and copurified with Gap b. As can be seen in FIGURE 2(C, gel 4), the RCA receptor isolated from the Empigen BB extract of the surface-labeled NILpy cells contains a glycoprotein receptor that comigrates with Gap b in addition to another, higher-mobility glycoprotein band (marked with an arrow and called Gap bT). It is possible that Gap bT represents a tumor-associated antigen or virus-coded surface antigen.

TABLE 4

RELATIVE AREA PROPORTIONS OF METHYLATED SUGARS IN INTACT AND IN NEURAMINIDASE AND β-GALACTOSIDASE-TREATED GLYCOPEPTIDES

	Fucos	Galactose			Mannose*			*N*-acetylglucosamine		
Glycopeptide	2,3,4-Trimethyl	2,3,4,6-Tetramethyl	2,4,6-Trimethyl	2,3,4-Trimethyl	3,4,6-Trimethyl	2,3,6-Trimethyl	2,4-Dimethyl	3,4,6-Trimethyl	3,6-Dimethyl	3-Mono-methyl
Intact	0.5	2.1	0.5	+	0.7	+	0.6	—	4.0	0.5
Sialidase, β-galactosidase-treated glycopeptide†	+				+	+	+	+	+	+

*The recovery of mannose derivatives seems to be lower than expected. Further studies are necessary to obtain quantitative data of this derivative.

†Too little material was available to obtain conclusive area ratios.

TABLE 5

CARBOHYDRATE COMPOSITION OF GAP b

Sugar	Gap b (% of dry wt sample)
Fucose	0.51
Mannose	2.36
Galactose	0.95
Glucose	0.25
N-acetylglucosamine	2.54
N-acetylgalactosamine	0.33

GLYCOPROTEIN 170K

A trypsin-resistant, transformation-sensitive membrane glycoprotein with a subunit molecular weight of 170,000 (170K) has been detected in NIL cells and was deleted in NILpy cells. The glycoprotein was not efficiently labeled by surface labeling procedures, either lactoperoxidase-catalyzed radioiodination or galactose oxidase-NaB^3H_4, and was not degraded readily by mild treatment with trypsin. This glycoprotein, is in contrast to Gap a, which was labeled very effectively by surface labeling methods and is highly susceptible to trypsin treatment (see FIGURE 7). Galactoprotein a metabolically labeled with [^{3}H]GlcN disappeared after incubating monolayer cells with 1 μg/ml of trypsin at 37°C for 3 hr. Galactoprotein a labeled by radioiodination can be removed by 0.1 μg/ml of trypsin; in contrast, 170K was not degraded by trypsin under these conditions. However, tryptic digestion with an increased quantity of trypsin (>2 μg/ml) will remove all the label associated with 170K. These results suggest that 170K is a surface membrane component but that it is less exposed than is Gap a to the ex-

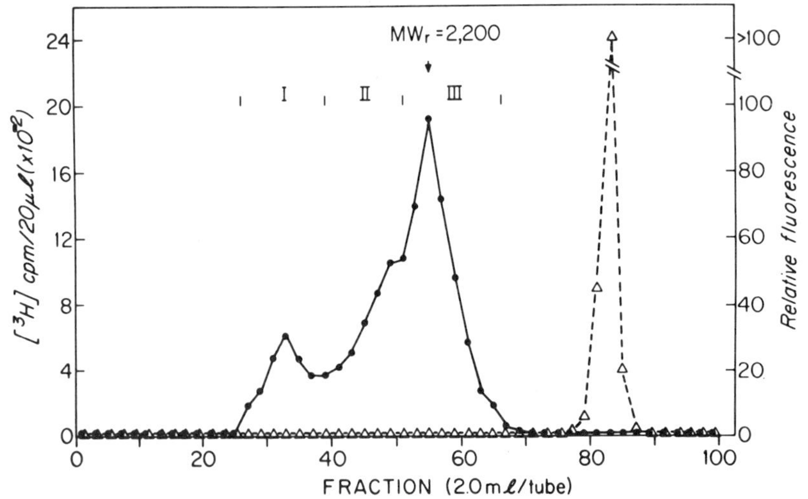

FIGURE 6. Surface-labeled glycopeptides from Gap b separated on a Biogel P-10 column. The solid line indicates radioactivity, and the dotted line indicates the peak for amino acid determined by fluorescamine assay.

ternal environment and/or is less sensitive to trypsin digestion. Both Gap a and 170K were insensitive to collagenase digestion (10 U/ml for 3 hr at 37°C). Glycoprotein 170K and Gap a were deleted from the cell surface by oncogenic transformation, but the deletion mechanism is not known.

Glycoprotein 170K can be readily extracted together with the minor 100K glycoprotein by treatment of intact NIL cells with a cold buffer solution that contains 20 mM Tris buffer (pH 7.6), 340 mM sucrose, 1 mM ATP, 1 mM EDTA, and

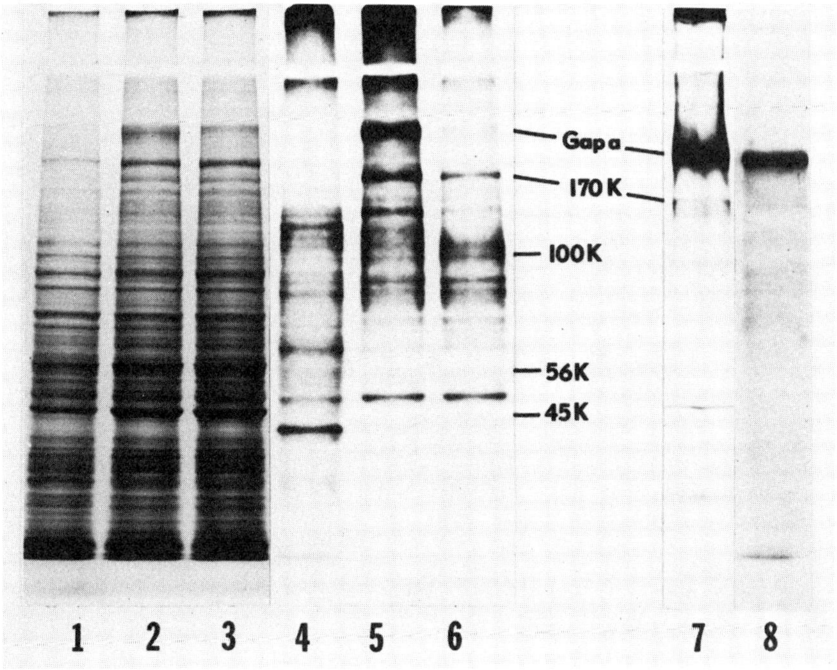

FIGURE 7. Electrophoresis of NIL and NILpy cells on pH 7.6 polyacrylamide gradient (5–14%) slab gels in the presence of 0.1% sodium dodecyl sulfate. NIL and NILpy cells were labeled metabolically with [^{3}H]GlcN or surface labeled by the galactose oxidase-NaB^3H_4 or lactoperoxidase-catalyzed ^{125}I method. 1, Protein stain of metabolically labeled NILpy cells; 2, protein stain of metabolically labeled NIL cells; 3, protein stain of metabolically labeled NIL cells digested with trypsin (1 μg of trypsin/ml of phosphate-buffered saline, 37°C, 3-hr incubation); 4, fluorograph of gel 1; 5, fluorograph of gel 2; 6, fluorograph of gel 3. Each sample applied to gels 1–3 contained 10^5 cpm of tritium label. 7, NIL cell surface labeled by the galactose oxidase-NaB^3H_4 method; 8, NIL cell surface labeled by lactoperoxidase-catalyzed ^{131}I. Mobility of known bands designated by labels (Gap a, 170K) in the margin.

0.2 mM TLCK (called sucrose-ATP extract; see FIGURE 1). Neither Gap a nor Gap b was extracted under these conditions. These extraction conditions are a modification of a procedure utilized for extraction of actin from homogenates of alveolar macrophages.[28] The 170K and 100K glycoproteins in the extract were purified by absorption onto an RCA column, followed by specific elution with 0.5 M galactose.

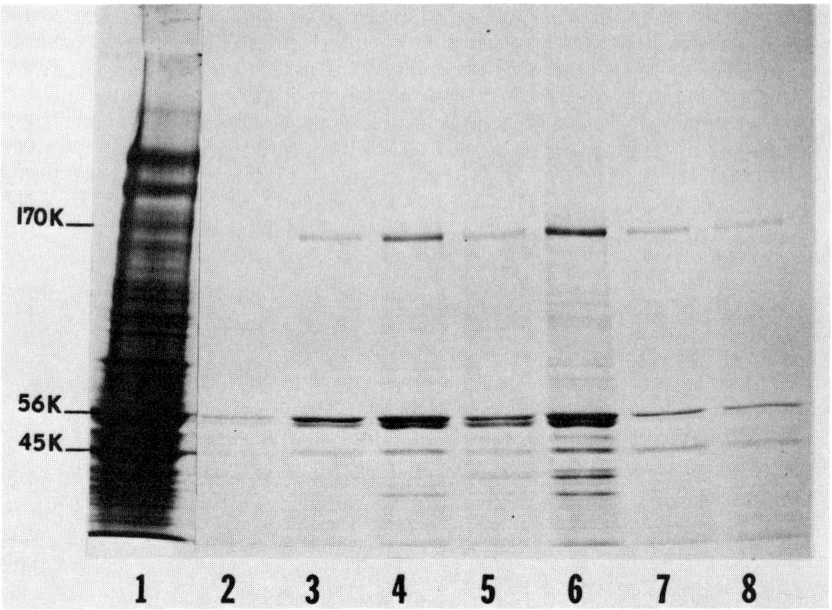

FIGURE 8. Effect of variations in polymerization conditions on precipitation of 56K protein as detected by sodium dodecyl sulfate polyacrylamide gel electrophoresis. Galactose oxidase-NaB^3H_4-labeled NIL cells (15 ml packed) were harvested with a rubber policeman and suspended in 4 vol (60 ml) of ice-cold 20 mM Tris-HCl buffer (pH 7.6) that contained 0.34 M sucrose, 1 mM ATP, 1 mM EDTA, and 0.2 mM TLCK, labeled "sucrose-ATP buffer." After 60 min on ice, the cell suspension was centrifuged at 800*g* for 10 min, and the supernatant was removed. The pellet was reextracted as above, and the second supernatant was pooled with the first. The combined supernatants were recentrifuged 120,000*g* for 60 min at 0–5°C, and the high-speed supernatant was removed and labeled "sucrose-ATP extract." Aliquots (8.8 ml) of the sucrose-ATP extract, on ice, were mixed with concentrated solutions of one or more of the following materials to the final concentrations indicated and a final volume of 9.2 ml: $MgSO_4$, 2 mM; KCl, 75 mM; $CaCl_2$, 2 mM; EGTA, 1 mM; GTP, 1 mM. Each solution was then incubated at either 4° or 37°C for 60 min and recentrifuged at 120,000*g* for 60 min at either 0–5° or 25°C, respectively. The supernatants were discarded, and each pellet was resuspended in 0.1 ml of ice-cold 20 mM Tris-HCl (pH 7.6) that contained 0.5 mM ATP and 2 mM EDTA, sonicated, and recentrifuged at 27,000*g* for 20 min at 0–4°C. An equal volume of each supernatant was applied to the gel.

Variations in polymerization condition of the sucrose-ATP extract are indicated for each gel: 1, NIL cell; 2, incubation at 37°C, no additions; 3, 37°C, KCl, Mg; 4, 37°C, KCl, Mg, Ca; 5, 37°C, GTP, Mg, EGTA; 6, 37°C, GTP, Mg, Ca; 7, 4°C, KCl, Mg; 8, 4°C, GTP, Mg, EGTA. Mobilities of 170K and 56K are designated by labels in the margin.

DO GAP a AND 170K ASSOCIATE WITH CYTOSKELETAL PROTEINS?

A possible connection of cell surface glycoproteins with cytoskeletal proteins has been suggested in the modified Singer-Nicolson model of cell surface membranes.[29] The alleged transmembrane control of motility, metabolism, and proliferation of cells can be operated through such a connection.[30] Although direct biochemical evidence for such a model has not been provided, some morphologic studies indicate that such a connection may indeed exist. The receptor for *Canavalia ensiformis* lectin (con A) can redistribute on the NRK cell surface after addition of fluorescent conjugated con A. The resultant linear arrays of con A receptors were superimposed on actomyosin-like filaments, as revealed by indirect immunofluorescence with antimyosin antibody.[31] The cytoplasmic actomyosin-like fibers were disrupted in transformed cells.[32]

In our studies, the majority of membrane and cytoplasmic proteins were extracted by sequential treatment of cells with sucrose-ATP-EDTA, followed by zwitterionic detergent Empigen BB, whereas a part of the cytoskeletal proteins, including an "actin-like protein," and essentially all Gap a remained unextracted. The subsequent urea extraction solubilized an actin-like protein, which was copurified with Gap a on affinity chromatography with RCA-PAHA (see FIGURE

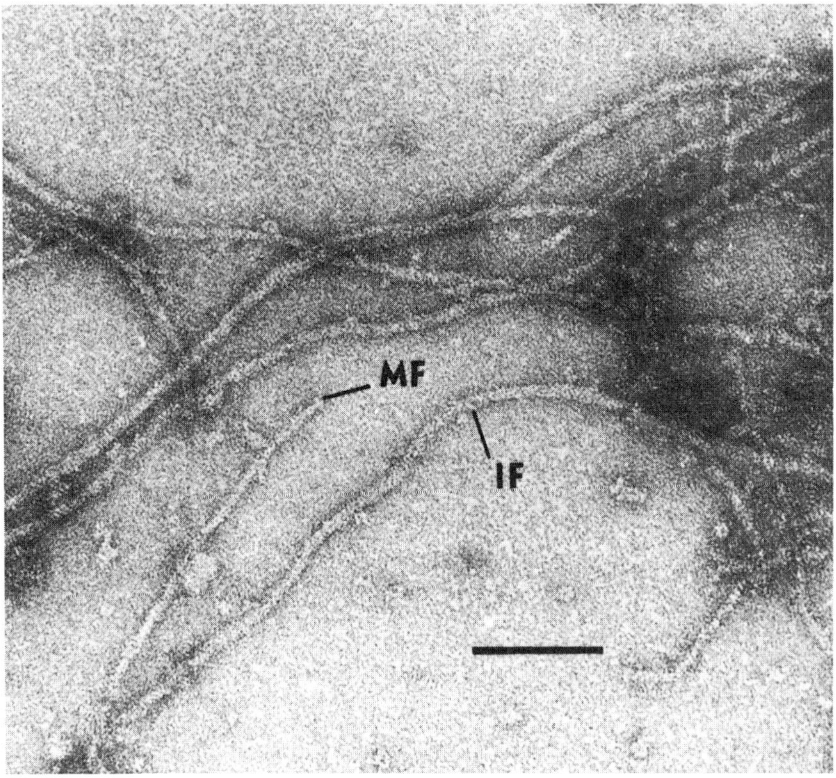

FIGURE 9. Electron micrograph of polymerized 56K protein (intermediate filament protein). The sucrose-ATP extract of NIL cells was polymerized in the presence of GTP, magnesium, and calcium. The polymerized material was gently resuspended in 20 mM Tris HCl buffer (pH 7.6) that contained 1 mM Ca and Mg by repeated pipetting, followed by centrifugation at 800*g* for 5 min. The supernatant was applied to carbon-coated specimen grids and negatively stained with 1% uranyl acetate, as described by Huxley.[38] Grids were observed in an AEI Corinth 500 transmission electron microscope. The filaments labeled IF (intermediate filaments) measured 105 Å in diameter. The filament labeled MF (microfilament) measured 69 Å in diameter. Bar measures 1000 Å. ×153,000.

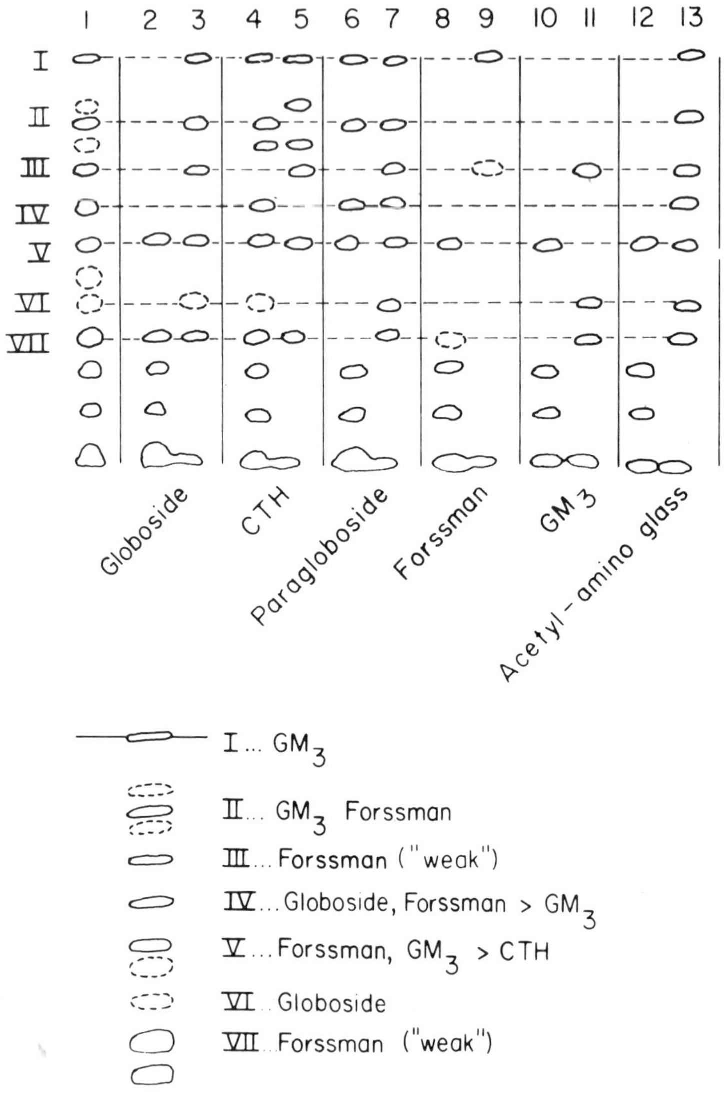

FIGURE 10. Sodium dodecyl sulfate polyacrylamide gel electrophoresis pattern of ^{35}S-labeled 3T3 cell membrane proteins (CMP) released by ethidium bromide and the interaction of CMP with various glycolipid-glass columns.

The pattern of total ^{35}S-labeled CMP is shown in lane 1, and the patterns of unbound CMP when treated with various glycolipid columns and with *N*-acetylamino-glass columns are shown in lanes 2, 4, 6, 8, 10, and 12. The patterns of CMP proteins eluted from respective columns with sodium iodide are shown in lanes 3, 5, 7, 9, 11, and 13. The kind of glycolipid-glass is identified as indicated.

Bands that were not recovered from the glycolipid-glass column can be seen if the total bands (lane 1) are compared to unbound and to those eluted with sodium iodide. Remarkably, bands 1, 2, and 4 were totally missed when CMP was treated with a GM_3-glass column,

2,D). The proportions of Gap a and actin present in the fraction coeluted from the column with 0.5 M galactose varied significantly from one preparation to another. Recently, Ali and Hynes[33] reported that cytochalasin B increased quantities of LETS protein released into the media, suggesting an association of LETS protein with actin-containing microfilaments. Previously, Kuusela *et al.*[34] observed that cell surface antigen (SFA), which is identical to Gap a or LETS, and actin-like material were coprecipitated by anti-Gap a antiserum, indicating a possible association of Gap a with actin. The question remains, however, as to whether actin may interact nonspecifically with anti-Gap a antibody. In fact, actin can be coprecipitated not only by anti-Gap a but also by antigloboside and antiganglioside antibodies.[39] However, actin present in the Empigen BB extract was not copurified with Gap b through the RCA-PAHA-column, excluding the possibility that actin-like protein may interact nonspecifically with the RCA-PAHA column. The experiment that demonstrated copurification of Gap a and actin on the RCA-PAHA column was performed with the extracts using a high concentration of urea; therefore, both Gap a and actin were denatured in these experiments. Further extensive study is necessary to draw an unequivocal conclusion concerning the specificity of this interaction.

A possible interaction of glycoprotein 170K and IFP was suggested by a few observations, as described below. When the sucrose-ATP extract of NIL cells was allowed to stand at room temperature for several hours or frozen, a turbid material formed that could be pelleted by centrifugation. The pellet contained significant quantities of two components, glycoproteins 170K and 56K. Precipitation was enhanced by addition of GTP and Mg^{2+}, tubulin polymerization conditions. Addition of 2 mM Ca^{2+} was found to further enhance the polymerization of 170K glycoprotein and 56K protein (see FIGURE 8). Surprisingly, Ca^{2+} is a reported inhibitor of microtubule formation.[35]

If the sucrose-ATP extract was immediately applied on the RCA-PAHA column, followed by elution with galactose, 170K and 100K were purified and were free from 56K protein. However, if the sucrose-ATP extract was first subjected to the polymerization conditions, followed by RCA-PAHA column chromatography, 56K was eluted by galactose, along with 170K and 100K. Therefore, pretreatment of the sucrose-ATP extract under polymerization conditions was required for coabsorption of 170K glycoprotein and 56K protein on the RCA-PAHA column.

After polymerization, the pelleted material was examined by transmission electron microscopy. As seen in FIGURE 9, the polymerized product contained long filaments of 105 Å diameter. On the basis of this observation, we shall henceforth consider the 56K protein to be identical to IFP.

---

and those bands were not recovered. Bands 2 and 4 were not recovered from the Forssman glass column, whereas essentially all of the bands were recovered from globoside, CTH, paragloboside, and *N*-acetylamino-glass columns, except for a few weaker bands. Therefore, bands 1 and 2 were strongly associated with hematoside (GM3), and bands 2 and 4 could be strongly associated with globoside and Forssman antigen. A summary of a putative correlation (*bottom*) between membrane proteins and glycolipids follows: bands 1 and 2 for GM3; 2, 3, 5, and 7 for Forssman; 4 and 6 for globoside. It is noteworthy that CTH and paragloboside did not show interaction with CMP of 3T3 cells. Both CTH and paragloboside are absent in 3T3 cells, whereas globoside, Forssman antigen, and GM3 are present in 3T3 cells.

The possible interaction of 170K and IFP may be due to nonspecific interactions and therefore requires further study to evaluate any possible biologic significance.

Do Cell Surface Proteins or Glycoproteins Interact with Membrane Glycolipids of Their Own Cells?

The label of Gap a in NIL cells introduced by the galactose oxidase-NaB^3H_4 method was greatly enhanced in the presence of a small quantity (10 μg/ml) of con A and RCA-I but was greatly suppressed by a larger quantity of these lectins (100 μg/ml). The label in Gap b of NIL cells was not effected by lectins. It was considered, therefore, that these lectins interacted with NIL cells through Gap a but not through Gap b. In contrast, the label in Gap b of transformed NILpy cells was greatly suppressed by RCA and con A; that is, these lectins interacted with NILpy cells through Gap b.[3]

A possible correlation between the exposure of Gap a and crypticity of cell surface glycolipid was suggested by the following observation: the label in globoside [GalNAc-(β1 $\rightarrow$ 3)-Gal-(α1 $\rightarrow$ 4)-Gal-(β1 $\rightarrow$ 4)-Glc $\rightarrow$ cer] through galactose oxidase-NaB^3H_4 was found to be greatly enhanced when cells were interacted with RCA through Gap a. The label of globoside was enhanced 300%, whereas the label in other glycolipids was not significantly altered.[3] It is possible, therefore, that globoside and Gap a of the NIL cell surface interacted with each other on the normal cell surface. When RCA is applied on a cell surface that reacts with Gap a strongly, the normal interaction between Gap a and globoside may be hampered; thus, globoside may become highly exposed. This finding may suggest that the Gap a polypeptide of NIL cells may include a specific affinity toward a specific glycolipid of NIL cells, and such "autorecognition" of glycolipid and glycoprotein could be an important basis for intercellular interactions.

Consequently, a possible interaction between glycolipids and proteins extracted from the same cells have been studied. Various glycolipids were attached covalently on amino-glass particles by the procedure described previously.[36] Cells were labeled metabolically in a medium that contained [^{35}S]methionine (10 μCi/ml), and the cell surface proteins were extracted with phosphate-buffered saline that contained ethidium bromide (10 μg/ml) for 30 min at 37°C. The supernatants were removed and dialyzed extensively against water and lyophilized. After extraction, cells were still viable. The ^{35}S-labeled proteins consisted of at least 13 bands separated by sodium dodecyl sulfate polyacrylamide gel electrophoresis (FIGURE 10). The protein mixtures were incubated with globoside-, CTH (αGal1 $\rightarrow$ 4βGal1 $\rightarrow$ 4βGlc $\rightarrow$ cer), paragloboside, Forssman antigen, and hematoside covalently attached, respectively, to amino-glass particles. The loss of the protein bands after interaction with glycolipid glass was observed by autoradiography of the sodium dodecyl sulfate polyacrylamide gel electrophoresis bands. Some of the bands were nonspecifically bound to *N*-acetyl amino-glass and could be eluted with 1 M NaI. Specific bands were lost and were not recovered, even by elution with guanidine hydrochloride or with 1% sodium taurochenodeoxycholate, indicating that the interaction was enormously strong. Globoside, CTH, and paragloboside columns did not show much interaction; however, Forssman antigen and hematoside columns showed extensive interaction, with four or five bands (see FIGURE 10).[39] The glycolipid-ektoprotein interaction could be a molecular basis of intercellular interaction, as has been predicted.[37]

REFERENCES

1. GAHMBERG, C. G. & S. HAKOMORI. 1973. J. Biol. Chem. **248:** 4311–4317.
2. GAHMBERG, C. G. & S. HAKOMORI. 1973. Proc. Nat. Acad. Sci. USA **70:** 3329–3333.
3. GAHMBERG, C. G. & S. HAKOMORI. 1975. J. Biol. Chem. **250:** 2447–2451.
4. CARTER, W. G. & S. HAKOMORI. 1977. Biochem. Biophys. Res. Commun. **76:** 299–307.
5. HYNES, R. O. 1973. Proc. Nat. Acad. Sci. USA **70:** 3170–3174.
6. WICKUS, G. G., P. E. BRANTON & P. W. ROBBINS. 1974. *In* Control of Proliferation in Animal Cells. B. Clarkson & R. Baserga, Eds.: 541–546. Cold Spring Harbor Laboratory Press. Cold Spring Harbor, N.Y.
7. VAHERI, A. & E. RUOSLAHTI. 1974. Int. J. Cancer **13:** 579–586.
8. YAMADA, K. M. & J. A. WESTON. 1974. Proc. Nat. Acad. Sci. USA **71:** 3492–3496.
9. KESKI-OJA, J., D. F. MOSHER & A. VAHERI. 1976. Cell **9:** 29–35.
10. HYNES, R. O. 1976. Biochim. Biophys. Acta **458:** 73–107.
11. HYNES, R. O. & J. M. BYE. 1974. Cell **3:** 113–120.
12. GAHMBERG, C. G. & S. HAKOMORI. 1974. Biochem. Biophys. Res. Commun. **59:** 283–291.
13. GAHMBERG, C. G. & S. HAKOMORI. 1975. J. Biol. Chem. **250:** 2438–2446.
14. PEARLSTEIN, E. 1976. Nature (London) **262:** 497–499.
15. ENGVALL, E. & E. RUOSLAHTI. 1977. Int. J. Cancer **20:** 1–5.
16. MOSESSON, M. W., A. N. CHEN & R. M. HUSEBY. 1975. Biochim. Biophys. Acta **386:** 509–524.
17. YAMADA, K. M., S. S. YAMADA & I. PASTAN. 1975. Proc. Nat. Acad. Sci. USA **72:** 3158–3162.
18. HYNES, R. O. & A. DESTREE. 1977. Proc. Nat. Acad. Sci. USA **74:** 2855–2859.
19. KESKI-OJA, J., D. F. MOSHER & A. VAHERI. 1977. Biochem. Biophys. Res. Commun. **74:** 699–706.
20. SHEN, L. C. & V. GINSBURG. 1972. Methods Enzymol. **28:** 195–201.
21. HAMMOND, K. S. & D. S. PAPERMASTER. 1976. Anal. Biochem. **74:** 292–297.
22. KOIDE, N. & T. MURAMATSU. 1974. J. Biol. Chem. **249:** 4897–4904.
23. TARENTINO, A. L., T. H. PLUMMER & F. MALEY. 1974. J. Biol. Chem. **249:** 818–824.
24. TAI, T., S. ITO, K. YAMASHITA, T. MURAMATSU & A. KOBATA. 1975. Biochem. Biophys. Res. Commun. **65:** 968–974.
25. ENGVALL, E. & E. RUOSLAHTI. 1977. Int. J. Cancer **20:** 1–5.
26. FUKUDA, M. N. & G. MATSUMURA. 1976. J. Biol. Chem. **251:** 6218–6225.
27. HAKOMORI, S. 1964. J. Biochem. (Tokyo) **55:** 205–208.
28. HARTWIG, J. H. & T. P. STOSSEL. 1975. J. Biol. Chem. **250:** 5696–5705.
29. NICOLSON, G. L. 1976. Biochim. Biophys. Acta **457:** 57–108.
30. NICOLSON, G. L. 1976. Biochim. Biophys. Acta **458:** 1–72.
31. ASH, J. F. & S. J. SINGER. 1976. Proc. Nat. Acad. Sci. USA **73:** 4575–4579.
32. TOH, B. H. & G. C. HARD. 1977. Nature (London) **269:** 695–697.
33. ALI, I. U. & R. O. HYNES. 1977. Biochim. Biophys. Acta **471:** 16–24.
34. KUUSELA, P., E. RUOSLAHTI & A. VAHERI. 1975. Biochim. Biophys. Acta **379:** 295–303.
35. WEISENBERG, R. C. 1972. Science **177:** 1104–1105.
36. LAINE, R. A., G. YOGEESWARAN & S. HAKOMORI. 1974. J. Biol. Chem. **249:** 4460–4466.
37. HAKOMORI, S., S. KIJIMOTO & B. SIDDIQUI. 1972. *In* Membrane Research. C F. Fox, Ed.: 253–277. Academic Press, Inc. New York, N.Y.
38. HUXLEY, H. E. 1963. J. Mol. Biol. **7:** 281–308.
39. LINGWOOD, C. & S. HAKOMORI. Unpublished results.

IMMUNOCHEMICAL AND COLLAGEN-BINDING PROPERTIES OF FIBRONECTIN*

Erkki Ruoslahti and Eva Engvall

Division of Immunology
City of Hope National Medical Center
Duarte, California 91010

Fibronectin (also called SFA,[1,2] LETS protein,[3,4] galactoprotein a,[5] CSP,[6] Z protein,[7] CIg,[8] CAF,[9] and CAP[10,11]), a cell surface and plasma protein,[1,2] has many interesting biologic properties, such as nonexistence in surfaces of transformed cells[4,5,7,12-14] and interaction with fibrin[15] and collagen.[16] These properties have made it the subject of much current research. It still remains a protein in search of a function, especially in regard to its plasma form, but the evidence available suggests that it may function in the attachment of cells to the extracellular matrix.[9,10]

In this paper, we will briefly summarize some of our past work and present some new data on the immunochemical properties and detection of fibronectin and its interaction with structural proteins, such as collagen and fibrin.

Antibodies to Fibronectin and Immunologic Comparison of the Fibroblast and Plasma Forms

In collaborative studies with Vaheri and others,[1,2] antibodies to material released from the surface of cultured chicken[1] and human[2] fibroblasts by treatment with insolubilized papain were prepared. The rationale of this approach was the demonstration that the histocompatibility antigens could be solubilized from lymphocyte plasma membranes by use of papain.[17] We expected to obtain antibodies to several surface components, but the antisera reacted with only one component in the material released by papain and in the cell extracts obtained by treatment with a combination of urea and Triton® X-100. An immunologically identical component was present in homologous serum and plasma. We initially named this antigen surface fibroblast antigen (SFA). Internal labeling of fibroblast cultures, followed by immunologic analysis showed that SFA was synthesized by fibroblasts and was not derived from serum.

Three properties of fibronectin explain our unexpected success in obtaining specific antibodies to it using what could be expected to be a crude mixture of cell surface and culture media components. Fibronectin (SFA) is by far the main component of fibroblast surfaces.[1,2] About 1% of the total cellular protein consists of fibronectin, and surface labeling shows almost exclusive incorporation of the label in fibronectin.[3,18] Fibronectin is extremely sensitive to certain proteases with respect to its release from the cell surface,[3,18-20] but it tends to cleave in large fragments (mol wt, ~200,000 daltons for trypsin[18]). There seems to be a "core" that is relatively resistant to proteases. The supposedly insoluble papain used in our early experiments did, however, leach considerable amounts of papain, and this leaching resulted, during the dialysis, in continued proteolysis of the soluble material released from the monolayers. This phenomenon may have led to the destruction of other contaminating components. This hypothesis is supported by the

*Supported by Grants CA 22108 and CA 16434 from the National Cancer Institute.

0077-8923/78/0312-0178 $01.75/0 © 1978, The New York Academy of Sciences

observation that inactivation of papain with iodoacetamide after digestion resulted in immunogen that gave rise to antibodies to papain[11] and calf serum proteins, in addition to fibronectin.[14] One of the reactive components in calf serum was cross-reactive bovine fibronectin (see below). We subsequently prepared our antiserum against antigen that was partially purified from the papain digest by removal of soluble papain with immunoadsorbents to which antipapain antibodies were attached, followed by gel filtration on Sephadex® G-200.[21] Most of the antigenic activity was present in a fraction with a molecular weight of about 50,000 daltons. It is interesting that this material was indistinguishable from the intact molecules, indicating that it contained all of the antigenic determinants of fibronectin.

The antigenic determinants of fibronectin appear to be extremely stable. Preparations treated with sodium dodecyl sulfate (SDS) and tested in the presence of SDS react identically with native fibronectin whether tested against antibodies to SDS-treated or native fibronectin.[52] Others have prepared antisera against antigen obtained from SDS gel bands and obtained antibodies that react with the intact molecule.[4] These results indicate that the structure of fibronectin is unusually stable and suggest that it may be composed of some kind of a repeating unit.

Antibodies to fibronectin, whether prepared by immunization with fibroblast-derived or plasma fibronectin, show identical reactivity against the plasma and cellular forms of the same species.[1,2,23] The immunologic identity of the fibroblast and plasma forms of fibronectin is strong evidence in favor of the contention that they are the same protein. However, as demonstrated by the immunologic identity of papain-digested fibronectin with the intact protein, this result does not exclude the possibility that the plasma form could be proteolytically modified from the cellular form.

Recent biochemical studies[24] performed in our laboratory have provided further evidence in support of the similarity of the fibroblast-derived and plasma fibronectins. Comparison of fibronectin obtained from spent culture medium and from plasma by gel electrophoresis, sugar and amino acid compositions, and peptide maps revealed no significant difference between them. The cellular and plasma forms also share an affinity to collagen and exhibit a similar pattern of specificity when tested against genetically different types of collagen (see below).

Isolation of Fibronectin with Antibodies

Fibronectin does not lend itself well to immunochemical purification with immunoadsorbents. This goal has been achieved but with considerable losses in terms of yield.[24] At the radiochemical level, it was possible to analyze the polypeptide composition of internally labeled fibroblast fibronectin by means of immunoprecipitation.[24] Three polypeptides were found in immunoprecipitates obtained when ^{35}S-labeled fibroblast extracts were first incubated with antiserum to fibronectin and the immune complexes precipitated with anti-γ-globulin serum. These polypeptides had molecular weights of about 210,000, 145,000, and 45,000 daltons. It became clear that the 210,000-dalton polypeptide corresponded to the lactoperoxidase-iodinatable transformation-sensitive protein.[18] The nature of the other two polypeptides is not clear. The 145,000-dalton polypeptide was found to have a very fast turnover, with short pulses of [^{35}S]methionine. This polypeptide appeared as one of the main labeled bands in autoradiography of gels.[14] Like the 210,000-dalton component, it was absent in transformed cells. Surface labeling by lactoperoxidase-catalyzed iodination did not label this protein. It is unlikely that this polypeptide would be present in the immunoprecipitates because of heterogeneity of the antibodies used. Extensive testing of the antibodies did not reveal any evidence of reactivity with more than one component in fibroblast extracts or serum, and antibodies prepared against fibroblast fibronectin and purified with affinity chromatography on plasma fibronectin gave the same polypeptides in the immunoprecipitates.

In view of the recent demonstration of binding of fibronectin to collagen,[16] it is possible that the 145,000-dalton polypeptide is a collagen precursor that becomes associated with the fibronectin immunoprecipitate because of interaction with fibronectin, not with the antibodies. In similar, more recent experiments, it was also found that fibronectin immunoprecipitates contained polypeptides other than polypeptide with an approximate molecular weight of 210,000 daltons, and some evidence that they were collagen precursors was presented.[26] The 45,000-dalton component comigrated with actin, but there is no evidence that actin would be specifically associated with fibronectin. A collagenous polypeptide with a molecular weight of 45,000 daltons has been described.[27]

Immunologic Interspecies Cross Reactions of Fibronectin

Strong antisera to fibronectin show extensive interspecies cross reactions[21] (see Figure 1). In immunodiffusion experiments, antihuman fibronectin precipitates all mammalian fibronectins, except for that of the species of origin of the antiserum. This finding, and the fact that the immunized animals remain quite healthy, suggests that immunization with a fibronectin from another species does

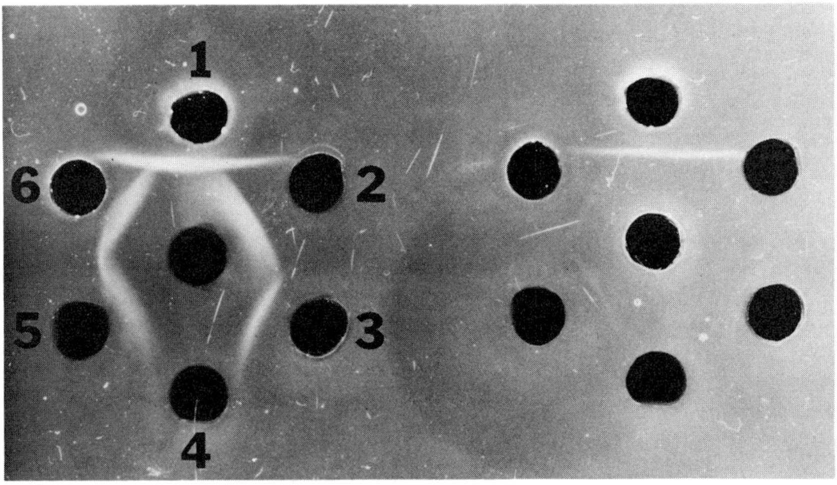

Figure 1. Immunodiffusion in agarose shows cross reactivities of fibronectins of various species. Rabbit antihuman fibronectin, when unabsorbed (*left*, center well), reacts strongly against human fibronectin (1) and cross reacts with bovine (2), rat (3), chicken plasma (5), and fibroblast (6) fibronectins but not with Torpedo fish fibronectin (4). The antiserum does not distinguish between bovine and rat fibronectins. Absorption of the antiserum with bovine fibronectin (*right*, center well) weakens its reactivity against human and chicken fibronectins and abolishes its reactivity against bovine and rat fibronectins (right side of the Figure).

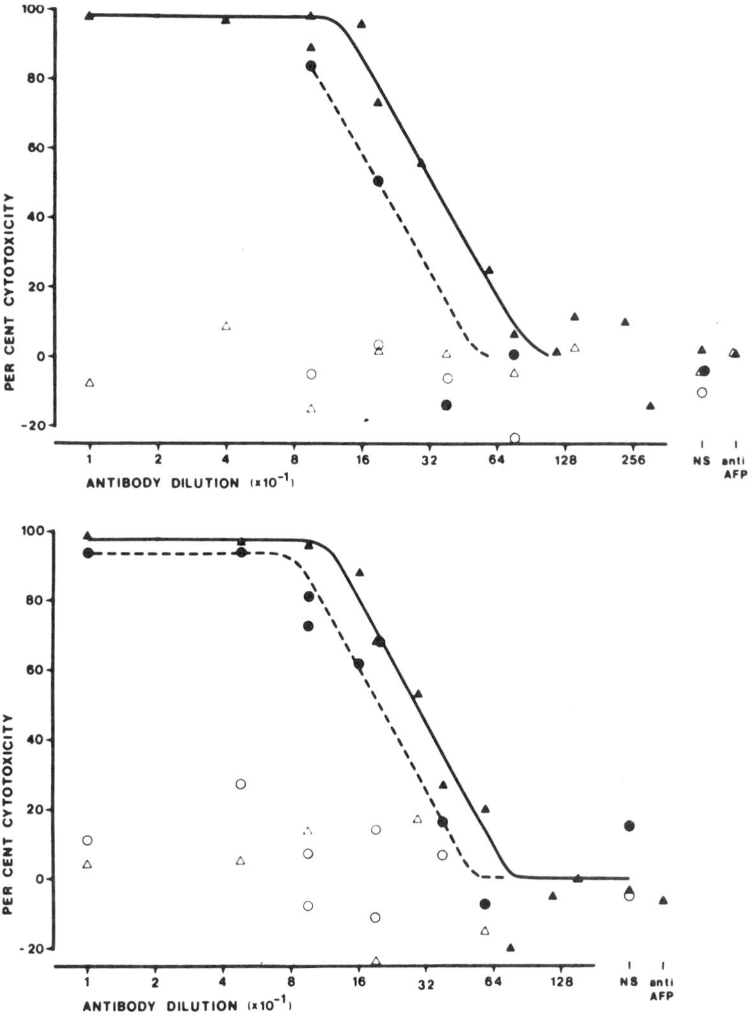

FIGURE 2. Complement-dependent cytotoxic activity of sheep antibodies to human fibronectin on human fibroblasts S64N and H894N tested separately or in mixture with HeLa cells (*top*) or H894A melanoma cells (*bottom*). ▲, Effect on fibroblasts tested separately; △, effect on HeLa cells (*top*) and melanoma cells (*bottom*) tested separately; ●, effect on fibroblasts tested in mixture with tumor cells; ○, effect on tumor cells tested in mixture with fibroblasts. Percentage cytotoxicity means percentage reduction of cell survival with active compared to inactive complement. Normal human serum (NS, diluted 1:20) from a healthy adult contains fibronectin and inhibits the cytotoxicity. The immunoglobulin fraction of a sheep anti-α-fetoprotein serum was used in a concentration of 1 mg/ml as a further control. (From Ruoslahti & Sjörgen.[29] by permission of *International Journal of Cancer*.)

not induce autoantibodies. With other proteins, such induction can occur.[28] It could be added, parenthetically, that attempts to demonstrate autoantibodies to fibronectin in patients with such antibodies against other connective tissue components have so far not revealed antifibronectin.[52] Quite remarkably, antisera to mammalian fibronectin also precipitate chicken fibronectin.[25] It is unusual that homologous proteins from species so widely separated show cross reactivity. It could be that the cross reactivity is due to similarity of the carbohydrate and not the protein portion. Evidence against this possibility has been presented.[21] The cross reactivity, like the rest of the immunologic reactivity of fibronectin, is sensitive to extensive pronase digestion but not to periodate treatment. Furthermore, fibronectin, both fibroblast and plasma derived, contains relatively little (4–5%) carbohydrate.[24]

The cross reactivities include reactivity of most antisera against bovine fibronectin. This fact is of importance in experiments with cell cultures. The fetal bovine serum used in the culture medium contains considerable amounts of fibronectin. In our work, we have used antibodies against the fibronectin studied and have always absorbed the antisera with bovine serum (which contains bovine fibronectin). Unfortunately, other laboratories often have not done the same. The interpretation of such experiments as measurement of fibronectin in rat or mouse cells grown in the presence of bovine fibronectin by use of antibodies to human fibronectin is difficult.

Cytotoxicity of Antibodies to Fibronectin

Antifibronectin extensively absorbed with red and white blood cells has been found to be cytotoxic to cultured fibroblasts in the presence of complement. This cytotoxicity is selective; melanoma cells and HeLa cells, which were found to be fibronectin negative in immunofluorescence studies, were not affected by antifibronectin. Furthermore, when mixtures of fibroblasts and the tumor cells were treated with antibodies, fibroblasts were eliminated selectively from the cultures (Figure 2). This finding suggests an important application that has not yet been attempted. Contaminating fibroblasts could be eliminated from primary cultures of tumor cells or epithelial cells negative for fibronectin. Antisera, probably reacting against fibronectin, have also been found to be selectively cytotoxic against fibroblasts in mixtures of fibroblasts with sarcoma cells, which lack surface fibronectin.[30] The fact that cytotoxicity can be achieved emphasizes the close association of fibronectin with the cell surface in fibroblasts.

Collagen-Binding Properties of Fibronectin

We have shown that human fibronectin has affinity to collagen and can be purified by affinity chromatography on gelatin-Sepharose[16] (gelatin is essentially pure, denatured collagen).

By means of a modification of the enzyme-linked immunoadsorbent assay ELISA,[31,32] we found that fibronectin from serum, plasma, or fibroblasts was bound to microtiter wells coated with calf skin collagen (mostly type-I collagen) or gelatin.[16] Gelatin had the higher capacity of the two (Figure 3). Denaturation of collagen by heating increased its activity to a level comparable to the gelatin activity. This result suggested that fibronectin interacts preferentially with unwound collagen. The specificity of the fibronectin-collagen interaction was evident from the fact that various other proteins (α_1-acid glycoprotein, serum albumin, ovalbumin, α-fetoprotein, fetuin) or glycosaminoglycans (chondroitin sulfates A, B, and C, heparin, hyaluronic acid) did not show such binding. The binding of fibronectin to collagen is inhibited by gelatin and vice versa. Both are inactivated by bacterial collagenase. None of the control substances listed above exhibits any inhibition.

More recent studies performed with this inhibition test have provided further insight into the specificity of the fibronectin-collagen interaction. We have found[23] that all four types of human collagen tested (I, II, III, and AB)[33] are active (inhibit binding of fibronectin to gelatin). The denatured forms are more active than are the native ones. Of the latter, type III is the most active, and type AB is the least active, both native and denatured forms. Further control proteins found to be negative include keratin and elastin (a gift from Dr. W. Gray).

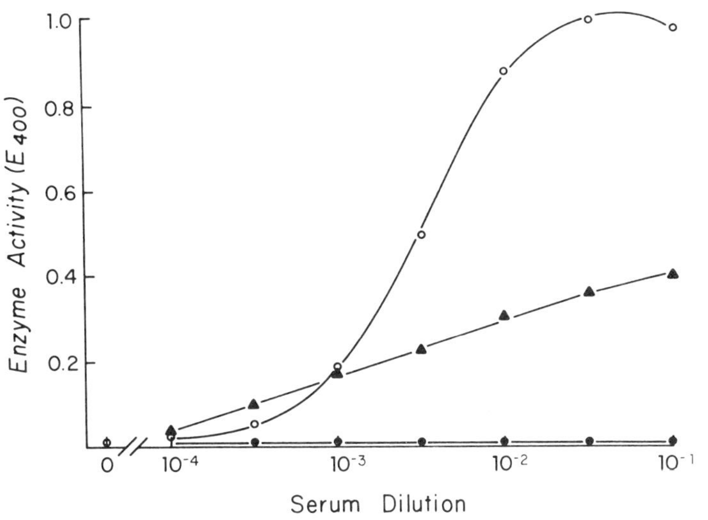

FIGURE 3. Binding of serum fibronectin to microtiter plate wells[16] coated with gelatin (○), native collagen (▲), and elastin (●).

It is known that fibronectin binds to fibrin and fibrinogen.[15,34] This interaction is manifest in the coprecipitation of fibronectin with cryofibrinogen and other components in the cryoprecipitate that forms when plasma is kept at a low temperature (hence the name "cold-insoluble globulin"). If blood is allowed to clot, a large fraction of fibronectin is incorporated into the fibrin clot,[15] and it can become cross-linked to fibrin.[8] Furthermore, fibronectin can be chromatographed on fibrin powder, to which it binds at low temperature.[15,34] Binding of fibronectin to fibrinogen can also be demonstrated by ELISA. However, on a weight basis, it is much less active than is collagen or gelatin. The binding of fibronectin to fibrinogen can be inhibited by collagen, and fibrinogen shows some inhibition of binding of fibronectin to collagen.[23] These results suggest that the same site in fibronectin is involved in both bindings.

It has recently been reported that human fibronectin binds to heparin and that addition of heparin to plasma causes precipitation of fibronectin, among other plasma components.[35] We also find decreased fibronectin levels in heparin plasma as compared to citrated plasma.[52] However, binding tests performed with the enzyme assay methods described above have not revealed any binding of fibronectin to heparin-coated microtiter plate wells, and heparin does not inhibit the binding of fibronectin to collagen.[16] It appears that the interaction of fibronectin with heparin is less clear-cut than is collagen binding and does not depend on the same binding site.

Chicken fibroblast fibronectin has been reported to bind to formalinized sheep red blood cells and to agglutinate them.[36] This agglutination was dependent on divalent metal ions and was inhibited by various sugars and compounds that contain amino groups. The collagen binding of fibronectin is unaffected by these substances and thus seems to be an unrelated property of fibronectin.

We have made use of the affinity of fibronectin to collagen in an affinity chromatography procedure that allows isolation of electrophoretically pure fibronectin from human plasma in one step with excellent yields[16] (FIGURE 4). We have now applied this method to the isolation of fibronectins from the plasma of several other mammalian species,[37] chicken plasma, and also from extracts of chicken fibroblasts.[23] Our recent results show that this method can be further improved by first passing the sample from which fibronectin is to be isolated on gelatin-Sepharose through a column of plain Sepharose. This step will remove a minor contaminant that binds to Sepharose and agglutinates rabbit red blood cells.[52] We have also isolated[23] a Torpedo fish plasma protein that displays similarities with the mammalian and chicken fibronectins. The same affinity chromatography procedure that results in isolation of mammalian and chicken fibronectins on gelatin-Sepharose was utilized. Apart from the affinity to collagen, implied by the fact that the fish protein could be purified on gelatin-Sepharose, it was similar to fibronectin in SDS gel electrophoresis in that it had a subunit of

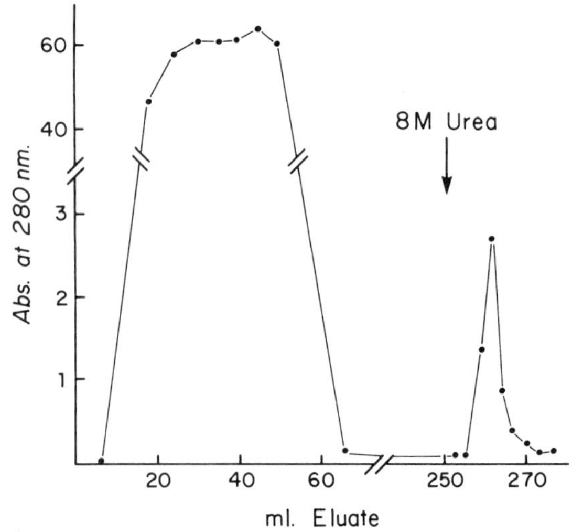

FIGURE 4. Affinity chromatography of human plasma fibronectin on gelatin-Sepharose. The second peak, obtained by elution with 8 M urea, contains the electrophoretically homogeneous fibronectin. (From Engvall & Ruoslahti.[16] By permission of *International Journal of Cancer.*)

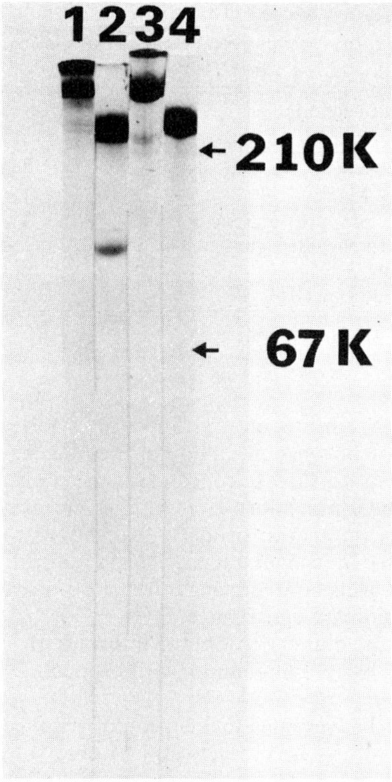

FIGURE 5. Polyacrylamide gel electrophoresis in the presence of sodium dodecyl sulfate. Comparison of the polypeptide composition of Torpedo fish (1 & 2) and human (3 & 4) fibronectin. Gels 1 and 3 were run without reducing agent in the sample and show the dimer forms. A minor component is present in the Torpedo fibronectin preparation. The positions of molecular weight markers have been indicated.

the same size and ran as a dimer if not reduced (FIGURE 5). This evidence allows the tentative conclusion that this protein is a homolog of the mammalian and chicken fibronectins. A minor polypeptide of lower molecular weight than fibronectin was also present in the Torpedo fibronectin preparation. Torpedo fibronectin did not cross react with other fibronectins (FIGURE 1). This finding, and the fact that we have been able to detect and isolate fibronectin from the plasma of every mammalian species studied, indicates that the affinity of fibronectin to collagen is shared by a variety of species, ranging from human to fish. No fibronectin-like protein was found in lobster hemolymph when gelatin-Sepharose columns were employed. Lobster fibrinogen has a polypeptide composition similar to that of fibronectin in that it is a dimer of a polypeptide chain with a molecular weight slightly above 200,000 daltons.[38] Clotting of lobster hemolymph is affected by direct cross-linking of fibrinogen by a transglutaminase. This phenomenon is reminiscent of the cross-linking of fibronectin to fibrin and to itself[8,39] by a transglutaminase. Finally, comparison of the amino acid compositions of lobster fibrinogen[40] and human fibronectin[24,41] reveal some similarities.

TABLE 1
SUMMARY OF CONDITIONS UNDER WHICH FIBRONECTIN ELUTES FROM GELATIN-SEPHAROSE

Fibronectin Is Eluted with	Fibronectin Is Not Eluted with
Urea, 2–4 M	EDTA, 10 mM
Sodium thiocyanate, 2 M*	Triton X-100, 1%
Sodium iodide, 4 M*	Tween® 20, 0.05%
Ethylene glycol, 50% in phosphate-buffered saline	sodium chloride, 5 M
Collagenase treatment	potassium chloride, 3 M
Purified antibodies to fibronectin†	

*Lower concentrations were not tested.
†Measured as release of bound [^{125}I]fibronectin from gelatin-Sepharose.

It appears that if lobster fibrinogen and fibronectin indeed are related, lobster fibrinogen does not share fibronectin's affinity for collagen.

Out results demonstrate that fibronectin is the only major component of plasma that binds to gelatin-Sepharose. It also is the only major component of fibroblast extracts that exhibits this property, as demonstrated by the purification of homogeneous fibronectin from extracts of chicken fibroblasts. This finding suggests strongly that the fibronectin-collagen interaction is not fortuitous. It is likelier that it has a profound biologic significance.

We have used gelatin columns to obtain information on the nature of fibronectin-collagen binding by studying elution of fibronectin with various treatments (TABLE 1).

The use of gradients has revealed that elution of fibronectin begins at urea concentrations of about 2 M and is complete at 4 M urea (FIGURE 6). Consequently, we now use 4 M urea instead of 8 M urea, as was originally reported.[16] Fibronectin was also successfully eluted with a glycine hydrochloride buffer (pH 2.6) and with high concentrations of sodium iodide, sodium thiocyanate, and magnesium chloride. Ethylene glycol, which is thought to disrupt hydrophobic interactions, was also effective in eluting fibronectin from gelatin columns. This has been found to be the case with another collagen-binding protein, collagen glycosyltransferase.[42] We have also found that treatment of gelatin columns, to which fibronectin was bound, with bacterial collagenase released fibronectin. This elution method may be useful when it becomes necessary to use the mildest possible elution conditions. Antibodies to fibronectin prevent binding of fibronectin to gelatin-Sepharose and cause elution of already bound fibronectin. This point is discussed below.

INTERACTION OF ANTIFIBRONECTIN WITH FIBRONECTIN-COLLAGEN COMPLEXES: EFFECT ON RADIOIMMUNOASSAY

Despite the importance of quantitation of fibronectin for experiments on its expression, many laboratories have employed methods, such as estimation of density of gel bands and immunofluorescence, that are semiquantitative at best. Purification of fibronectin on gelatin columns has allowed us to greatly simplify and improve on previous radioimmunoassay procedures. We have reported the details of the method as applied to human and mouse fibronectins.[37] The salient points of the procedure are that fibronectin is iodinated directly from a solution that contains 2 M urea and that after iodination, the labeled protein is refractionated on gelatin-Sepharose.

More than 90% of the radioactivity in the labeled fibronectin obtained binds to antifibronectin and to gelatin. Labeled fibronectin obtained with other procedures, such as cold-insoluble globulin purification[8] or immunoadsorbence,[24] shows lower binding to antibodies and especially to gelatin.

We have used the labeled fibronectin and the radioimmunoassay to study the effects of fibronectin antibodies on collagen binding of fibronectin and vice versa.

The main findings have been that collagen, which is likely to be present in fibronectin samples, does not affect the quantitation of fibronectin in radioimmunoassays. The explanation of this observation is that antibodies inhibit the binding of fibronectin to collagen and even reverse such binding (TABLE 1).[37] This fact suggests that it is possible to obtain antibodies to the collagen-binding site of fibronectin. However, the antibodies could attach to areas other than the collagen-binding site if such binding activity results in a change of the conformation of the fibronectin molecule so as to make it unable to bind collagen. Steric hindrance could also be involved. In any event, it seems that the antibodies must have a higher affinity for fibronectin than does collagen. This hypothesis is in agreement with the fact that fibronectin can be eluted from gelatin columns with 2–4 M urea (see above), while antigen-antibody complexes usually require higher concentrations of urea to dissociate.[43]

Possible Biologic Significance of the Collagen- and Fibrin-Binding Properties of Fibronectin

Immunofluorescence reveals that fibronectin is especially abundant in the basement membranes,[44] and fibronectin could be expected to display preferential binding to basement membrane collagens. However, we found that type-AB collagen, which may be a basement membrane collagen,[45] is less active than are the

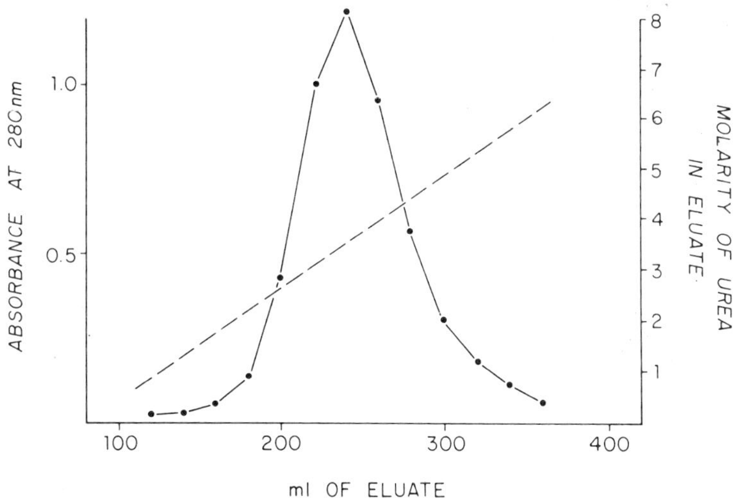

FIGURE 6. Elution of human plasma fibronectin found gelatin-Sepharose by use of a urea gradient. Fibronectin from 400 ml of human plasma was bound to an 80-ml column of gelatin-Sepharose. The column was washed with phosphate-buffered saline and eluted with a linear gradient of 0–8.5 M urea in 0.05 M Tris HCl buffer (pH 7.5). The volume of the gradient was 400 ml.

other collagens and that type III was the most active of the native collagens. Recent evidence[46,47] suggests that there are several other kinds of basement membrane collagens that have not yet been characterized. Once they become available, it will be important to analyze their capacity to bind fibronectin.

Since the specialized functions of the different collagens, if any, are not known, it is difficult to assess the significance of the higher activity of native type-III collagen in fibronectin binding. This phenomenon may have little influence *in vivo*. Because our results[16,33] and those obtained in studies on the specificity of cell attachment protein[48] show that collagens are active, it may be the type and amount of collagen available at a given location, rather than the relative binding activities of the different collagens, that determine which collagen interacts with fibronectin. That such an interaction does occur *in vivo* is only inferred, not proven.

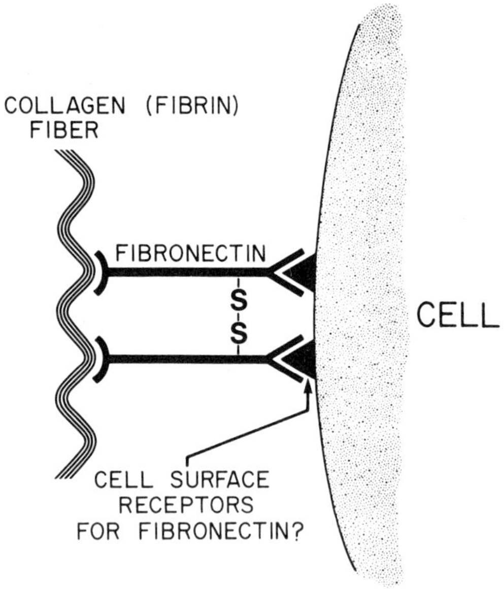

FIGURE 7. A hypothetical model for the arrangement of fibronectin as an adhesion molecule between the cell surface and the extracellular matrix.

At present, the codistribution of collagen and fibronectin observed by immunofluorescence on the surface of cultured fibroblasts[49] and the detachment of cell surface fibronectin by treatment of cells with collagenase[19] are the only evidence available to suggest that fibronectin does bind to collagen under conditions that approximate the *in vivo* situation.

Since fibronectin is a cell surface component, it may be that fibronectin attaches cells to the extracellular collagen matrix. There is direct evidence from *in vitro* experiments that such attachment occurs. A fibroblast surface component, cell attachment factor (CAF),[9] and the plasma protein CAP[10] mediate the attachment of cultured fibroblasts to collagen-coated plates. Both are apparently identical to fibronectin.[9,11]

We would like to propose the model presented in FIGURE 7 as a basis for further experimentation. The two subunits of fibronectin, in addition to their linkage by a disulfide bond(s), are well-established facts (FIGURE 5[8,50,51]). It is clear from the evidence[16] discussed above that fibronectin binds to collagen in a test tube, and it is reasonable to assume that this binding occurs *in vivo*. This hypothesis accounts for the binding to a collagen fiber shown in the model. It is not known how many binding sites fibronectin has for collagen. Although plasma fibronectin exhibits two bands when analyzed on SDS gels after reduction,[37] we have assumed that the subunits are similar with respect to collagen binding and have therefore drawn two binding sites. The evidence for another kind of binding site with an affinity for the cell surface is less well established, and the presence of a specific cell surface receptor for fibronectin as depicted is entirely hypothetical. There is some evidence from work on membrane and plasma factors that mediate cell attachment[9,10] that two binding phenomena may be involved.

The adhesion phenomenon requires the addition of divalent cations and is inhibited by amino group-containing compounds, such as lysine. Lysine was found to affect the attachment of cells to collagen-coated plates previously treated with CAF, but it did not prevent the attachment of CAF to the collagen-coated plates.[9] The collagen binding of fibronectin does not require the addition of calcium or magnesium, is not inhibited by 1 mM EDTA,[16] and is not affected by lysine.[52] These results support the idea of a different binding site. The putative binding of fibronectin to heparin is also calcium dependent,[35] and heparin has no inhibitory effect on collagen binding.[16] It is possible that the heparin binding and the apparent affinity for cell surfaces are related. All of this evidence is suggestive of the presence of a separate binding site for cell surfaces in fibronectin, but until proven or disproven by binding experiments utilizing purified reagents, this other binding site must be considered hypothetical. We hope that this part of the model will provoke further experimentation.

CONCLUDING REMARKS

Fibronectin has an extremely stable immunochemical structure both in terms resistance to alteration by chemical modification of the molecule and in terms of similarity of fibronectins of different species. Fibronectin, subjected to treatment with SDS, retains its immunologic reactivity, and cross reactivity is found between mammalian and avian fibronectins. The evidence available indicates that the antigenic determinants reside in the protein part of the glycoprotein molecule. Antibodies against fibronectin isolated from fibroblasts or plasma show complete immunologic identity between the fibronectins from these two sources. Recent chemical evidence and shared functional properties, such as collagen binding, strongly support the immunochemical evidence on the similarity of the plasma and fibroblast fibronectins.

Fibronectin has affinity for collagen and can be purified from plasma of different mammalian species, chicken plasma, and extracts of cultured chicken fibroblasts by use of affinity chromatography on gelatin coupled to Sepharose particles. A fibronectin-like protein has been isolated from the plasma of Torpedo fish by this method.

The collagen-binding properties of fibronectin studied by use of genetically different types of collagen and their derivatives demonstrate that heat-denatured collagens of types I–III are equal in their binding capacity and more active than are the native forms or denatured type-AB collagen. Native type-III collagen is more active than are the other native collagens.

Fibronectin binds to fibrinogen and fibrin. Comparison of this binding to collagen binding shows that fibrinogen inhibits binding of fibronectin to collagen but is not very active. The binding of fibronectin to fibrinogen is inhibited by collagen. This finding suggests that the interaction of fibronectin with collagen and fibrinogen depends on the same binding site.

It is proposed that cell surface fibronectin mediates attachment of cells to the collagenous extracellular matrix and to a temporary fibrin matrix in a wound. Since transformed cells lack fibronectin on their surface, their capacity to grow uninhibitedly and metastasize may, at least partly, depend on the resultant lack of the supportive, and possibly growth inhibitory, attachment to the surrounding extracellular matrix.

Acknowledgment

We thank Aulikki Pekkala for technical assistance.

References

1. Ruoslahti, E., A. Vaheri, P. Kuusela & E. Linder. 1973. Biochim. Biophys. Acta **322:** 352–358.
2. Ruoslahti, E. & A. Vaheri. Nature (London) **248:** 789–791.
3. Hynes, R. O. 1973. Proc. Nat. Acad. Sci. USA **70:** 3170–3174.
4. Hynes, R. O. 1976. Biochim. Biophys. Acta **458:** 73–107.
5. Gahmberg, C. G. & S. Hakomori. 1973. Proc. Nat. Acad. Sci. USA **70:** 3329–3333.
6. Yamada, K. & J. Weston. 1974. Proc. Nat. Acad. Sci. USA **71:** 3492–3496.
7. Robbins, P. W., G. G. Wickus, P. E. Branton, B. J. Gaffney, C. B. Hirschberg, P. Fuchs & P. M. Blumberg. 1974. Cold Spring Harbor Symp. Quant. Biol. **39:** 1173–1180.
8. Mosher, D. F. 1975. J. Biol. Chem. **250:** 6614–6621.
9. Pearlstein, E. 1976. Nature (London) **262:** 497–500.
10. Klebe, R. J. 1974. Nature (London) **250:** 248–251.
11. Jilek, F. & H. Hörmann. 1977. Hoppe-Seyler's Z. Physiol. Chem. **358:** 133–136.
12. Stone, K. R., R. E. Smith & W. K. Joklik. 1974. Virology **58:** 86–100.
13. Hogg, N. M. 1974. Proc. Nat. Acad. Sci. USA **71:** 489–492.
14. Vaheri, A. & E. Ruoslahti. 1974. Int. J. Cancer **13:** 579–586.
15. Ruoslahti, E. & A. Vaheri. 1975. J. Exp. Med. **141:** 497–501.
16. Engvall, E. & E. Ruoslahti. 1977. Int. J. Cancer **20:** 1–5.
17. Cresswell, P., M. J. Turner & J. L. Strominger. 1973. **70:** 1603.
18. Keski-Oja, J., A. Vaheri & E. Ruoslahti. 1976. Int. J. Cancer **17:** 261–269.
19. Blumberg, P. & P. W. Robbins. 1975. Cell **6:** 137–147.
20. Zetter, B. R., L. B. Chen & J. M. Buchanan. 1976. Cell **7:** 407–412.
21. Kuusela, P., E. Ruoslahti, E. Engvall & A. Vaheri. 1976. Immunochemistry **13:** 639–642.
22. Vaheri, A. & E. Ruoslahti. 1975. J. Exp. Med. **142:** 530–535.
23. Engvall, E., E. Ruoslahti & E. J. Miller. 1978. J. Exp. Med. In press.
24. Vuento, M., M. Wrann & E. Ruoslahti. 1977. FEBS Lett. **82:** 227–231.
25. Kuusela, P., E. Ruoslahti & A. Vaheri. 1975. Biochim. Biophys. Acta **379:** 295–303.
26. Adams, S. L., M. E. Sobel, B. H. Howard, K. Olden, K. M. Yamada, B. deCrombrugghe & I. Pastan. 1977. Proc. Nat. Acad. Sci. USA **74:** 3399–3403.
27. Mayne, R., M. S. Vail, E. J. Miller, S. H. Blose & S. Chacko. 1977. Arch. Biochem. Biophys. **181:** 462–469.
28. Ruoslahti, E., H. Pihko & M. Seppälä. 1974. Transplant. Rev. **20:** 38–60.
29. Ruoslahti, E. & H. O. Sjögren. 1976. Int. J. Cancer **18:** 375–378.
30. Rieber, M. & E. Romano. 1976. Cancer Res. **36:** 3568–3573.

31. ENGVALL, E., K. JONSSON & P. PERLMANN. 1971. Biochim. Biophys. Acta **251:** 427–434.
32. ENGVALL, E. & P. PERLMANN. 1972. J. Immunol. **109:** 129–135.
33. MILLER, E. J. 1976. Mol. Cell. Biochem. **13:** 165–191.
34. STEMBERGER, A. & H. HÖRMANN. 1976. Hoppe-Seyler's Z. Physiol. Chem. **357:** 1003–1005.
35. STATHAKIS, N. E. & M. W. MOSESSON. J. Clin. Invest. **60:** 855–865.
36. YAMADA, K. M., S. S. YAMADA & I. PASTAN. 1975. Proc. Nat. Acad. Sci. USA **72:** 3158–3162.
37. RUOSLAHTI, E., M. VUENTO & E. ENGVALL. 1978. Biochim. Biophys. Acta. In press.
38. FULLER, G. M. & R. F. DOOLITTLE. 1971. Biochemistry **10:** 1311–1315.
39. MOSHER, D. 1977. Biochim. Biophys. Acta **491:** 205–210.
40. FULLER, G. M. & R. F. DOOLITTLE. 1971. Biochemistry **10:** 1305–1311.
41. MOSESSON, M. W., A. B. CHEN & R. M. HUSEBY. 1975. Biochim. Biophys. Acta **386:** 509–524.
42. ANTTINEN, H., R. RISTELI & K. KIVIRIKKO. 1977. Eur. J. Biochem. **78:** 11–17.
43. RUOSLAHTI, E. 1976. Scand. J. Immunol. Suppl. **3:** 39–49.
44. LINDER, E., A. VAHERI, E. RUOSLAHTI & J. WARTIOVAARA. 1975. J. Exp. Med. **142:** 41–49.
45. BURGESON, R. E., F. A. EL ADLI, I. I. KAITILA & D. HOLLISTER. 1976. Proc. Nat. Acad. Sci. USA **73:** 2579–2583.
46. KEFALIDES, N. A. (Ed.) 1977. First International Symposium on Basement Membrane Collagens 1976. Academic Press, Inc. New York, N.Y. In press.
47. FESSLER, L. J. & J. H. FESSLER. 1978. Annu. Rev. Biochem. In press.
48. KLEINMAN, H. K., E. B. MCGOODWIN & R. J. KLEBE. 1976. Biochem. Biophys. Res. Commun. **72:** 426–432.
49. BORNSTEIN, P. & J. F. ASH. 1977. Proc. Nat. Acad. Sci. USA **74:** 2480–2484.
50. KESKI-OJA, J., D. E. MOSHER & A. VAHERI. 1977. Biochem. Biophys. Res. Commun. **74:** 699–706.
51. HYNES, R. O. & A. DESTREE. 1977. Proc. Nat. Acad. Sci. USA **74:** 2855–2859.
52. RUOSLAHTI, E. & E. ENGVALL. 1977. Unpublished results.

LABELING, SOLUBILIZATION, AND PEPTIDE MAPPING OF FIBROBLAST SURFACE GLYCOPROTEIN

R. Colin Hughes and Roderick Nairn

National Institute for Medical Research
Mill Hill, London NW7 1AA, England

One necessary first step toward determining the structure of biologically active surface components is to solubilize the surface membrane by using conditions that do not destroy biologic activity and then to separate and characterize the active molecules. Although integral membrane proteins require detergents for solubilization, other cell surface components classified as peripheral are often released by manipulation of the ionic conditions or pH of the suspending medium.[1-3] However, the concept[4] of integral and peripheral components is only a generalization, and exceptions might be expected to these guidelines, which were formulated from data obtained from a limited sample of membrane components from a few cell types. The fibroblast surface-associated large external temperature-sensitive (LETS) glycoprotein has been extracted from cells with urea alone[5] or mixtures of urea and detergents.[6] However, the yields obtained and the effects of other solubilizing reagents have not been studied systematically. The experiments detailed in this report describe the effects of various detergents and other solubilization agents on the iodinated surface components of BHK cells, including the high-molecular-weight LETS component, and extend previous communications of ourselves[7] and other workers.[8] In addition, primary structure comparisons by peptide analysis of the radiolabeled LETS component of BHK cells[7] and the equivalent components of other cells are presented. Cell surface glycoproteins that show similar biologic properties on a variety of cell lines might be expected to have a similar molecular structure. Some early evidence[9] indicated little antigenic cross reaction between avian and mammalian components, although this claim has been modified more recently.[10] However, since antigenic cross reactivity need not necessarily indicate molecular identity, we have attempted to employ another approach to define the extent of structural homology between these surface components.

Materials and Methods

General Methods

BHK 21 C13 cells were grown[11] to confluency (approximately 2×10^7 cells) on 60 mm diameter plastic petri dishes in modified Eagle's medium plus 10% tryptose phosphate broth and 10% fetal calf serum. Primary chick embryo fibroblasts were passaged normally at 1:4 split ratios in the supplemented Eagle's medium and used within five to 10 passages. Human embryo fibroblasts were a gift of Dr A. Meager, Department of Biological Sciences, University of Warwick, Coventry, England and were maintained with a 1:2 split ratio in the same growth medium, except that tryptose phosphate was replaced with 10% nonessential amino acid mixture (Gibco-Biocult Ltd., Hounslow, Middlesex, England). Conditions of lactoperoxidase-catalyzed radioiodination of cell monolayers, harvesting cells,

0077/8923/78/0312-0192 $01.75/0

and washing the cell pellets have been described previously.[11,12] Electrophoresis on polyacrylamide disc or slab gels (7.5%) with a stacking gel (3.6%) was performed in buffered sodium dodecyl sulfate. Samples were dissociated at 90°C with 1% sodium dodecyl sulfate-1% β-mercaptoethanol before electrophoresis. Procedures for gel slicing and radioactive counting have been described.[11]

Extraction of Cell Pellets

This procedure was performed at 2°C for 30 min in 10 mM Tris HCl (pH 8.2) supplemented with the appropriate extractant unless otherwise indicated. The cell pellet was suspended in 1 ml of extractant in a conical centrifuge tube, and a sample (0.3 ml) was taken and frozen immediately. The remaining suspension was mixed frequently throughout the extraction period and then centrifuged at 1000*g* for 5 min at 2°C. The supernatant was removed by means of a hyperdermic syringe and transferred to 0.5 ml of cellulose nitrate tubes for centrifugation at 35,000 rpm for 100 min ($\sim 10^7 g$/min) in a Beckman type 40 rotor fitted with appropriate adaptors. The $10^7 g$/min supernatant ("soluble fraction") was removed, mixed with 2 vol of cold absolute alcohol, 100 μg of bovine serum albumin was added, and the protein was allowed to precipitate at −20°C for 2 days. These fractions were centrifuged at 2°C and 10,000*g* for 30 min, the ethanol supernatants were collected, and the precipitate was washed once more with cold absolute alcohol by centrifugation. Urea (Aristar, Hopkins and Williams Ltd., Chadwell Heath, Essex, England) -containing solutions were prepared fresh; Nonidet®P40 (BDH Chemicals Ltd., Poole, Dorset, England), Triton® X-100 (Koch-Light Ltd., Colnbrook, Bucks., England), sodium deoxycholate (E. Merck, Darmstadt, Federal Republic of Germany), and other extractants were stable and generally stored for up to 1 month at −2°C.

Elution of Iodinated Polypeptides from Gels

Slices (2 mm wide) of disc gels that contained the particular iodinated band to be analyzed were homogenized by 12 strokes in a Dounce homogenizer with projections on the pestle (Uniform MTG3/P, Jencons Scientific Ltd., Hemel Hempstead, Herts, England) in three separate portions of 0.1% sodium dodecyl sulfate, Tris HCl buffer (pH 8.3). The clear supernatants collected by centrifugation at 10,000*g* for 30 min, which contained 70–80% of the total ^{125}I radioactivity, were dialyzed against water at 2°C for 2–3 days and freeze dried.

Proteolysis

Unless otherwise indicated, the freeze-dried material was dissolved in 0.2 M NH_4HCO_3 and digested either with 100 μg of trypsin (type III, Sigma Chemical Co., Kingston upon Thames, England) or 100 μg of thermolysin (type X, Sigma) for 16 hr at 37°C. For purposes of exact comparison, thermolytic digestion was performed at 37°C rather than at 55°C, where it is more efficient. After digestion, the solution was centrifuged at 1000*g* for 5 min. The supernatant, which contained essentially all (>90%) of the radioactivity, was removed, frozen, and freeze dried.

Peptide Analysis

The freeze-dried material was taken up in a pyridine-acetic acid-water mixture (1:10:989, pH 3.5). Then, 10 μl of a 1% solution of the dye markers methyl green and ϵ-2,4-dinitrophenyl lysine were added. Samples of 50–100 μl ($\sim 10^5$ cpm of ^{125}I radioactivity) were applied to full sheets of Whatman no. 3 chromatography paper (45 × 57 cm) and electrophoresed at pH 3.5.[13] Conditions were 3 kV and 50 mA for 90 min. By this time, methyl green and ϵ-dinitrophenyl lysine had migrated 33–36 and 4–5 cm, respectively. The dried papers were then chromatographed for 16–20 hr with butan-1-ol-acetic acid-water (4:1:5, upper phase). At the end of chromatography, the paper was dried and exposed to Kodak x-ray film (Kodirex Auto process AP54, Kodak Ltd., London, England) for 4–6 weeks.

Iodination of Material Eluted from Gels

This procedure was performed by a modification of the chloramine T method.[14] Freeze-dried samples of gel eluates were reconstituted in water (0.4 ml) and protein precipitated by addition of acetone (1.6 ml). After centrifugation, the pellet was washed with 80% acetone, and to the washed protein pellet were then added 0.04 ml of 0.1% sodium dodecyl sulfate (SDS) in 50 mM Tris HCl (pH 7.0), 100 μCi of ^{125}I, and 10 μl of chloramine T (2.5 mg/ml in phosphate-buffered saline). Sodium metabisulfate (0.05 ml, 2.5 mg/ml) was added after 5 min at room temperature (18°C), and the samples were cooled to 2°C for 15 min. Then, bovine serum albumin (1 mg/ml), followed by 50% trichloroacetic acid (0.2 ml) were added, and precipitated protein was collected by centrifugation. The pellets were washed with 0.1 M HCl and acetone, followed by acetone alone, dried in a vacuum, and taken for peptide analysis.

RESULTS

Labeled Products

The major surface proteins labeled by lactoperoxidase-catalyzed iodination of BHK cells are resolved by SDS polyacrylamide gel electrophoresis into three areas of nominal molecular weights: 250K (LETS glycoprotein component), 120–140K, and 75–100K.[11] The extent of labeling of the 250K component relative to that of other iodinated components varies in separate experiments two- to threefold, even when conditions of growth and labeling are kept as constant as practicable. The extent of labeling also depends on cell type. Chick embryo fibroblasts appear, for example, to have 40–50 times more label in the 250K fraction than do BHK cells iodinated under comparable conditions, indicating that chick cells have more of this surface component or that the chick component contains a large excess of iodinatable tyrosine residues. In all cases, monoiodotyrosine was the only iodinated amino acid detected under the conditions described, that is, at carrier-free iodide concentrations. This was shown by precipitating iodinated cells with 20% trichloroacetic acid and treating the washed precipitated proteins with pronase (*Streptomyces griseus* protease). The radiolabeled amino acid in the digest was identical to authentic monoiodotyrosine by paper chromatography and clearly distinguishable from diiodotyrosine and other iodinated amino acid derivatives.[26]

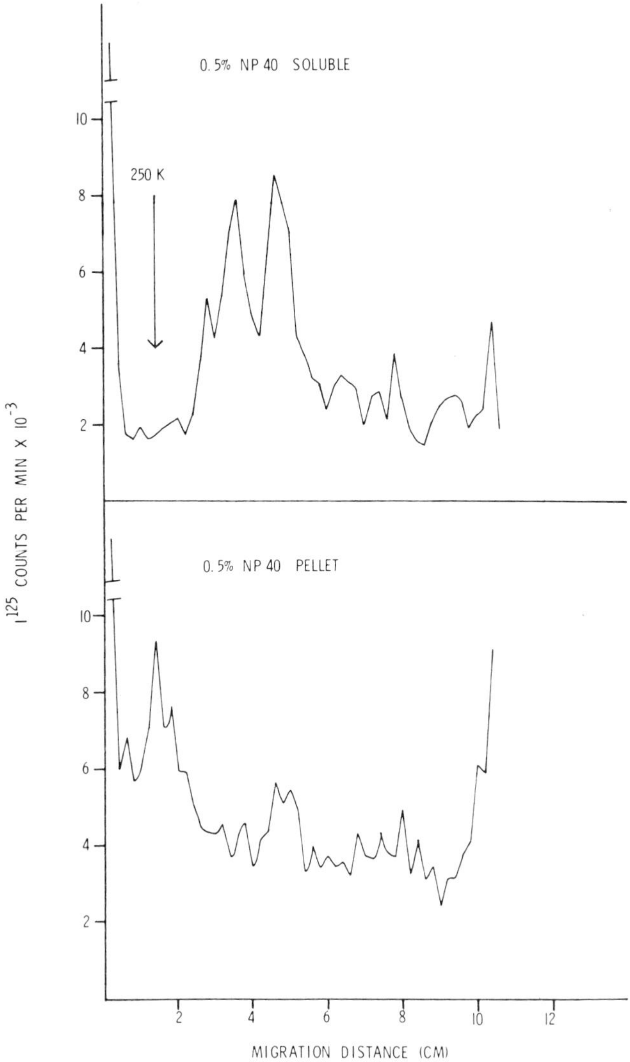

FIGURE 1. Sodium dodecyl sulfate polyacrylamide gel electrophoresis of ^{125}I-labeled polypeptides of lactoperoxidase-catalyzed iodinated BHK cells. Electrophoresis is from left to right.

Solubilization of Iodinated BHK Cell Surface Components

The definition of soluble was taken to be material nonsedimentable at $10^7 g$/min. Nonionic detergents readily solubilized iodinated components other than the LETS component (FIGURE 1). Almost complete extraction of iodinated components, except LETS glycoprotein, was obtained after 30–45 min of extraction with 0.5% Nonidet P40 or 1% Triton X-100. The ionic detergent sodium deoxy-

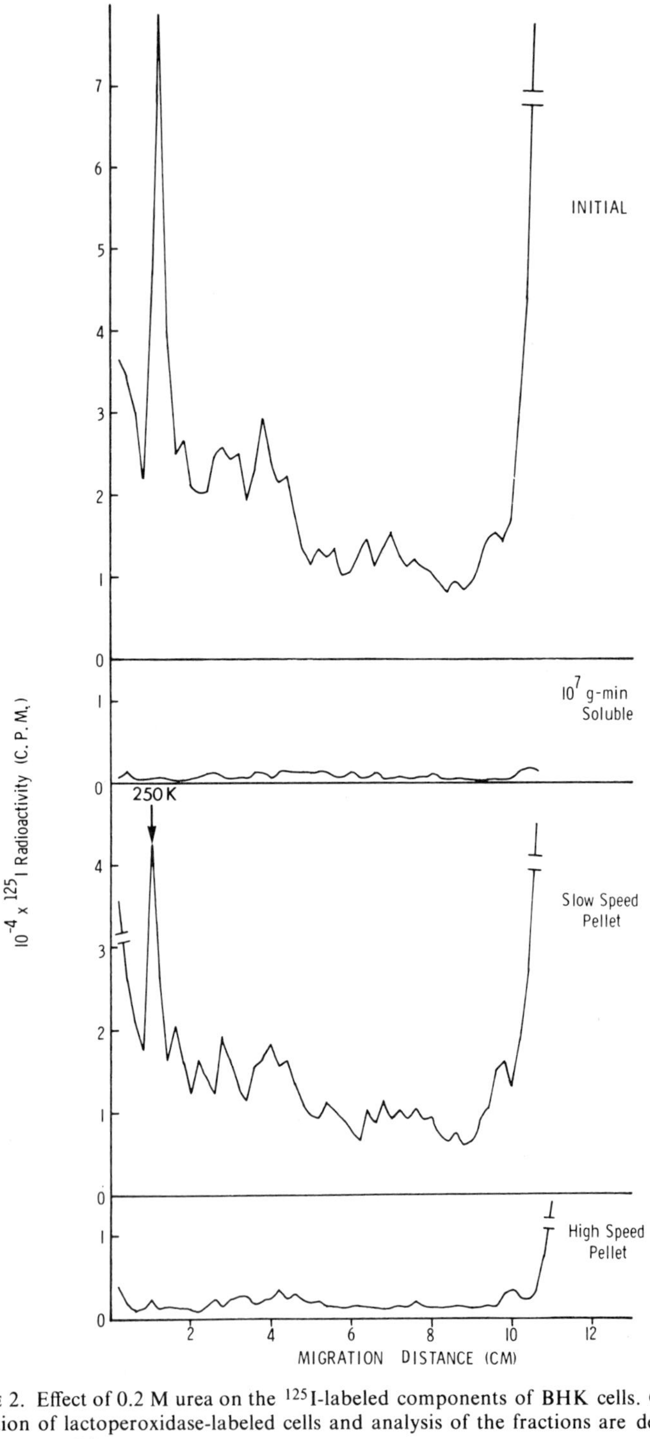

FIGURE 2. Effect of 0.2 M urea on the ^{125}I-labeled components of BHK cells. Conditions of extraction of lactoperoxidase-labeled cells and analysis of the fractions are described in MATERIALS AND METHODS.

cholate (1%) extracted most of the iodinated species, except LETS glycoprotein, after 15 min at 2°C and essentially all, except LETS glycoprotein, by 30 min. After 30 min, some degradation of LETS glycoprotein was observed by SDS polyacrylamide gel electrophoresis of the extracted residue, and this phenomenon was not prevented by addition of the protease inhibitor phenylmethylsulfonyl

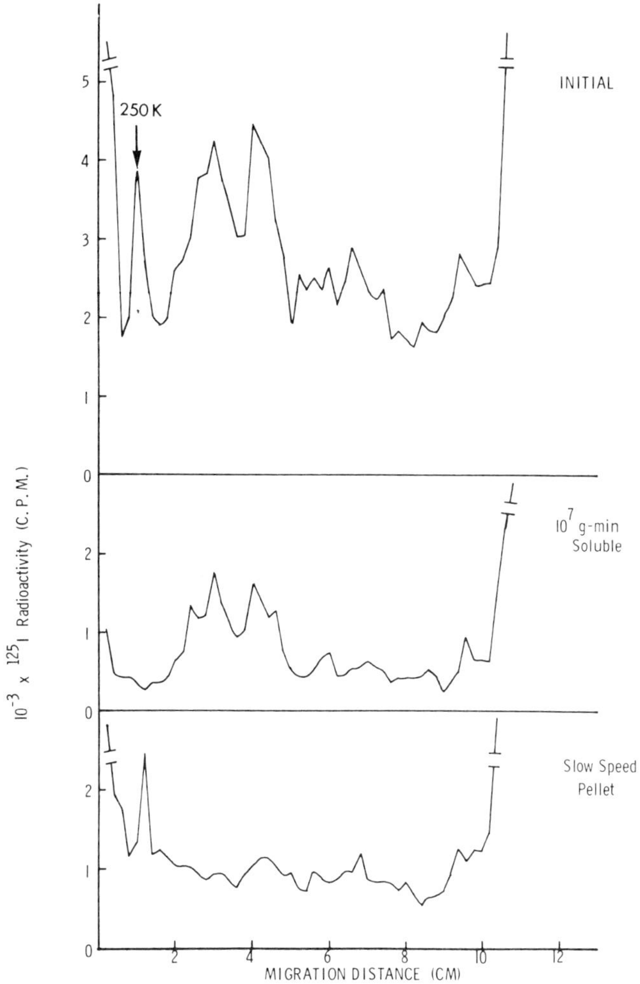

FIGURE 3. Solubilization of ^{125}I-labeled components of BHK cells with 0.2 M urea-1% Triton X-100.

fluoride (PMSF) at 2 mM final concentration. In contrast to other reports with chick cells, extraction[5] of BHK cells with 0.2 M urea removed no detectable LETS glycoprotein and indeed did not appear to extract any major iodinated surface component (FIGURE 2). Much higher concentrations (8–9.5 M) of urea extracted all of the iodinated components, including LETS glycoprotein, particularly when

TABLE 1

EXTRACTION OF IODINATED BHK CELLS*

Extractant	Slow-Speed Pellet†	High-Speed Pellet‡	High-Speed Supernatant§	Alcohol Supernatant¶
Urea, 0.2 M	86‖	1	1	ND**
PMSF, 2 mM	65	1	1	1
Urea, 0.2 M; Triton X-100, 1%	46	1	1	ND
Urea, 0.2 M; Triton X-100, 1%; PMSF, 2 mM	43	1	1	1

*BHK cells were iodinated and extracted under standard conditions with the stated reagents. The various fractions were analyzed by SDS polyacrylamide gel electrophoresis.

†Radioactivity not extracted.

‡$10^7 g$/min pellet.

§$10^7 g$/min supernatant.

¶Supernatant from the cold alcohol precipitation and wash.

‖All data are counts per minute (% of total) and were estimated as the summed counts in the resolving gel, less background and the activity at the dye front, divided by the total counts in the resolving gel of the initial starting sample.

**Not determined.

incorporated in 10 mM cyclohexylaminopropane sulfonic acid (CAPS), 0.15 M NaCl, 1 mM $CaCl_2$ (pH 11) buffer.[27] A mixture of urea (0.2 M) and Triton X-100 (1%) did not solubilize the LETS component, although it did solubilize the other iodinated polypeptides (FIGURE 3). The urea-Triton X-100 mixtures used by others[10] successfully to extract LETS component from chick cells contained 2 mM PMSF to control endogenous proteolysis during extraction. Phenylmethylsulfonyl fluoride is made up in 2% propan-2-ol, and we checked that at this concentration propan-2-ol had no solubilizing effect by itself or in conjunction with the 0.2 M urea-1% Triton X-100 reagent. The latter results closely resembled those obtained with 1% Triton X-100 alone or in combination (FIGURE 3) with 0.2 M urea. To be certain in all of these extractions that no iodinated polypeptides, including LETS component, was being lost, any high-speed pellet ($10^7 g$/min) and the supernatant from the cold alcohol precipitation step were analyzed by SDS polyacrylamide gel electrophoresis (TABLE 1). Although there was considerable radioactivity present, most of it appeared to be free radioiodine, and little, if any, was associated with polypeptide material.

The LETS component was also resistant to extraction with 0.2% EDTA (FIGURE 4). In agreement with this result, lactoperoxidase-labeled BHK cells can be released from monolayer cultures with 0.2% EDTA without loss of any iodinated polypeptide species, including LETS glycoprotein. No extraction of iodinated polypeptides was obtained with NaCl (0.1–2 M) or 0.14 M salt plus 10% ethanol. The latter solution extracts a large 250K molecular weight surface glycoprotein of *Paramecium aurelia*.[15] Although high-salt (1 M NaCl) released no radioactive polypeptides into the $10^7 g$/min supernatant, incubation with 1–2 M NaI did solubilize the iodinated components, including LETS glycoprotein (FIGURE 5), probably due to the chaotropic effect of the iodide ion. However, although 11 and 40% of the total counts incorporated into LETS glycoprotein were found in the $10^7 g$/min supernatant after treatment with 1 or 2 M NaI, respectively, there was a considerable amount of iodinated LETS glycoprotein in the high-speed pellet (FIGURE 5 & TABLE 2). This is material released into a form not sedimentable at $5 \times 10^3 g$/min but sedimentable at $10^7 g$/min.

Peptide Analysis

The results so far showed that the LETS component of BHK cells could not be solubilized by most reagents. Therefore, we utilized the high solubilizing activity of SDS and polyacrylamide gel electrophoresis to prepare material for analysis. The LETS component, resolved by SDS polyacrylamide gel electrophoresis from the other iodinated cell surface components, was analyzed after elution from the resolving gel and removal of SDS. By means of these procedures,

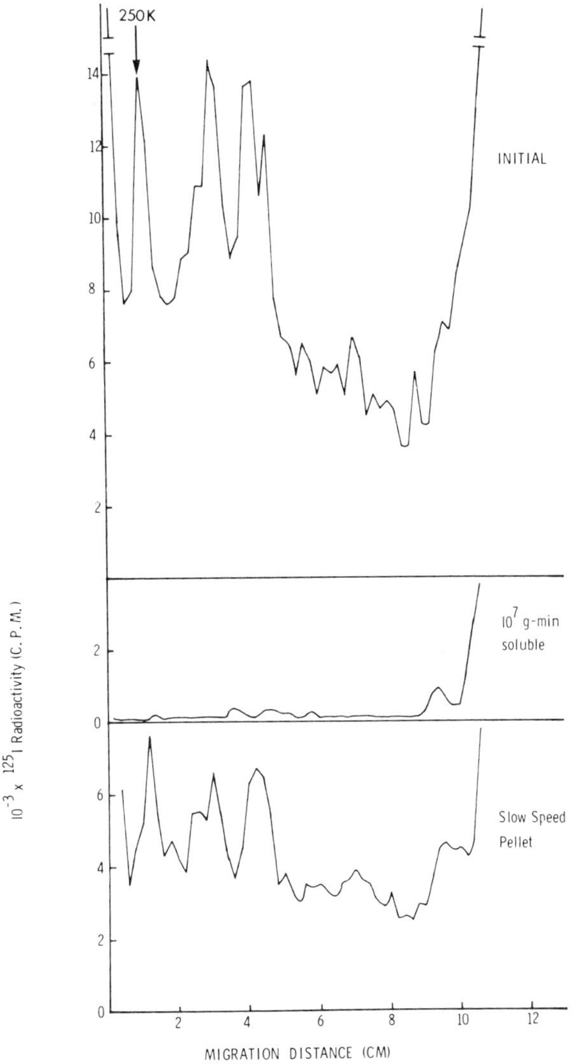

FIGURE 4. Effect of 0.2% EDTA on the ^{125}I-labeled components of BHK cells.

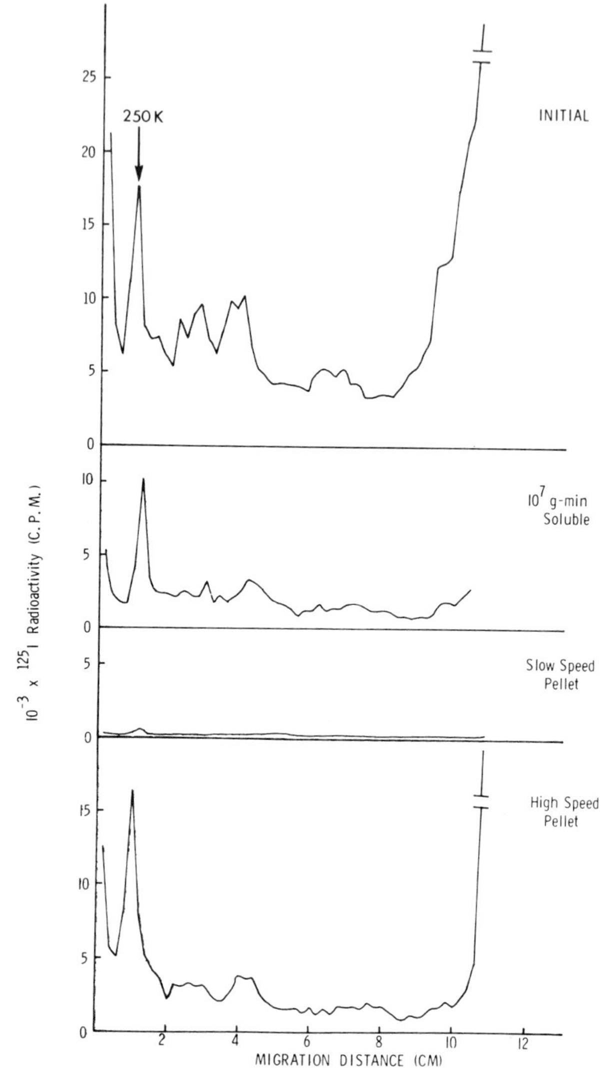

FIGURE 5. Solubilization of the ^{125}I-labeled components of BHK cells with 2 M sodium iodide.

"fingerprints" of the mixture of iodinated tyrosine-containing peptides produced by protease digestion of LETS glycoproteins were obtained. Thermolysin was found in trial experiments to give a more efficient digestion than that obtained with trypsin, as expected from its broader specificity. Perhaps the ability of thermolysin to catalyze hydrolysis of hydrophobic residues, such as leucine, isoleucine, valine, and phenylalanine, which comprise about 21% of the total polypeptide content of hamster LETS glycoprotein,[27] induces better fragmentation, although the

TABLE 2

EXTRACTION OF THE IODINATED LETS GLYCOPROTEIN FROM BHK CELLS

Extractant	Total cpm Extracted (%)*	LETS Glycoprotein cpm Extracted (%)†
Urea, 0.2 M; Triton X-100, 1%; PMSF, 2 mM	39	0
NaI, 1 M	21	11
NaI, 2 M	55	46

*Total counts in the resolving gel of the soluble fraction, namely, $10^7 g$/min supernatant, less background and the activity that migrated at the dye front, expressed as percentage of total counts per minute summed in the same way for starting sample.

†Counts per minute of the LETS glycoprotein peak in the soluble fraction expressed as percentage of total counts per minute in the LETS glycoprotein peak of the starting sample.

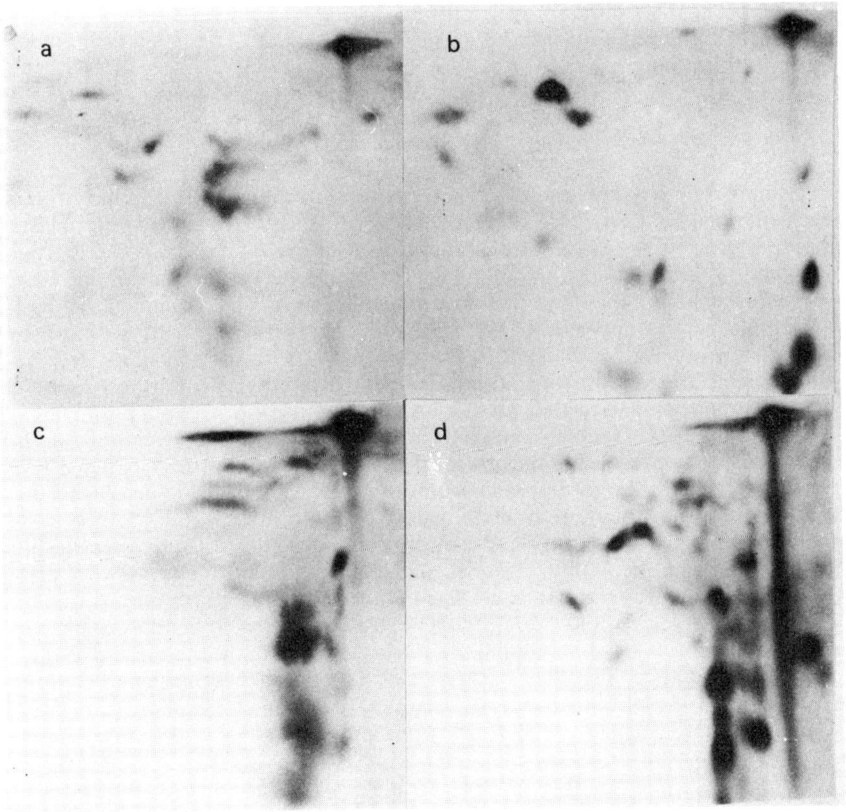

FIGURE 6. Comparison of the thermolysin peptide maps of LETS glycoproteins labeled by lactoperoxidase-catalyzed iodination of intact cells and isolated by polyacrylamide gel electrophoresis. a, BHK cells; b, chick embryo fibroblasts; c, human embryo fibroblasts. The trypsin peptide map of chick cells LETS glycoprotein is shown for comparison (d). Samples were loaded at the top right-hand side of each map. Electrophoresis is from right to left.

glycoprotein is not particularly rich in nonpolar residues and contains 10% lysine plus arginine.

FIGURE 6(a) shows a two-dimensional peptide map prepared from the radio-iodinated component of BHK cells, eluted from SDS gel slices (~ 10^5 cpm) and treated with thermolysin. In several independent experiments, a consistent pattern emerged of approximately 15–20 radioactive peptides. The radioactivity that remained at the origin was identified, in part, as glycopeptide fragments, by elution from the chromatogram with 1% Nonidet P40 and adsorption onto a small column of concanavalin A-Sepharose (5 ml capacity, 5 mg of lectin). Up to 30% of the total ^{125}I radioactivity was washed through the column with 1% Nonidet P40, but a majority adsorbed tightly, revealing that a carbohydrate moiety was present. The relative intensities of the radioactive peptides resolved by "fingerprinting" tended to be rather similar between experiments and, since the only detectable iodinated amino acid derivative was identified as monoiodotyrosine, may give a general indication of the abundance and/or accessibillty of tyrosine residues of each peptide. However, other work suggests caution in interpreting the degree of labeling in these terms, because the reactivity of tyrosine residues to lactoperoxidase-catalyzed iodination, even in soluble proteins, varies by at least one order of magnitude.[16]

Peptide Mapping of LETS Components from Other Cells

Examples of two-dimensional peptide maps prepared from LETS glycoprotein that contained 10^5 cpm of [^{125}I]lactoperoxidase-labeled chick embryo fibroblasts, after purification by SDS polyacrylamide gel electrophoresis and thermolysin (FIGURE 6, b) or trypsin (FIGURE 6, d) digestion, are shown. Sixteen peptide spots were reproducibly found, and five or six of them were very prominent. A basis hydrophilic peptide appeared to be readily released by thermolysin, as shown by controlled proteolysis of chick LETS glycoprotein (FIGURE 7, a–c arrows), recalling the extremely mild conditions of proteolysis by chymotrypsin (with a specificity very similar to that of thermolysin), or trypsin necessary to remove preferentially the LETS component from cells.[17] However, the relation of the proteolysis site(s) that leads to release of the basic peptide from isolated LETS glycoproteins and the glycoprotein from attachment to the cell surface, for example, by trypsin, remains to be investigated. The pattern of peptides (Figure 6, b) is highly characteristic and not particularly complex and is, therefore, easily distinguishable from the pattern obtained under comparable conditions with BHK cell LETS component (FIGURE 6, a). The number of iodinated peptides is similar. Therefore, the elevated (40–50-fold) labeling of this iodinated band in chick cells presumably reflects a greater abundance of this component at the surface of these cells. Human embryo fibroblasts at passage 15 produced a peptide map (FIGURE 6, c) that showed peptides migrating more slowly during electrophoresis than the peptides of BHK or chick cell LETS glycoprotein.

Reiodination of Chick LETS Glycoprotein

The radioactive peptides obtained by thermolysin treatment of LETS glycoprotein labeled on intact cells presumably derive from tyrosine-containing segments that are fully exposed at the cell surface and reactive to lactoperoxidase. To detect other tyrosine residues, chick embryo fibroblasts were iodinated with lactoperoxidase, solubilized in SDS, and the iodinated LETS component was isolated from polyacrylamide gels after electrophoresis. Since little material was present, the labeled glycoprotein was reiodinated by the chloramine T procedure, which is more efficient than lactoperoxidase labeling. It was necessary to remove residual SDS by acetone precipitation, although 0.1% SDS was useful to resuspend the small protein precipitate obtained. The thermolysin peptide map (FIGURE 7, d) of the reiodinated glycoprotein shows remarkably few (~7) additional peptides relative to the maps obtained for surface-labeled material. The two maps are not strictly comparable, becuase the extent of labeling of a particular tyrosine residue

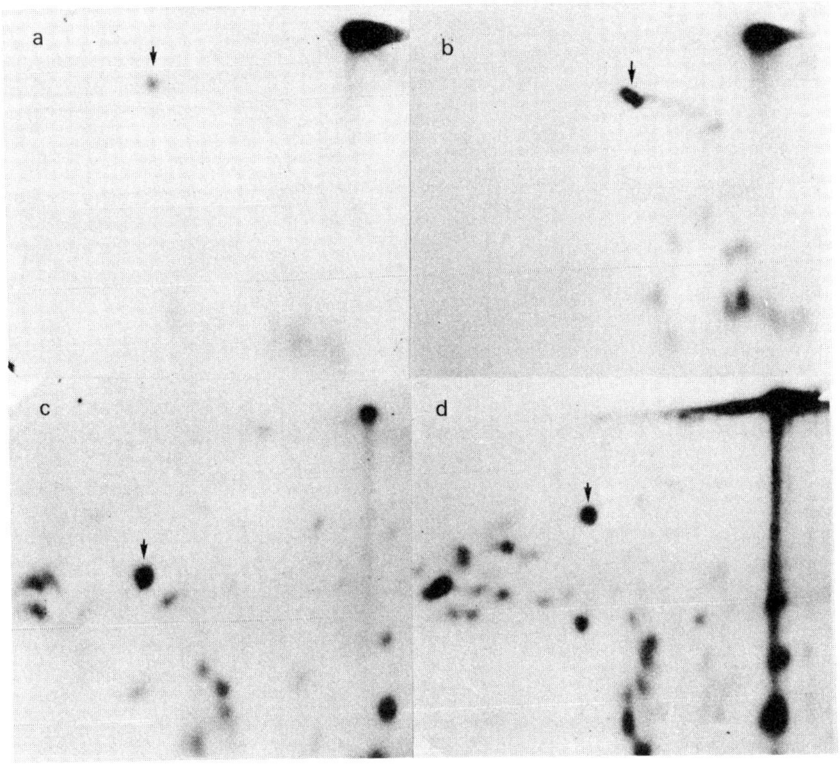

FIGURE 7. Peptide maps of the LETS glycoprotein of chick cells obtained by use of different amounts of thermolysin: a, 10 μg, b, 50 μg; c, 100 μg. The thermolysin map of chick embryo fibroblasts LETS glycoprotein reiodinated by the chloramine T procedure after isolation from polyacrylamide gels is also shown (d).

need not be the same for both iodination methods, and particular peptides may show a greater intensity after one iodination procedure than with another.

DISCUSSION

The results of the solubilization experiments described here and elsewhere[7] suggest a novel arrangement of LETS glycoprotein at the surface of BHK cells. Chelating agents, which deplete bridging ions, low ionic strength, which depletes stabilizing ions, and urea, which affects hydrogen bonding and water structure, are reagents that affect polar interactions and that have no solubilizing effect on the LETS glycoprotein of BHK cells. Detergents, known to bring about solubilization of integral proteins,[1,2,12] perhaps by disturbing lipid-protein associations by binding hydrophobically to sites on proteins normally occupied by lipid, are also ineffective. The latter observation is difficult to reconcile with the suggestion[18] that the LETS glycoprotein may be an integral membrane component or even a transmembrane glycoprotein unless other interactions are postulated. Thus, although nondenaturing detergents, such as Triton X-100, Nonidet P40, or sodium deoxycholate, solubilize all of the other iodinated polypeptides, the LETS component is not released into a form nonsedimentable after $10^7 g$/min centrifugation. In contrast, 1% deoxycholate has been used[19] to solubilize LETS glycoprotein from isolated BHK cell membranes. We suggest that the LETS component present in plasma membrane fractions may already be in a form "loosened" from some normal interactions with other membrane-associated structural components. The Finnish group have shown, for example, that in chick cells, LETS glycoprotein may copurify with actin and have proposed some interaction between LETS glycoproteins and microfilaments.[6] Alternatively, perhaps weak interactions are formed in cell pellets disrupted by nonionic detergents, while the much smaller amounts of actin usually associated with purified membrane fractions are insufficient to interfere seriously with solubilization of LETS glycoprotein. In fact, it is not clear from the published data[19] what proportion of LETS glycoprotein was solubilized from isolated BHK cell membranes, and it may represent only a minor portion, perhaps that released during cell homogenization and trapped inside membrane vesicles. In connection with the latter possibility, it may be relevant that sodium iodide released LETS glycoprotein from BHK cells into a form sedimentable at $10^7 g$/min and into true solution.

Combinations of reagents, such as urea-Triton X-100, that are capable of disrupting both polar and nonpolar interactions were also ineffective in extracting LETS glycoprotein. This observation suggests that the molecule is associated with the cell surface by interactions that will require other combinations of reagents to be disrupted. Recently, the LETS glycoprotein has been isolated[27] by extraction into 8 M urea-CAPS buffer (pH 11) and fractionated in excellent yield by collagen affinity column chromatography.[20] This material will be a convenient source for further structural analysis and confirmation of the preliminary conclusions described here.

Our preliminary comparisons of the radioiodinated LETS glycoproteins from hamster, chick, and human fibroblasts by two-dimensional mapping of tyrosine-containing peptides indicate sequence differences. To some extent, these comparisons might be invalidated if the iodinated band eluted from the gel slices for analysis contained other lactoperoxidase-labeled polypeptides, in addition to LETS glycoprotein. Any difference observed might then be due to the constituents of the mixture being present in different relative amounts in the different cells. However, reiodination of gel eluates that contained the LETS glycoprotein fraction from chick cells produced few additional radioactive peptides compared with the pattern of peptides obtained from the surface-labeled fraction, arguing against the presence of a mixture of polypeptides at the 250K molecular weight position on gels. These results also suggest that the bulk of the chick LETS glycoprotein is accessible to lactoperoxidase in intact cells and are in contrast to another proposal that in murine L cells, LETS glycoprotein is a transmembrane component, with the bulk of the molecule on the inside of the bilayer.[18]

The sequence differences suggested by the peptide maps are perhaps surprising if, as seems likely, LETS glycoprotein has been conserved to perform a similar function in different species. Presumably, whatever the role of LETS glycoprotein turns out to be, this role does not require strict conservation of polypeptide sequence. In this respect, LETS glycoprotein differs from muscle and nonmuscle actin, which show a high degree of evolutionary conservation. Peptide maps of actin from mammalian and avian muscle are indistinguishable, while those of frog, fish, and scallop differ in only one, two, and eight peptides, respectively.[21] Similarly, human HLA antigens of different specificities gave essentially identical peptide maps when radioiodinated tyrosine-containing peptides were compared.[22] By contrast, comparison of tryptic peptides from radioactive arginine- or lysine-labeled murine H-2K or H-2D glycoproteins revealed[23] extensive differences, even in products of allelic genes (only eight of nine of 20–26 peptides showed similar positions). However, it is likely that peptide map comparisons exaggerate actual structural differences of similar polypeptide sequences. A single amino acid change could alter the charge or chromatographic behavior of a peptide that contains many conserved amino acids. This is confirmed for H-2 antigens by sequence data that demonstrate approximately 65% sequence homology with allelic products over the first 27 residues from the N-terminus.[24] Thus, although the peptide maps of LETS glycoprotein reveal little homology, this method of analysis may be exaggerating any differences there might be. Certainly, additional mapping and amino acid sequence data are required to establish the degree of homology between species that immunologic probes suggest. The extensive antigenic cross reactions among SF (LETS) antigens of the chicken, Japanese quail, mouse, rat, guinea pig, hamster, and human[10] would not be inconsistent with nonidentical peptide maps or extensive sequence differences. Sequence differences of the order of 40% appear to be necessary[25] before there is a loss of immunologic cross reactivity in many proteins.

Summary

We have studied the major glycoprotein (LETS) species of nominal molecular weight 250K labeled by lactoperoxidase-catalyzed radioiodination of cultured fibroblast monolayers of primary human embryo, chick embryo, and hamster BHK cells. The LETS glycoprotein is not solubilized from iodinated cells by nonionic detergents or sodium, deoxycholate, which preferentially extracts the other major iodinated glycoproteins, by chelating agents, high salt concentrations, or modest concentrations of urea. Higher concentrations of urea and 2 M NaI partially solubilized LETS glycoprotein. Fingerprints of peptides that contained [^{125}I]tyrosine residues obtained by thermolysin degradation of LETS glycoproteins of different species show differences, indicating an incomplete interspecies homology among this group of surface-associated glycoproteins.

References

1. Helenius, A. & K. Simons. 1975. Biochim. Biophys. Acta **415:** 29–79.
2. Butters, T. D. & R. C. Hughes. 1974. Biochem. J. **140:** 469–478.
3. Yu, J., D. A. Fischman & T. L. Steck. 1973. J. Supramol. Struct. **1:** 233–248.
4. Singer, S. J. & G. L. Nicolson. 1972. Science **175:** 720–731.
5. Yamada, K. M. & J. A. Weston. 1974. Proc. Nat. Acad. Sci. USA **71:** 3492–3496.
6. Kuusela, P., E. Ruoslahti & A. Vaheri. 1975. Biochim. Biophys. Acta **379:** 295–303.

7. Nairn, R. & R. C. Hughes. 1976. Biochem. Soc. Trans. **4:** 165–167.
8. Hynes, R. O., A. T. Destree & V. Mautner. 1976. J. Supramol. Struct. **2** (Suppl. 1): 189–201.
9. Ruoslahti, E. & A. Vaheri. 1974. Nature (London) **248:** 789–791.
10. Vaheri, A., E. Ruoslahti, E. Lindner, J. Wartiovaara, J. Keski-Oja, P. Kuusela & O. Saksela. 1976. J. Supramol. Struct. **4:** 63–70.
11. Meager, A., R. Nairn & R. C. Hughes. 1975. Virology **68:** 41–57.
12. Butters, T. D. & R. C. Hughes. 1975. Biochem. J. **150:** 59–69.
13. Katz, A. M., W. J. Dreyer & C. B. Anfinsen. 1959. J. Biol. Chem. **234:** 2897–2900.
14. Hunter, W. M. & F. C. Greenwood. 1962. Nature (London) **194:** 495–496.
15. Hansma, H. G. & C. Kung. 1975. Biochem. J. **152:** 523–528.
16. Gow, J. & A. C. Wardlaw. 1975. Biochem. Biophys. Res. Commun. **67:** 43–49.
17. Teng, N. N. H. & L. B. Chen. 1975. Proc. Nat. Acad. Sci. USA **72:** 413–417.
18. Hunt, R. C. & J. C. Brown. 1975. J. Mol. Biol. **97:** 413–422.
19. Smart, J. E., E. Pearlstein & M. D. Waterfield. 1974. Biochem. Soc. Trans. **2:** 1316–1317.
20. Ruoslahti, E. & E. Engvall. 1977. Int. J. Cancer **2:** 1–5.
21. Carsten, M. E. & A. M. Katz. 1964. Biochim. Biophys. Acta **90:** 534–541.
22. Cunningham-Rundles, C., C. Jersild, A. Svejgaard, B. Dupont & R. A. Good. 1975. Proc. Nat. Acad. Sci. USA **72:** 5081–5085.
23. Brown, T. L., K. Kato, J. Silver & S. G. Nathenson. 1974. Biochemistry **13:** 3174–3178.
24. Silver, T. & L. Hood. 1976. *In* Contemporary Topics in Molecular Immunology. H. N. Eisen & R. A. Reisfeld, Eds. Vol. 5: 35–68. Plenum Publishing Corporation. New York and London.
25. Prager, E. M. & A. C. Wilson. 1971. J. Biol. Chem. **246:** 7010–7017.
26. Nairn, R. 1976. Ph.D. Thesis. University of London.
27. Pena, S. D. J. & R. C. Hughes. 1978. Cell Biol. Int. Rep. In press.

INTERACTION OF AN L-CELL SURFACE PROTEIN WITH THE CYTOSKELETON*

Richard C. Hunt

Department of Biochemistry
University of Oxford
Oxford OX1 3QU, England

Jay C. Brown

Department of Microbiology
University of Virginia Medical School
Charlottesville, Virginia 22901

Proteins on the external surface of the plasma membrane are thought to be involved in intercellular recognition, specific adhesion, and interaction with growth-controlling substances in the environment. All of these phenomena involve transmembrane events in which the external protein interacts directly or indirectly with structures within the cell. Studies with drugs, such as colchicine and cytochalasin B, have shown that external receptors are probably under the control of microtubules and microfilaments, which, together with other structures, constitute the cytoskeleton of the cell. However, the way in which external proteins and internal cytoskeletal components interact is not known. (For a review of transmembrane phenomena, see Reference 12.)

External proteins have been detected on fibroblast surfaces by the use of non-penetrating reagents. Of these external proteins, the most prominently labeled is the high-molecular-weight (210,000–250,000 daltons) transformation-sensitive protein, which has variously been called LETS, SFA, CSP, and so on.[3,6,14,16]

We have previously shown that mouse L-929 cells, a transformed cell line, possess a surface protein similar to LETS with respect to its mobility on polyacrylamide gels.[4] In intact cells, this protein, the 240K protein (originally called band 1 because of its position on polyacrylamide gel electrophoresis), is labeled by lactoperoxidase-catalyzed iodination, which labels external proteins, and is sensitive to protease digestion. In addition, it is exposed on the cell surface only during the G_1 period of the cell cycle.[5] The fraction of this protein that is incorporated into inside-out vesicles derived from the L-cell plasma membrane has been shown to span the membrane.[4]

In this paper, we report our studies on the interaction of the transmembrane 240K protein with the cytoskeleton. The 240K protein is isolated along with the cytoskeleton after detergent extraction of the cells, but this association is broken down by incubation of the cells with cytochalasin B. The interaction between the 240K protein and components of the cytoskeleton is further emphasized by the finding that disulfide cross-links can be formed between this protein and a protein that has a molecular weight of 45,000 daltons, which is probably actin, the monomer of the microfilament system.

*Supported in part by a grant from the British Medical Research Council and by Grant HD 06390 from the National Institutes of Health.

0077-8923/78/0312-0207 $01.75/0

MATERIALS AND METHODS

Cells. Mouse L-929 cells were grown as monolayers in 75 cm^2 plastic flasks (Falcon) with 25 ml of Eagle's basal medium (Gibco) that contained 5% fetal calf serum (Gibco) and antibiotics.

Polyacrylamide Gel Electrophoresis. Electrophoresis in 5, 6, or 8.5% sodium dodecyl sulfate (SDS) polyacrylamide gels was performed with the system of Laemmli.[11] Cells, cytoskeletons, and membrane fractions were dissolved in either 5% SDS and 2% 2-mercaptoethanol (reducing gels) or in 5% SDS that contained 0.125 mM 5,5′-dithiobis(2-nitrobenzoic acid)(nonreducing gels). Molecular weights were estimated with reference to the mobilities of standard proteins. These proteins included myosin, thyroglobulin, bovine serum albumin, ovalbumin, and Sendai virus proteins.

Sendai virus consists of five major proteins that range in molecular weight from 35,000 to 88,000 daltons. Under nonreducing conditions and after cross-linking by oxidation or dimethyl 3,3′-dithiobispropionimidate, two proteins form homooligomers. One of these homooligomers (G_1) is an external glycoprotein that forms dimers and tetramers with molecular weights of 144,000 and 288,000 daltons, respectively.[17]

Lactoperoxidase-Catalyzed Iodination. Cells were washed in Dulbecco's phosphate-buffered saline (PBS) and iodinated with lactoperoxidase (Sigma), glucose oxidase (Sigma), and Na ^{125}I (Radiochemical Centre, Amersham).[6]

Inside-out Plasma Membrane Vesicles. These vesicles were prepared as described previously.[4]

Preparation of Cytoskeletons. Cell monolayers were washed with PBS and incubated for 15 min with 5 ml of 2% Triton® X-100 (Sigma) in PBS. The extract was poured off, and the remaining residue attached to the flask was washed with 2% Triton X-100 in PBS. The residue (the cytoskeleton fraction) was then dissolved for electrophoresis or fixed for electron microscopy.

RESULTS

The Two Surfaces of the L-929 Cell Plasma Membrane

Radioactivity is incorporated into several proteins when lactoperoxidase-catalyzed iodination is employed to label the external surface of mouse L-929 cells; most prominent among these labeled proteins is an external protein with an apparent molecular weight of approximately 240,000 (± 10,000) daltons, which will be referred to as the 240K protein. FIGURE 1 (*top*) shows an optical density scan of an autoradiogram of a 6% SDS polyacrylamide gel on which the externally labeled proteins have been separated.

To investigate the inside surface of the plasma membrane, inside-out vesicles were prepared by the polystyrene latex bead method.[4] Cells were removed from the flask surface by scraping with a rubber policeman and were suspended in growth medium with approximately 10^4 polystyrene beads per cell. After a 1-hr incubation, the ingested beads were isolated, and the surrounding membrane (an inside-out plasma membrane vesicle) was labeled by lactoperoxidase-catalyzed iodination. The results are shown in FIGURE 1 (*bottom*). In contrast to the spectrum of labeled proteins on the outer surface of the cell, the spectrum of proteins on the membranes surrounding the beads was more complex, with about 14 iodinated bands readily resolved. Some of these bands may be serum components that ad-

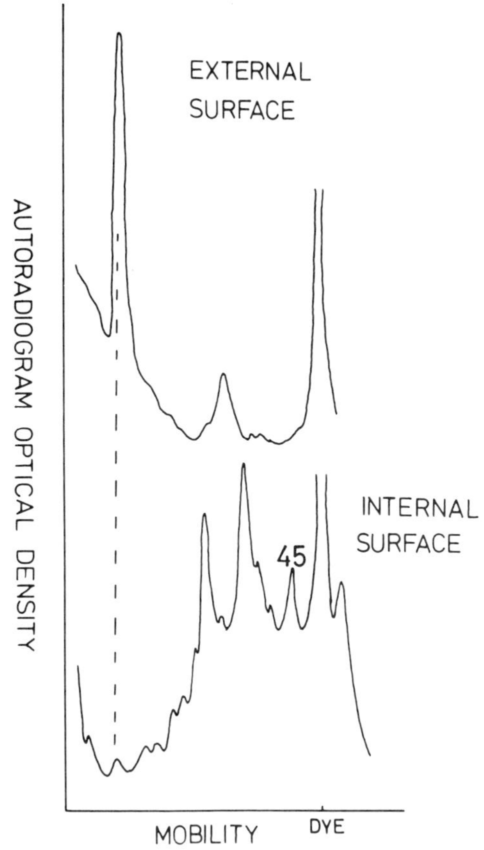

FIGURE 1. Lactoperoxidase-catalyzed iodination of the external and internal surfaces of the plasma membrane of mouse L-929 cells. *Top:* Optical density scan of an autoradiogram of a 6% SDS polyacrylamide gel after separation of ^{125}I-labeled proteins from intact cells. *Bottom:* Optical density scan of an autoradiogram of a 6% SDS polyacrylamide gel after separation of proteins labeled by iodination of inside-out membrane vesicles made by the polystyrene-latex bead method.

here to uningested polystyrene beads during the incubation of cells with beads, but most bands were associated specifically with plasma membrane vesicles. Among these bands was one with a molecular weight of approximately 45,000 daltons. Tryptic peptide analysis revealed that this band is actin, a protein that has frequently been observed to be associated with plasma membranes. A minor band comigrates with the external 240K protein. This band may represent the internal end of the 240K transmembrane protein.

Properties of the 240K External Surface Protein of L-929 Cells

To examine further the manner in which the externally iodinated proteins associate with the plasma membrane, labeled cells were extracted with a variety of

TABLE 1

EXTRACTION OF 240K PROTEIN FROM L-929 CELLS*

Extractant†	Concentration	Solubility‡
PBS		–
EDTA	0.025 M in PBS	–
Triton X-100	5.0% in PBS	–
Tween® 20	1.0% in PBS	–
Nonidet® P-40	1.0% in PBS	–
Deoxycholate (Na)	0.5% in PBS	–
LIS	0.02 M in 0.05 M Tris HCl (pH 7.6)	–
LIS	0.2 M in 0.05 M Tris HCl (pH 7.6)	+
SDS	2.0% in water	+++
Urea	0.1 M in PBS	–
Urea	8 M in PBS	+++

*Cells were labeled by lactoperoxidase-catalyzed iodination and were extracted with the above solutions for 15 min at room temperature. The supernatant was decanted from the monolayer and centrifuged to remove insoluble material. Sodium dodecyl sulfate-2-mercaptoethanol solution was added to the extracted material to give a final concentration of 2% SDS, 2% 2-mercaptoethanol, and the solution was boiled. The insoluble material was also dissolved in SDS-2-mercaptoethanol and boiled. Both the soluble and insoluble materials were analyzed by polyacrylamide gel electrophoresis.

†PBS, Dulbecco's phosphate-buffered saline; LIS, lithium diiodosalicylate.

‡–, 240K protein completely insoluble; +, 240K protein less than 30% solubilized; +++, 240K protein completely solubilized.

agents that solubilize either peripheral or integral membrane proteins. Peripheral proteins, those that are not intimately associated with the lipid bilayer, are usually extracted by agents that break down electrostatic interactions (high salt, chelating agents), whereas integral proteins, those embedded in the lipid bilayer, require more drastic treatment, such as the use of detergents or denaturing agents.

TABLE 1 shows the effects of a variety of agents on the solubilization of the iodinated 240K external protein of L-929 cells. Cells were washed with PBS and iodinated as a monolayer on the surface of the culture flask, washed again, and incubated for 15 min with 5 ml of the appropriate extracting solution at room temperature. The 240K protein was not extracted by agents that usually remove peripheral proteins (e.g., EDTA) nor was it solubilized by nonionic detergents, as would be expected for an integral protein. Only the anionic detergent SDS and 8 M urea, a denaturing agent, completely solubilized the protein, although it was to some extent solubilized by 0.2 M lithium diiodosalicylate (LIS). FIGURE 2 shows the effect of Triton X-100 (2% in PBS) on the solubilization of iodinated proteins from the L-929 cell surface. Of all of the labeled proteins, only the 240K protein is not extracted by this detergent. Similar results were found with the other nonionic detergents used. These results indicate that the 240K protein does not fall into either the peripheral or integral membrane protein class.

Cytoskeletal Fraction

The fraction of the cells that remained attached to the culture flask surface after incubation with 2% Triton X-100 was examined by electron microscopy. This technique revealed the insoluble residue to contain recognizable nuclear material surrounded by a network of filaments, interspersed among which were ribosomes (FIGURE 3). The preparation was therefore characteristic of a cell cytoskeleton.

The protein composition of the Triton X-100-insoluble fraction was investigated by means of polyacrylamide gel analysis. Triton X-100 extraction solubilized about 90% of the total cell protein, and Coomassie blue staining of gels on which the residual proteins were analyzed demonstrated a simple pattern. Three major protein classes were observed (FIGURE 4). Several proteins ran with the tracking dye on 6% gels, and their identities are unknown; one protein ran with a molecular weight of about 45,000 daltons and comigrated with actin, the monomer of the microfilament system; a third protein migrated with an apparent molecular weight of 60,000 daltons. The latter protein may be the monomer of the 100 Å filament system (further details of this protein will be published elsewhere). Despite the fact that all of the 240K protein remained in the Triton X-100-insoluble fraction, only minor proteins could be detected in this region of Coomassie blue-stained gels, indicating that the 240K protein must represent a very small fraction of the total cytoskeletal protein.

Effect of Cytochalasin B on 240K Protein

The preceding data show that the 240K protein is isolated along with the cytoskeleton. Several possibilities may account for this finding. The nuclear material, cytoskeleton, and external 240K protein system could form, for example, concentric spheres in the cell that might be preserved after removal of the plasma membrane by detergent extraction. In this case, there would be no physical con-

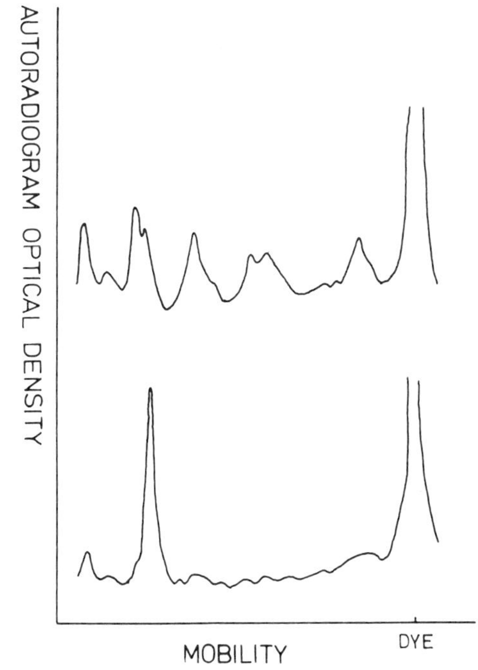

FIGURE 2. Extraction of surface proteins labeled by iodination of L-929 cells. *Top:* Proteins extracted by 2% Triton X-100 in PBS. Optical density scan of an autoradiogram of a 6% SDS polyacrylamide gel. *Bottom:* Proteins that are not soluble in 2% Triton X-100. Optical density scan of an autoradiogram of a 6% SDS polyacrylamide gel.

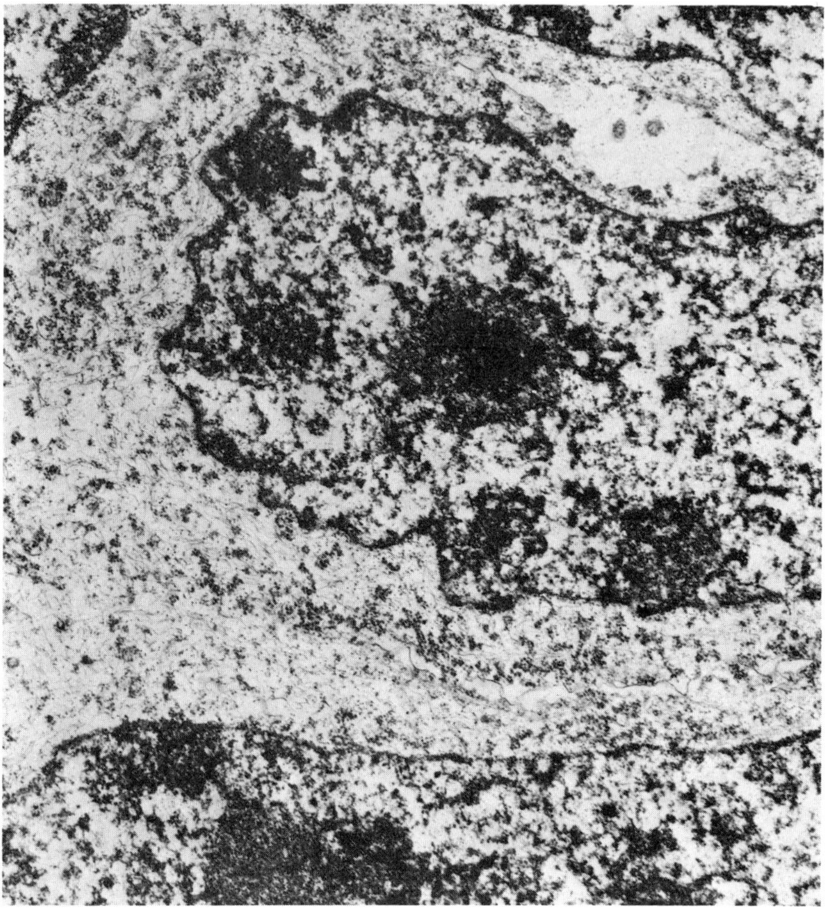

FIGURE 3. Electron micrograph of L-929 cells after extraction with 2% Triton X-100. ×7000.

nection between them, and the cytoskeleton and nuclear material would simply be trapped inside the sphere of cell surface material. Alternatively, the 240K protein system and the cytoskeleton might be independently attached to the substratum, the surface of the culture flask. A third possibility is that the 240K system and the cytoskeleton are isolated together because some interaction between them remains after the membrane is dissolved. This situation could result from either a direct interaction between the two systems or the presence of a transmembrane bridging protein. If the third possibility were true, breakdown of the cytoskeleton might result in an alteration in the properties of the 240K protein.

To test this possibility, cells were labeled and incubated for 1 hr in the presence of 0–100 μg of cytochalasin B per milliliter of PBS. The cytochalasin B-containing solution was then decanted, and the cells were extracted as before with 2% Triton X-100. The cytochalasin B incubation solution, the Triton X-100 extract, and the cytoskeleton residue were then analyzed on 6% SDS polyacrylamide gels. No evidence for solubilization of the 240K protein by cytochalasin B solution during the 1-hr incubation could be found, although incubation of cells for several days in the presence of similar concentrations of cytochalasin B in growth medium did indicate shedding into the medium. However, incubation for 1 hr in the presence of 10 μg of cytochalasin B per milliliter of PBS did affect Triton X-100 extraction of the 240K protein. In FIGURE 5, the iodinated proteins that remained associated with the Triton-insoluble residue are shown. Preincubation with cytochalasin B causes the 240K protein to be extracted by the detergent and thus to assume one of the properties of an integral membrane component. Similar experiments were performed in which iodinated cells were first extracted with Triton X-100 to obtain cytoskeletal preparations. The cytoskeletons were then incubated with cytochalasin B solutions. Some of the 240K protein was solubilized by the cytochalasin B-containing solution (data not shown).

Interaction of 240K Protein with a 45,000-Dalton Protein

The results of experiments with cytochalasin B indicate that the 240K protein may be insoluble in nonionic detergents because of an interaction with the cytoskeleton. When the integrity of the cytoskeleton is destroyed, at least some of the 240K protein can be dissolved along with the plasma membrane and its associated proteins. Possible interactions of 240K proteins with other polypeptides were therefore investigated further. It has been shown that the major lactoperoxidase-iodinated, external protein migrates on polyacrylamide gels with an apparent

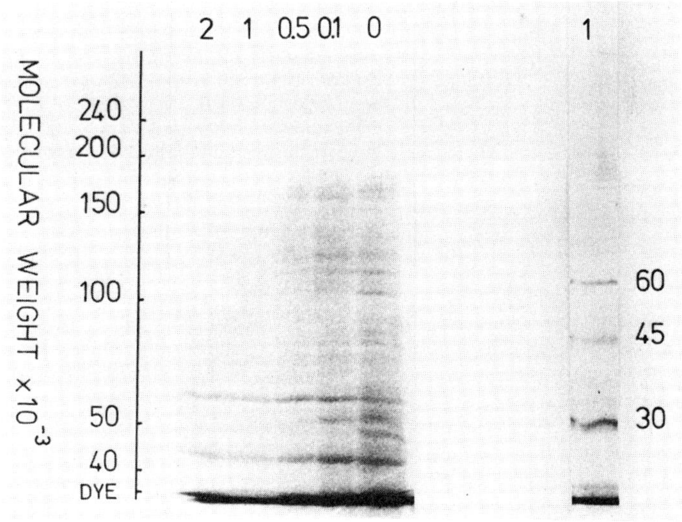

FIGURE 4. Protein composition of the cytoskeletal fraction. *Left:* Six percent SDS polyacrylamide gel of the Triton X-100-insoluble fraction after extraction of cells with PBS, 0.1%, 0.5%, 1, and 2% Triton X-100. *Right:* Eight and one-half percent SDS polyacrylamide gel of the cytoskeletal fraction. Cells were extracted with 1% Triton X-100 in PBS. Gels were stained with Coomassie blue.

molecular weight of 240,000 (± 10,000) daltons after it has been dissolved in SDS and 2-mercaptoethanol (FIGURE 6, A). Under nonreducing conditions, however, the pattern of iodinated proteins is more complex. Much more material fails to enter the stacking gel. More iodinated protein is also found at the interface between the stacking and separating gels. This material probably represents aggregates with molecular weights in excess of 330,000 daltons. Other new bands appear in the separating gel under nonreducing conditions; for example, a shoulder appears on the 240K band, and a new band (band X) runs between the stacking/separating gel interface and the 240K position (FIGURE 6, A).

To determine the nature of the protein with which the iodinated 240K protein forms disulfide bridges, the molecular weight of band X was investigated. Unfortunately, it is difficult to determine molecular weights in this region of the gel, because apparent molecular weights of unknown proteins are determined with

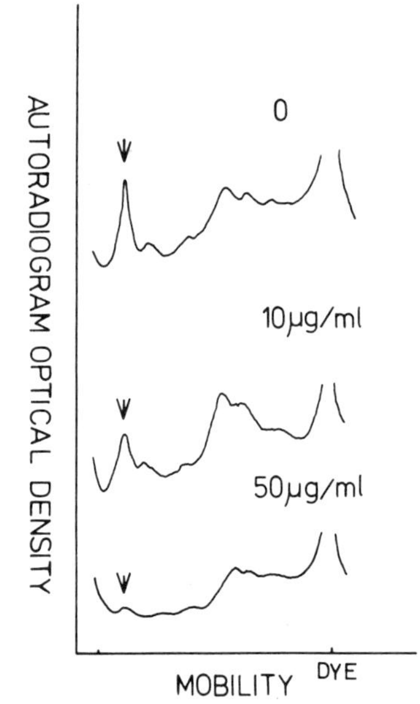

FIGURE 5. Effect of preincubation of L-929 cells with cytochalasin B on the Triton X-100 solubility of the 240K protein. L cells were washed in PBS, iodinated, and then incubated with 0–50 μg of cytochalasin B per milliliter of PBS for 1 hr. They were then extracted with Triton X-100 (2% in PBS), and the residue was analyzed on 6% SDS polyacrylamide gels. The gels were autoradiographed, and the film was scanned with a densitometer. Iodinated proteins that remained in the cytoskeleton after extraction with Triton X-100. Cells were preincubated in PBS (*top*); in 10 μg of cytochalasin B per milliliter of PBS (*middle*), or in 50 μg of cytochalasin B per milliliter of PBS (*bottom*) prior to extraction with Triton X-100. The arrows indicate the position of the 240K protein.

reference to the molecular weights of known standards. The largest standard usually available is myosin, which has a molecular weight of 200,000 daltons. For accurate apparent molecular weight determinations, two standard proteins must bracket the unknown protein.

When the mobility of standard proteins is plotted against the logarithm of the molecular weight, a linear relationship is observed up to 200,000 daltons, and if it is assumed that linearity extends above this point (as is usually done), the apparent molecular weight of band X is 285,000 (± 5000) daltons. To confirm this estimate, a protein with a molecular weight higher than 280,000 daltons was sought and found in cross-linked Sendai virus. The tetramer of this external viral glycoprotein has a molecular weight of 288,000 and migrates slightly behind band X (FIGURE 6, C). These results confirm the linearity of the logarithmic molecular weight versus mobility curve and indicate that the molecular weight of band X

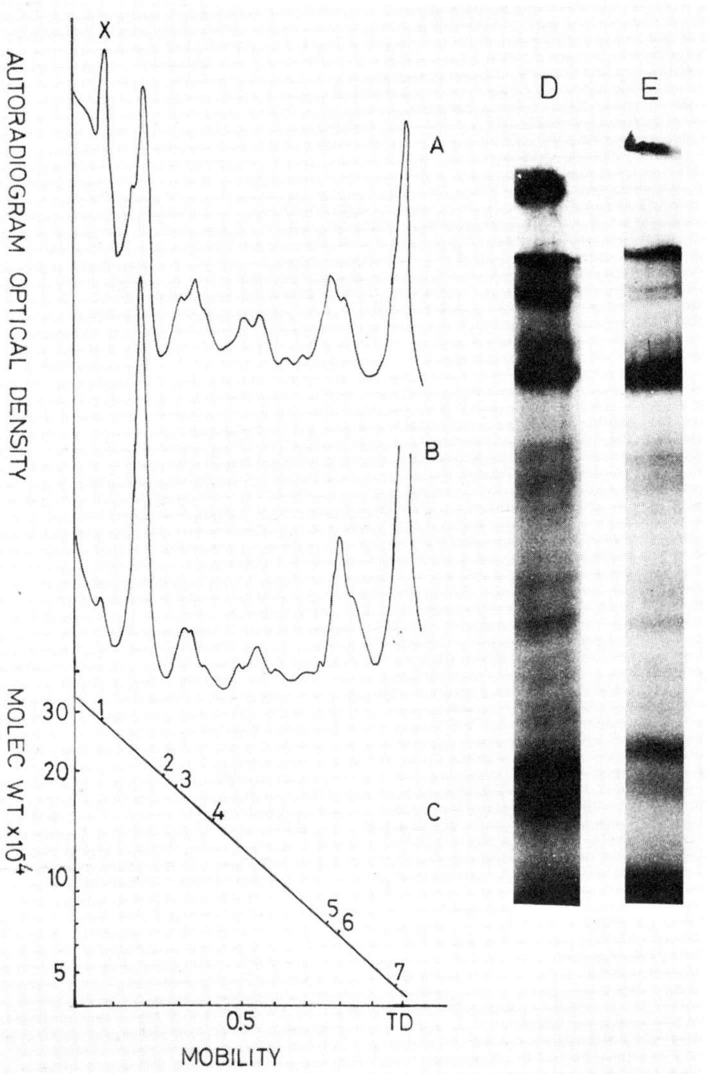

FIGURE 6. Iodinated proteins on the surface of L-929 cells analyzed on 5% SDS polyacrylamide gels in the presence and absence of a disulfide bridge reducing agent. A: Optical density scan of an autoradiogram of a 5% SDS polyacrylamide gel of unreduced iodinated surface proteins. B: Optical density scan of an autoradiogram of a 5% SDS polyacrylamide gel of reduced (with 2-mercaptoethanol) iodinated surface proteins. C: Molecular weight versus mobility plot of standard proteins separated on 5% SDS polyacrylamide gels. Protein 1, tetramer of Sendai virus G_1 protein (molecular weight, 288,000); 2, myosin (200,000); 3, thyroglobulin (167,000); 4, dimer of Sendai virus G_1 protein (144,000); 5, monomer of Sendai virus G_1 protein (72,000); 6, bovine serum albumin (68,000); 7, ovalbumin (46,000). D: Photograph of gel scanned in A shows material at top of the gel. E: Photograph of gel scanned in B.

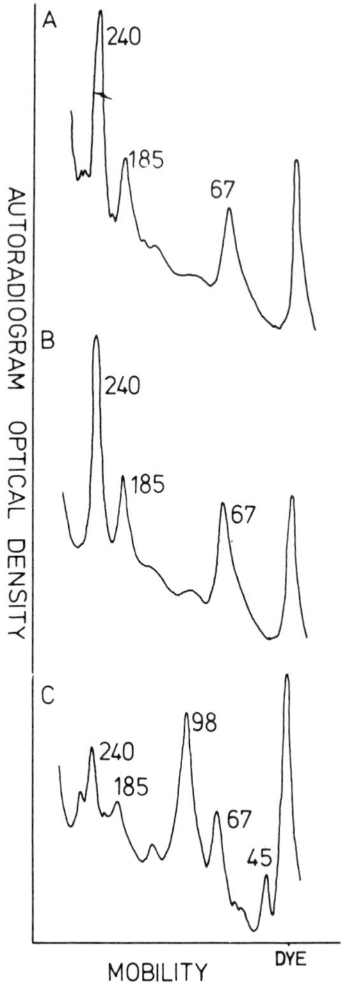

FIGURE 7. Reduction of the iodinated 285K protein complex from the surface of L-929 cells. A: Cells were iodinated, and the proteins were separated without a reducing agent on a polyacrylamide gel. The 285K region of the gel was cut out and soaked in SDS-2-mercaptoethanol solution to reduce and elute the proteins. The eluted proteins were separated on a 6% SDS polyacrylamide gel that was dried, autoradiographed, and scanned. B: Cells were iodinated, and cytoskeletons were prepared from them by Triton X-100 extraction. The proteins were then solubilized and separated without reduction, and the 285K region of the gel was treated as in A. C: Cytoskeletons were prepared by Triton X-100 extraction and iodinated. They were then solubilized without reduction, and the proteins were separated on a 5% SDS polyacrylamide gel. The 285K region of the gel was then treated as in A.

lies between 280,000 and 285,000 daltons. If, therefore, the 240K protein is a component of the band X complex, it must be disulfide bridged to a protein with a molecular weight of about 45,000 daltons.

Two-dimensional electrophoresis was utilized to obtain information on the proteins that formed band X and the higher-molecular-weight complexes that were observed at the top of the stacking and separating gels. In all cases, the major iodinated component of these complexes was the 240K protein (data not shown).

In other experiments, the nature of the proteins that formed disulfide bridges with the 240K protein in band X was investigated by carefully cutting the 285K region from the gel, reducing it by soaking the gel slice for 24–48 hr in a solution of 10% SDS-10% 2-mercaptoethanol, and rerunning the material under reducing conditions on a 6% polyacrylamide gel. The results are shown in FIGURE 7. Reduction of band X from iodinated intact cells yielded proteins with molecular weights

TABLE 2
REDUCTION OF THE 285K DISULFIDE-BRIDGED COMPLEX BY 2-MERCAPTOETHANOL

	Molecular Weight		Apparent Molecular Weight of Unlabeled Disulfide-Bridged Proteins
	Unreduced	After Reduction	
Iodinated intact cell*	285K	240K	45K
		185K	100K
		67K	218K
Iodinated intact cell extracted with 2% Triton X-100 to yield	285K	240K	45K
		185K	100K
cytoskeletal fraction†		67K	218K
Iodinated cytoskeleton‡	285K	240K	45K
		185K	100K
		98K	187K
		67K	218K
		45K	240K

*Cells were labeled by lactoperoxidase-catalyzed iodination and analyzed on 5% SDS polyacrylamide gels in the absence of 2-mercaptoethanol. The 285K band was cut out, and the material was reduced by 2-mercaptoethanol. The proteins were analyzed under reducing conditions on 6% SDS polyacrylamide gels.

†Cells were labeled and extracted with 2% Triton X-100 in PBS. The resultant cytoskeleton was analyzed in the absence of 2-mercaptoethanol on 5% SDS polyacrylamide gels. The 285K band was cut out, reduced, and reanalyzed on 6% SDS polyacrylamide gels.

‡Cytoskeletons were prepared by extracting cells with 2% Triton X-100 in PBS. The cytoskeletons were labeled and analyzed in the absence of 2-mercaptoethanol. The 285K band was cut out and reanalyzed under reducing conditions.

of approximately 240,000, 185,000, and 67,000 daltons, suggesting that they were disulfide bridged to proteins with molecular weights of 45,000, 100,000, and 218,000 daltons, respectively (TABLE 2). However, no iodinated proteins of these molecular weights were detected on reduction (FIGURE 7, A).

Similar results were obtained when iodinated cells were extracted with Triton X-100 to prepare cytoskeletons. The band X of the cytoskeletons gave reduced proteins with molecular weights of 240,000, 185,000, and 67,000 daltons (FIGURE 7, B).

When cells were extracted with Triton X-100 and the cytoskeletons labeled by lactoperoxidase-catalyzed iodination, many labeled proteins were observed, as expected. The distribution of radioactivity in the polyacrylamide gel resembled the Coomassie blue-stained material (as shown in FIGURE 4). Thus, in these preparations, band X represented a minor proportion of the incorporated radioactivity; all proteins in the cytoskeleton were labeled under these conditions. A more complex range of proteins than that found with iodinated intact cells was obtained on elution and reduction of the band-X region of the gel of labeled cytoskeletons (FIGURE 7, C). In addition to the bands found when band X of iodinated intact cells was reduced, bands with molecular weights of 98,000 and 45,000 daltons, plus several minor bands, appeared. These other labeled proteins are likely to be interior to the plasma membrane, because they are inaccessible to the iodinating reagents in intact cells. The 45,000-dalton protein is clearly a candidate for the protein with which the 240K protein forms disulfide bridges, and the 98,000-dalton protein a candidate for the protein that forms disulfide bridges with the 185,000-dalton protein. If this is so and the 45,000- and 98,000-dalton bands are internal proteins, the 240K and 185K proteins must span the plasma membrane.

Discussion

On the external surface of mouse L-929 cells (a transformed cell line) is a class of proteins that have a molecular weight of approximately 240,000 daltons (the 240K protein). Like LETS and SFA, this protein is very sensitive to digestion with proteases[6] and is not extracted from the cell by nonionic detergents.[1,8] The detergent-insoluble residue consists of microfilaments plus nuclear material and ribosomes, a composition that is supported by polyacrylamide gel analysis, which reveals that actin (the major protein of microfilaments) is one of the major cytoskeletal proteins. Coisolation of the 240K protein with the cytoskeleton may not necessarily indicate an association between them, but the existence of such a relationship is supported by the experiments with cytochalasin B, a drug that breaks down microfilaments. After incubation of the cells with cytochalasin B, the 240K protein is lost from the insoluble fraction after detergent extraction.

The possible association of the 240K protein directly or indirectly with the microfilament system has been investigated by identifying the proteins with which the 240K protein forms disulfide bonds after lactoperoxidase-catalyzed iodination. In the absence of a reducing agent, SDS polyacrylamide gel electrophoresis of iodinated surface proteins revealed a pattern quite different from that observed when 2-mercaptoethanol was included in the solubilization mixture. Additional high-molecular-weight complexes were found, some of which failed to enter either the stacking or separating gel and one of which migrated with an apparent molecular weight of 285,000 daltons. This finding suggested that the 240K protein formed disulfide bonds with a protein(s) with a total molecular weight of 45,000 daltons. The 45,000-dalton protein (which comigrated with actin) in the 285,000-dalton region of the gel, which was labeled only when detergent-treated cells were iodinated, is a possible candidate for the protein with which the 240K protein forms disulfide bonds. If this 45,000-dalton protein is that protein, these results imply a direct association between an external surface protein and the cytoskeleton. They also imply that the 240K protein molecules that cross-link to actin span the plasma membrane. Such an orientation of high-molecular-weight proteins in the plasma membrane of L-929 cells has already been suggested by experiments performed with inside-out membrane vesicles.[4]

The observation that the 240K protein forms disulfide bridges with a 45,000-dalton protein after lactoperoxidase-catalyzed iodination does not necessarily imply that these complexes exist *in vivo*. This is because the glucose oxidase component of the iodination system generates hydrogen peroxide, which can oxidize sulfhydryl groups and induce the formation of disulfide bridges between adjacent proteins. Even though the pattern of iodination is not changed in the presence of sulfhydryl blocking agents, this hypothesis cannot be completely ruled out. Nevertheless, the generation of such bonds cross-linking the 240K and 45,000-dalton proteins indicates that they are in close proximity within the cell, and therefore this observation does not alter our conclusion of an association between them.

The conclusion that these high-molecular-weight proteins interact with the cytoskeleton is supported by other evidence. First, cells that have a poorly developed cytoskeleton, such as transformed cells, also contain little external high-molecular-weight protein.[3,6] Second, the high-molecular-weight protein of chick embryo fibroblasts has been observed to be associated with ridges on the external surface of the cell. Underlying these ridges are bundles of actin-containing microfilaments.[10] Third, two reports have demonstrated that a high-molecular-weight protein can be isolated along with actin.[2,10] Immune precipitation of the SFA complex from the surface of chick embryo fibroblasts coprecipitates proteins with molecular weights of 230,000, 150,000, and 45,000 daltons.[10] Interestingly, immune precipitation of the SFA shed into the medium by these cells only coprecipitates the 230,000- and 150,000-dalton proteins. In a more recent report, Carter and Hakomori[2] have shown that affinity chromatography on castor bean lectin columns isolates galactoprotein a (equivalent to LETS) of hamster NIL cells, together with actin. Since actin is not glycosylated, it must bind to the lectin because of its association with galactoprotein a. These two reports suggest more than a casual indirect association of high-molecular-weight external proteins with the microfilaments; indeed, they suggest that some of the external protein is directly associated with the cytoskeleton.

Our finding of a disulfide bridge between an external high-molecular-weight protein and a 45,000-dalton protein in mouse L-929 cells is in contrast to results obtained with hamster NIL cells[7] and chick embryo fibroblasts.[9] These reports demonstrated a high-molecular-weight protein that exists as disulfide-bridged dimers. The results reported here for mouse L-929 cells, however, are more similar to those that have recently been reported for another transformed mouse cell line. On the surface of LM cells (a derivative of L-929 cells) is an iodinatable 250,000-dalton protein that forms disulfide bridges to a protein(s) with a total molecular weight of 90,000 daltons (a dimer of a 45,000-dalton protein?).[13] In both L-929 and LM cells, the lower-molecular-weight protein was not labeled by lactoperoxidase-catalyzed iodination of intact cells, suggesting that the protein may exist internal to the plasma membrane.

Acknowledgments

We thank Mrs. F. Croall for expert technical assistance and Mr. F. Caddick for preparing the photographs.

References

1. Brown, S., W. Levinson & J. A. Spudlich. 1976. Cytoskeletal elements of chick embryo fibroblasts revealed by detergent extraction. J. Supramol. Struct. **5:** 119–130.
2. Carter, W. G. & S. Hakomori. 1977. Isolation and partial characterisation of galactoprotein A (LETS) and galactoprotein B from hamster embryo fibroblasts. Biochem. Biophys. Res. Commun. **76:** 229–308.
3. Hogg, N. M. 1974. A comparison of membrane proteins of normal and transformed cells by lactoperoxidase labeling. Proc. Nat. Acad. Sci. USA **71:** 489–492.
4. Hunt, R. C. & J. C. Brown. 1975. Identification of a high molecular weight transmembrane protein in mouse L cells. J. Mol. Biol. **97:** 413–422.
5. Hunt, R. C., E. Gold & J. C. Brown. 1975. Cell cycle dependent exposure of a high molecular weight protein on the surface of mouse L cells. Biochim. Biophys. Acta **413:** 453–458.

6. Hynes, R. O. 1973. Alteration of cell surface proteins by viral transformation and proteolysis. Proc. Nat. Acad. Sci. USA **70:** 3170–3174.
7. Hynes, R. O. & A. Destree. 1977. Extensive disulfide bonding at the mammalian cell surface. Proc. Nat. Acad. Sci. USA **74:** 2855–2859.
8. Hynes, R. O., A. T. Destree & V. Mautner. 1976. Spatial organisation at the cell surface. *In* Membranes and Neoplasia: New Approaches and Strategies. V. T. Marchesi, Ed.: 189–201. Alan R. Liss, Inc. New York, N.Y.
9. Keski-Oja, J., D. F. Mosher & A. Vaheri. 1977. Dimeric character of fibronectin—a major cell surface associated glycoprotein. Biochem. Biophys. Res. Commun. **74:** 699–706.
10. Kuusela, P., E. Ruoslahti & A. Vaheri. 1975. Polypeptides of a glycoprotein antigen (SF) present in serum and surface of normal but not transformed chicken fibroblasts. Biochim. Biophys. Acta **379:** 295–303.
11. Laemmli, U. K. 1970. Cleavage of structural proteins during assembly of the head of bacteriophage T4. Nature (London) **227:** 680–685.
12. Nicolson, G. L. 1976. Transmembrane control of receptors on normal and tumor cells. 1. Cytoplasmic influence over cell surface components. Biochim. Biophys. Acta **457:** 57–108.
13. Takemoto, L. J., T. Miyakawa & C. F. Fox. 1977. Analysis of membrane protein topography of Newcastle disease virus and cultured mammalian fibroblasts. *In* Cell Shape and Surface Architecture. J. P. Revel, U. Henning & C. F. Fox, Eds.: 605–614. Alan R. Liss, Inc. New York, N.Y.
14. Vaheri, A. & E. Ruoslahti. 1974. Disappearance of a major cell type specific surface glycoprotein antigen after transformation of fibroblasts with Rous sarcoma virus. Int. J. Cancer **13:** 579–586.
15. Wartiovaara, J., E. Linder, E. Ruoslahti & A. Vaheri. 1974. Distribution of fibroblast surface antigen. J. Exp. Med. **140:** 1522–1533.
16. Yamada, K. M. & J. A. Weston. 1974. Isolation of a major cell surface glycoprotein from fibroblasts. Proc. Nat. Acad. Sci. USA **71:** 3492–3496.
17. Moore, N. F. & R. C. Hunt. 1978. In preparation.

CELL SURFACE COAT OF HAMSTER FIBROBLASTS

John M. Graham

Department of Biochemistry
St. George's Hospital Medical School
London SW17 ORE, England

Richard O. Hynes

Center for Cancer Research
Massachusetts Institute of Technology
Cambridge, Massachusetts 02139

Charles Rowlatt and Jane K. Sandall

Department of Cellular Pathology
Imperial Cancer Research Fund
London WC2A 3PX, England

The ^{125}I-lactoperoxidase system has been used by many workers to label the surface proteins of intact tissue culture cells.[1-4] When intact hamster embryo fibroblasts (NIL 8) are labeled in this manner, analysis of the whole cells by sodium dodecyl sulfate (SDS) polyacrylamide gel electrophoresis reveals that either six or seven proteins have incorporated ^{125}I. A high-molecular-weight protein (>200,000 daltons) is considerably more heavily labeled than any other protein, and this protein is absent from hamster sarcoma virus-transformed hamster embryo fibroblasts (NIL 8-HSV). Hynes[4] called this protein "large, external transformation-sensitive" (LETS) protein.

Graham *et al.*[5] showed, by fractionation of [^{125}I]peroxidase-treated cells, that while the majority of these labeled proteins were associated with the plasma membrane, the LETS protein was concentrated in a fraction that was called ultra-heavy fraction (UHF) on the basis of its high density ($\rho \sim 1.250$). Because of its composition (14% carbohydrate, 85% protein), relative absence of plasma membrane marker enzymes, and appearance in the electron microscope, it was tentatively identified as "cell surface coat." Subsequently,[6] a simple and rapid fractionation procedure was devised for the isolation of this high-density material that could be adapted to either small-scale (swing-out rotors) or large-scale (zonal rotors) preparations.

A technique for demonstrating the presence of the glycocalyx at the surface of cells by electron microscopy, devised by Luft,[7] involves simultaneous staining and fixation of the cells with ruthenium red, glutaraldehyde, and/or osmium tetroxide. In a previous paper, we reported briefly that in the absence of these fixatives, monolayers of hamster embryo fibroblasts could be stained with ruthenium red, and the deep red stain could be used as a visual marker for the high-density fraction in density gradients.[6] We have now studied these stained preparations from both NIL 8 and NIL 8-HSV cells by electron microscopy, attempted to obtain a value for the yield of ruthenium red-stained material from the two types of cell, and extended our observations on the protein compositions of both the unstained and stained UHF from these cells.

0077-8923/78/0312-0221 $01.75/0 © 1978, The New York Academy of Sciences

MATERIALS AND METHODS

Ruthenium Red Staining

We use the term "stained" to describe material that colors red with ruthenium red but that has not received the standard ruthenium red-glutaraldehyde-osmium tetroxide treatment for electron microscopy.[7] NIL 8 and NIL 8-HSV cells were grown in 9-cm plastic Petri dishes in Dulbecco's modification of Eagle's medium supplemented with 10% calf serum and subcultured by trypsinization. At confluency, the medium was removed from 30–40 dishes, and the monolayer was washed twice with cold 0.25 M sucrose-5 mM Tris (pH 7.4). The monolayer was then stained in the dark with 0.2 mg/ml of ruthenium red in 0.25 M sucrose-5 mM Tris (pH 7.4, 2 ml per dish) at 4°C for 20 min. Ruthenium red is light sensitive; solutions should be made up fresh and protected from the light. After removal of the staining solution, the monolayer was *carefully* washed twice with cold 0.25 M sucrose-5 mM Tris (pH 7.4) to remove as much of the unbound ruthenium red as possible and was then scraped with a rubber policeman into 2 ml of this medium supplemented with 0.2 mM Mg^{2+}. The cells were colored a deep red.

Fractionation

Stained or unstained cells were sedimented at 500*g* for 10 min, resuspended in a total volume of 30 ml of 0.25 M sucrose-5 mM Tris (pH 7.4) – 0.2 mM Mg^{2+} and then homogenized by nitrogen cavitation as described previously.[5,6] Nuclei were removed from the homogenate (1000*g* for 5 min), and the supernatant was made 1 mM with respect to EDTA, 47.5% (w/w) with respect to sucrose, and a total volume of approximately 50 ml. Each tube of a 3 × 25-ml swing-out rotor was loaded with 2 ml of 60% sucrose, 16 ml of the sample, 2 ml of 40% sucrose, and 4 ml of 20% sucrose [all solutions contained 5 mM Tris (pH 7.4) and 1 mM EDTA]. Centrifugation at $100{,}000g_{av}$ for 12 hr resolved two fractions: the bulk of the membranous material collected at the top of the gradient in the 20–40% sucrose region, while the dense UHF gathered at the 47.5/60% interface. With ruthenium red-stained cells, the UHF was colored a deep red, and the membrane band (A) was white-yellow.

The two bands of material were so well separated that they could be removed with a Pasteur pipette without any risk of cross-contamination. To purify the UHF, the latter was collected and recentrifuged in an identical gradient (in one tube) for 3–4 hr. After removal from the second gradient, the fraction was diluted with an equal volume of 5 mM Tris (pH 7.4), sedimented at $80{,}000g_{av}$ for 40 min, and either processed for electron microscopy or polyacrylamide gel electrophoresis or assayed for various enzymes. Stained material was also assayed for ruthenium red content. Fractions from the second gradient were denoted as A_1 and UHF_1.

Electron Microscopy

Material for electron microscopy, either pellets from the cell fractions or cells growing on plastic culture dishes (stained or unstained), was routinely fixed in 2.5% glutaraldehyde-0.1 M sodium cacodylate buffer (pH 7.2) at 20°C, washed in several changes of the cacodylate buffer, postfixed in 1% aqueous osmium tetroxide, and rinsed in water. After graded ethanol dehydration, the samples were embedded in Araldite® and hardened at 60°C (or 50°C in the case of the plastic dishes). An intermediate propylene oxide stage between the 100% ethanol and the Araldite was included for the pellets but omitted with the plastic dishes. The dishes were separated from the hardened Araldite and replaced by a further layer of Araldite to allow sectioning perpendicular to the culture monolayer. Thin sections were poststained with 25% uranyl acetate in methanol and lead citrate[8] and examined in a Philips EM301G electron microscope. For comparison, cultures of NIL 8 cells were also treated with ruthenium red (0.5 mg/ml) in both glutaraldehyde and osmium tetroxide fixatives according to the method of Luft.[7]

^{125}I Labeling

In some experiments, the cell monolayer was labeled with ^{125}I by use of the lactoperoxidase-glucose oxidase technique.[1] The membrane fraction and UHF were then analyzed for protein composition by SDS polyacrylamide gel electrophoresis and ^{125}I-labeled proteins were detected by autoradiography of the gels as described previously.[2]

Assays

5′-Nucleotidase and Na^+/K^+-ATPase were measured as described by Avruch and Wallach[9]; alkaline phosphatase was measured by use of *p*-nitrophenyl phosphate as the substrate[6]; the protein content of unstained material was assayed by the method of Lowry *et al.*[10] Galactosyl transferase was measured by use of both *N*-acetylglucosamine[11] and desialylated degalactosylated fetuin[12] as the acceptors. The assay mixture with *N*-acetylglucosamine included 20 μg of uridine diphosphate (UDP) [^{14}C]galactose, 3 μmol of $MnCl_2$, 10 μl of 2-mercaptoethanol, 4 μmol of *N*-acetylglucosamine, 0.8 μmol of Hepes (pH 7.0), and 20 μl of the sample (20–60 μg of protein) in a total volume of 80 μl. *N*-acetylglucosaminyl [^{14}C]galactose was separated on a Dowex®-I column.[11] The assay mixture with the glycoprotein acceptor included 10 μl of 1% Triton® X-100, 5 μmol of $MnCl_2$, 0.4 μg of UDP [^{14}C]galactose, 1 μg of desialylated degalactosylated fetuin, 5 μmol of Tris (pH 7.0), and 20 μl of the sample (20–60 μg of protein) in a total volume of 170 μl. The controls contained either desialylated fetuin or fetuin. After incubation at 37°C for 15 min, the protein was precipitated with 3 ml of 1% phosphotungstic acid in 0.5 N HCl (plus 2 mg of bovine serum albumin) at 0°C; the precipitate was washed twice with 4 ml of 10% cold trichloroacetic acid, extracted twice with chloroform-methanol-diethyl ether (2:1:1), solubilized in 1 ml of hyamine, and counted.

The ruthenium red content of a fraction was measured spectrophotometrically at 560 nm after solubilizing the material in 1% SDS-0.1 M sodium hydroxide. Spectra of ruthenium red-stained material show that the wavelength of maximum absorption (λ_{max}) near 560 nm tends to shift to lower wavelengths compared to ruthenium red in solution. Addition of sodium hydroxide to 0.1 M almost abolishes this shift. Estimation of the protein and enzyme content of ruthenium red-stained fractions is discussed under RESULTS.

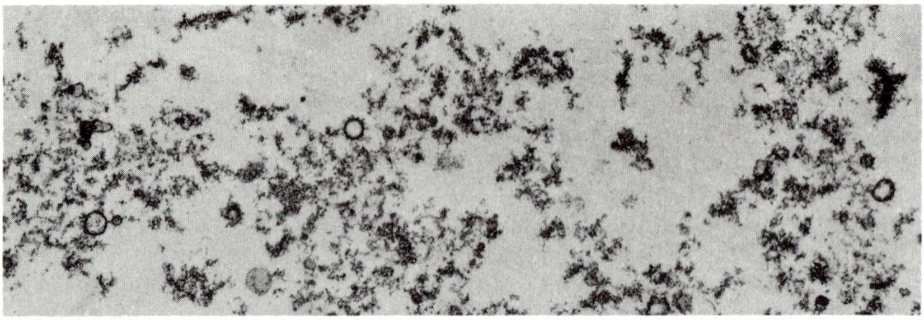

FIGURE 1. Electron micrograph of a section from an Araldite-embedded pellet of the UHF from NIL 8 cells stained with ruthenium red. The pellet consists predominantly of flocculated filamentous material, with very little contamination with membranous structures. × 16,000.

Enzymes and Chemicals

Lactoperoxidase was obtained from Calbiochem (London, England), glucose oxidase from Worthington Biochemicals, carrier-free sodium ^{125}I-iodide from the Radiochemical Centre (Amersham, England), and ruthenium red from Fluka (Buchs, Switzerland; batch 84071). All other chemicals were Analar grade.

RESULTS

Structure of Ultra-Heavy Fraction

FIGURE 1 is an electron micrograph that shows the appearance of the high-density ruthenium red-stained material from NIL 8 cells. The material is composed predominantly of an amorphous granular substance with a few membrane profiles. FIGURE 2 (*left*) is a higher-magnification electron micrograph of the material from NIL 8 cells. The structure of this stained material is very similar to that of the LETS protein-containing UHF from NIL 8 cells observed previously[5]; however, there were considerably more membranous structures in the unstained fraction.

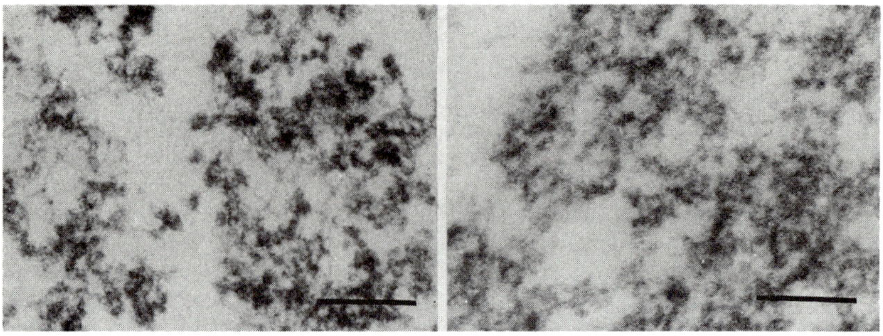

FIGURE 2. The UHF from ruthenium red-pretreated cultures at higher magnifications to demonstrate that the substructure is similar in material from NIL 8 cells (*left*) NIL 8-HSV cells (*right*). Bars = 100 nm.

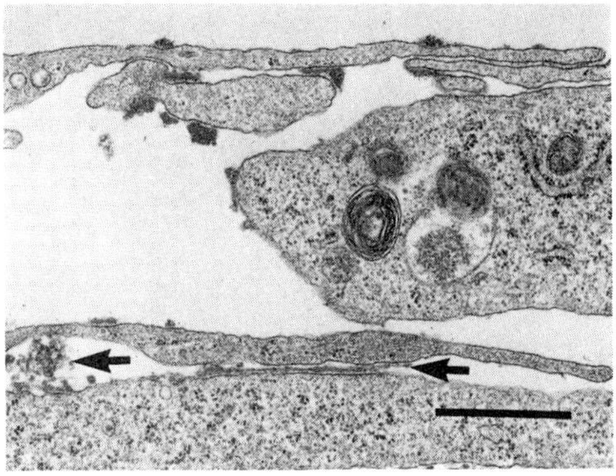

FIGURE 3. Vertical section through the full thickness of an NIL 8 cell culture stained with ruthenium red before fixation *in situ*. Aggregates of filamentous material are associated with the superficial cells. Normally occurring strands of extracellular material are indicated by arrows. Bar = 100 nm.

Electron micrographs of intact NIL-HSV cells presented in that paper[5] suggested that these cells possess very little surface coat material. However, when monolayers of these cells, stained with ruthenium red, were washed prior to harvesting, it was noted that even the third and fourth washings of cold 0.25 M sucrose-5 mM Tris (pH 7.4) were colored red, while with NIL 8 cells, the second washing was almost colorless. Apparently, ruthenium red-stained material was being continuously washed off the NIL-HSV cells. This material did not consist of whole cells, but it could be sedimented at 100,000 g_{av} for 30 min. Moreover, the sedimented material banded at the isopycnic density of the UHF in a sucrose gradient. It would appear, therefore, that the cell coat material of NIL-HSV cells is less well bound to the cell surface and may be partially lost during washing of the cells. Only if the washing of ruthenium red-stained NIL-HSV cells is performed very gently may this loss be minimized. When the washing is performed carefully, not only can a red-stained fraction be isolated from NIL-HSV cells (FIGURE 2, *right*) that is structurally similar to that from NIL 8 cells (FIGURE 2, *left*) but also cell coat material can be distinguished at the surface of intact monolayers of these cells. The material shown in FIGURES 1 and 2 is very similar in structure to that observed at the surface of both NIL 8 cells (FIGURES 3 & 4) and NIL 8-HSV (FIGURES 5 & 6). These figures are electron micrographs of vertical sections through intact monolayers of cells that have been stained with ruthenium red; they demonstrate the presence of filamentous and aggregated material on the surface of the cell. In comparison with the fractionated UHF, however, the cell

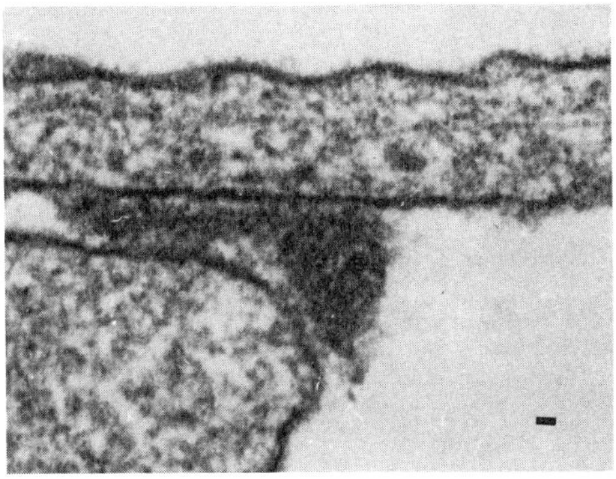

FIGURE 4. Higher magnification of part of the specimen in FIGURE 3 to illustrate the normal surface filamentous material and the detail in the aggregated material. The aggregates are separated from the plasma membrane by surface filaments. Bar = 100 nm.

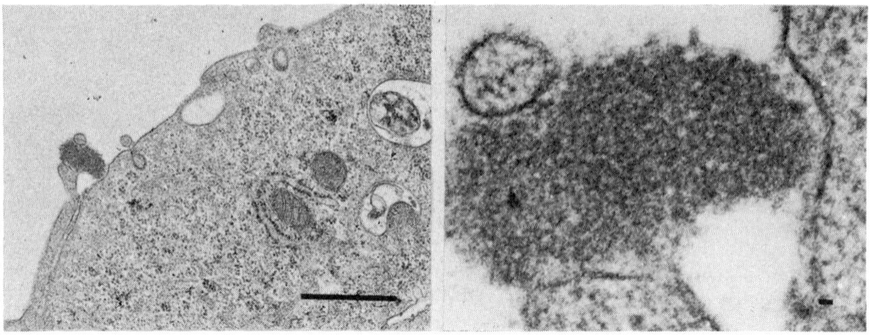

FIGURES 5 (*left*) & 6 (*right*). Vertical sections through the upper surface of a cell culture of NIL 8-HSV cells that had been stained with ruthenium red to illustrate aggregation of material and the normal surface filamentous material. Bars = 100 nm.

surface aggregates appear more condensed. The more diffuse nature of the fractionated material may be the result of nitrogen cavitation of the cells; during this process, the rapid formation of bubbles of nitrogen gas in the medium apparently causes the glycocalyx to be stripped off the surface membrane of the cell; at the same time, it may lead to a disaggregation and partial breakdown of the glycocalyx itself.

When either NIL 8 or NIL 8-HSV cells are stained with ruthenium red, they are colored a deep red. However, this reaction alone, when performed prior to fixation for electron microscopy, does not cause an increase in electron density of the surface material, even though it is subsequently treated with osmium tetroxide [compare FIGURES 7 and 8 (NIL 8 and NIL 8-HSV cells not stained with ruthenium red) with FIGURES 4 and 5]. In contrast, when ruthenium red is added along with both glutaraldehyde and osmium tetroxide fixatives,[7] the electron density of the surface material is enhanced, and its volume is apparently increased (FIGURE 9).

Yield of Ruthenium Red-Stained Ultra-Heavy Fraction

To evaluate the protein content of a ruthenium red-stained fraction, chemical methods cannot be used because of interference from the dye, and the latter is not readily removed from the sample once it is bound. High concentrations of sodium chloride (1–2 M) remove approximately 75% of the bound ruthenium red from

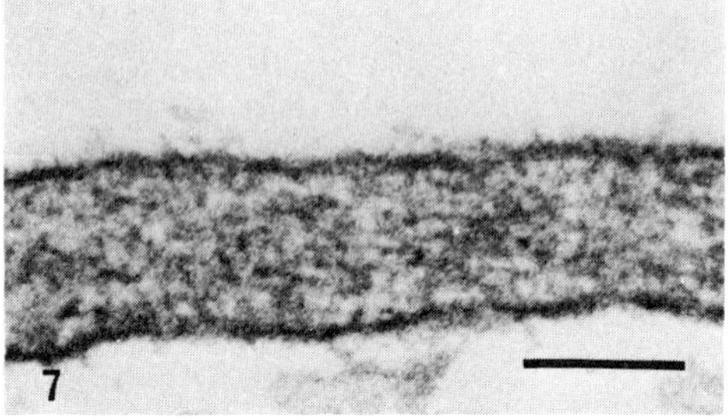

FIGURE 7. Similar field from an NIL 8 cell culture that had not been stained with ruthenium red before fixation. The surface filamentous material and the extracellular strands have the same density as material that has been stained with ruthenium red. Bar = 100 nm.

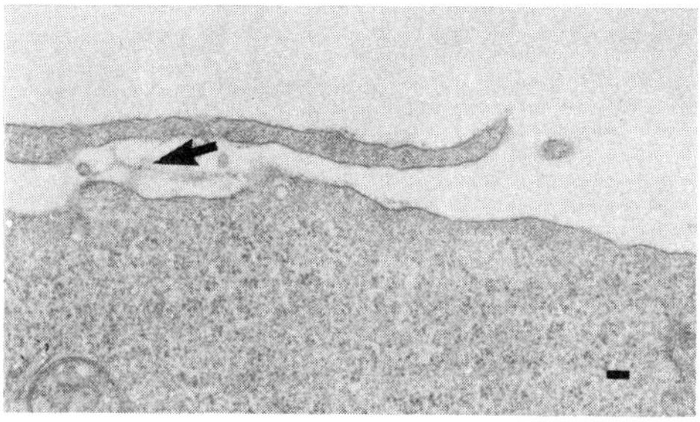

FIGURE 8. Similar field to that in FIGURE 6 from a culture of NIL 8-HSV cells that had not been stained with ruthenium red before fixation. The extracellular strands seen in the NIL 8 cell cultures are also present (arrow) in the transformed cultures. Bar = 100 nm.

stained UHF, but this reagent also solubilizes approximately 15% of the protein. An approximate value for the protein content can be most readily obtained by measurement of absorption at 280 nm. Since, however, ruthenium red itself shows significant absorption at 280 nm, the contribution of the dye to the total absorption must be taken into account. In the presence of 1% SDS-0.1 M NaOH, unstained UHF, or the membrane fraction, exhibits negligible absorption at 560 nm, so that any absorption measured at this wavelength with stained fractions must be due to the ruthenium red. FIGURE 10 shows calibration curves for the absorption of ruthenium red solution (in 1% SDS-0.1 M NaOH) at 560 and 280 nm. From the 560-nm absorption of a stained fraction, the 280-nm absorption due to ruthenium red can be calculated (from FIGURE 10) and subtracted from the observed 280-nm absorption of the sample to yield a value for the protein absorption alone. Since we have used the Lowry method for protein assay in all previous experiments, the 280-nm absorptions for various concentrations of unstained postnuclear supernatant, total membrane fraction, and UHF were plotted against the protein content (micrograms) as determined by the method of Lowry *et al.*[10] with bovine serum albumin used as the standard. From FIGURE 11, therefore, the 280-nm protein absorption of stained fractions can be converted to micrograms of protein.

To test the validity of this approach, a suspension of UHF was divided into two parts: to one part was added an equal volume of a ruthenium red solution, so that the final concentration was 25 μg/ml; to the other was added an equal volume of water, and both were then made 1% and 0.1 M with respect to SDS and NaOH, respectively. Measurements at 280 and 560 nm were made on a series of dilutions of the stained and unstained samples. FIGURE 12 shows good agreement between the $A_{280\,\mathrm{nm}}$ versus UHF concentration plot of the unstained material and that

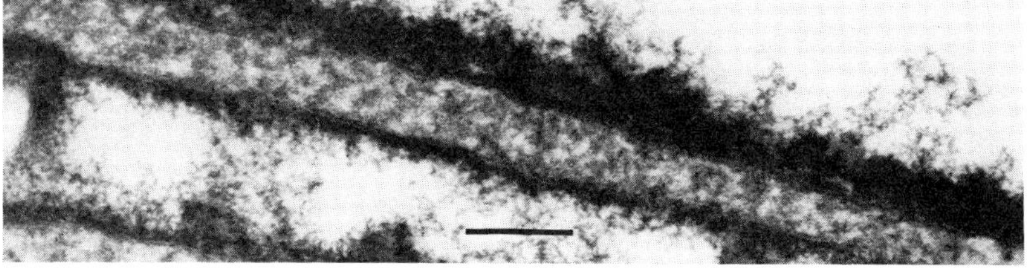

FIGURE 9. Surface cells in an NIL 8 culture treated *in situ* with ruthenium red, according to Luft,[7] to illustrate the greater electron density in the surface material. Bar = 100 nm.

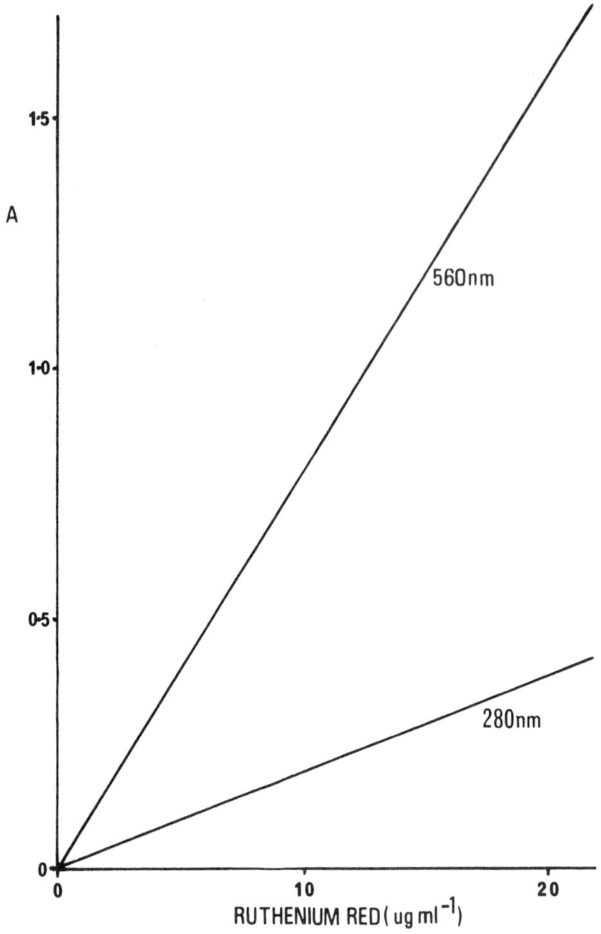

FIGURE 10. Absorption of solutions of ruthenium red in 1% SDS-0.1 M NaOH at 280 and 560 nm.

calculated for the stained sample. In a second experiment, similar measurements were made on the original stained and unstained samples and on various mixtures of the two (i.e., at varying protein-ruthenium red ratios but constant protein concentration). TABLE 1 compares the protein content of the samples as estimated by the Lowry procedure on the unstained material with the values obtained from absorption measurements made on the stained samples. The method yields an estimated value accurate to about ±9% of the true value.

The yield of UHF from nonruthenium red-stained NIL 8 cells represents approximately 0.65% of the total cellular protein; for NIL-HSV cells, the value is approximately 0.45%. This difference may reflect the apparently poorer adhesion of the cell coat in the transformed cell. From ruthenium red-treated NIL 8 and NIL-HSV cells, the recoveries were approximately 0.4 and 0.3%, respectively. The nuclear pellet from homogenates of stained cells, however, is significantly colored, suggesting that much of the surface coat is not recovered. The pellet resisted all attempts at resuspension by Dounce homogenization or ejection through the fine-gauge needle of a syringe. Under the phase contrast microscope, this material revealed large aggregates of nuclei and partially broken cells, cell

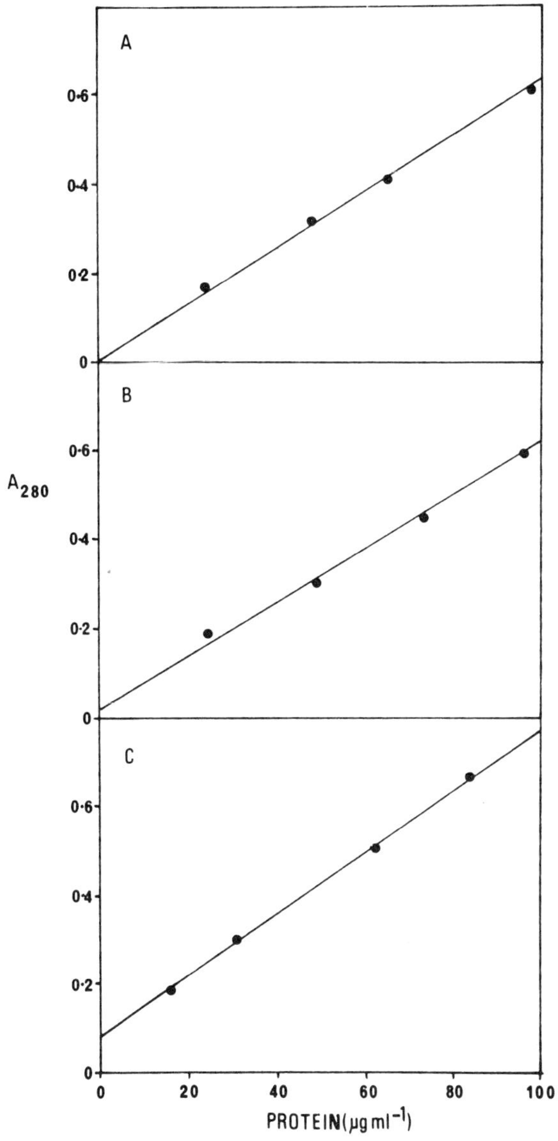

FIGURE 11. Relation between absorbance at 280 nm (A_{280}) and protein concentration as determined by the method of Lowry *et al.*[10] A, membrane fraction; B, homogenate; C, UHF.

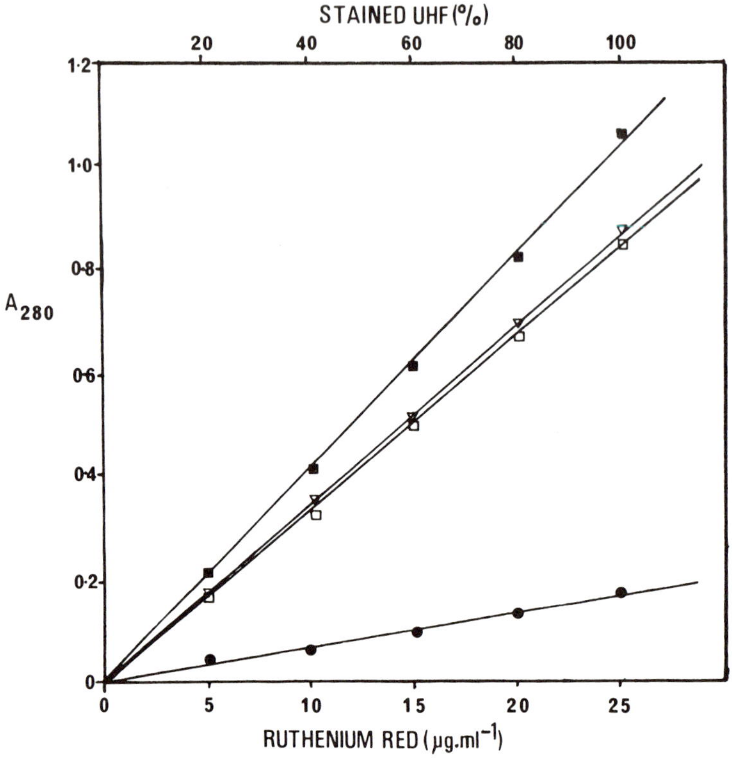

FIGURE 12. Relation between absorbance at 280 nm (A_{280}) and concentration of ruthenium red and concentrations of stained and unstained fractions. UHF, either stained with ruthenium red at 25 µg/ml or unstained, was diluted to give five samples that contained 20, 40, 60, 80, and 100% of the original protein concentration; for the stained fraction, the concentrations were 5, 10, 15, 20, and 25 µg/ml of ruthenium red. Ruthenium red in aqueous solution (●), unstained UHF (▽), stained UHF (■), stained fraction corrected for absorbance due to ruthenium red (□).

TABLE 1

ESTIMATION OF PROTEIN CONTENT OF RUTHENIUM RED-STAINED UHF

Ruthenium Red-Stained UHF (%)	$A_{560\,nm}$	$A_{280\,nm}$	$A^{RR*}_{280\,nm}$	$A^{corr\dagger}_{280\,nm}$	Protein (µg)
0	—	—	—	—	48.5
20	0.270	0.500	0.065	0.435	51.0
40	0.510	0.570	0.120	0.450	53.0
60	0.720	0.550	0.170	0.380	43.0
80	1.020	0.635	0.245	0.390	44.0
100	1.380	0.740	0.330	0.410	47.0

*Calculated absorption of ruthenium red.
†Corrected for ruthenium red contribution.

TABLE 2

ALKALINE PHOSPHATASE AND 5′-NUCLEOTIDASE OF FRACTIONS FROM UNSTAINED NIL 8 AND NIL 8-HSV CELLS

	Alkaline Phosphatase*		5′-Nucleotidase*	
Fraction	NIL 8	NIL 8-HSV	NIL 8	NIL 8-HSV
Homogenate	0.045	0.092	0.56	0.72
A	0.065	0.129	3.22	2.05
UHF	0.128	0.200	1.56	1.50
A_1	0.115	0.150	9.81	2.43
UHF_1	0.130	0.223	1.21	0.54

*Activity is expressed in micromoles of substrate hydrolyzed per hour per milligram of protein.

debris, and amorphous material. It appears that treatment with ruthenium red renders the cells less susceptible to breakage by nitrogen cavitation, and the nuclei aggregate almost irreversibly with some of the ruthenium red-stained material. It seems unlikely that the coloration of the nuclear pellet is due to entry of ruthenium red into the cells because the cell monolayers appear to be intact after staining, as determined by electron microscopy, and the membrane fraction that can be recovered from homogenized cells is either unstained or only very lightly colored. Thus, although the UHF from ruthenium red-stained cells is far less contaminated by membranes compared to that from untreated cells, the yield from NIL 8 and NIL 8-HSV cells was reduced by 30%.

Enzyme Composition of Ultra-Heavy Fraction

Unstained Material

Previously,[6] we had reported the existence of 5′-nucleotidase and alkaline phosphatase activities in unstained UHF from NIL 8 cells; we have now confirmed and extended these results. Both of these enzymes are present in the UHF from NIL 8 and NIL 8-HSV cells, and the enzymatic activities persist in this

TABLE 3

PROPERTIES OF ALKALINE PHOSPHATASE IN FRACTIONS FROM NIL 8 AND NIL 8-HSV CELLS

	Alkaline Phosphatase*			
	Tris Buffer		Borate Buffer	
Fraction	pH 9.8	pH 8.8	pH 9.8	pH 8.8
NIL 8				
Membrane	0.064	0.104	0.076	0.084
UHF	0.140	0.148	0.111	0.096
NIL 8-HSV				
Membrane	0.054	0.120	0.045	0.071
UHF	0.168	0.210	0.210	0.210

*Activity is expressed in micromoles of *p*-nitrophenol phosphate hydrolyzed per hour per milligram of protein.

TABLE 4

EFFECT OF RUTHENIUM RED (RR) ON ENZYMATIC ACTIVITIES OF NIL 8 CELL FRACTIONS

Fraction	Na^+/K^+-ATPase*		5′-Nucleotidase*	
	−RR	+RR	−RR	+RR
Membrane	0.57	< 0.01	5.3	3.3
UHF	< 0.01	< 0.01	1.5	1.0

*Enzymatic activities expressed in micromoles of substrate reacted per hour per milligram of protein.

fraction when it is repurified through a second gradient (TABLE 2). However, in NIL 8 cells, at least part of the 5′-nucleotidase is associated, in high concentration, with the low-density fraction (A_1); that is, it may not be a true component of UHF. For the alkaline phosphatase, however, the situation is altered, because its activity is higher in the UHF than in the membrane fraction from both normal and transformed fibroblasts. Although assays for this enzyme are normally performed at pH 9.8 (often with a borate buffer), the enzyme in NIL 8 cells has a pH optimum of about pH 8.8 and appears to be more active in the presence of Tris (TABLE 3). Conversely, in the NIL 8-HSV cell, the alkaline phosphatase is more active in borate buffer at pH 9.8 and is unaffected by this buffer type at pH 8.8, while the reverse is true for the membrane fraction. There is also a significant redistribution of enzymatic activity in favor of the UHF in the transformed cell (TABLE 3).

Stained Material

In view of the reduced membrane contamination of ruthenium red-stained UHF, it was of interest to study the enzyme content of this material. To check the effect of ruthenium red on two enzymes (assayed radiometrically), a membrane fraction and UHF were prepared from unstained NIL 8 cells. Each fraction was suspended in 0.25 M sucrose-5 mM Tris (pH 7.4), divided, and a solution of ruthenium red added to one half (final concentration, 0.2 mg/ml). After maintaining the suspensions at 4°C for 10 min in the dark, they were sedimented at $100,000 g_{av}$ for 30 min and washed twice in 0.25 M sucrose-5 mM Tris (pH 7.4) before resuspension in this medium. Na^+/K^+-ATPase and 5′-nucleotidase were

TABLE 5

5′-NUCLEOTIDASE COMPOSITION OF FRACTIONS ISOLATED FROM RUTHENIUM RED-STAINED NIL 8 CELLS

Fraction	5′-Nucleotidase Activity*	
	Observed	Corrected
Homogenate	0.45	0.64
Membrane	2.80	4.00
UHF	1.10	1.57

*Enzymatic activity expressed in micromoles of AMP hydrolyzed per hour per milligram of protein.

assayed in each fraction, and the unstained fractions were assayed for protein. TABLE 4 shows that ruthenium red abolished Na^+/K^+-ATPase activity in the membrane fraction, while 5′-nucleotidase activity was reduced by approximately 30% in both membrane and UHF fractions. Since colorimetric assays (e.g., alkaline phosphatase) of this material are difficult, we have restricted enzyme assays on stained UHF to the 5′-nucleotidase. TABLE 5 shows the 5′-nucleotidase content of fractions isolated from NIL 8 cells stained with ruthenium red. Clearly, despite the reduction of membrane contamination in the stained UHF (compared to unstained), there is no significant reduction in 5′-nucleotidase. It thus seems likely that this enzyme is a true constituent of the surface coat material; moreover, 5′-nucleotidase has been shown to be an exoenzyme.[13]

TABLE 6

GALACTOSYL TRANSFERASE OF FRACTIONS FROM NIL 8 AND NIL 8-HSV CELLS

Fraction	Galactosyl Transferase: Acceptor, *N*-acetylglucosamine* pH 7.0	Galactosyl Transferase: Acceptor, *N*-acetylglucosamine* pH 6.4	Galactosyl Transferase: Acceptor, Desialylated Degalactosylated Fetuin†
NIL 8			
Homogenate	26	ND‡	63
A	154	108	277
UHF	40	ND	ND
A_1	80	65	ND
UHF_1	32	44	130
NIL 8-HSV			
Homogenate	24	ND	51
A	120	ND	135
UHF_1	0–10	ND	72

*Activity expressed in micrograms of galactose transferred to *N*-acetylglucosamine per hour per milligram of protein.

†Activity expressed in nanograms of galactose transferred to desialylated degalactosylated fetuin per hour per milligram of protein.

‡Not determined.

Galactosyl Transferase

In view of interest in the possible existence of surface glycosyl transferase enzymes,[14-16] and because our previous attempts[17] to detect such activities in the plasma membrane were equivocal, we assayed the unstained UHF for this enzyme. TABLE 6 demonstrates that the UHF from NIL 8 cells contains a significant amount of galactosyl transferase activity (using *N*-acetylglucosamine as the acceptor of galactose from UDP galactose), and this activity is not diminished by repurification of the UHF through a second gradient. Moreover, the pH optimum for the enzyme is shifted down slightly in the UHF compared to the membrane fraction. In the UHF from NIL-HSV cells (TABLE 6), enzymatic activity is about five- to 10-fold lower compared to that in untransformed cells. It is also interesting that, compared to the membrane enzyme, the UHF enzyme is able to utilize desialylated degalactosylated fetuin considerably better than *N*-acetylglucosamine (TABLE 6); for NIL 8 cells, the membrane-UHF activity ratio is 4.8 for *N*-acetylglucosamine and 2.0 for the glycoprotein acceptor. With NIL

8-HSV cells, the enzyme in the UHF is always at a lower level compared to that in normal cells; sometimes no activity is detectable with *N*-acetylglucosamine as the acceptor.

SDS Polyacrylamide Gel Electrophoresis

Unstained Ultra-Heavy Fraction

Although the UHF from NIL 8 cells contains only one ^{125}I-labeled protein (the LETS protein), the Coomassie blue staining pattern of SDS polyacrylamide gels reveals that this fraction also contains several other proteins that are not labeled. Furthermore, the pattern of protein bands (FIGURE 13) in the repurified UHF is quite distinct from that of either of the two membrane (low-density) fractions obtained in the first (A) and second (A_1) isopycnic sucrose gradients. Comparison of tracks b–d with track a (the postnuclear supernatant from which all subsequent fractions are derived) indicates that bands 1, 2 (LETS protein), 3, 11, 22, 27, 33, 38, and 39 are specifically enriched in the NIL 8 UHF (track d); the other prominent bands, notably 14, 15, 20, 23, 28–30, 35, and 36, are also major

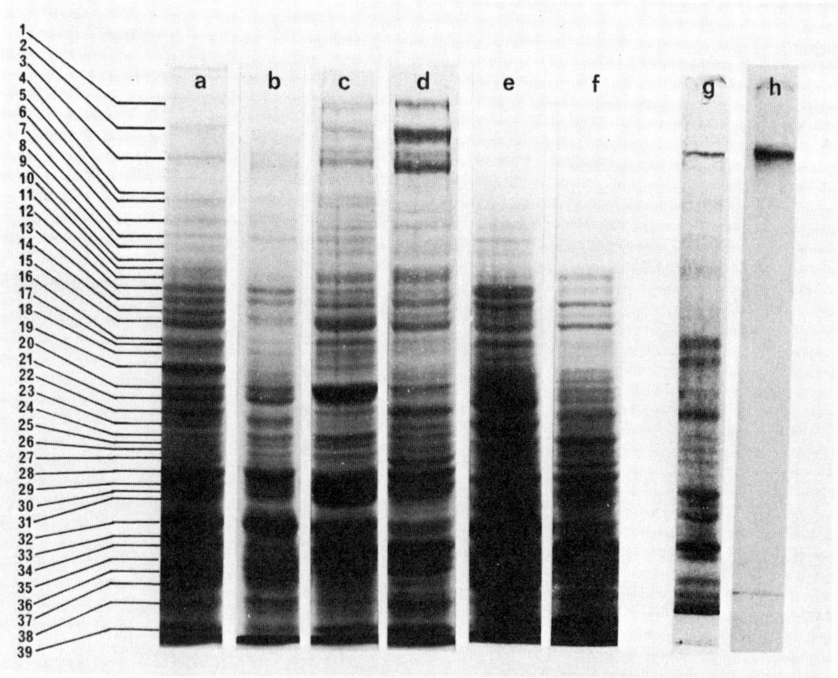

FIGURE 13. Sodium dodecyl sulfate polyacrylamide gel electrophoresis of fractions from NIL 8 and NIL 8-HSV cells. a, NIL 8 postnuclear supernatant; b, NIL 8 membrane fraction (A); c, NIL 8 membrane fraction (A_1); d, NIL 8 repurified UHF_1; e, NIL 8-HSV membrane fraction; f, NIL 8-HSV UHF; g & h, NIL 8 ruthenium red-stained ^{125}I-labeled UHF. Tracks a–g are from Coomassie blue-stained gels; track h is an autoradiograph of track g. The molecular weights of bands 3, 15, 23, and 36 are approximately 200,000, 130,000, 70,000, and 40,000 daltons, respectively.

bands in either A (track b) or A_1 (track c). In addition, there are minor bands in the separated fractions that are not apparent in the postnuclear supernatant. Regions 12–15 of the postnuclear supernatant are considerably more complex than would be suggested from track a; A (track b) contains two minor bands that run between bands 14 and 15, and one of these bands appears in the UHF. Regions 23 and 24 are similarly more complex than would be suggested from track a; a minor band between these two is apparent in both A and A_1. It is clear, moreover, that material A_1 (track c), derived during repurification of the UHF, is significantly different from A (track b) in its protein composition and thus does not represent merely a nonspecific association of membrane material with the UHF. The pattern of bands in the UHF from NIL 8-HSV cells (track f) shows a broad similarity to that from NIL 8 cells, except for bands 1–3, which are almost totally absent; bands 22 and 38 are much reduced, while a band in the region of 24 is more prominent. Track e (the NIL 8-HSV membrane fraction) displays many differences from both the NIL 8-HSV UHF and the corresponding NIL 8 membrane fraction A (track b).

Ruthenium Red-Stained Ultra-Heavy Fraction

In one series of experiments, the intact NIL 8 cell monolayer was labeled with ^{125}I by use of the lactoperoxidase-glucose oxidase technique prior to staining with ruthenium red. Sodium dodecyl sulfate polyacrylamide gels of ^{125}I-labeled fractions were analyzed by autoradiography. Clearly, the stained high-density fraction is heavily enriched in the [^{125}I]LETS protein (track h); the latter is absent from the membrane fraction.

Although the stained NIL 8 UHF was run on a gel different from that used for unstained UHF, it is possible to compare the band patterns by reference to a post-nuclear supernatant run on the same gel (not shown). Bands 1 and 3 are notably absent in the stained UHF, as are bands in the regions 11–14. We have tentatively assigned the major bands as 2, 20, 22, 27 or 28, 33, 35, and 38 or 39.

Discussion

Staining of NIL 8 hamster embryo fibroblasts with ruthenium red appears to aid the resolution of the UHF from membranes of these cells; electron micrographs reveal that fewer membrane profiles are found than had been reported previously.[5] Some reaction occurs in the extracellular material during staining that allows a cleaner separation of the UHF from other cellular components. Under the staining conditions employed, no significant cellular rupture occurs, and endocytosis is prevented. Since in intact cells ruthenium red stains the cell coat specifically,[18] the UHF, identified in gradients by its deep red color, must be derived from the cell surface coat. The results presented in this paper demonstrate clearly that in NIL 8 hamster embryo fibroblasts, the LETS protein, which, by definition, is extremely trypsin sensitive and therefore located externally on the surface of the cell, is heavily concentrated in this fraction.

Although the stained UHF visualized in electron micrographs appears more disperse than does the extracellular material of intact monolayers of both NIL 8 and NIL 8-HSV cells, the overall similarity of structure is clear. There is no obvious difference between the stained UHF isolated from normal fibroblasts and that from transformed cells. Interestingly, the electron density of the cell surface material from unstained monolayers is very similar to that of our stained cells;

only when osmic acid and glutaraldehyde are applied simultaneously with ruthenium red does intensification of the electron density occur. Luft[7] has described this process in detail. He has postulated two models in which successive accretion of osmium-containing molecules occurs at the site of the initial ruthenium red localization. We cannot determine whether the additional material observed on the cell surface after Luft's treatment (FIGURE 9), as compared with our stained or unstained cell surfaces, represents a real retention of cell surface material or the accumulation of reaction products during processing. Under the conditions used, ruthenium red was able to react with all extracellular material in the cultures that had not been washed off in the initial rinses. This material includes the fibrillar extracellular substance, shown in FIGURES 3, 7, and 8 (see also Reference 19), normally not considered to be the immediate cell coat, which may be preelastin. The separation of the filamentous aggregates from the immediate vicinity of the cell membranes (FIGURES 4 & 6) may be significant.

Investigations into the amount and origin of the surface coat at the surfaces of normal and virus-transformed cells have yielded much conflicting evidence. Data obtained from measurements made on the layer of glycoprotein deposited on the surface of the culture vessel suggested that the coat material of normal mouse fibroblasts (Balb/c 3T3) was thicker than that of transformed derivatives.[20] Transformation of human blood lymphocytes is also associated with a decrease in the lanthanum-stained cell coat.[21] With ruthenium red, however, Martinez-Palomo *et al.*[22] reported an increase in thickness in transformed derivatives of hamster cells, but other workers[23] were not able to demonstrate any clear-cut difference with this stain. Ellipsometry measurements[24] on a variety of cell lines have generally indicated increased coat synthesis in transformed variants, although Goggins *et al.*[25] found that SV-40-transformed 3T3 cells displayed decreased mucopolysaccharide synthesis.

Our results with unstained cells suggest a reduction in the amount of cell coat in NIL 8-HSV cells compared to NIL 8 cells. Observations on the ruthenium red-stained material did not confirm this finding due to the recovery problem. Quantitative comparisons of the UHF between the two cell types, however, must be made with caution, because the stained NIL 8-HSV extracellular material is clearly far less well bound to the surface than that of NIL 8. Shigematsu and Dmochowski[26] have also emphasized that the cell coat of virus-transformed cells may be more diffuse than that of normal cells. On the other hand, Temmink and Spiele[27] discovered that part of the osmic acid-ruthenium red-stained material at the surface of 3T3 cells could be removed by washing, unlike that of SV40-3T3 cells. Furthermore, they proposed that all of this material was serum derived; yet, the protein profile of the UHF (stained or unstained) does not resemble that of normal serum.

The enzymatic profile of the UHF is interesting, inasmuch as the functional role of the cell coat is not well understood. Of the three enzymes that we have detected in the UHF, the most problematical is the 5′-nucleotidase: although it persists in this fraction during purification of the UHF, the level of this enzyme progressively decreases; UHF levels of alkaline phosphatase, however, particularly that from NIL 8-HSV cells, are higher than that of any other enzyme, under optimum conditions of assay for that fraction. The very high activity of the 5′-nucleotidase in fraction A_1 from both types of cells suggests that either the membrane components of the unpurified UHF contain the enzyme or that fraction A_1 also contains some specific part of the extracellular material, which is resolved from the rest in the second isopycnic gradient. In the ruthenium red-stained UHF from NIL 8 cells, 5′-nucleotidase levels are as high, or even slightly higher than, that of the unstained fraction; yet, electron microscopy has shown that the stained fraction is considerably impoverished in the membranous components, suggesting that the enzyme is a true component of the UHF.

The association of alkaline phosphatase with the UHF is supported by findings in electron micrographs of cells stained histochemically for this enzyme.[28] Benham *et al.*[28] also suggest that the plasma membrane enzyme is different from that of the surface coat. It is interesting that in reports on alkaline phosphatase levels in normal and virus-transformed tissue culture cells, no general trend can be distinguished.[29-31] We have found that during long periods of continuous culture of NIL 8 cells, the actual levels of the enzyme in all fractions fluctuate quite widely. Thus, in TABLE 2, the values for alkaline phosphatase are taken from a single experiment; although the specific activity of the enzyme in a fraction may be as much as 50% higher or 30% lower than the values given, the relative specific activities in each fraction remain constant. With NIL 8-HSV cells, however, these fluctuations are far less marked. What is important is that examination of a particular fraction from each type of cell need not indicate any significant difference between the normal and the virus-transformed cell; on the other hand, examination of both cell fractions consistently reveals the more pronounced enrichment of the alkaline phosphatase in the UHF of NIL 8-HSV cells compared to that of NIL 8 cells. From data on the effect of buffer type and pH on the alkaline phosphatase, it would appear that the enzyme is altered in the transformed cell; the significance of this alteration is not known.

The results on galactosyl transferase shed further light on the problem of the composition of fraction A_1. In NIL 8 cells, it appears that when *N*-acetylglucosamine is used as the acceptor, the pH optimum of the enzyme is 6.4 in the UHF compared to 7.0 in the membrane fraction (A). The activity in fraction A_1 is more like that of the membrane fraction (A) than that of the UHF, suggesting that this material does contain membrane components. Interestingly, the activity in the UHF from NIL 8-HSV cells is considerably lower than that from normal cells, when both *N*-acetylglucosamine and desialylated degalactosylated fetuin are used as galactose acceptors. In some experiments, no activity could be detected with the simple molecule with the NIL 8-HSV UHF; differences in the membrane fraction of the two types of cell were, however, less marked. When the glycoprotein acceptor was used, the level in both membrane and UHF fractions is approximately 50% lower in the transformed cell. The other important feature of these results is that compared to the membrane fraction (in both types of cell), the UHF utilizes the glycoprotein acceptor more efficiently than the simple molecule. This difference in efficacy would be expected if the UHF enzyme functions in the lengthening of preexisting carbohydrate chains.

Sodium dodecyl sulfate polyacrylamide gel electrophoresis has demonstrated that although the UHF from NIL 8 cells contains only one major iodinated protein (LETS), it possesses numerous other unlabeled proteins. Why these proteins are not labeled is not clear. If at least some of these proteins are real components of the cell coat, they must somehow be unavailable to the iodination reaction, probably because of protection by either the LETS protein itself or some other component. Alternatively, if the LETS protein is the only true component, the other bands represent contaminants derived from the internal membranes and/or cytosol during homogenization. If this is the case, only certain specific proteins become adsorbed, because the protein pattern of the UHF is unlike that of the membrane fraction or the postnuclear supernatant (membrane, cytosol, and UHF). Clearly, in NIL 8-HSV cells, this situation is untenable, because the UHF derived from these cells does not contain the LETS protein. Furthermore, although the protein profile of this UHF is similar to that of NIL 8 cells, there are some notable differences between the two, particularly the absence of bands 1–3.

The complexity of the surface structure is emphasized by the fact that fraction A_1, derived from the unpurified UHF, has a protein pattern quite distinct from that of either the total membrane fraction or the purified UHF. Thus, it does not simply represent the release of membranes nonspecifically trapped in the UHF. It is either a specific membrane fraction associated with the cell coat or a particular component of the UHF or possibly both. Likewise, the ruthenium red-stained UHF may represent only a particular subfraction of the total coat material rather than a highly purified form of this cellular component.

It is tempting to speculate that at least some of the unlabeled proteins in the UHF may reside in parts of the surface membrane overlaid and protected by the surface coat; other "naked" parts of the plasma membrane, on the other hand, would contain labeled proteins other than the LETS protein. The UHF would, therefore, contain surface coat together with parts of the plasma membrane normally associated with the coat in the intact cell.

Summary

The UHF fraction from NIL 8 hamster embryo fibroblasts contains the LETS protein and several other major proteins. It exhibits three enzymatic activities in significant amounts: 5′-nucleotidase, alkaline phosphatase, and galactosyl transferase. The latter two appear to be different from the membrane-bound enzymes. This fraction is heavily stained with ruthenium red, a dye specific for the cell coat in intact cells. A comparable fraction from hamster sarcoma virus-transformed cells exhibits a similar overall protein composition but lacks at least three major proteins, including the LETS protein. Compared to NIL 8 cells, the distribution of alkaline phosphatase in fractions from these cells is different, and the level of galactosyl transferase in the UHF is much reduced.

Acknowledgments

We thank Marilyn C. Fripp and Jagdish Rao for excellent technical assistance and Helen Withers Green for typing the manuscript.

References

1. Hynes, R. O. 1973. Alteration of cell surface proteins by viral transformation and by proteolysis. Proc. Nat. Acad. Sci. USA **70:** 3170–3174.
2. Wickus, G. G., P. E. Branton & P. W. Robbins. 1974. Rous sarcoma virus transformation of the chick cell surface. *In* Cold Spring Harbor Conferences on Cell Proliferation. B. Clarkson & R. Baserga, Eds. Vol. 1:541–546. Cold Spring Harbor Laboratory. Cold Spring Harbor, N.Y.
3. Morrison, M., R. E. Gates & C. T. Huber. 1974. The use of lactoperoxidase as a molecular and macromolecular probe. *In* Membrane Transformations in Neoplasia. Miami Winter Symposia. J. Schultz & R. E. Block, Eds. Vol. 8: 33–49. Academic Press, Inc. New York, N.Y.
4. Hynes, R. O. & K. C. Humphryes. 1974. Characterization of the external proteins of hamster fibroblasts. J. Cell Biol. **62:** 438–448.
5. Graham, J. M., R. O. Hynes, E. A. Davidson & D. F. Bainton. 1975. The location of proteins labeled by the ^{125}I-lactoperoxidase system in the NIL 8 hamster fibroblast. Cell **4:** 353–365.

6. GRAHAM, J. M. & R. O. HYNES. 1975. Isolation and characterization of a cell surface fraction from hamster-embryo fibroblasts. Biochem. Soc. Trans. **3:** 761–763.
7. LUFT, J. H. 1971. Ruthenium red and violet: chemistry, purification, methods of use for electron microscopy and mechanism of action. Anat. Rec. **171:** 347–368.
8. REYNOLDS, E. S. 1963. The use of lead citrate at high pH as an electron dense stain in electron microscopy. J. Cell Biol. **17:** 208–212.
9. AVRUCH, J. & D. F. H. WALLACH. 1971. Preparation and properties of plasma membrane and endoplasmic reticulum fragments from rat fat cells. Biochim. Biophys. Acta **233:** 334–347.
10. LOWRY, O. H., N. J. ROSEBROUGH, A. L. FARR & R. J. RANDALL. 1951. Protein measurement with the Folin phenol reagent. J. Biol. Chem. **193:** 265–275.
11. FLEISCHER, B. & S. FLEISCHER. 1970. Preparation and characterization of Golgi membranes from rat liver. Biochim. Biophys. Acta **219:** 301–319.
12. BOSMANN, H. B. 1971. Mechanism of cellular drug resistance. Nature (London) **233:** 566–569.
13. DE PIERRE, J. W. & M. L. KARNOVSKY. 1974. Ecto-enzymes of the guinea pig polymorphonuclear leukocyte. I—Evidence for an ecto-adenosine monophosphatase, -adenosine triphosphatase and -*p*-nitrophenyl phosphatase. J. Biol. Chem. **249:** 7111–7120.
14. ROTH, S., E. J. MCGUIRE & S. ROSEMAN. 1971. Evidence for cell surface glycosyltransferases—their potential role in cellular recognition. J. Cell Biol. **51**:536–547.
15. ROTH, S. & D. WHITE. 1972. Intercellular contact and cell surface galactosyl transferase activity. Proc. Nat. Acad. Sci. USA **69:** 485–489.
16. KEENAN, J. W. & D. J. MORRE. 1975. Glycosyl transferases: do they exist at the surface of mammalian cells? FEBS Lett. **55:** 8–13.
17. CHANDRABOSE, K. A., J. M. GRAHAM & I. A. MACPHERSON. 1976. Glycolipid glycosyl transferases of a hamster cell line in culture. II. Subcellular distribution and the effects of culture age and density. Biochim. Biophys. Acta **429:** 112–122.
18. LUFT, J. H. 1971. Ruthenium red and violet: fine structural localization in animal tissue. Anat. Rec. **171:** 369–415.
19. FRANKS, L. M. & P. D. WILSON. 1970. "Spontaneous" neoplastic transformation *in vitro:* the ultrastructure of the tissue culture cell. Eur. J. Cancer **6:** 517–523.
20. CULP, L. A. 1974. Substrate-attached glycoproteins mediating adhesion of normal and virus-transformed mouse fibroblasts. J. Cell Biol. **63:** 71–83.
21. ANTEUNIS, A. & M. VIAL. 1975. Cytochemical and ultra-structural studies concerning the cell coat glycoproteins in normal and transformed human blood lymphocytes. Exp. Cell Res. **90:** 47–55.
22. MARTINEZ-PALOMO, A., C. BRAISLOVSKY & W. BERNHARD. 1969. Ultrastructural modification of the cell surface and intercellular contacts of some transformed cell strains. Cancer Res. **29:** 925–937.
23. VORBRODT, A. & H. KOPROWSKI. 1969. Ruthenium red-stained coat of normal and SV 40-transformed cells. J. Nat. Cancer Inst. **43:** 1241–1248.
24. POSTE, G. & I. A. MACPHERSON. 1973. The synthesis of surface coat materials in normal and transformed cells. Exp. Cell Res. **77:** 264–270.
25. GOGGINS, J. F., G. S. JOHNSON & I. PASTAN. 1972. Effect of dibutyryl cyclic AMP on the synthesis of sulphated-mucopolysaccharides by transformed fibroblasts. J. Biol. Chem. **247:** 5759–5768.
26. SHIGEMATSU, T. & L. DMOCHOWSKI. 1973. Studies on the acid mucopolysaccharide coat of viruses and transformed cells. Cancer **31:** 165–175.
27. TEMMINK, J. H. M. & H. SPIELE. 1974. Ruthenium red staining of normal and transformed murine fibroblasts *in vitro*. J. Cell Biol. **61:** 548–553.
28. BENHAM, F., D. C. COTTELL, L. M. FRANKS & P. D. WILSON. 1977. Alkaline phosphatase in human bladder tumor cell lines. J. Histochem. Cytochem. **25:** 266–274.
29. SELA, B.-A. & L. SACHS. 1974. Alkaline phosphatase activity and the regulation of growth in transformed mammalian cells. J. Cell. Physiol. **83:** 27–34.
30. DOELL, R. G. & B. J. MATHIESON. 1970. Absence of alkaline phosphatase in rat thymic lymphoma induced by murine leukaemia virus. Cancer Res. **30:** 2456–2457.
31. NOSE, K., T. TAKAOKA & H. KATSUTA. 1973. Two different activities of alkaline phosphatase in cultured mammalian cells. Arch. Biochem. Biophys. **155:** 1–8.

IDENTIFICATION AND CHARACTERIZATION OF NORMAL AND MALIGNANT HUMAN BLOOD LEUKOCYTES BY SURFACE GLYCOPROTEIN PATTERNS*

Carl G. Gahmberg† and Leif C. Andersson‡

†Department of Bacteriology and Immunology
and ‡ Transplantation Laboratory
Department of Surgery IV
University of Helsinki
00290 Helsinki 29, Finland

Numerous reports have shown that cell surface carbohydrate-containing macromolecules are altered in malignant cells (for recent reviews, see References 1–3), and these structures are implicated in the regulation of cell growth and cell-cell interactions.[4-6]

The analysis of cell surface glycoconjugates has recently been facilitated by the development of methods to introduce specific radioactive labels into the oligosaccharide portions of surface-exposed glycoproteins and glycolipids. The galactose oxidase-NaB^3H_4 and periodate-NaB^3H_4 labeling techniques are relatively simple to use, only a few cells are needed, and it is usually not necessary to isolate surface membranes for subsequent studies.

When intact cells are treated with galactose oxidase terminal galactosyl/*N*-acetylgalactosaminyl residues are oxidized to the corresponding C-6 aldehydes.[7] The large size of the enzyme molecule inhibits its penetration into the cell, and only surface-exposed glycoproteins and glycolipids are oxidized. Because sialic acids often are linked to penultimate galactosyl residues, it is usually necessary to treat the cells with neuraminidase to obtain efficient oxidation. The cells are then washed, and the aldehydes are reduced with tritiated sodium borohydride back to the original sugars.[8-10]

With low concentrations of periodate (0.1–2 mM), sialic acids are specifically oxidized; after reduction with tritiated borohydride, radioactive 5-acetamido-3,5-dideoxy-L-arabino-2-heptulosonic acid or 5-acetamido-3,5-dideoxy-L-arabino-2-octulosonic acid is formed.[11,12] This method has previously been used to label sialoglycoproteins of erythrocytes,[13] but it has not been considered to be specific for the cell surface of other cells. We recently showed, however, that if oxidation is performed for a short time at 0°C, there is no appreciable oxidation of intracellular glycoproteins; only surface-exposed sialic acids are oxidized.[14]

These methods have been successfully applied to the analysis of surface glycoproteins of various different cells[15-17] and envelope viruses.[18,19] For example, on normal fibroblasts, fibronectin (galactoprotein a) is strongly labeled by the galactose oxidase-NaB^3H_4 method,[15] and transformed fibroblasts do not incorporate label into this protein.

We have recently began to study surface glycoproteins of mouse and human lymphoid cells.[16,20] Such cells are very suitable for analysis with surface labeling

*Supported by the Academy of Finland, the Finnish Cancer Society, and the Sigrid Jusélius Foundation.

0077–8923/78/0312–0240 $01.75/0

techniques, because the cells exist in suspension and there is a large background information about their surface marker characteristics and functions. Very little, however, is known about the molecular organization and structure of their cell surfaces.

We now report that by use of surface carbohydrate labeling techniques, we can distinguish between different types of human blast leukemias on the basis of their surface glycoprotein patterns and can also evaluate their degree of differentiation. This is relevant because, although a proportion of human leukemias can be typed according to their characteristic morphology and immunologic and functional markers, many leukemic cells do not exhibit surface characteristics by which their cellular relationships can be ascertained at present.[21]

Materials and Methods

Chemicals and Enzymes

The sources of the chemicals used were as follows: sodium metaperiodate, Merck AG, Darmstadt, Federal Republic of Germany; acrylamide and *N*,*N*′-methylenebisacrylamide, Eastman Chemical Company, Rochester, New York; Triton® X-100, British Drug Houses Ltd., Poole, England; 2-aminoethylisothiouronium bromide (AET), Sigma, St. Louis, Missouri; 2,5-diphenyloxazole (PPO) and 1,4-bis[2-(5-phenyloxazolyl)]benzene (POPOP), New England Nuclear, Boston, Massachusetts; phytohemagglutinin (PHA), Difco, Detroit, Michigan; and concanavalin A (con A), Pharmacia, Uppsala, Sweden. Tritiated sodium borohydride (8.6 Ci/mmol) and [^{14}C]formaldehyde (2.0 mCi/mmol) were obtained from the Radiochemical Centre, Amersham, England. Tritiated sodium borohydride was handled as described previously.[16] *Vibrio cholerae* neuraminidase (500 U/ml) was obtained from Behringwerke, Marburg-Lahn, Federal Republic of Germany; galactose oxidase was purchased from Kabi AB, Stockholm, Sweden. The galactose oxidase preparation was dissolved in 0.15 M NaCl-0.01 M sodium phosphate (PBS, pH 7.4) to a concentration of 200 U/ml and kept at 4°C. The enzymes did not contain proteolytic activity when measured as described,[8] and the galactose oxidase did not contain neuraminidase activity.[8]

Initial Fractionation of Normal Blood Cells

The following main populations of normal human blood leukocytes were isolated: granulocytes, monocytes, T and B lymphocytes, and null cells. These cell populations were purified from the buffy coats from blood units of 400 ml each obtained from the Finnish Red Cross Blood Transfusion Service. The buffy coats were diluted with 2 vol of PBS, and platelets and mononuclear cells were separated from granulocytes and erythrocytes by one-step Ficoll-Isopaque® (density 1.077, Pharmacia) gradient centrifugation at 400*g* for 40 min at 22°C.

Purification of Granulocytes

The pellets obtained after Ficoll-Isopaque centrifugation, which contained granulocytes and erythrocytes, were suspended in 20 ml of PBS. Ten milliliters of 0.15 M NaCl that contained 6% dextran were added (6% Macrodex®, Leiras,

Turku, Finland), and the cell suspension was kept for 40 min at 37°C. The granulocyte-rich buffy coat was then collected, and contaminating erythrocytes were lysed by incubation with a Tris-buffered 0.84% aqueous solution of ammonium chloride. The leukocytes were then washed three times with PBS. This procedure yielded a cell population that contained more than 97% granulocytes as judged from May-Grünwald-Giemsa (MGG) -stained cytocentrifuged smears. Cellular viability was nearly 100%, as evaluated by the trypan blue dye exclusion test.

Purification of Monocytes

The blood monocytes were separated from the cell population recovered at the interphase after Ficoll-Isopaque centrifugation. The platelets were depleted by three washes with PBS at 200*g* for 10 min. The mononuclear cell population was mixed at a ratio of 1:50 with AET-treated sheep red blood cells (sRBC).[22] After incubation for 15 min at 37°C, the mixture was centrifuged for 10 min at 200*g*, and the T cells were allowed to rosette with the AET-sRBC for 1 hr on ice. The pellet was then gently suspended into cold PBS that contained 30% newborn calf serum, and the rosette-forming cells were separated from the nonrosette-forming cells by one-step Ficoll-Isopaque density gradient centrifugation. The cells recovered from interphase were mainly monocytes and non-T lymphocytes. The monocytes were then purified from the contaminating lymphocytes by 1*g* velocity sedimentation. By this procedure, the cells are fractionated according to cellular size.[23] The cells were then suspended in PBS that contained 5% newborn calf serum and layered on top of a linear gradient of 15–30% newborn calf serum in PBS. After sedimentation for 4 hr at 4°C, the gradient was drained into fractions of 20 ml, and the cellular content of each fraction was analyzed from MGG-stained smears. The early fractions, which contained mainly monocytes, were pooled. The cellular population thus obtained contained 90–95% monocytes as judged by conventional morphologic criteria. Viability always exceeded 98%. Slight contamination was caused by occasional lymphoblasts, myeloid precursor cells, and some granulocytes.

Purification of T Lymphocytes

The platelet-depleted mononuclear cell population recovered after Ficoll-Isopaque gradient centrifugation was passed over a human immunoglobulin (Ig) -rabbit antihuman Ig column as described by Wigzell *et al.*[24] The column-passed lymphocyte population contained less than 1% surface Ig-positive lymphocytes, as judged by staining with fluorescein-conjugated polyvalent sheep antihuman Ig obtained from Professor Astrid Fagraeus, Stockholm, Sweden. Viability of the T-cell population was always greater than 98%.

Purification of Non-T Lymphocytes, Null Cells, and B Lymphocytes

The mononuclear cells obtained after Ficoll-Isopaque centrifugation were suspended into RPMI 1640 culture medium supplemented with 10% normal human AB plasma. Carbonyl iron was added, and the suspension was incubated for 1 hr at 37°C. The majority of the phagocytic cells were then removed with a magnet. The phagocyte-depleted cell population was then allowed to form rosettes with AET-sRBC as described above. The T cells were depleted by centrifugation of the rosette-forming cell suspension on a Ficoll-Isopaque gradient. The cell suspension recovered from interphase contained mainly lymphocytes and larger monocyte-like cells. The non-T lymphocytes were further purified from this population by velocity sedimentation as described above. The later fractions obtained after sedimentation for 4 hr contained mainly small lymphocytes as judged from MGG smears. These fractions were pooled, and the cell population thus obtained was contaminated with less than 3% nonlymphocytic cells. This cell population was characterized with regard to surface markers. Two percent formed rosettes with AET-sRBC, and 61% were positive by surface Ig staining. The remaining lymphocytes in this population represented null cells. Cellular viability exceeded 95%.

This lymphocyte suspension was further fractionated by passage over an Ig-anti-Ig column at 4°C to minimize Fc binding. The cells that passed, of which more than 95% exhibited morphologic features of small lymphocytes, did not bind AET-sRBC (<3%) and lacked surface Ig (<1%). These cells were designated null cells. The lymphocytes retained by the column were eluted mechanically by shaking the glass beads in PBS. Eighty-six percent of the eluted lymphocytes stained positively for surface Ig, and less than 2% bound AET-sRBC. This population was designated blood B lymphocytes.

Stimulation and Isolation of T Lymphoblasts

The mononuclear cells recovered from the buffy coats after Ficoll-Isopaque gradient centrifugation were further enriched for T cells by passage over a nylon wool column, as described by Julius *et al.*[25] The passaged lymphocyte population was cultivated at a density of 10^6 cells/ml of RPMI 1640 culture medium supplemented with 10% normal human AB plasma or with 5% fetal calf serum for 6 days in the presence of optimal dilutions of the T-cell mitogens PHA or con A. A similarly purified lymphocyte population was mixed with mitomycin C-treated allogeneic mononuclear blood cells and cultivated for 6 days. At the end of the culture period, the blast cells were purified from the mitogen-stimulated and the mixed lymphocyte cultures by 1g velocity sedimentation as described.[23] The blast cell populations thus obtained contained more than 90% lymphoblasts as judged from MGG-stained cytocentrifuged smears. The blast cells from the mitogen cultures and from the mixed lymphocyte cultures formed rosettes with sRBC at a frequency of 80–98%.

Isolation of Blast Cells from Patients with Leukemia and Acute Infectious Mononucleosis

Leukemic cells were obtained from patients with newly diagnosed acute lymphocytic leukemia (ALL) or acute myeloid leukemia (AML) and from patients with chronic myeloid leukemia (CML) undergoing blast crisis. One patient with acute monoblastic leukemia and one with acute infectious mononucleosis were also included. Nucleated cells were isolated from heparinized blood or bone marrow aspirates by one-step Ficoll-Isopaque density gradient centrifugation. If the cell population recovered from interphase was contaminated with more than 10% normal, nonleukemic cells, the cell population was further purified by centrifugation on a linear Ficoll-Isopaque density gradient as described.[23] The cells used for surface labeling always contained more than 93% morphologically uniform leukemic blasts.

Cultivation of Human Leukemic Cell Lines

The acute lymphocytic leukemia cell line JM[26] and the chronic myeloid leukemia cell line K562[27] were grown in RPMI 1640 medium supplemented with 10% newborn calf serum.

Labeling of Cell Surface Glycoproteins by the Galactose Oxidase-NaB^3H_4 Technique

A total of 20–50 × 10^6 cells were used for each labeling experiment. The cells were washed three times in PBS and suspended in 1 ml of Dulbecco's PBS that contained Ca^{2+} and Mg^{2+}. In most cases, 25 μl of neuraminidase and 25 μl of galactose oxidase were added to the cell suspensions. In some experiments, neuraminidase was omitted; sometimes, both enzymes were left out. The cells were incubated at 37°C for 30 min, washed twice in PBS, and suspended in 0.5 ml of PBS. Then, about 0.5 mCi of NaB^3H_4 in 0.05 ml of 0.01 N NaOH was added, and the tubes were incubated at room temperature for 30 min. The cells were washed three times in PBS and to the pellets was added, on ice, 0.2–0.5 ml of PBS, which contained 2 mM phenylmethylsulfonyl fluoride (as a protease inhibitor) and 1% Triton X-100. After 5 min, the nuclei were removed by centrifugation at 2000 rpm for 5 min, and the supernatants were recovered. Aliquots were counted for radioactivity,[28] and the rest of the samples were boiled in the sample buffer of Laemmli.[29]

Labeling of Cell Surface Sialoglycoproteins by the Periodate-NaB^3H_4 Technique

From 20 to 50 × 10^6 cells were suspended in 1 ml of PBS, and 10 μl of 0.1 M sodium metaperiodate were added to give a final concentration of 1 mM. After 10 min on ice in the dark, the cells were washed twice by centrifugation, and 0.5 mCi of NaB^3H_4 was added. The cells were subsequently handled in the same way as for the galactose oxidase-treated cells.

Polyacrylamide Slab Gel Electrophoresis

Polyacrylamide slab gel electrophoresis was performed on 8% acrylamide gels in the presence of sodium dodecyl sulfate according to the method of Laemmli.[29] After electrophoresis, the gels were fixed in 20% sulfosalicylic acid and treated for fluorography as described.[30] They were then covered with Kodak X-Omat® R film and exposed at −70°C for 1–3 days. The apparent molecular weights of the glycoproteins were calculated according to the method of Weber and Osborn.[31] Protein standards were bovine thyroglobulin, ovalbumin, human hemoglobin (Sigma), transferrin, and human serum albumin (Kabi). They were labeled with [^{14}C]formaldehyde according to the technique of Rice and Means.[32]

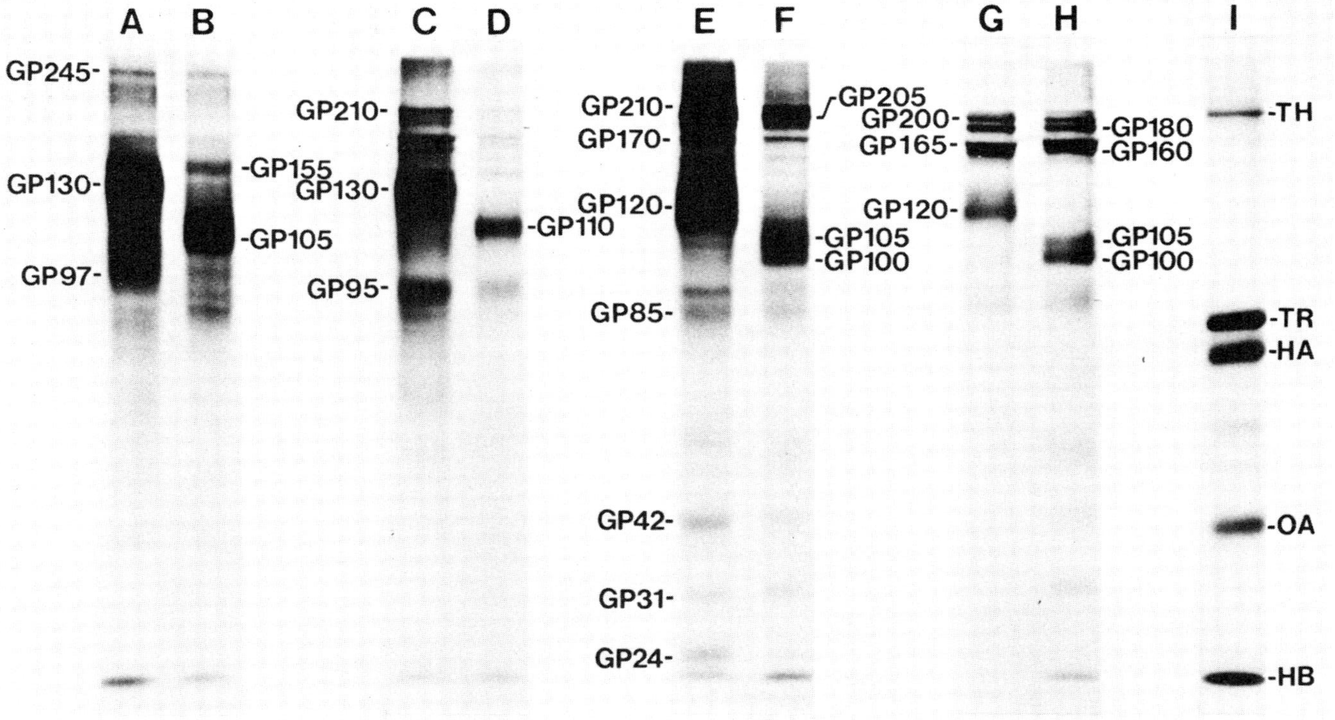

FIGURE 1. Fluorographic patterns obtained after polyacrylamide slab gel electrophoresis of normal blood cells labeled with $NaB^3 H_4$ after neuraminidase plus galactose oxidase (NE + GO) or periodate treatment. A, Granulocytes labeled after treatment with NE + GO. The apparent molecular weights of the major labeled proteins are indicated: GP245 represents molecular weight 245,000 daltons, for example. B, Granulocytes labeled after treatment with periodate. C, Monocytes labeled after treatment with NE + GO. D, Monocytes labeled after treatment with periodate. E, Non-T cells (B cells and null cells) labeled after treatment with NE + GO. F, Non-T cells labeled after treatment with periodate. G, T lymphocytes labeled after treatment with NE + GO. H, T lymphocytes labeled after treatment with periodate. I, ^{14}C-labeled standard proteins: TH, thyroglobulin; TR, transferrin; HA, human albumin; OA, ovalbumin; HB, hemoglobin.

Results: Surface Glycoproteins

Granulocytes

The following major surface glycoproteins were labeled in granulocytes after treatment with neuraminidase and galactose oxidase: GP245, GP230, GP165, GP155, GP130 (major band), GP105, GP97, GP85, GP80, GP62, GP50, and GP42 (Figure 1,A). When the surface sialic acids were labeled after periodate treatment, a fluorographic pattern similar to that seen after neuraminidase and galactose oxidase treatment were obtained (Figure 1,B). The main differences were seen in the area GP155–105: GP130 was not visualized after periodate treatment, whereas GP105 was more strongly labeled than after neuraminidase and galactose oxidase treatment.

Monocytes

The following major surface glycoproteins were labeled in monocytes after treatment with neuraminidase and galactose oxidase: GP210, GP205, GP200, GP170, GP165, GP155, GP130 (major component), GP95, GP85, GP80, and GP42 (Figure 1,C). Labeling of the sialic acid residues after periodate treatment yielded a pattern similar to that seen after application of neuraminidase and galactose oxidase (Figure 1,D). The main differences were recorded in the area GP165–110: GP110 was strongly labeled after periodate treatment, whereas the same protein exhibited the mobility of GP130 when labeled after administration of neuraminidase and galactose oxidase.

Non-T Lymphocytes

The following major glycoproteins were labeled on non-T lymphocytes after treatment with neuraminidase and galactose oxidase: GP210, GP205, GP180, GP170, GP155, GP130, GP120, GP95, GP85, GP50, GP42, GP31, and GP24 (Figure 1,E). Labeling after treatment with periodate yielded a similar pattern. The major differences were in areas GP105 and GP87, which were strongly labeled in periodate-treated cells; however, apparently similar proteins displayed the mobility of GP130 and GP120 in the pattern obtained after neuraminidase and galactose oxidase application (Figure 1,F).

T Lymphocytes

GP200, GP180, GP165, GP160, GP130, GP120, GP97, GP95, GP85, GP50, and GP42 were the major labeled glycoproteins in T cells pretreated with neuraminidase and galactose oxidase (Figure 1,G). Labeling after treatment with periodate yielded a pattern similar to that observed following administration of neuraminidase and galactose oxidase (Figure 1,H). There was, again, the typical change in the mobilities of two major bands. After application of neuraminidase and galactose oxidase, they were recorded as GP130 and GP120 but as GP105 and GP97 after periodate treatment. GP31 was labeled after periodate treatment but not after administration of neuraminidase and galactose oxidase.

Blood B Lymphocytes

FIGURE 2(B) shows the surface glycoprotein pattern of normal blood B lymphocytes labeled after neuraminidase and galactose oxidase treatment. The dominant bands are GP210, GP120, and GP110. Surface immunoglobulin, which should migrate in the 50,000 molecular weight region, is not seen. GP31 and GP24 are visualized only after prolonged exposure.

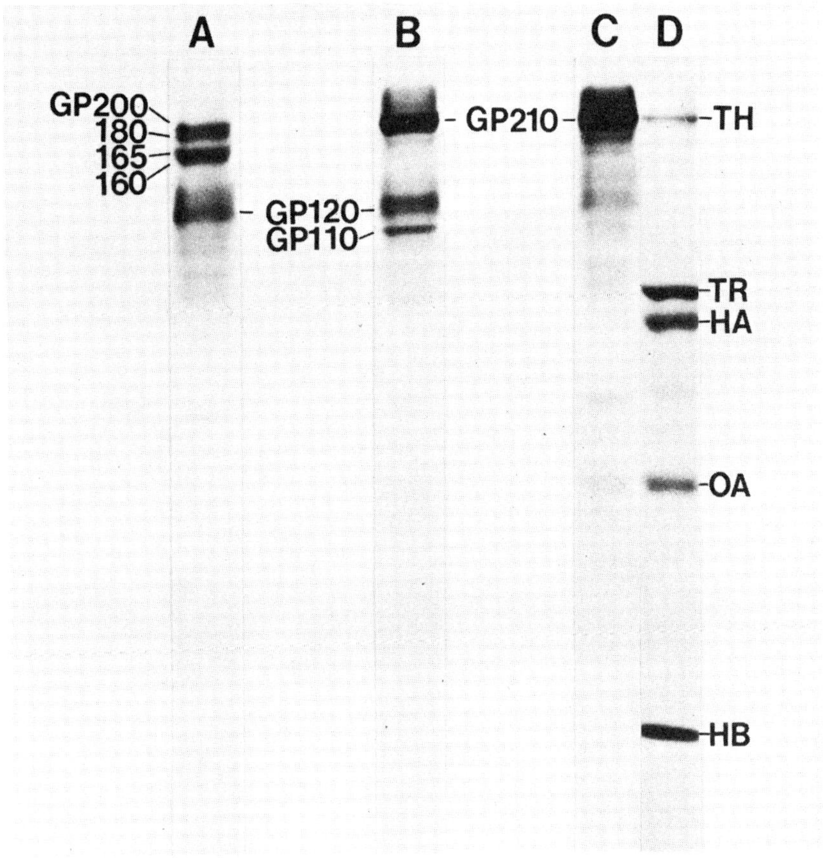

FIGURE 2. Fluorographic patterns of surface glycoproteins of normal blood T and B lymphocytes and null cells labeled after treatment with neuraminidase and galactose oxidase. A, Pattern of normal T lymphocytes; B, blood B lymphocytes; C, blood null cells; D, ^{14}C-labeled standard proteins, same abbreviations as in FIGURE 1 (I).

Blood Null Lymphocytes

The most strongly labeled band in null cells after neuraminidase and galactose oxidase treatment was GP210. GP120 was more weakly labeled than in T or B lymphocytes (FIGURE 2,C).

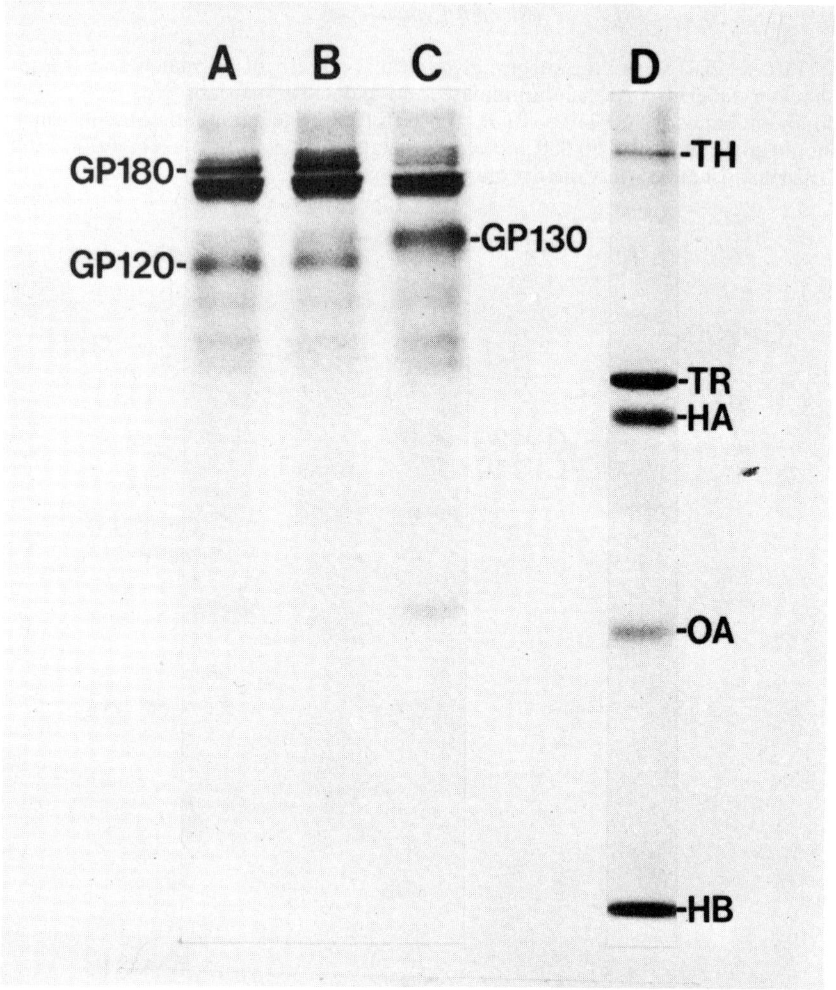

FIGURE 3. Fluorographic patterns of T-lymphoblast surface glycoproteins labeled after treatment with neuraminidase and galactose oxidase. A, Con A-stimulated cells; B, PHA-stimulated cells; C, cells isolated from mixed lymphocyte culture; D, ^{14}C-labeled standard proteins, same abbreviations as in FIGURE 1 (I).

Normal T Lymphoblasts

The surface glycoprotein patterns of lymphoblasts obtained after stimulation with con A or PHA and labeled after treatment with neuraminidase and galactose oxidase are shown in FIGURE 3(A & B). GP180, GP165, GP160, and GP120 are relatively strongly labeled. GP200 is weaker than in unstimulated T cells. The lymphoblasts isolated from mixed lymphocyte cultures contain relatively weak GP200, GP180, and GP165 bands. GP120 is not clearly seen; instead, these cells contain a strong GP130 band.

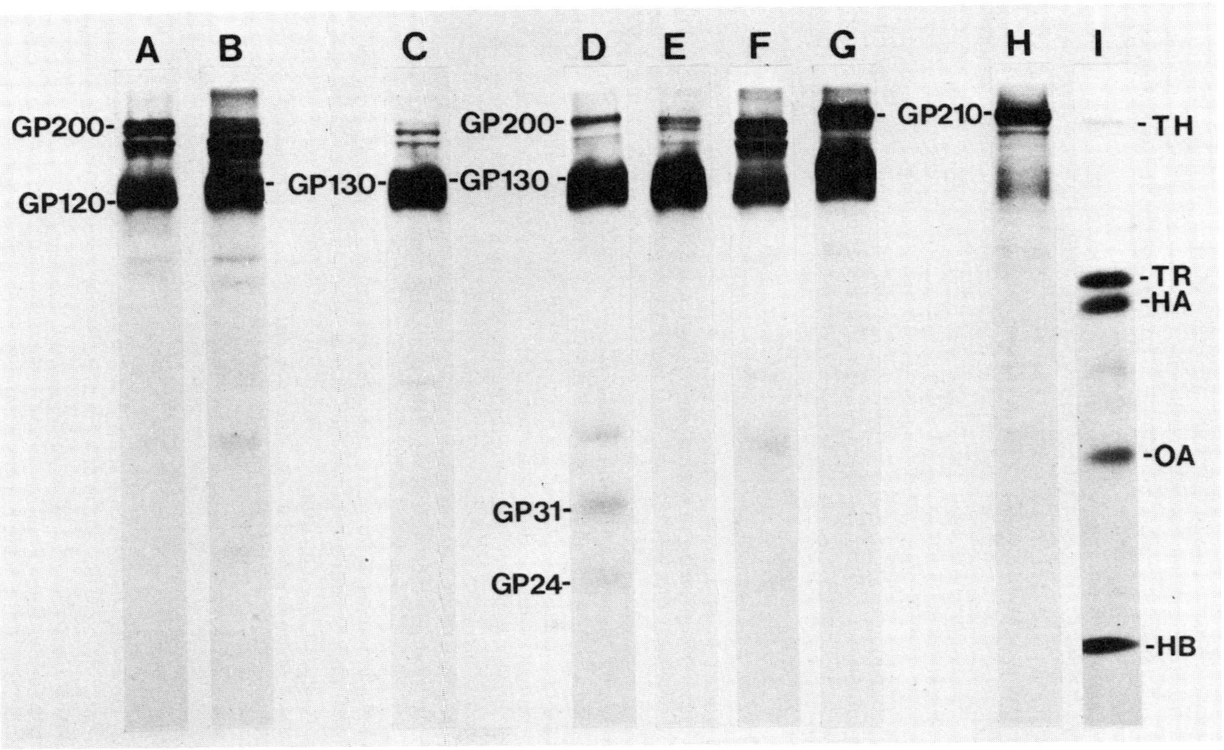

FIGURE 4. Fluorographic patterns of surface glycoproteins of blast cells from patients with acute lymphocytic leukemia labeled after treatment with neuraminidase and galactose oxidase. A, Pattern of normal T cells; B, lymphoblasts from a patient with infectious mononucleosis; C, ALL cell line JM; D–G, leukemic cells from patients with ALL; H, normal blood null cells; I, ^{14}C-labeled standard proteins, same abbreviations as in FIGURE 1 (I).

Blast Cells from Patients with Acute Lymphocytic Leukemia or Acute Infectious Mononucleosis

The surface glycoproteins of T lymphocytes and those of lymphoblasts from patients with infectious mononucleosis labeled after administration of neuraminidase and galactose oxidase are shown in FIGURE 4(A & B). GP130 is relatively strongly labeled in FIGURE 4(B). Compared to these patterns, the glycoproteins of continuous cell line JM (FIGURE 4,C) are different. GP130 is strongly labeled, and GP200, GP180, and GP165 are relatively weakly labeled. The glycoproteins from the ALL patients shown in FIGURE 4(D & E) are similar, except that the cells in FIGURE 4,D, contain glycoproteins GP31 and GP24. Both of these cells are null cells. FIGURE 4(F) shows the pattern of the cells from another patient with ALL. These cells have retained the capacity to form rosettes with sRBC. Actually, the pattern is similar to that of normal T cells, with about equally strong GP200, GP180, GP165, and GP160 bands. The cells in FIGURE 4,G, exhibit a strong GP210 band, which is characteristic for normal blood null cells (FIGURE 4,H). In addition, the cells in FIGURE 4,G, demonstrate a very strong and broad GP130 band.

Myeloid and Monoblastic Leukemia Cells

The surface glycoproteins of normal granulocytes are shown in FIGURE 5(A); those of continuous cell line K562, obtained from a patient with chronic myeloid leukemia, are illustrated in FIGURE 5(B). The surface glycoproteins of the latter partially resemble those of AML and CML cells (FIGURE 5,C–I). The glycoproteins of patients with AML shown in FIGURE 5(C & D) contain a strong GP210 band and a more weakly labeled GP205 band. GP130 is the dominant band in all cells from patients with myeloid leukemia. The surface glycoproteins shown in FIGURE 5(E; AML) resemble granulocytes in that they lack GP210. FIGURE 5(F) shows glycoproteins from a patient with AML, which contain GP31 and GP24. The glycoproteins of cells from patients with CML (FIGURE 5,G–I) are similar to those of cells from patients with AML. The cells in FIGURE 5(G & H) contain GP210, but those in FIGURE 5(I) do not.

Surface glycoproteins of normal monocytes are shown in FIGURE 5(J); those of cells from a patient with monoblastic leukemia are illustrated in FIGURE 5(K). The glycoproteins of the cells from the patient with monoblastic leukemia clearly resemble those of monocytes but are different. GP130 is weakly labeled in the leukemic cells, and GP165 and GP95 are not present.

DISCUSSION

Human leukemias have been grouped according to the morphologic and histochemical characteristics of the malignant cells and according to the clinical picture. During the past few years, however, new methods, employing different immunologic and functional markers, have come into use. By use of such techniques, about 20% of the acute human lymphocytic leukemias have been shown to exhibit T-cell characteristics, such as the ability to form rosettes with sRBC.[33] On the other hand, leukemic cells from patients with chronic lymphocytic leukemia usually are B-cell derived and contain surface immunoglobulin. However,

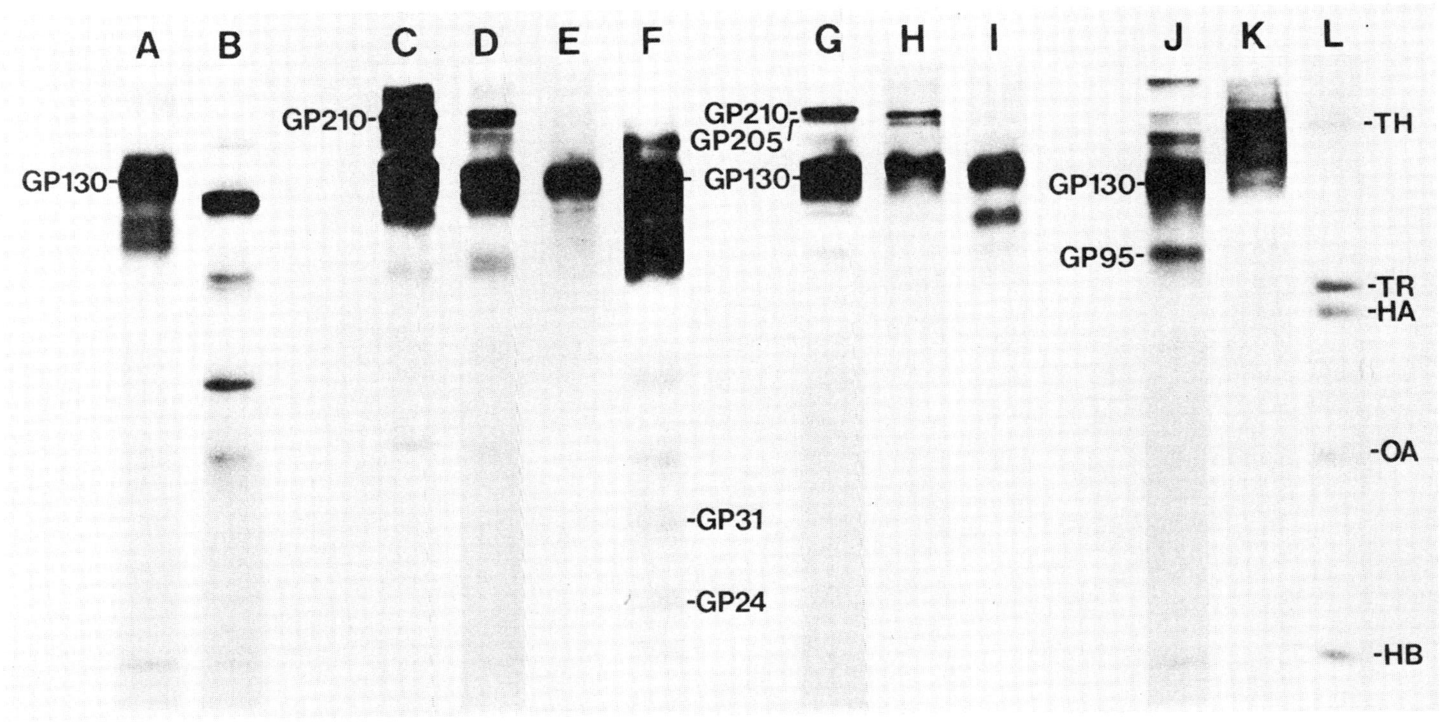

FIGURE 5. Fluorographic patterns of surface glycoproteins of blast cells from patients with acute myeloid leukemia, chronic myeloid leukemia in blast crisis, and acute monoblastic leukemia. The cells were labeled after treatment with neuraminidase and galactose oxidase. A, Normal granulocytes; B, chronic myeloid leukemia line K562; C–F, leukemic cells from patients with AML; G–I, leukemic cells from patients with CML; J, normal monocytes; K, blasts from a patient with acute monoblastic leukemia; L, ^{14}C-labeled standard proteins, same abbreviations as in FIGURE 1 (I).

in ALL, a large proportion of the leukemic cells lack all commonly used markers; they are referred to as null cells. Because treatment and prognosis of acute blast leukemias depend on the type of leukemia and the degree of differentiation,[34] and because these parameters cannot always be ascertained with present techniques, methods that could provide this information would be of great clinical importance.

Here we report a new approach to both classify human blast leukemias and to estimate their degree of differentiation. By selective radioactive labeling of cell surface glycoproteins, combined with slab gel electrophoresis in the presence of sodium dodecyl sulfate, followed by fluorography, an analytic tool of high sensitivity was developed.[10] We have previously shown that cells from patients with chronic lymphocytic leukemia have unique surface glycoprotein patterns, which are different from those of T and B lymphocytes.[20] Recently, we analyzed a large panel of established human lymphoid cell lines and found that cells from patients with Burkitt's lymphoma, non-Burkitt lymphocytic lymphomas, myelomas, and acute lymphocytic leukemias all exhibited surface glycoprotein patterns that were characteristic for the different malignancies. In addition, they were clearly different from those of benign lymphoblastoid cell lines.[35,36]

Here we report the surface glycoprotein patterns of leukemic cells, from patients with blast leukemias. The patterns of ALL cells resembled those of normal T or null lymphocytes. The ALL cells contained a very strong GP130 band and, in this respect, resembled lymphoblasts isolated from the blood of patients with acute infectious mononucleosis.[37] In contrast to this finding, GP180, GP165, and GP160 were reduced. GP200 was also reduced in con A- or PHA-stimulated lymphoblasts and in lymphoblasts from mixed lymphocyte cultures; in addition, GP180 was weak. In ALL cells, the largest glycoprotein, GP210/200, was the dominant labeled glycoprotein in the high-molecular-weight region. GP210 is characteristic of B cells[16,36,38] but is also strongly labeled in null cells (FIGURE 2,B). It was present in one of the ALL cells studied here (FIGURE 4,G), which indicates that this cell is more closely related to null cells and might have a better prognosis.[34] Other ALL cells contain GP200 and may be more differentiated than this ALL cell. One of them resembles normal T cells, with all four high-molecular-weight glycoproteins present (FIGURE 4,F), and this cell actually has retained the capacity to bind sRBC. The strong GP130 band is also a feature of lymphoblasts from mixed lymphocyte cultures and is present in much lower levels in con A- or PHA-stimulated cells (FIGURE 3). The corresponding protein is highly sialylated in ALL cells, because it is very poorly labeled after treatment with galactose oxidase only (data not shown). When labeled by the periodate method, the apparent molecular weight of GP130 in normal cells is 105,000 daltons. Because of limited amounts of leukemic cells, we did not label these cells after periodate oxidation. It is reasonable to assume, however, that the apparent molecular weight of this protein would be lower. As expected, the surface glycoprotein patterns of ALL cells closely resemble those of cultured cell lines, considered to be of the ALL type (cf. Reference 35).

Surface glycopeptides are known to be more sialylated in malignant fibroblasts[39] and leukemic cells[40] than in normal cells. Such glycopeptides may at least partially derive from GP130.

The surface glycoprotein patterns of blast cells from patients with AML or CML in blast crisis resemble each other. They were, however, distinctly different from those of normal granulocytes. Also, these leukemic cells contained a dominant GP130 band; this band, however, appeared to be different from that of ALL cells. In AML and CML cells, it was usually more narrow on gels. Many of the cells from myeloid leukemia patients contained the high-molecular-weight glycoproteins GP210 and GP205; GP210 was the dominant band. This finding may indicate that the cells of these cases of myeloid leukemia are more closely related to null cells or stem cells than the others. The GP95–97 region always contained relatively strong, diffusely labeled glycoproteins in the cells from AML and CML patients. This observation is in contrast to the ALL cells, which demonstrated an almost total lack of labeled components in this region. Normal T cells and T blasts contained labeled glycoproteins in this region, but these bands were sharp and relatively weak.

The glycoprotein pattern of the monoblastic leukemia patient was clearly different from all others. The pattern resembles that of monocytes, and GP130 was not strongly labeled.

From the surface glycoprotein patterns observed, one can draw tentative conclusions regarding the degree of differentiation of the various normal cells from the common stem cell. Granulocytes must have diverged early from the stem cell line, because they are very different from other cells and, for example, do not contain clearly labeled GP200, GP180, GP165, and GP160 bands. Monocytes contain not only these glycoproteins but also the typical GP130 band of granulocytes. Therefore, these cells probably have diverged after the granulocytes. T lymphocytes show similarities to monocytes in the high-molecular-weight region but also contain GP120, which obviously is different from GP130. B lymphocytes contain GP210 but do not have GP180, GP165, or GP160. In addition, they contain the Ia antigens GP31 and GP24. The major glycoproteins of null cells resemble those of B cells, containing a dominant GP210 but a very weak GP120 band. The malignant cells contain unique components, but it is characteristic that some of the glycoproteins normally confined to other types of leukocytes are evidently derepressed by malignancy.

The identity and function of most labeled bands are at present unknown. Exceptions are GP31 and GP24, which carry the human counterpart of the mouse Ia antigens.[41,42,48] These glycoproteins, which occur in pairs of similar strength, are present in normal B cells and null cells, but the pair is absent from T cells. Most of the ALL cells lack these glycoproteins with only one exception (FIGURE 4,D). Normal granulocytes lack Ia bands, but, interestingly, some of the myeloid leukemia cells contain them (FIGURE 5,F). This observation confirms the results of Winchester *et al.*,[43] who, working with fluorescent anti-Ia antisera, recently reported that AML and CML blasts often were positive.

GP42 is the heavy chain of the HL-A antigen.[36,44,45] As expected, it is found in most cells but is relatively weakly labeled.

GP210 resembles fibronectin in some respects: it has a similar molecular weight and, at least the mouse counterpart, is very sensitive to proteases. B lymphocytes, which contain this protein, are stimulated by trypsin to divide.[46] In addition, GP210 is present at lower levels in actively growing B cells.[36] However, it does not react with antifibronectin antiserum[48] and obviously, is not transformation sensitive. Burkitt's lymphoma cells, for example, also contain this protein.[36]

Surface Ig is not clearly labeled in any cell by the NaB^3H_4 method. The reason for this result is unknown. The polypeptide portion is known to be accessible to lactoperoxidase-catalyzed iodination.[47] It might be that the carbohydrate moiety is embedded within the polypeptide portion and therefore is not accessible to the labeling reagents. Obviously, many of the other surface proteins play important roles in the physiologic functions and antigenic properties of white blood cells, and their further isolation and characterization will be a major subject of our future research.

Summary

We have labeled surface glycoproteins of normal and malignant human blood leukocytes by the galactose oxidase-NaB^3H_4 and periodate-NaB^3H_4 labeling techniques. The labeled glycoproteins were separated by slab gel electrophoresis in the presence of sodium dodecyl sulfate and visualized by fluorography. The different types of normal blood cells could be distinguished by their surface glycoprotein patterns. The surface glycoproteins of cells from patients with acute lymphoblastic, myeloid, or monoblastic leukemia were different from those of normal cells. The leukemic cells could be classified by their surface glycoprotein patterns with respect to their relationships to normal blood cells, and an estimation of their degree of differentiation was obtained.

Acknowledgments

The skillful technical assistance of Anneli Asikainen and Liisa Räisänen is acknowledged. We thank Pekka Vuopio, M.D. and Martti Siimes, M.D. for blood samples from leukemia patients.

References

1. Nicolson, G. L. 1976. Biochim. Biophys. Acta **458**(1): 1–72.
2. Gahmberg, C. G. 1977. *In* Dynamic Aspects of Cell Surface Organization. G. Poste & G. L. Nicolson, Eds. Vol. 3: 371–421. North Holland Publishing Company. Amsterdam, The Netherlands.
3. Vaheri, A. & D. F. Mosher. 1978. Biochim. Biophys. Acta. In press.
4. Yen, P. H. & C. E. Ballou, 1973. J. Biol. Chem. **248:** 8316–8318.
5. Merrell, R., D. I. Gottlieb & L. Glaser. 1975. J. Biol. Chem. **250:** 5655–5659.
6. Bowles, D. J. & D. E. Hanke. 1977. FEBS. Lett. **82:** 34–38.
7. Avigad, G., D. Amaral, C. Asensio & B. L. Horecker. 1962. J. Biol. Chem. **237:** 2736–2743.
8. Gahmberg, C. G. & S.-I.Hakomori. 1973. J. Biol. Chem. **248:** 4311–4317.
9. Steck, T. L. & G. Dawson. 1974. J. Biol. Chem. **249:** 2135–2142.
10. Gahmberg, C. G. 1976. J. Biol. Chem. **251:** 510–515.
11. Van Lenten, L. & G. Ashwell. 1971. J. Biol. Chem. **246:** 1889–1894.
12. Suttajit, M. & R. J. Winzler. 1971. J. Biol. Chem. **246:** 3398–3404.
13. Liao, T.-H., P. M. Gallop & O. O. Blumenfeld. 1973. J. Biol. Chem. **248:** 8247–8253.
14. Gahmberg, C. G. & L. C. Andersson. 1977. J. Biol. Chem. **252:** 5888–5894.
15. Gahmberg, C. G. & S. Hakomori. 1973. Proc. Nat. Acad. Sci. USA **70:** 3329–3333.
16. Gahmberg, C. G., P. Häyry & L. C. Andersson. 1976. J. Cell Biol. **68:** 642–653.
17. Gahmberg, C. G., G. Myllylä, J. Leikola, A. Pirkola & S. Nordling. 1976. J. Biol. Chem. **251:** 6108–6116.
18. Luukkonen, A., C. G. Gahmberg & O. Renkonen. 1977. Virology **76:** 55–59.
19. McLellan, W. L. & J. T. August. 1976. J. Virol. **20:** 627–636.
20. Andersson, L. C., C. Wasastjerna & C. G. Gahmberg. 1976. Int. J. Cancer **17:** 40–46.
21. Bennett, J. M., D. Catovsky, M.-T. Daniel, G. Flandrin, D. A. G. Galton, H. R. Gralnick & G. Sultan. 1976. Brit. J. Haematol. **33:** 451–458.
22. Pellegrino, M. A., S. Ferrone, M. P. Dierich & R. A. Reisfeld. 1975. Clin. Immunol. Immunopathol. **3:** 324–333.
23. Häyry, P. & L. C. Andersson. 1976. Scand. J. Immunol. **5**(Suppl. 5): 31–44.
24. Wigzell, H., Y. G. Sundqvist & T. O. Yoshida. 1972. Scand. J. Immunol. **1:**75–87.

25. JULIUS, M. H., E. SIMPSON & L. A. HERZENBERG. 1973. Eur. J. Immunol. **3:** 645–649.
26. SCHWENK, H.-U. & U. SCHNEIDER, 1975. Blut **31:** 299–306.
27. LOZZIO, C. B. & B. B. LOZZIO. 1975. Blood **45:** 321–334.
28. BRAY, G. A. 1960. Anal. Biochem. **1:** 279–285.
29. LAEMMLI, U. K. 1970. Nature (London) **227:** 680–685.
30. BONNER, W. M. & R. A. LASKEY. 1974. Eur. J. Biochem. **46:** 83–88.
31. WEBER, K. & M. OSBORN. 1969. J. Biol. Chem. **244:** 4406–4412.
32. RICE, R. H. & G. E. MEANS. 1971. J. Biol. Chem. **246:** 831–832.
33. BROWN, G., M. F. GREAVES, T. A. LISTER, N. RAPSON & M. PAPAMICHAEL. 1974. Lancet **2:** 753–755.
34. BELPOMME, D., G. MATHÉ & A. J. S. DAVIES. 1977. Lancet **1:** 555–558.
35. ANDERSSON, L. C., C. G. GAHMBERG, K. NILSSON & H. WIGZELL. 1977. Int. J. Cancer **20:** 702–707.
36. NILSSON, K., L. C. ANDERSSON, C. G. GAHMBERG & H. WIGZELL. 1977. Int. J. Cancer **20:** 708–716.
37. ANDERSSON, L. C. & C. G. GAHMBERG. 1978. Clin. Immunol. Immunopathol. In press.
38. TROWBRIDGE, J. S., P. RALPH & M. J. BEVAN. 1975. Proc. Nat. Acad. Sci. USA **72:** 157–161.
39. BUCK, C. A., M. C. GLICK & L. WARREN. 1971. Science **172:** 169–171.
40. VAN BEEK, W. P., L. A. SMETS & P. EMMELOT. 1973. Cancer Res. **33:** 2913–2922.
41. KLARESKOG, L., L. SANDBERG-TRÄGARDH, L. RASK, J. B. LINDBLOM, B. CURMAN & P. A. PETERSON. 1977. Nature (London) **265:** 248–251.
42. GIPHART, M. J., J. F. KAUFMAN, A. FUHS, D. ALBRECHTSEN, B. G. SOLHEIM, J. W. BRUNING & J. L. STROMINGER. 1977. Proc. Nat. Acad. Sci. USA **74:** 3533–3536.
43. WINCHESTER, R. J., G. D. ROSS, C. J. JAROWSKI, C. Y. WANG, J. HALPER & H. E. BROXMEYER. 1977. Proc. Nat. Acad. Sci. USA **74:** 4012–4016.
44. BRIDGEN, J., D. SNARY, M. J. CRUMPTON, C. BARNSTABLE, P. GOODFELLOW & W. F. BODMER. 1976. Nature (London) **261:** 200–205.
45. TERHORST, C., R. ROBB, C. JONES & J. L. STROMINGER. 1977. Proc. Nat. Acad. Sci. USA **74:** 4002–4006.
46. VISCHER, T. L. 1974. J. Immunol. **113:** 58–62.
47. VITETTA, E. S., S. BAUR & J. W. UHR. 1971. J. Exp. Med. **134:** 242–264.
48. GAHMBERG, C. G. & L. C. ANDERSSON. 19xx. Unpublished results.

PART IV. ROLE OF THE HIGH-MOLECULAR-WEIGHT GLYCOPROTEIN IN CELLULAR MORPHOLOGY AND ADHESION

TRANSFORMATION-SENSITIVE CELL SURFACE PROTEIN: ISOLATION, CHARACTERIZATION, AND ROLE IN CELLULAR MORPHOLOGY AND ADHESION

Kenneth M. Yamada, Kenneth Olden, and Ira Pastan

Laboratory of Molecular Biology
National Cancer Institute
Bethesda, Maryland 20014

The major protein found on the cell surface of fibroblastic cells explanted into tissue culture is a glycoprotein known as CSP (cell surface protein),[1] LETS (large, external, transformation sensitive) protein,[2] fibronectin,[3] or by several other names (listed in Reference 4). This glycoprotein is a major protein of such explanted cells, constituting 1–3% of total cellular protein.[1,5] Quantities of this cell surface protein are often partially decreased on established, permanent cell lines.[5] In addition, many laboratories have reported decreased amounts or an absence of this protein from the cell surface after malignant transformation, although several exceptions exist to this generality (listed in References 6 and 7).

We have been investigating this glycoprotein's structure, composition, function, mobility, and regulation with purified preparations of CSP from chick embryo fibroblasts. Where applicable, we shall include recent data obtained with material from human fibroblasts. Our goal is to understand the role and the mechanism of action of this glycoprotein and how its loss after transformation affects cellular behavior.

ISOLATION AND PURIFICATION

Because this class of glycoprotein is exposed on the outer surface of the plasma membrane, an ideal method of purification would be to detach it from the surface of cells that remain intact and thus achieve a substantial one-step purification. Low concentrations of urea (0.2–1.0 M) appear to be the most gentle, effective means of releasing CSP from chick fibroblasts in biologically active form.[1,8] In contrast, a wide range of concentrations of chelating agents, salts, nonionic and ionic detergents, guanidine, and dithiothreitol are much less effective in releasing CSP or result in substantial cell lysis and yield contaminated preparations of CSP.[9,47]

Although the CSP released from chick cells by 1.0 M urea is relatively pure by sodium dodecyl sulfate (SDS) gel electrophoresis,[8] we routinely purify it further by use of ammonium sulfate and resuspend it at pH 11 to resolubilize the protein.[10] We often fractionate it further on molecular sizing columns.[11] The isolation and purification protocol is summarized in FIGURE 1. Each step has been evaluated by substituting other conditions, such as alternative times, media, and reagents. Use of the conditions indicated by asterisks results in significantly greater recoveries or purity than do substitutions of these conditions. Particularly important precautions are to use tertiary rather than secondary cultures to minimize procollagen contamination and to avoid drying of monolayers (to avoid cell lysis). Optimal

0077–8923/78/0312–0256 $01.75/0

cell density is also important to increase yields[1] while avoiding cell lysis that occurs in overcrowded cultures.

CSP can also be affinity purified by adsorption to formalin-fixed erythrocytes[8] or immobilized gelatin[12] and can be eluted with urea or alkaline pH. In tests of biologic activity (see below), CSP assayed immediately after extraction from cells, after ammonium sulfate fractionation, or after either type of affinity purification has the same hemagglutinating[8] activity (end points at 1–2 μg/ml). The protein therefore appears to be purified to constant specific activity by all of these methods.

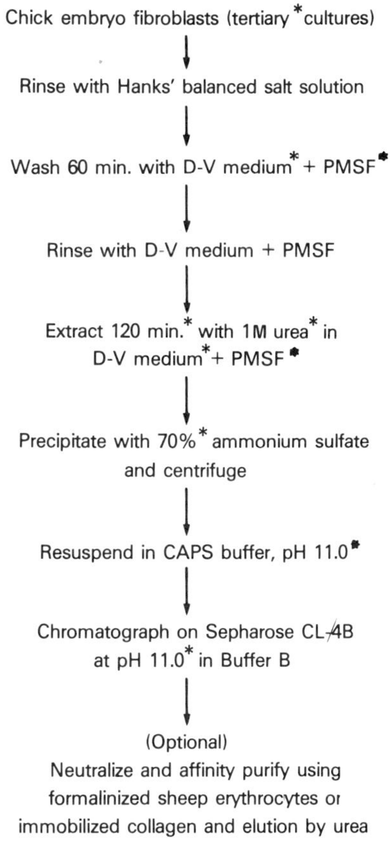

FIGURE 1. Isolation and purification of CSP. Asterisks indicate factors that decrease purity or recovery of CSP if modified. D-V, Dulbecco-Vogt's modified Eagle's; PMSF, phenylmethanesulfonyl fluoride; CAPS, cyclohexylaminopropane sulfonic acid; buffer B, 0.1 M NaCl, 1 mM $CaCl_2$, 0.05 M CAPS (pH 11.0).

Other criteria of purity include homogeneity after disulfide bond reduction on SDS or 8 M urea gels and migration as a single peak (after reduction) on molecular sizing columns[11] and by ultracentrifugation criteria.

Extraction of fibroblasts from other species, for example, human diploid fibroblasts, with 1–2 M urea results in enriched, but only partially pure, CSP. Such preparations can be purified to homogeneity by affinity chromatography on immobilized gelatin columns. Pure human CSP (LETS protein) has a hemagglutinating activity similar to that of chick CSP (end points at 1–2 μg/ml).

STRUCTURE

Isolated CSP exists in the form of disulfide-linked dimers and multimers, with only minimal amounts of the 220,000-dalton monomer.[11] A similar subunit organization exists in the CSP located on the surface of cells in culture.[3,11,13] On molecular sizing columns, native (not reduced) CSP migrates as three protein peaks, which were shown by analysis on SDS polyacrylamide gels to consist of peaks of CSP monomer, dimer, and multimer.[11] After disulfide reduction of each size class, only monomer is present. These results indicate that the CSP we isolate is not linked via disulfide bonds to other cell surface or extracellular matrix proteins and that CSP does not consist of smaller subunits. Even strong denaturing conditions, such as boiling in SDS plus 8 M urea at pH 11, will not dissociate the CSP monomer.[11]

The monomeric and multimeric forms of CSP are highly asymmetric. The Stokes radius of the 220,000-dalton monomer is 110 Å, regardless of whether the protein is reduced.[11] This result indicates that the protein is either unfolded or rodlike rather than globular and that intrachain disulfide bonds do not impose major constraints on the extended configuration of this protein. Such a highly asymmetric shape may contribute to CSP's adhesive properties (discussed below) by permitting noncovalent interactions between adjacent extended polypeptides. Such interactions are thought to occur in many artificial adhesives composed of repeating polymers.

The interchain disulfide bonds of CSP appear to be confined to a short peptide of approximately 10–15,000 daltons at an end of each monomer (FIGURE 2). Limited proteolysis by trypsin had been previously shown to cleave this peptide from CSP,[1] and analysis by use of SDS gels without reduction of CSP reveals that this cleavage results in formation of a major product with an apparent molecular weight of 205,000 daltons, regardless of whether the sample is reduced (FIGURE 2). In some preparations, we also see a 230,000-dalton component that apparently corresponds to the CSP monomer coupled to the short peptide via disulfide bonds (FIGURE 2). Similar results have been reported recently after treatment of the immunologically similar[14] plasma glycoprotein cold-insoluble globulin with the protease plasmin.[15,45] The localization of interchain disulfide bonds to one end of each subunit may result in a maximizing of the asymmetry of the whole molecule, permitting the formation of very long polymeric structures.

COMPOSITION AND CHEMICAL PROPERTIES

The amino acid composition of the major form of isolated CSP, the dimer, is not unusual[11] (TABLE 1). Cell surface protein does not contain an unusual number of hydrophobic residues and is not a simple polyelectrolyte. Cell surface protein contains slightly more proline and glycine than do many proteins, but its composition is quite different from that of collagen, another extracellular protein. In addition, CSP is not hydrolyzed by even high concentrations of protease-free bacterial collagenase (at enzyme/substrate ratios of 1:20).[11]

The amino acid composition of CSP is very similar, although not identical, to that of cold-insoluble globulin.[16,17] It displays some similarity to the microfibillar protein of smooth muscle cells[18] and lobster "fibrinogen."[46] The amino terminus of CSP is blocked and is not removed by pyroglutamate amino peptidase, suggesting that it may be acetylated.[11] In contrast, the end group of cold-insoluble globulin appears to be pyrrolidone carboxylic acid.[16]

The carbohydrate composition of the CSP dimer suggests that it contains oligosaccharides linked via *N*-glycosidic bonds to asparaginyl residues, because substantial amounts of mannose and *N*-acetylglucosamine are present.[11] In contrast, no *N*-acetylgalactosamine is present, suggesting that no *O*-glycosidic bonds are present. When CSP is labeled by incorporation of radioactive glucosamine,

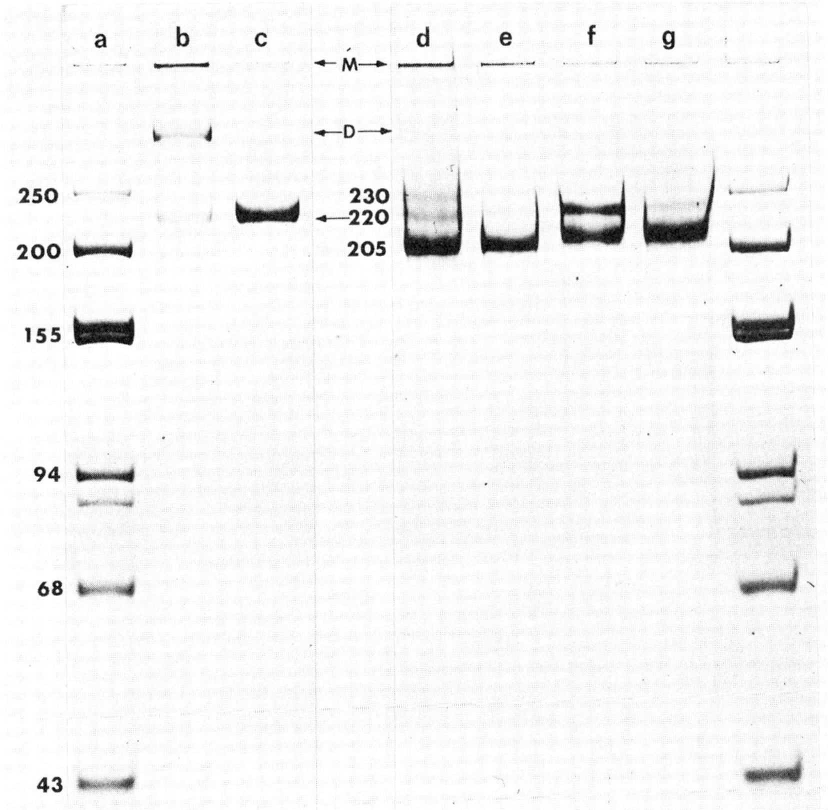

FIGURE 2. Interchain disulfide link region identified by tryptic cleavage. Purified CSP was incubated with a 1:400 (w/w) ratio of crystalline trypsin to CSP (approximately 1:40 molar ratio) in 1 mM $CaCl_2$, 0.03 M NaCl, 20 mM sodium phosphate (pH 7.0) for the indicated times, then lyophilized and electrophoresed on 5% SDS slab gels.[1] a, Protein molecular weight standards; b, original untreated CSP, not reduced, 3 μg; c, original CSP reduced with 0.1 M dithiothreitol, 3 μg; d, CSP trypsinized for 1 min, not reduced, 8 μg; e, trypsinized for 2 min, not reduced, 8 μg; f, trypsinized for 1 min, reduced, 8 μg; g, trypsinized for 2 min, reduced, 8 μg. M, multimer; D, dimer; 220, 220,000-dalton apparent molecular weight monomer; 205, 205,000-dalton tryptic cleavage product.

and then subjected to alkaline borohydride hydrolysis to cleave *O*-glycosidic bonds, virtually no label is released.[19] Moreover, the total incorporation of sugar precursors into CSP is reduced 86–98% by treatment with tunicamycin, a specific inhibitor of the synthesis of dolichol phosphate-linked carbohydrates that are required for the glycosylation of asparagine residues. This simplicity of CSP's

TABLE 1

AMINO ACID COMPOSITION OF CSP COMPARED TO THOSE OF OTHER PROTEINS (RESIDUES PER 1000)[11]

	CSP	CIg	Microfibrillar Protein	Lobster Fibrinogen B	α1 Collagen	Average Protein
Aspartic acid	94.1	94	87.6	99.0	41	104
Threonine	100.3	102	75.3	71.2	18	63
Serine	74.0	74	101.0	81.9	27	76
Glutamic acid	117.5	116	130.0	109.1	78	94
Proline	76.9	79	69.9	52.5	123	48
Cysteine	18.4	26	10.2	13.1	0	28
Glycine	106.2	83	131.0	61.7	329	77
Alanine	47.5	42	65.5	54.6	127	84
Valine	65.6	81	61.9	69.0	13	69
Methionine	12.7	10	10.7	16.9	9	17
Isoleucine	37.5	45	35.6	51.2	7	48
Leucine	55.4	56	57.9	96.0	20	75
Tyrosine	36.6	44	33.0	31.8	2	36
Phenylalanine	18.8	26	27.0	41.0	12	36
Tryptophan	36.0	28	—	19.4	0	13
Lysine	34.1	36	37.0	41.8	29	69
Histidine	15.6	22	22.4	41.6	2	22
Arginine	53.4	53	44.1	47.5	51	41
Hydroxyproline	0	—	0	—	107	—

carbohydrate structure (primarily asparagine-linked oligosaccharides) has permitted studies of the role of CSP's carbohydrate by use of tunicamycin (see below).

In analyses of total cellular concanavalin A-binding proteins of chick cells, we found that CSP was the major receptor for this lectin.[20] The mannose in CSP apparently accounts for its capacity to bind the lectin concanavalin A,[20,21] because treatment with tunicamycin abolishes this activity.[19]

Cell surface protein is an unusual protein. It is insoluble at physiologic pH under a variety of conditions of salt and chelating agents and in a variety of detergents other than strong denaturing agents.[11] This property had interfered with studies of its structure and function. Although solubilized in strong denaturing agents, such as 8 M urea, 6 M guanidine, or SDS, CSP could not be conveniently used in biologic experiments after these treatments. In our experience, the most useful solubilizing agent is alkaline pH.[10,11] Cell surface protein is completely solubilized at pH 11, and CSP preparations can be stored frozen at −60°C at this pH for months, defrosted, and used immediately for biologic experiments after neutralization. This insolubility under physiologic conditions represents a form of noncovalent self-association and can account for CSP's propensity to form large aggregates on the cell surface (see below).

A tentative model for the structure of CSP is presented in FIGURE 3. Cell surface protein is an asymmetric molecule, with subunits linked by interchain disulfides present on at least one end of each chain. It has not been established whether the interchain disulfide bridges are near the amino or carboxyl ends of the peptide chains. Cell surface protein's carbohydrate consists primarily of asparagine-linked oligosaccharides that terminate in sialic acid or fucose. These core oligosaccharides probably are the sites of concanavalin A binding.

FUNCTION: CELLULAR ADHESION

One obvious possible role for CSP is that of an adhesive protein, because it is a major cell surface protein, and because a major function of the cell surface is to participate in cell-cell adhesion. In addition, since the adhesiveness of cells in tissues,[22] and to tissue culture substrata,[23,24] is often reduced in malignant cells, it was possible that some of this decrease could have been due to the reductions in CSP reported after transformation.

We evaluated the function of this glycoprotein by means of three approaches: direct measurement of CSP's effect on cell-cell adhesiveness in model systems that measure the aggregation of dissociated cells,[8] reconstitution experiments in which we attempted to restore the isolated and purified protein to cells from which it is missing,[10,25,26] and antibody inhibition experiments that employed anti-CSP to determine the effects of complexing and, presumably, inactivation of CSP.[27]

We found that CSP is an agglutinin that will aggregate formalin-fixed sheep erythrocytes (FIGURE 4) at concentrations as low as 1 μg/ml of CSP.[8] The hemagglutination reaction was prevented by treatment with low concentrations of proteases (1 μg/ml of trypsin or chymotrypsin) or of chelating agents (10–15 μM EDTA or EGTA).[8] These findings were consistent with an adhesive role for CSP, because these agents are known to disrupt cellular adhesion.

The hemagglutination reaction is also inhibited by urea, hypertonic solutions of some salts, amino sugars, and several other amines. We have been unable to identify any monosaccharide specificity for this reaction.[8]

Cell surface protein also increases the aggregation of living mammalian or

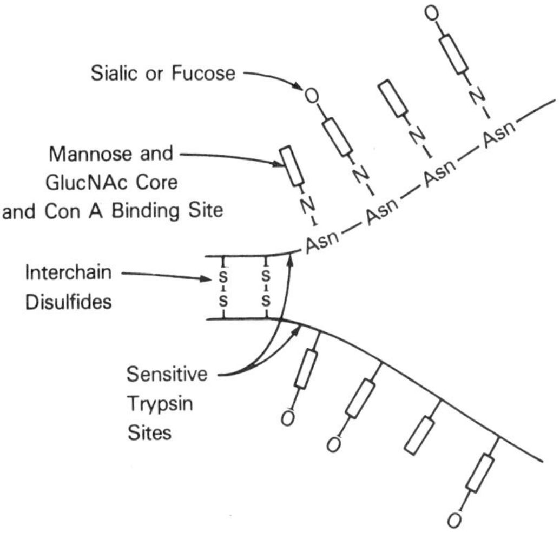

FIGURE 3. Hypothetical model for structure of the CSP dimer. Note the asymmetric structure and localization of interchain disulfide bridges to a terminal portion of each chain. Rectangles represent core oligosaccharides of mannose and *N*-acetylglucosamine (GlucNAc) residues; if these oligosaccharides are analogous to those of other glycoproteins, they should terminate in the galactose and/or the sialic acid or fucose residues found on CSP by either chemical or isotopic labeling procedures.

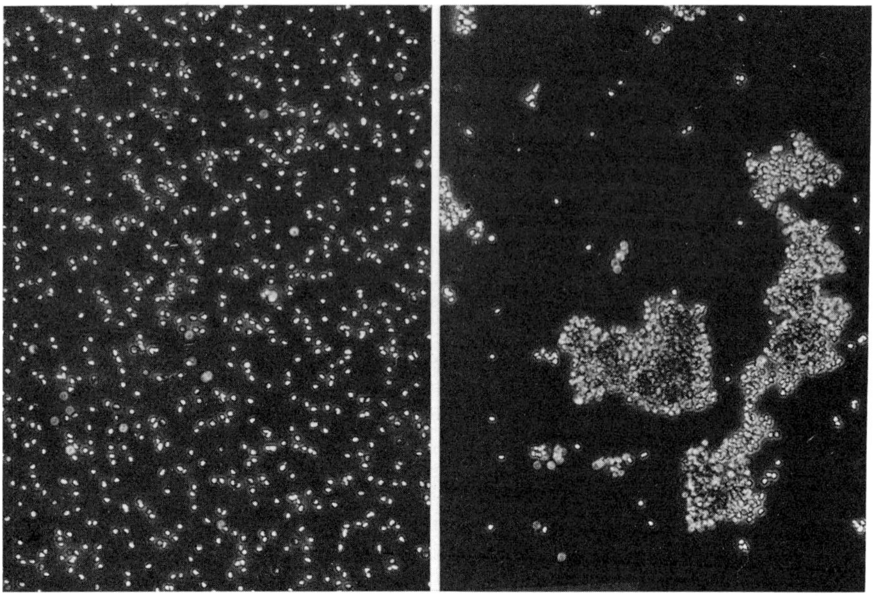

FIGURE 4. Agglutination of erythrocytes by CSP. Formalin-fixed sheep erythrocytes were incubated with purified CSP at a final concentration of 25 μg/ml in PBS for 2 hr, then photographed by means of phase-contrast microscopy. *Left,* Untreated control; *right,* CSP-treated erythrocytes. × 150.

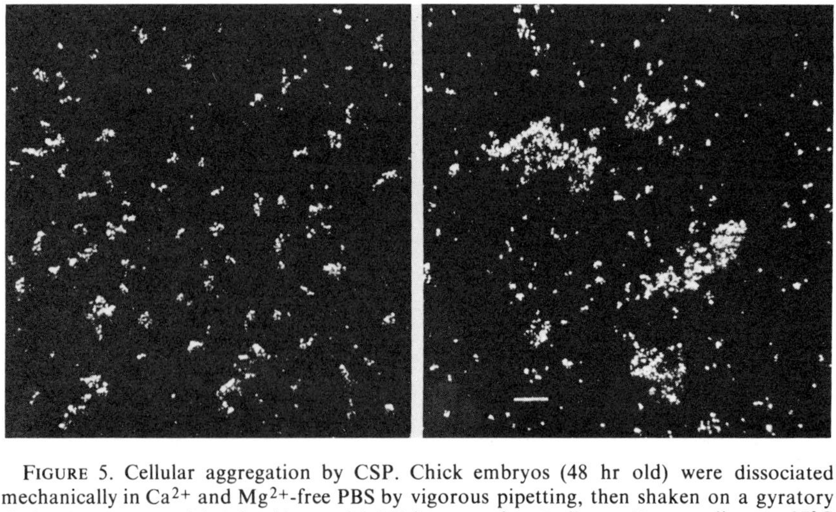

FIGURE 5. Cellular aggregation by CSP. Chick embryos (48 hr old) were dissociated mechanically in Ca^{2+} and Mg^{2+}-free PBS by vigorous pipetting, then shaken on a gyratory shaker (New Brunswick Scientific, model G2) in serum-free Dulbecco-Vogt medium at 37°C for 60 min. *Left,* Control; *right,* plus 50 μg/ml of CSP. Bar = 50 μm.

chicken cells. For example, addition of purified CSP to dissociated cells shaken on a gyratory shaker in a modification of the cell aggregation assay of Moscona[28] promotes cell aggregation within 30–60 min (FIGURE 5). This aggregation reaction is not affected by inclusion of DNase (100 μg/ml) during the assay but is completely inhibited by inclusion of crystalline trypsin (1 μg/ml). In addition, in the cell-cell adhesion assay described by Roseman *et al.*,[29] CSP at 50 μg/ml increases the initial rate of attachment of ^{14}C-labeled BHK cells to monolayers of BHK cells by 2–2.5-fold during the first 30 min of the assay.

Pearlstein has reported that partially purified CSP from BHK cells increases the attachment of transformed BHK cells to collagen.[30] We have confirmed this finding for SV40-3T3 cells with highly purified chick CSP in the Klebe[31] cell-collagen attachment assay.

TABLE 2

CSP EFFECTS ON MORPHOLOGY OF TRANSFORMED CELLS *in Vitro*

	Flattening	Cell Process Elongation	Alignment
Chicken			
Schmidt-Ruppin RSV chick fibroblasts	++*	++	+
Bryan high-titer RSV chick fibroblasts	—	+	—
Mouse			
L929	++	+	—
SV40 BALB/C 3T3 (SVT2)	++	++	++
SV40 BALB/C 3T3 tumor (SV1)	+	++	++
Harvey sarcoma virus BALB/C 3T3	+	++	—
Kirsten sarcoma virus BALB/C 3T3	+	+	+
Methylcholanthrene BALB/C 3T3	++	++	+
Rat			
Kirsten sarcoma virus NRK	++	++	++
Harvey sarcoma virus NRK	++	++	—
Moloney sarcoma virus NRK	++	+	—
Schmidt-Ruppin RSV NRK	+	+	—
Hamster			
Chinese hamster ovary (CHO)	++	+	+
Human			
SV40 WI38 (VA13)	—	+	—

*++, Marked effect similar to that illustrated in FIGURE 6 or 7; +, moderate effect; —, questionable or undetectable effect.

RECONSTITUTION EXPERIMENTS: ROLE OF CELL SURFACE PROTEIN IN CELL-SUBSTRATUM ADHESION AND MORPHOLOGY

A second approach to evaluate CSP's function is to reconstitute it on the surfaces of cells from which it is missing.[25] We have interacted purified CSP with a wide variety of transformed fibroblastic cells[10,26] (TABLE 2). The CSP is added to standard culture media and spontaneously attaches to the cell surface; approximately 15% attaches to the cell layer, and 5% attaches to the substratum.[10] The CSP on the cell surface remains intact according to SDS gel electrophoresis, and it remains on the cell surface and is accessible to hydrolysis by low concentrations of trypsin.[10] It is organized in the same fibrillar patterns as is found on untransformed fibroblasts[27] (see also Reference 32 for similar results).

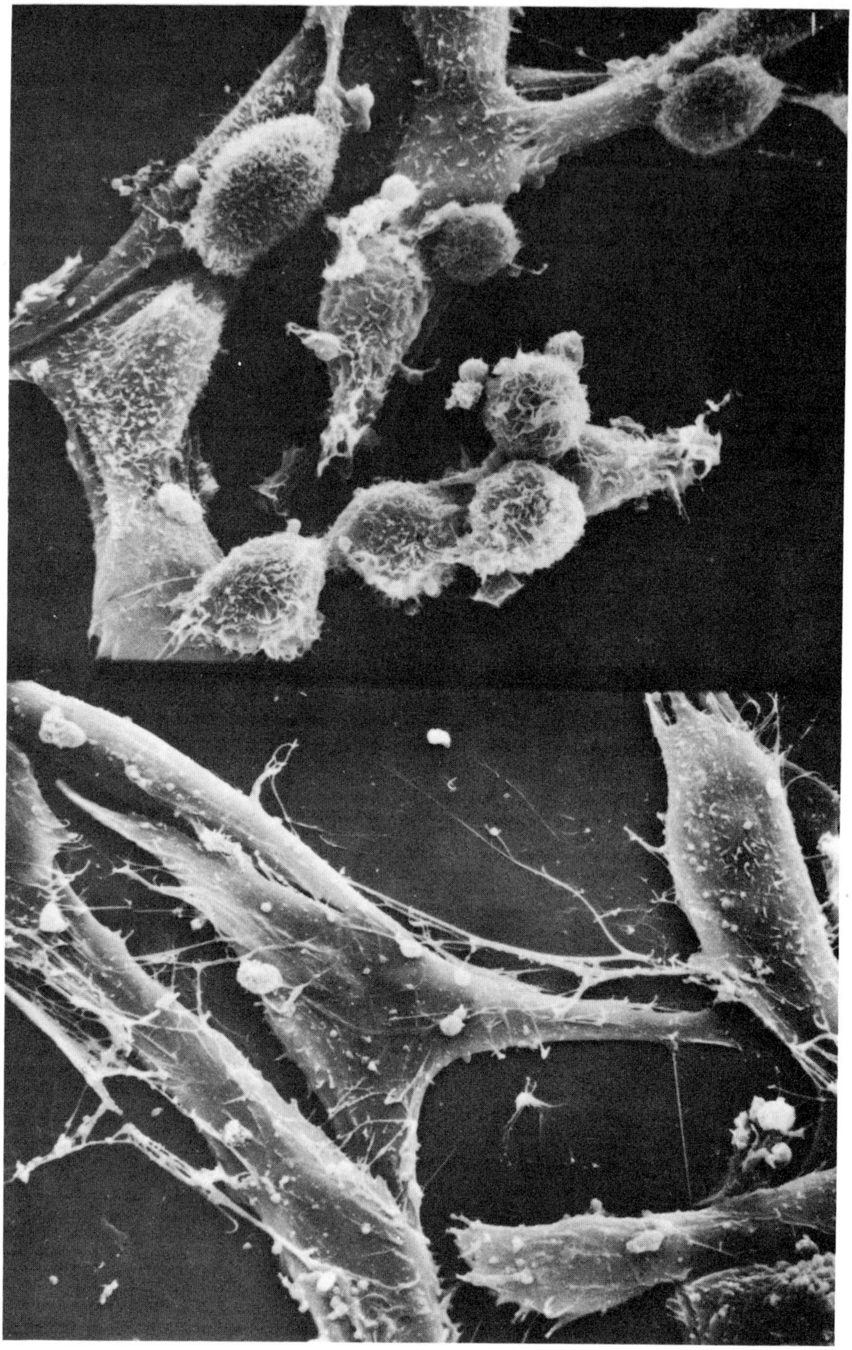

Cell surface protein-treated cells exhibit an increased adhesiveness to the substratum, as assessed by two different assays that measure the susceptibility of cells to detachment from the substratum.[10,33] In addition, the cells undergo striking alterations in morphology and assume a flattened, elongated shape similar to that of untransformed fibroblasts (FIGURE 6).[10,26] Cell surface architecture also reverts from a convoluted plasma membrane covered by numerous microvilli, blebs, and ruffles (lamellipodia) to a much smoother, flat surface that is similar in appearance to that of untransformed fibroblasts.[26]

These alterations in morphology are accompanied by changes in the intracellular organization of microfilament bundles, which reappear in transformed cells treated for 24–48 hr with CSP.[33] In general, the presence of microfilament bundles in cultured cells appears to correlate best with the extent of adhesion of the cells to their substrata.[33]

Cell Surface Protein and Cellular Motility

Reconstitution experiments have also implicated CSP in cellular motility. The AD6 mutant cell line is defective in cell-substratum adhesion. Pouyssegur *et al.*[43] found that it has a substantially decreased rate of locomotion, as determined by time lapse photomicroscopy. Cell surface protein treatment increases the rate and directionality of AD6 locomotion to normal.[43] Cells of the tumor cell line SV1 are also not highly motile, and these cells migrate out only slowly from aggregates of cells onto a substratum (FIGURE 7). Treatment of these aggregates with exogenous CSP added to culture media results in a marked increase in the rate and extent of cellular migration from the aggregates (FIGURE 7).

Cell surface protein may act to increase the motility of these cells by several mechanisms that result from increased cell-substratum adhesiveness, including increased forward motility of cells due to more effective traction to the substratum, or increased cell polarity, as evidenced by formation of microfilament bundles; this cytoskeletal rearrangement might produce more effective directional motility of cells.

Cell Surface Protein and Cell Social Interactions

Cell surface protein also alters cell-cell interactions. Fibroblastic cells characteristically align in parallel arrays as they crowd together when cultures become confluent, and this behavior is usually lost after malignant transformation (FIGURE 8). The crisscrossed, piled-up arrangement of transformed cells can be reverted by CSP treatment. Parallel arrays of cells are observed in approximately half of the transformed cell lines we have tested[10] (FIGURE 8 & TABLE 2). This restoration of alignment is also reflected in a decrease in overlapping of cells, which can be quantitated by Abercrombie's nuclear overlap ratio. Cell surface

---

FIGURE 6. Effects of CSP on tumor cell morphology. Mouse SV1 tumor cells were cultured on glass coverslips with or without addition of CSP to regular culture medium for 48 hr, then fixed and processed for scanning electron microscopy as described previously. We thank Dr. Sarkis Ohanian for these scanning electron micrographs. *Top,* Control SV1 cells; *bottom,* SV1 cells treated with 25 μg/ml of CSP. × 1700.

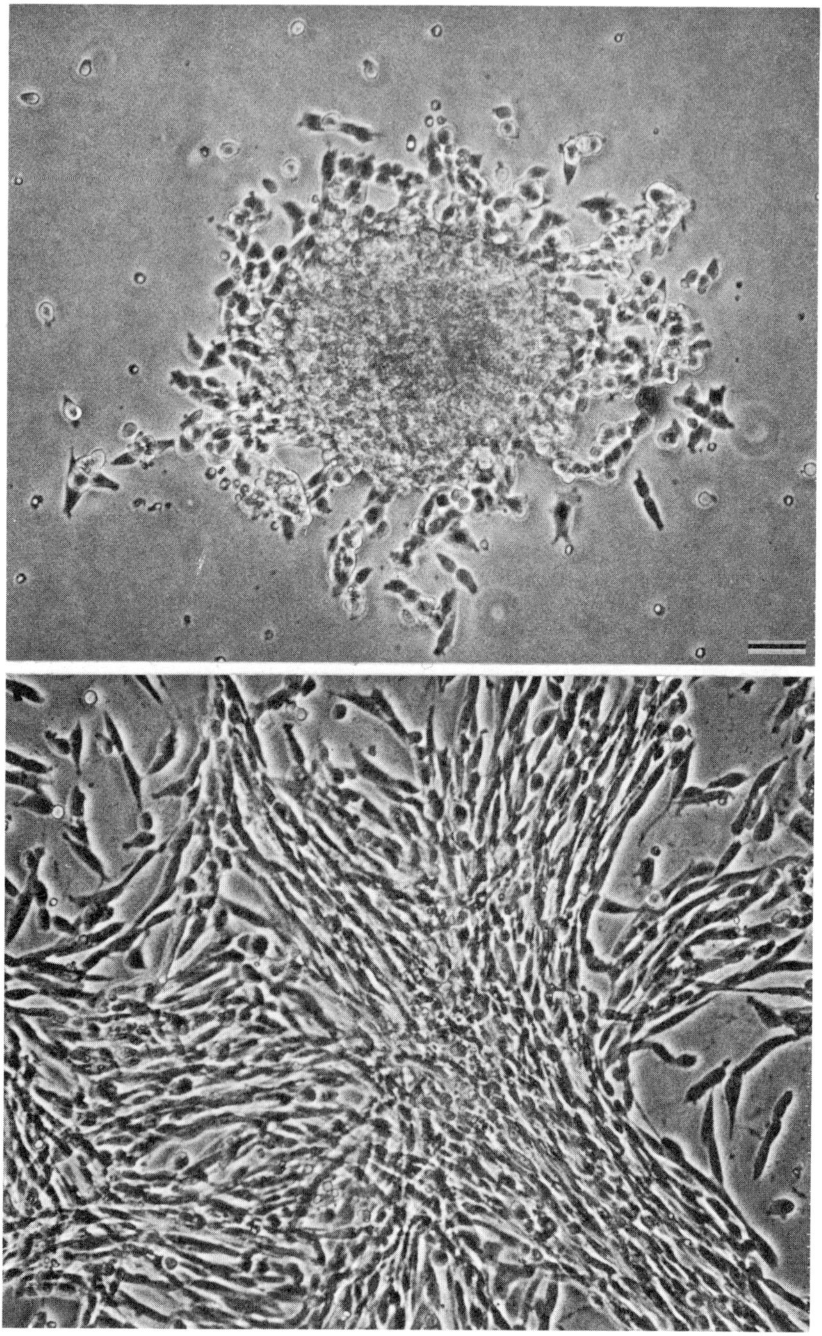

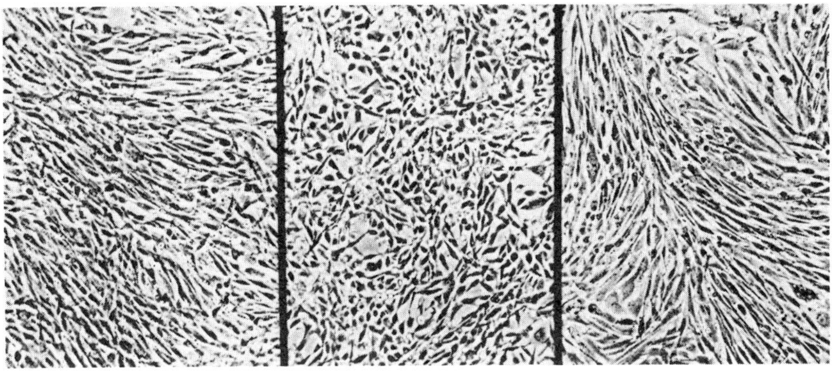

FIGURE 8. Effects of CSP on cellular alignment. All cells were cultured for 48 hr in regular culture medium. *Left,* Culture of untransformed chick embryo fibroblasts; *center,* culture of chick cells transformed by the Schmidt-Ruppin strain of Rous sarcoma virus; *right,* transformed chick cells incubated with 50 μg/ml of CSP for 48 hr. × 70.

protein treatment of an SV40-transformed 3T3 cell line results in a fivefold decrease in overlapping by this criterion.[10]

Cell surface protein is clearly distinct from another cell factor identified by Weston and Hendricks that also affects cell overlapping.[34] The latter factor is also released from cultured chick cells by low concentrations of urea. However, we have been able to demonstrate[35] that the two proteins are different by several criteria: the Weston-Hendricks factor is released entirely during the first 2 hr of extraction into serum-free medium with or without 0.2 M urea, whereas CSP is released continuously; CSP is removed by membrane ultrafilters that permit passage of the Weston-Hendricks factor; CSP is inactive in the assay system used to identify the Weston-Hendricks factor.[35] These results indicate that multiple factors will be involved in controlling the alignment of fibroblasts at confluence and in preventing cellular overlapping.

Cell Surface Protein Versus Other Adhesive Molecules

To begin to analyze the adhesive mechanism of action of CSP, we have compared the effects of CSP on the morphology and cellular alignment of the mouse tumor cell line SV1 with those of two other substances that increase adhesion of cells to a substratum, collagen and polylysine. Cells plated on tissue culture dishes precoated with either rat tail collagen or a purer preparation of rat skin type-I collagen (a generous gift from Dr. Hynda Kleinman) flatten and spread on these substrata more than on control dishes (FIGURE 9,a–c).

If the dishes are, instead, precoated with CSP, the cells also flatten, and the spreading is generally greater than on collagen (FIGURE 9,d). It is important to

FIGURE 7. Effects of CSP on cellular migration from cellular aggregates. Aggregates of SV1 cells were formed by allowing dissociated cells to settle and aggregate for 3 hr in regular culture medium[26] at 37°C. The aggregates were then transferred to 35-mm dishes that contained glass coverslips and cultured for another 24 hr in (*top*) regular medium or (*bottom*) medium plus 50 μg/ml of CSP. Bar = 50 μm.

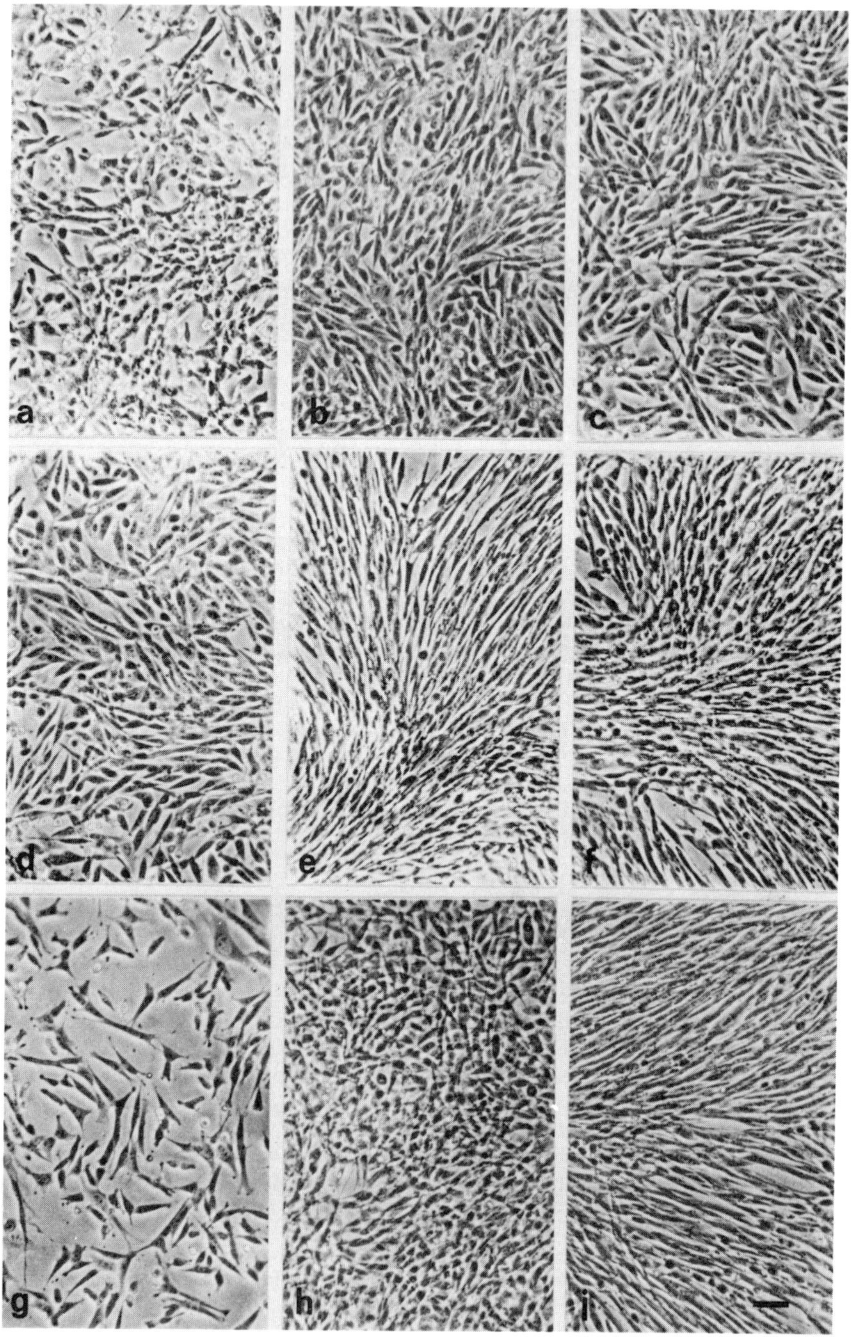
a
b
c
d
e
f
g
h
i

note that precoating of dishes with either of these glycoproteins does not result in alignment of cells at confluence, which is characteristic of the CSP reconstitution experiments described above (in which CSP is added directly to culture media). A second difference is that cells that spread on CSP- or collagen-coated dishes are less elongated than are such CSP-treated cells. However, if CSP is added to the culture medium of cells cultured on collagen-coated substrata, the cells form particularly striking parallel arrays at confluence (FIGURE 9,e). This alignment is even more striking than that which results from CSP treatment of cells cultured on uncoated dishes (FIGURE 9,f).

Precoating of dishes with the adhesive polycation polylysine also results in cell flattening, but not elongation. The effects are most striking at 24 hr (FIGURE 9,g) and decrease in magnitude thereafter (FIGURE 9,h). No cellular alignment is present in cells on polylysine-treated dishes unless CSP is added to the medium (FIGURE 9,i)

Direct addition of collagen, polylysine, or a variety of other proteins or glycoproteins to culture media does not result in cellular flattening, elongation, or alignment.[26]

In this regard CSP has two separable functions: it directly increases adhesiveness to the substratum, resulting in cellular spreading, and it also causes cellular elongation and alignment of cells at confluence. These effects of CSP are enhanced on collagen substrata.

Cell Surface Protein and Cellular Growth and Metabolism

Purified CSP that is highly effective in agglutinating erythrocytes and in altering morphology and alignment of transformed cells does not decrease the growth rate of chick, rat, or mouse cells transformed by oncogenic viruses nor does it alter their saturation densities.[10] Cell surface protein likewise does not suppress the elevated rates of glucose and amino acid transport characteristic of transformed chick and mouse cells.[36] These results strongly suggest that CSP plays no direct role in growth control or in the altered rates of nutrient transport that accompany transformation.

Cell surface protein also did not alter cyclic AMP levels of transformed mouse cells.[10] Conversely, treatment of the transformed mouse cell lines L929 or SV40-3T3 with dibutyryl cyclic AMP or the phosphodiesterase inhibitor methylisobutyl xanthine did not result in significant alterations in CSP levels, as determined by means of lactoperoxidase-mediated iodination with ^{125}I or by immunofluorescent

---

FIGURE 9. Comparison of the effects of CSP, collagen, and polylysine on transformed cellular morphology and alignment. Mouse SV1 tumor cells were plated at 2.5×10^4 cells/cm^2 and cultured for 48 hr in regular culture medium in Falcon tissue culture dishes coated with various substances to provide substrata of increased adhesiveness. Where indicated, CSP was added to the culture media at a final concentration of 50 μg/ml. a, Control, untreated plastic substratum; b, dish coated with rat tail collagen, then air-dried; c, dish coated with rat skin type-I collagen; d, dish coated with CSP by incubating with 1.5 ml of PBS that contained 50 μg/ml of CSP for 1 hr, then washing four times with PBS; e, dish coated with rat tail collagen, then the cells were cultured in the presence of 50 μg/ml of CSP; f, uncoated dish, with cells in the presence of 50 μg/ml of CSP; g, polylysine-coated dish, 1 mg/ml for 1 hr, then rinsed five times with PBS, 24 hr in culture; h, polylysine-coated dish after 48 hr; i, polylysine-coated dish with cells cultured with 50 μg/ml of CSP. Bar = 50 μm.

staining with anti-CSP. Cell surface protein and cyclic AMP therefore appear to function independently in regulating cellular shape and related phenomena.

The biologic activities of CSP are summarized on TABLE 3. It is likely that the primary function of CSP is in adhesion.

IMMUNOLOGIC STUDIES OF CELL SURFACE PROTEIN

We have used CSP purified by column chromatography or by SDS gel electrophoresis to obtain antibodies against CSP.[8,27] These antibodies appear to be monospecific by several different criteria. To ensure monospecificity, the anti-CSP antibodies have been affinity purified by use of CSP dimer coupled covalently to Sepharose 4B.[27] Untransformed chick fibroblasts treated with this antibody at concentrations of 50–100 μg/ml in regular culture medium rapidly become more rounded and assume a morphology that resembles that of many transformed cell lines. These alterations occur within 30 min, and the morphology of cells reverts to normal 2–3 hr after removal of the antibody from culture media.[27] Anti-CSP antibody also inhibits the action of CSP on transformed cellular morphology and in restoring alignment at confluence in reconstitution experiments[10] and also blocks the activity of CSP as an agglutinin.[8] These results are consistent with the notion that CSP is an adhesive protein that contributes substantially to cellular adhesion and morphology.

TABLE 3

SUMMARY OF CSP BIOLOGIC ACTIVITIES

Cell-cell aggregation (agglutination)
Formalinized sheep erythrocytes
Live cells (chick embryo and BHK)
Reversion of transformed cells
↑ Cell-substratum adhesion
(plastic, glass, collagen substrata)
Morphology more normal
↑ Alignment and ↓ overlapping
↓ Microvilli and ruffles
↑ Microfilament bundles
↑ Cell motility
No restoration of growth control or decrease in nutrient transport

LOCATION AND MOBILITY

By direct or indirect immunofluorescence, affinity-purified CSP stains cells in fibrillar patterns[4,5,27] of aggregated CSP similar to those described by others.[7,32,37] We see no evidence for staining of the plasma membrane itself that is greater than the weak background membrane staining observed with preimmune, nonimmune, or a control immune antiserum directed against the intracellular protein filamin. When cells are rendered permeable to the antibody by acetone treatment, granular perinuclear intracytoplasmic staining becomes visible. This staining is apparently due to pools of CSP newly synthesized prior to secretion, because this staining can be abolished and chased to the cell surface in pulse-chase experiments with protein synthesis inhibitors.[27] This finding may permit studies of the intracellular pathway of processing for CSP from its site of biosynthesis to its points of insertion onto the cell surface or into an extracellular matrix.

In contrast to procollagen, which is primarily released directly into culture media, CSP of chick fibroblasts is located primarily on the cell surface, although significant amounts are eventually shed into the medium.[38] As cells become dense, CSP is found in a meshwork of extracellular fibrils that remain closely associated with the cell layer.[5,7,27,32,37] This localization suggests that the protein may play an important role in the interaction of cells with elements of the extracellular matrix *in vivo*.

The lateral mobility of CSP on the cell surface can be studied with anti-CSP and with CSP labeled with fluorescein. Schlessinger *et al.*[39] found that purified fluorescein-labeled CSP can be reincorporated into the CSP present on chick fibroblasts and that the patterns of this incorporated CSP were identical to those of the original CSP (which could be identified in double-labeling experiments with rhodamine-labeled anti-CSP). This fluorescent CSP had a very low lateral mobility (less than 1 μm/min) as determined by the fluorescence photobleaching procedure.[39] Rhodamine-labeled anti-CSP bound to CSP on the cell surface was also found to be immobile by these criteria. Cell surface protein therefore forms a nonmoving, interconnected framework on the cell surface, superimposed on highly mobile lipid and glycoprotein molecules.

The possibility that CSP might restrain the movement of these other surface molecules was tested by measuring the mobility of a lipid analog, a ganglioside derivative, concanavalin A receptors, and general cell surface antigens in regions of the cell in which CSP was present or absent.[39] Cell surface protein had no detectable effect on the mobility of any of these components, except for concanavalin A receptors, of which a subset was immobilized. Since CSP is a concanavalin A receptor itself,[20] it is likely that the immobile molecules were CSP. Cell surface protein therefore appears to provide immobile adhesive points on the cell surface. These attachment sites do not interfere with the lateral mobility of other cell surface components.

Although CSP is relatively immobile in comparison to other surface components, it can eventually be redistributed by prolonged treatment with anti-CSP, forming large aggregates and supranuclear caps.[29] This "capping" is inhibited by metabolic inhibitors and resembles ligand-induced redistribution of other cell surface molecules. These findings imply a possible connection between intracellular movements and the distribution of this glycoprotein. It should be noted that the converse relationship also occurs[32,33]: added CSP affects the organization of intracellular microfilament bundles.

Biosynthesis, Secretion, and Turnover

We have studied CSP's synthesis and turnover by means of immunoprecipitation, protein staining of SDS gels for CSP, and labeling with radioactive leucine.[25,38] Leucine is particularly useful as a tracer, because it is abundant in CSP,[11] and because under appropriate conditions, it is not reutilized in studies of protein degradation.[25,38,40] Short pulses of radioactive leucine are incorporated with linear kinetics into CSP, and this incorporation constitutes 3% of the total incorporation into acid-precipitable material[38]; this value corresponds to the 3% of total chick cell protein constituted by CSP.[1] Once synthesized, the rate of transit of

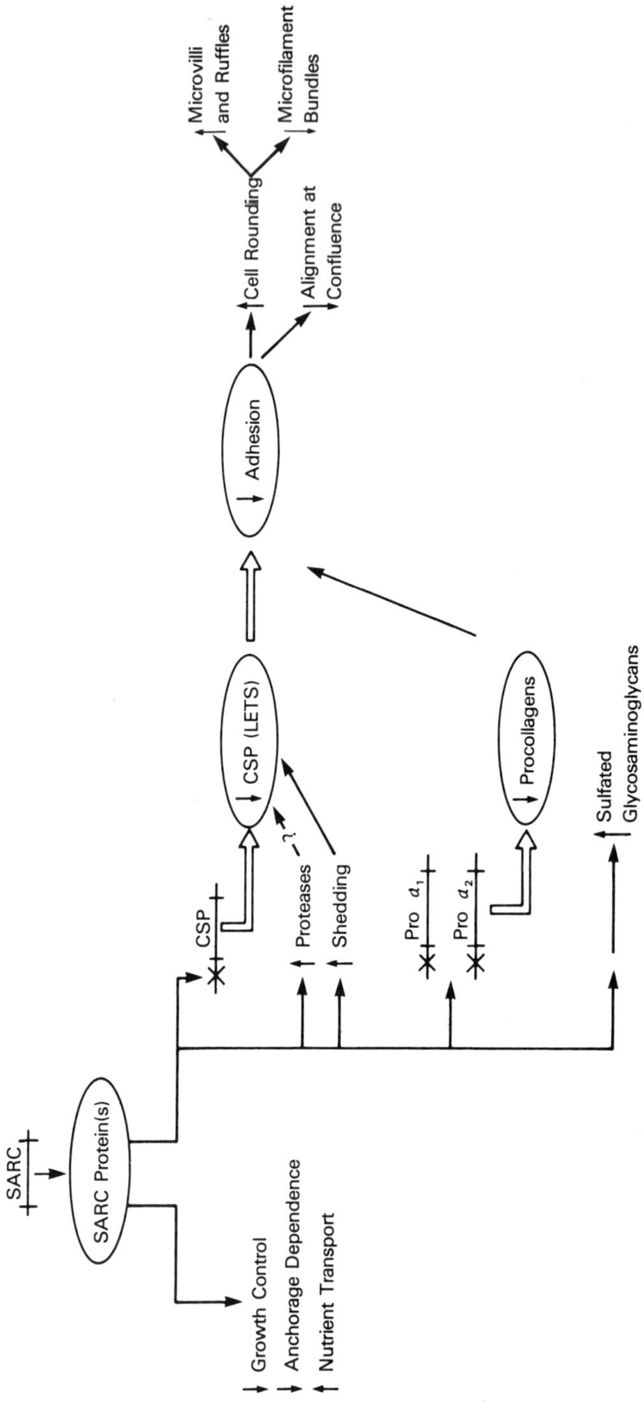

FIGURE 10. Proposed pathway for the effects of transformation on chick embryo fibroblasts.

CSP to the cell surface can be quantitated by the rate at which label in CSP becomes sensitive to tryptic hydrolysis as it appears on the cell surface. This transit time is approximately 30–45 min.[38] The resynthesis of CSP removed from the cell surface by trypsinization is detected by protein staining within 3 hr and continues for more than 24 hr, although the rate of accumulation becomes progressively slower.[25]

Cell surface protein is turned over on the cell surface in a process that involves loss of intact CSP into culture media both by sloughing and by direct proteolytic degradation.[38] We found that the overall turnover rate of CSP is 35–48 hr, which is similar to that of total cellular protein.[25,38] Approximately half of the turnover was attributable to loss into medium (where it remains undegraded) and half to degradation.[38]

Role of Cell Surface Protein's Carbohydrates

We examined the role of the carbohydrate moiety of CSP by use of tunicamycin, a specific inhibitor of dolichol phosphate-mediated protein glycosylation.[19] Although carbohydrates have been implicated in protein secretion by experiments that employed various inhibitors of glycosylation in other systems, we found that a 98% inhibition of CSP glycosylation did not block its secretion onto the cell surface.[19] Total amounts of CSP were reduced, however. This reduction was not due to a decrease in CSP biosynthesis or transit to the cell surface but, instead, could be explained by a two- to threefold increase in the rate of CSP degradation (accompanied by virtually no change in the rate of total protein degradation).[19] The carbohydrate on CSP therefore appears to function to help stabilize the molecule against proteolytic degradation.

Regulation of Cell Surface Protein

Quantitites of CSP are reduced by five- to sevenfold after transformation of chick embryo fibroblasts by four different strains of Rous sarcoma virus.[38] We investigated the mechanism of this decrease. The major effect of transformation was on the biosynthetic rate of CSP, which was decreased approximately fivefold in cells transformed by wild-type or temperature-sensitive viruses.[38] This decrease occurred regardless of the population density of the cells.

The decrease in biosynthesis was analyzed by Adams *et al.*[41] in a wheat germ *in vitro* translation system to evaluate total cellular mRNA. Translation of CSP was reduced fivefold in RNA extracted from transformed chick fibroblasts compared to RNA from untransformed cells, and this reduction was also found in temperature-sensitive mutants at the temperature permissive for transformation.[41] The decrease in CSP biosynthesis is therefore not due to significant alterations in mRNA translation but, instead, to a decreased level of translatable CSP mRNA. Interestingly, the decrease in translatable mRNA for CSP is accompanied by a 10-fold decrease in mRNA activity for another major class of extracellular proteins, the procollagens.[41]

Transfer of newly synthesized CSP to the cell surface is not detectably altered by transformation.[38] However, CSP turnover rates are increased 1.5-fold due both to increases in the rate of sloughing of intact CSP into culture medium and to increased degradation of CSP.[38] Increased loss from the cell surface has also been postulated for transformed human cells[42] and could be due to alterations in cell surface receptors for CSP. The increased rates of proteolytic degradation could be a result of increases in cell surface proteases in transformed cells.[44]

Transformation of chick fibroblasts by tumor viruses therefore alters quantities of CSP by acting at several different regulatory points. A hypothetical scheme for the mechanism of transformation of chick cells by Rous sarcoma virus is presented in FIGURE 10. This speculative diagram is the simplest interpretation of known effects of the "sarc" transforming gene on cell behavior and may omit other actions, such as other mechanisms that decrease cellular adhesion and possible mechanisms of metastasis not associated with the alterations listed here. Nevertheless, it suggests that knowledge of the mechanisms of regulation of CSP mRNA levels may lead to an understanding of the regulatory mechanisms of transformation. Whether these alterations in cell surface and extracellular macromolecules play roles in metastasis and invasiveness is also important to determine.

SUMMARY

Cell surface protein is the major cell surface glycoprotein of chick embryo fibroblasts. We have isolated and purified this glycoprotein and find that it is an adhesive protein that increases cell-cell and cell-substratum adhesiveness in a variety cellular adhesion assays.

Transformation of chick fibroblasts results in decreased quantities of CSP due primarily to a fivefold reduction in CSP biosynthesis, although increased proteolytic degradation and shedding from the cell surface also contribute. The decreased biosynthesis is apparently due to a fivefold reduction in translatable mRNA for CSP.

Reconstitution of isolated purified CSP on 14 transformed cell lines from several species results in reversion to a more normal fibroblastic morphology, adhesiveness, cell surface architecture, microfilament bundle organization, motility, and alignment at confluence. Cell surface protein does not restore growth control. The effects of CSP appear to be due to at least two actions, increased cell-substratum adhesion plus altered cell-cell interactions.

Untransformed chick cells treated with affinity-purified antibodies to CSP develop the rounded morphology characteristic of many transformed cells that are deficient in CSP (LETS protein).

Cell surface protein is found primarily in fibrillar aggregates on the cell surface. These CSP fibrils are relatively immobile and do not affect the mobility of other cell surface components. However, CSP can be eventually redistributed to caplike structures with anti-CSP.

Isolated CSP consists of highly asymmetric disulfide-linked dimers and multimers. The interchain disulfide bridges are confined to a short terminal fragment that is readily removed by trypsin. Cell surface protein and cold-insoluble globulin have similar compositions but differ in solubility and amino termini.

Cell surface protein contains primarily asparagine-linked oligosaccharides that appear to be responsible for CSP's concanavalin A receptor activity. Inhibition of CSP's glycosylation by treatment with tunicamycin results in decreased CSP due to marked increases in its degradation rate, without inhibition of synthesis or secretion.

Studies of this major cell surface glycoprotein have provided insight into the biochemical mechanisms of cellular adhesion, morphology, and social interaction and provide an approach to analyze the dynamics and regulation of protein synthesis, glycosylation, secretion, and turnover.

ACKNOWLEDGMENTS

We thank Dorothy Kennedy for invaluable assistance with the limited tryptic cleavage experiments and S. Adams, S. Alexander, L. Barak, G. Colonna, B. De Crombrugghe, H. Edelhoch, E. Elson, G. Hammes, K. Hendricks, B. Howard, S. Ohanian, R. Pratt, D. Schlesinger, J. Schlessinger, M. Sobel, W. Webb, J. Weston, M. Willingham, and S. Yamada for valuable collaboration.

REFERENCES

1. YAMADA, K. M. & J. A. WESTON. 1974. Isolation of a major cell surface glycoprotein from fibroblasts. Proc. Nat. Acad. Sci. USA **71:** 3492–3496.
2. HYNES, R. O. & J. M. BYE. 1974. Density and cell cycle dependence of cell surface proteins in hamster fibroblasts. Cell **3:** 113–120.
3. KESKI-OJA, J., D. F. MOSHER & A. VAHERI. 1977. Dimeric character of fibronectin, a major cell surface-associated glycoprotein. Biochem. Biophys. Res. Commun. **74:** 699–706.
4. YAMADA, K. M. & I. PASTAN. 1976. Cell surface protein and neoplastic transformation. Trends Biochem. Sci. **1:** 222–224.
5. YAMADA, K. M., S. S. YAMADA & I. PASTAN. 1977. Quantitation of a transformation sensitive, adhesive cell surface glycoprotein: decrease on several untransformed permanent cell lines. J. Cell Biol. **74:** 649–654.
6. HYNES, R. O. 1976. Cell surface proteins and malignant transformation. Biochim. Biophys. Acta **458:** 73–107.
7. VAHERI, A. & D. F. MOSHER. 1978. High molecular weight, cell surface-associated glycoprotein (fibronectin) lost in malignant transformation. Biochim. Biophys. Acta. In press.
8. YAMADA, K. M., S. S. YAMADA & I. PASTAN. 1975. The major cell surface glycoprotein of chick embryo fibroblasts is an agglutinin. Proc. Nat. Acad. Sci. USA **72:** 3158–3162.
9. HYNES, R. O., A. T. DESTREE & V. MAUTNER. 1976. Spatial organization at the cell surface. *In* Membranes & Neoplasia: New Approaches & Strategies.: 189–201. V. T. Marchesi, Ed. Alan R. Liss, Inc. New York, N.Y.
10. YAMADA, K. M., S. S. YAMADA & I. PASTAN. 1976. Cell surface protein partially restores morphology, adhesiveness, and contact inhibition of movement to transformed fibroblasts. Proc. Nat. Acad. Sci. USA **73:** 1217–1221.
11. YAMADA, K. M., D. H. SCHLESINGER, D. W. KENNEDY & I. PASTAN. 1977. Characterization of a major fibroblast cell surface glycoprotein. Biochemistry **16:** 5552–5559.
12. ENGVALL, E. & E. RUOSLAHTI. 1977. Binding of soluble form of fibroblast surface protein, fibronectin, to collagen. Int. J. Cancer **20:** 1–5.
13. HYNES, R. O. & A. DESTREE. 1977. Extensive disulfide bonding at the mammalian cell surface. Proc. Nat. Acad. Sci. USA **74:** 2855–2859.
14. RUOSLAHTI, E. & A. VAHERI. 1975. Interaction of soluble fibroblast surface antigen with fibrinogen and fibrin: identity with cold insoluble globulin of human plasma. J. Exp. Med. **141:** 497–501.
15. CHEN, A. B., D. L. AMRANI & M. W. MOSESSON. 1977. Heterogeneity of the cold-insoluble globulin of human plasma (CIg), a circulating cell surface protein. Biochim. Biophys. Acta **493:** 310–322.
16. MOSESSON, M. W., A. B. CHEN & R. M. HUSEBY. 1975. The cold-insoluble globulin

of human plasma: studies of its essential structural features. Biochim. Biophys. Acta **386:** 509–524.
17. MOSHER, D. F. 1975. Cross-linking of cold-insoluble globulin by fibrin-stabilizing factor. J. Biol. Chem. **250:** 6614–6621.
18. MUIR, L. W., P. BORNSTEIN & R. ROSS. 1976. A presumptive subunit of elastic fiber microfibrils secreted by arterial smooth-muscle cells in culture. Eur. J. Biochem. **64:** 105–114.
19. OLDEN, K., R. M. PRATT & K. M. YAMADA. 1978. Role of carbohydrates in protein secretion and turnover: effects of tunicamycin on the major cell surface glycoprotein of chick embryo fibroblasts. Cell **13:** 461–473.
20. OLDEN, K. & K. M. YAMADA. 1977. Direct detection of antigens in sodium dodecyl sulfate polyacrylamide gels. Anal. Biochem. **78:** 483–490.
21. BURRIDGE, K. 1976. Changes in cellular glycoproteins after transformation: identification of specific glycoproteins and antigens in sodium dodecyl sulfate gels. Proc. Nat. Acad. Sci. USA **73:** 4457–4461.
22. MCCUTCHEON, M., D. R. COMAN & F. B. MOORE. 1948. Studies on invasiveness of cancer: adhesiveness of malignant cells in various human adnocarcinomas. Cancer **1:** 460–467.
23. SANFORD, K. K., B. E. BARKER, M. W. WOODS, R. PARSHAD & L. W. LAW. 1967. Search for "indicators" of neoplasmic conversion *in vitro*. J. Nat. Cancer Inst. **39:** 705–718.
24. WEBER, M. J., A. H. HALE & L. LOSASSO. 1977. Decreased adherence to the substrate in Rous sarcoma virus-transformed chicken embryo fibroblasts. Cell **10:** 45–51.
25. YAMADA, K. M. & J. A. WESTON. 1975. The synthesis, turnover, and artificial restoration of a major cell surface glycoprotein. Cell **5:** 75–81.
26. YAMADA, K. M., S. H. OHANIAN & I. PASTAN. 1976. Cell surface protein decreases microvilli and ruffles on transformed mouse and chick cells. Cell **8:** 241–245.
27. YAMADA, K. M. 1978. Immunological characterization of a major transformation-sensitive fibroblast cell surface glycoprotein: localization, redistribution, and role in cell shape. J. Cell Biol. In Press.
28. MOSCONA, A. A. 1973. Cell aggregation. *In* Cell Biology in Medicine. E. Bittar, Ed. Wiley Interscience. New York, N.Y.
29. ROSEMAN, S., W. ROTTMANN, B. WALTHER, R. OHMAN & J. UMBREIT. 1974. Measurement of cell-cell interactions. Methods Enzymol. **32B:** 597–611.
30. PEARLSTEIN, E. 1976. Plasma membrane glycoprotein which mediates adhesion of fibroblasts to collagen. Nature (London) **262:** 497–500.
31. KLEBE, R. J. 1974. Isolation of a collagen-dependent cell attachment factor. Nature (London) **250:** 248–251.
32. ALI, I. U., V. MAUTNER, R. LANZA & R. O. HYNES. 1977. Restoration of normal morphology, adhesion and cytoskeleton in transformed cells by addition of a transformation-sensitive surface protein. Cell **11:** 115–126.
33. WILLINGHAM, M. C., K. M. YAMADA, S. S. YAMADA, J. POUYSSEGUR & I. PASTAN. 1977. Microfilament bundles and cell shape are related to adhesiveness to substratum and are dissociable from growth control in cultured fibroblasts. Cell **10:** 375–380.
34. WESTON, J. A. & K. L. HENDRICKS. 1972. Reversible transformation by urea of contact-inhibited fibroblasts. Proc. Nat. Acad. Sci. USA **69:** 3727–3731.
35. WESTON, J. A., K. M. YAMADA & K. L. HENDRICKS. 1978. Characterization of factor(s) in culture supernatants affecting cell social behavior. Submitted for publication.
36. YAMADA, K. M. & I. PASTAN. 1977. The relationship between cell surface protein and glucose and α-aminoisobutyrate transport in transformed chick and mouse cells. J. Cell. Physiol. **89:** 827–830.
37. CHEN, L. B., P. H. GALLIMORE & J. K. MCDOUGALL. 1976. Correlation between tumor induction and the large external transformation sensitive protein on the cell surface. Proc. Nat. Acad. Sci. USA **73:** 3570–3574.
38. OLDEN, K. & K. M. YAMADA. 1977. Mechanism of the decrease in the major cell surface protein of chick embryo fibroblasts after transformation. Cell **11:** 957–969.

39. SCHLESSINGER, J., L. S. BARAK, G. G. HAMMES, K. M. YAMADA, I. PASTAN, W. W. WEBB & E. L. ELSON. 1977. Mobility and distribution of a cell surface glycoprotein and its interaction with other membrane components. Proc. Nat. Acad. Sci. USA **74:** 2909–2913.
40. NEFF, N. T., A. ROSS, J. C. BARTHOLOMEW & M. J. BISSELL. 1977. Leucine in cultured cells: its metabolism and use as a marker for protein turnover. Exp. Cell Res. **106:** 175–183.
41. ADAMS, S. L., M. E. SOBEL, B. H. HOWARD, K. OLDEN, K. M. YAMADA, B. DE CROMBRUGGHE & I. PASTAN. 1977. Levels of translatable mRNAs for cell surface protein, collagen precursors, and two membrane proteins are altered in Rous sarcoma virus-transformed chick embryo fibroblasts. Proc. Nat. Acad. Sci. USA **74:** 3399–3403.
42. VAHERI, A. & E. RUOSLAHTI. 1975. Fibroblast surface antigen produced but not retained by virus transformed human cells. J. Exp. Med. **142:** 530–538.
43. POUYSSEGUR, J., M. WILLINGHAM & I. PASTAN. 1977. Role of cell surface carbohydrates and proteins in cell behavior: studies on the biochemical reversion of an N-acetylglucosamine-deficient fibroblast mutant. Proc. Nat. Acad. Sci. USA **74:** 243–247.
44. REICH, E., E. SHAW & D. B. RIFKIN. (Eds.) 1975. Proteases and Biological Control. Cold Spring Harbor Laboratory. Cold Spring Harbor, N.Y.
45. JILEK, F. & H. HÖRMANN. 1977. Cold-insoluble globulin, II. Plasminolysis of cold-insoluble globulin. Hoppe-Seyler's Z. Physiol. Chem. **358:** 133–136.
46. FULLER, G. M. & R. F. DOOLITTLE. 1971. Studies of invertebrate fibrinogen. I. Purification and characterization of fibrinogen from the spiny lobster. Biochemistry **10:** 1305–1311.
47. YAMADA, K. M. 1977. Unpublished results.

HIGH-MOLECULAR-WEIGHT GLYCOPROTEIN AS A MEDIATOR OF CELLULAR ADHESION*

Edward Pearlstein and Leslie I. Gold

Irvington House Institute
and Department of Pathology
New York University Medical Center
New York, New York 10016

Cell adhesion factor (CAF) is a major external glycoprotein expressed on the surface of a variety of adherent cells, but it is absent from the plasma membrane of many,[1-4] but not all,[5,6] transformed cells. We have previously demonstrated[7] the requirement for a component in partially purified CAF during fibroblast attachment and spreading on collagen-coated dishes pretreated with 8 M urea. Physicochemical evidence indicated that the active factor was CAF. In the present study, we have developed a reproducible isolation procedure for obtaining highly purified CAF from human plasma and have obtained direct evidence that CAF is responsible for mediating the attachment and spreading of fibroblasts on urea-treated collagen. Furthermore, inhibition of attachment by antisera specific for CAF has been used to provide additional support for these conclusions.

Pretreatment of collagen-coated dishes with 8 M urea denatures the triple helical structure of collagen[8] and allows a high efficiency of CAF binding to the primary structure.[9] To more closely mimic the *in vivo* situation, assay of the CAF dependence of fibroblast attachment to undenatured collagen has been attempted following incorporation of CAF into the collagen gel both during and after the induction of gelation by the elevation of pH. Although poor cellular spreading on the native collagen substratum was demonstrated, cellular attachment, albeit slower, remained CAF dependent.

There is evidence that cellular attachment and spreading on plastic tissue culture dishes involve at least two biochemically distinct processes. The attachment process necessitates divalent cations[7,10] and intact sulfhydryl groups,[11,12] and cellular spreading requires intact microtubules and microfilaments. In the present study, these requirements were also demonstrated to be important on collagen-coated dishes. On the collagen substratum, cellular spreading, but not attachment, can be inhibited by amino-containing compounds[7]; both attachment and spreading require ATP.

We have utilized pharmacologic agents to determine structural and metabolic requirements for normal and transformed cells during attachment and spreading on collagen in the presence of exogenously supplied CAF. In addition, normal and transformed cells were compared to ascertain whether previously described alterations in cytoskeletal structures following transformation affected CAF-mediated attachment and spreading. Requirements for adhesion and spreading were studied as separate events, because it has already been established that cellular attachment can be measured independently of spreading.[7] In our experiments, we were unable to demonstrate any qualitative difference between BHK and PyBHK cells in cellular adhesion or spreading within all experimental variations.

*Supported in part by United States Public Health Service Grants CA 22085 and AM 01431.

0077-8923/78/0312-0278 $01.75/0

Materials and Methods

Cell Cultures

Baby hamster kidney fibroblasts, BHK 21/C13, and a polyoma virus-transformed derivative, PyBHK, were obtained from the Imperial Cancer Research Fund, London, England. Cultures were maintained in Dulbecco's modified Eagle's medium (E4) (Gibco) that contained 10% calf serum, 4 mM glutamine, 100 U/ml of penicillin, and 100 μg/ml of streptomycin (Gibco), at 37°C and 5% CO_2. Cells were passaged twice weekly.

Isolation of Human Plasma Cell Adhesion Factor

A reliable and reproducible procedure for isolating CAF from outdated, citrated human plasma has been adapted from a combination of previously published procedures.[9,13,14] From 1 liter of plasma, 20–30 mg of CAF were routinely purified. To 1 liter of plasma at room temperature, 100 g of barium sulfate and 6.5 g of ϵ-amino caproic acid (ϵ-ACA) were added with constant stirring and then centrifuged in a Sorvall ultracentrifuge equipped with a GS-3 rotor at 5000 rpm for 10 min. The supernatant was cooled to 4°C and solid ammonium sulfate was added to a final concentration of 129 g/liter. The precipitate was again pelleted and washed three times with a total of 2000 ml of 25% saturated ammonium sulfate at 4°C. The washed precipitate was redissolved in 0.05 M sodium phosphate and 0.01 M sodium EDTA (pH 6.6) and again precipitated with ammonium sulfate (129 g/liter). This precipitate was spun as before, dissolved in 0.01 M Tris and 0.1 M sodium chloride (pH 7.4) and dialyzed overnight at 4°C against this buffer. Any precipitate formed during dialysis was discarded; the solution was then cooled to 0°C on ice and brought to a final concentration of 0.1 M ϵ-ACA and 7% ethanol in that order. The copious precipitate was pelleted and redissolved in 500 ml of 0.1 M NaCl, 0.01 M Tris HCl, and 1 mM EDTA (pH 6.8) and clotted at room temperature by the addition of 1 U/ml of thrombin (topical, Parke-Davis). After 3 hr, the clot was dispersed and removed by centrifugation. The clear supernatant was dialyzed overnight at 4°C against phosphate-buffered saline (PBS) and applied directly to a 2.5 × 40-cm gelatin-Sepharose 4B affinity chromatography column.[9] The column was washed with PBS until the optical density fell to less than 0.05. Bound CAF was eluted with 8 M urea in 0.05 M Tris buffer (pH 7.5).

Preparation of Collagen-Coated Plastic Petri Dishes

Tissue culture dishes (35 mm, Falcon) were coated according to the method of Klebe,[17] which involves covering the dishes with 1.5 ml of 0.25% collagen (acid soluble, type IV from calf skin, Sigma) dissolved in 0.2% acetic acid that contains 0.002% phenol red. The dishes were exposed to 1 atm of concentrated NH_4OH in a closed vessel for 20 min and air-dried overnight. Dishes that did not receive urea were washed twice with PBS and used directly. Those that received urea were washed with 8 M urea for 20 min, followed by three 5-min washes with distilled water and again air-dried.

In some experiments, CAF or an equal volume of PBS was added to the collagen solution immediately before the gelation. All subsequent steps were as described before.

Cell Attachment Assay

To measure cellular adhesion and spreading, purified human plasma CAF or 20% calf serum in Dulbecco's modified Eagle's medium (E4) was used as the source of CAF. An $E^{1\%}_{280\ \mathrm{nm}}$ of 13.5 was determined for the pure protein by use of the Lowry *et al.* method,[18] and protein concentration was estimated at 280 nm. Collagen-coated dishes were incubated for 1 hr at 37°C with CAF and washed twice with Dulbecco's PBS (Gibco) prior to addition of the test reagents and cells.

Cells were removed from monolayers with 0.25% trypsin-EDTA (Gibco) and washed several times with E4 that contained 0.02% bovine serum albumin (E4-BSA) or PBS, depending on individual experimental requirements. The cells were counted in a hemocytometer and suspended to a concentration of $0.5 \times 10^6/3.0$ ml in E4-BSA or PBS. Various concentrations of test reagents in 0.1 ml of E4-BSA were added to the collagen-coated dishes, followed by 3.0 ml of media that contained the cells. Cells were allowed to attach and spread at 37°C in 5% CO_2 for 1.5–2 hr. The duration of incubation depended on the time required for control dishes to satisfactorily settle and spread, because internal variations of the system invariably occurred. Attachment and spreading were assessed in the following way. After the incubation period, dishes were washed twice with PBS to remove nonadherent cells, and the dishes were then examined microscopically to determine the percentage of spread cells among the adherent population. Spreading was considered positive only if a polygonal morphology was apparent. At least three fields were examined on each dish, and a minimum of 200 cells were counted. Percentage spreading was determined as the number of polygonal cells per field divided by the total number of cells. Cells were then trypsinized off the dishes, and the total adherent cell number was determined by counting in a hemocytometer. For attachment, the final numbers that appear in the Tables were calculated as follows:

$$\%\ \text{of cells attached} = \frac{\text{total number attached}}{0.8\ (\text{total number plated})} \times 100,$$

as the recovery of total cells in the adherent and nonadherent population was 80%.

Cellular viability of nonadherent cells was quantitated by trypan blue dye exclusion (Gibco) to determine whether nonadherence resulted from a toxic effect of the drug rather than an actual effect on the process in question.

Anticell Adhesion Factor Antibody

Rabbits were immunized with approximately 1 mg of purified human plasma CAF per injection in Freund's complete adjuvant (v/v), initially in the foot pad and subsequently intradermally for 4 weeks. Five days after the final injection, serum was collected, and the IgG fraction was purified by DEAE column chromatography. Contaminating antibody to human IgM was removed by adsorption with a human IgM myeloma protein. Immunoelectrophoresis was performed in 1% agar in barbitol buffer according to the method of Grabar and Williams.[19] Antinormal human serum (anti-NHS) antibody was purchased from Meloy Laboratories.

Methods That Employ Anticell Adhesion Factor Antibody

Inhibition of cellular attachment by anti-CAF antibody was performed as described in the cellular attachment assay. Concentrations of serum γ-globulin ranged from 0.8 to 0.0125 mg/ml. Preimmune rabbit sera at the same concentrations were used as controls.

The role of CAF in cellular spreading was tested by use of anti-CAF antibody as follows: collagen-coated dishes were incubated for 1 hr with media that contained 10% calf serum, washed, and fibroblasts (2×10^5/3 cm dish) were allowed to attach in the presence of 0.1 M L-lysine, an inhibitor of cellular spreading.[7] After attachment, cells were washed twice to remove the inhibitor and immediately replenished with media that contained anti-CAF. The cells were again incubated at 37°C to assess inhibition of spreading by the antibody.

Pharmacologic Reagents

Sodium cyanide and sodium azide were obtained from Fisher; puromycin, cycloheximide, Triton® X-100, Triton X-15, Tween® 80, sodium deoxycholate (DOC), procaine hydrochloride, colchicine, vinblastine sulfate, and L-lysine were purchased from Sigma. Nonidet® P40 was acquired from Particle Data Laboratories, and dibucaine was obtained from ICN Pharmaceuticals. Lidocaine was a product of Astra Pharmaceuticals. Cytochalasin B was obtained from Calbiochemical.

Polyacrylamide Gel Electrophoresis

Protein samples were dialyzed against 0.1 M Tris HCl (pH 6.8) and adjusted to a final concentration of 2% sodium dodecyl sulfate and 10% glycerol. Prior to electrophoresis, the samples that were being reduced were adjusted to 0.1 M dithiothreitol. All samples were boiled for 2 min. Polyacrylamide gel electrophoresis (PAGE)[20,21] was performed in a slab gel[22] or a disc gel at the acrylamide monomer concentration indicated. After electrophoresis, the gels were stained with Comassie brilliant blue (0.5 g of Coomassie, 520 ml of H_2O, 473 ml of methanol, and 57 ml of glacial acetic acid) to visualize protein bands.

Iodination of Cell Adhesion Factor

Plasma CAF was iodinated by the chloramine T method.[23] Briefly, 2.5 mg of CAF in 2.5 ml of 0.05 M sodium phosphate (pH 7.0) were iodinated at 4°C, with constant stirring for 5 min, with 50 μCi of ^{125}I (New England Nuclear) and 0.125 ml of chloramine T at 200 μg/ml. The reaction was terminated with 0.125 ml of sodium metabisulfite (Fisher) at 200 μg/ml, and free iodide was removed by exhaustive dialysis against 0.1 M sodium phosphate (pH 7.0).

RESULTS

Purification of Cell Adhesion Factor

Purity and molecular composition of the CAF preparation were ascertained by PAGE. As shown in FIGURE 1 (a), when 100 μg of protein were electrophoresed, the only visible band possessed a molecular weight of 235,000 daltons, identical to the molecular weight of the CAF monomer. No additional stained protein band was found. This technique is capable of detecting at least 5 μg of protein, and we

FIGURE 1. Polyacrylamide gel electrophoresis of purified CAF. Samples were reduced with 0.1 M dithiothreitol prior to electrophoresis. Protein concentration was (a) 100 μg and (b) 20 μg. Monomer concentration was 7.5%.

therefore have less than 5% contamination with other proteins. From 1 liter of outdated human plasma, 20–30 mg of CAF could be isolated. FIGURE 1(b) shows 20 μg of CAF electrophoresed under reducing conditions. Only one protein band of molecular weight 235,000 daltons is visible.

When the purified protein was electrophoresed under nonreducing conditions (FIGURE 2, a) approximately 80% of the protein was present as a dimer, 15% as monomer, and the rest as aggregates of high molecular weight. For comparison, CAF reduced with 0.1 M dithiothreitol is shown in FIGURE 2(b).

Cell Adhesion Factor-Mediated Adhesion

As shown in FIGURE 3, highly purified CAF is capable of mediating cellular adhesion to collagen-coated dishes. A clear dose-response relationship is apparent, with 50% of cells adherent at approximately 10 μg/ml of CAF. These data are in agreement with previously published results.[7] Activity could be detected with as little as 0.5 μg/ml of CAF. The CAF extracted from the plasma membrane of BHK cells, chick embryo fibroblasts, or Omm cells, a human fibroblast-like cell line, exhibited similar specific activities when total protein concentration was adjusted for CAF content (as judged by PAGE and densitometry).

The CAF remained active for 4–6 weeks at 4°C in 0.1 M Tris phosphate buffer

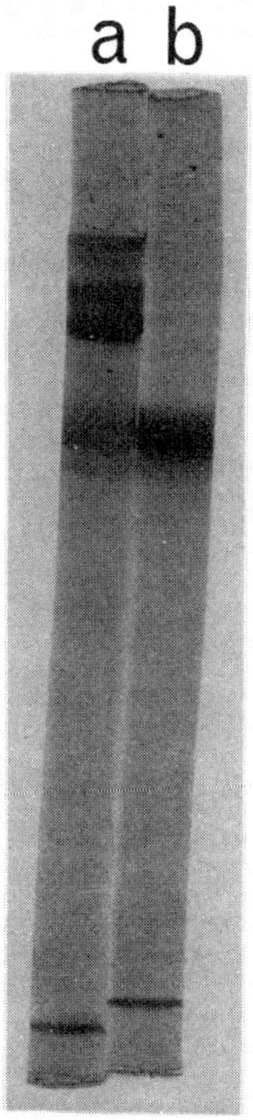

FIGURE 2. Polyacrylamide disc gel electrophoresis of purified CAF. a, Unreduced; b, reduced with 0.1 M dithiothreitol. Monomer concentration was 5%.

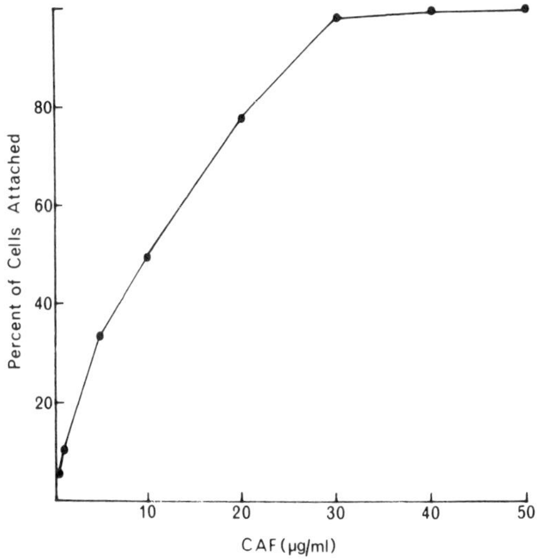

FIGURE 3. Adhesion of PyBHK cells to urea-washed collagen-coated dishes incubated with increasing concentrations of CAF prior to the addition of cells. Results are an average of four experiments.

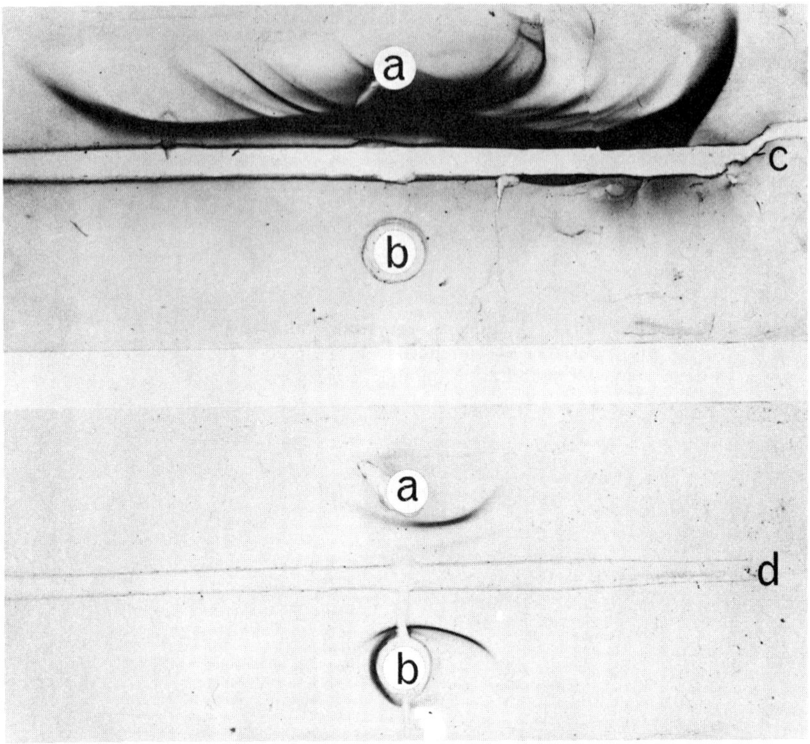

FIGURE 4. Immunoelectrophoresis in 1% agar gels of (a) NHS and (b) purified CAF. Samples were electrophoresed for 3 hr at 3 mA/slide and reacted for 2 days with (c) anti-NHS or (d) anti-CAF. Anode is to the right.

(pH 7.0) and for prolonged periods of time in the same buffer stored in liquid nitrogen.

Antibody Specificity

The specificity of the antiserum was established by immunoelectrophoresis. In FIGURE 4, normal human serum (well a) and purified CAF (well b) at a concentration of 3 mg/ml were electrophoresed, and trough c was filled with anti-NHS. Although CAF failed to react with this antiserum, the preparation contained no additional protein detectable with the antiserum. The same antigens that reacted with anti-CAF (trough d) showed a single precipitin line of β mobility.

Effect of Anticell Adhesion Factor Antibody on Cell Adhesion Factor-Mediated Attachment and Spreading

Inhibition of CAF-Mediated Attachment and Spreading with Anti-CAF Antibody

As shown in TABLE 1, when anti-CAF antibody and BHK or PyBHK cells were simultaneously added to CAF-treated collagen dishes, the antibody inhibited

TABLE 1
EFFECT OF ANTI-CAF ANTIBODY ON CAF-MEDIATED CELLULAR ATTACHMENT AND SPREADING ON COLLAGEN

Final Concentration of Antibody (IgG)	BHK Attachment	BHK Spreading	PyBHK Attachment	PyBHK Spreading
0.4 mg/ml	0*	0	0	0
0.2 mg/ml	0	0	0	0
0.16 mg/ml	100	0	90	50
0.14 mg/ml	100	50	90	50
0.12 mg/ml	100	50	90	50
0.1 mg/ml	100	100	100	100
0.08 mg/ml	100	100	100	100

*All data are percentages and represent averages of four experiments each. Total volume = 2.5 ml of E4-BSA.

attachment and spreading. Control dishes received cells plus the IgG fraction of sera obtained from nonimmunized rabbits, which at the same concentration was not inhibitory.

At 0.2 mg/ml of the IgG fraction of anti-CAF sera, adherence and spreading of both cell types was completely abolished. However, reduction of antibody concentration by only 0.04 mg/ml allowed complete attachment of the cells. Total spreading of both cell types required lowering of the IgG concentration to 0.1 mg/ml.

Inhibition of Cell Spreading by Anti-CAF Antibody

The use of L-lysine, which allows attachment but inhibits spreading, permitted analysis of CAF involvement in spreading as a separate entity not dependent on cellular attachment. Inhibition of spreading with L-lysine could be reversed within 1.5 hr in controls that did not contain antibody.[7] In contrast, dishes to which anti-CAF antibody was added, after removal of the inhibitor, revealed a graded inhibitory effect on cellular spreading, which varied directly with the concentration of anti-CAF antibody added. The percentage of cells that spread per dish is presented in TABLE 2. Anti-CAF antibody did not reverse cellular attachment.

TABLE 2
INHIBITION OF SPREADING BY ANTI-CAF ANTIBODY AFTER ATTACHMENT

	1.5 hr First Incubation 0.1 M Lysine	Spreading (%)	1.5 hr Second Incubation Anti-CAF ab	Spreading (%)
Control	none*	100	none	100
Control	+	0	none	100
Test	+	0	1 mg/ml	1.4
Test	+	0	0.33 mg/ml	10
Test	+	0	0.17 mg/ml	18
Test	+	0	0.03 mg/ml	80

*All results are averages of four experiments each.

Attachment to Undenatured Collagen

As an approach to understanding the *in vivo* significance of CAF-mediated adhesion to collagen, cells were tested for their ability to attach to fibrillar collagen that was not treated with urea. FIGURE 5 demonstrates that 100% of the cells were capable of attaching to urea-washed collagen dishes that received 50 μg/ml of CAF after 1 hr at 37°C. On untreated collagen, 6 hr were required for 80% of the cells to attach. This result was obtained whether the collagen was exposed to CAF during gelation of after the final air drying (see MATERIALS AND METHODS). In addition, spreading on the untreated collagen was considerably slower and never achieved more than 50%, even after 24 hr.

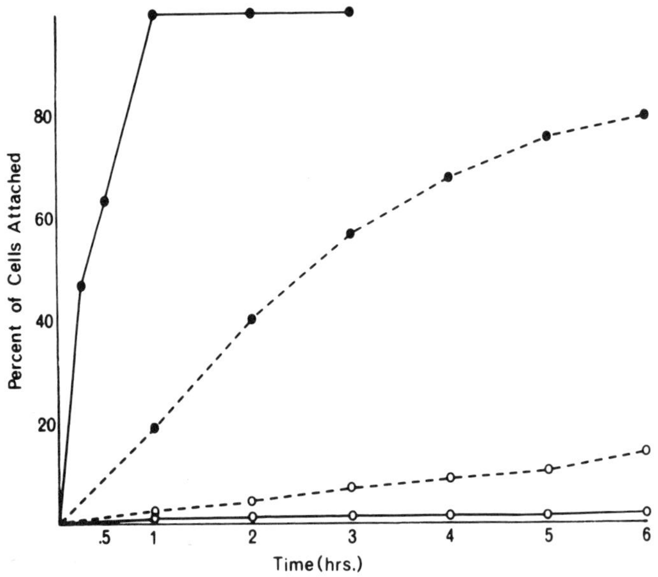

FIGURE 5. Collagen-coated dishes were treated with 8 M urea or PBS, washed, and then incubated with 50 μg/ml of CAF or PBS for 1 hr at 37°C. A total of 5 × 10^5 pyBHK cells were added in E4-BSA, and the percentage of cells attached was determined at successive time intervals. Results are an average of four experiments. ○—○, Urea; ●—●, urea plus CAF; ○- - -○, PBS; ●- - -●, PBS plus CAF.

Although attachment was impeded on untreated collagen, it nevertheless was still CAF dependent, as can be seen from the comparison of cellular attachment with and without CAF (FIGURE 5).

One explanation for the decreased rate of attachment and spending on undenatured collagen may be a decrease in the number of available binding sites for CAF. This hypothesis has been proposed[9] and is confirmed by the binding of ^{125}I-labeled CAF to both forms of collagen. TABLE 3 shows that untreated collagen binds approximately half of the amount of CAF per unit as does urea-treated collagen. Labeled CAF retains the ability to mediate attachment and spreading.

TABLE 3

BINDING OF ^{125}I-LABELED CAF TO COLLAGEN-COATED DISHES*

Collagen Washed with	Dish 1 (cpm)	Dish 2 (cpm)
Urea	6938	6466
PBS	3764	3816

*A total of 10^4 cpm were initially placed on the washed dish, then incubated for 1 hr at 37°C, washed twice with PBS, and bound counts per minute removed for counting by collagenase treatment (1 mg/ml for 0.5 hr at 37°C).

Effects of Metabolic Inhibitors and Inhibitors of Protein Synthesis on Attachment and Spreading

Metabolic Inhibitors

If metabolic inhibitors were added in the presence of E4 that contained 0.45% glucose, no effect on cellular attachment or spreading was observed. However, washing the cells three times before the incubation period with glucose-free PBS depleted the glucose pool sufficiently to permit inhibition of attachment with both sodium azide and sodium cyanide at nontoxic concentrations. It appears (see TABLE 4) that the process of spreading is more sensitive to these inhibitors. Concentrations of 10 mM and above of sodium cyanide proved toxic as assessed by trypan blue dye exclusion. Sodium azide at a concentration of 10 mM was not toxic and resulted in total inhibition of cellular attachment and spreading.

Protein Synthesis Inhibitors

Both cycloheximide and puromycin at concentrations that completely inhibited [^{3}H]-L-leucine incorporation into trichloroacetic acid-precipitable material proved

TABLE 4

EFFECTS OF METABOLIC INHIBITORS AND INHIBITORS OF PROTEIN SYNTHESIS ON CELLULAR ADHESION AND SPREADING

			BHK		PyBHK	
Group	Reagent	Concentration	Attachment	Spreading	Attachment	Spreading
I. Metabolic inhibitors	sodium azide	10 mM	0*	0	0	0
		1 mM	75	0	100	0
		0.1 mM	100	75	100	75
	sodium cyanide	1 mM	37	0	66	0
		0.1 mM	100	75	75	50
II. Inhibitors of protein synthesis	puromycin	10^{-4} M	100	100	100	100
		10^{-5} M	100	100	100	100
	cyclo-heximide	10^{-5} M	100	100	100	100
		10^{-6} M	100	100	100	100

*All data are percentages.

TABLE 5

EFFECTS OF DETERGENTS ON CELLULAR ATTACHMENT AND SPREADING

Group	Reagent	Concentration (%)	BHK Attachment	BHK Spreading	PyBHK Attachment	PyBHK Spreading
Detergents*	Triton X-100	0.001	75†	75	96	96
		0.0005	94	94	100	100
		0.0001	100	100	100	100
	Triton X-15	0.1	50	50	50	50
		0.01	100	100	74	74
		0.005	ND‡	ND	100	67
	Tween 80	0.01	75	75	100	100
	Nonidet P40	0.0005	62	37	ND	ND
		0.001	75	62	100	100
	DOC§	0.001	88	71	66	50
		0.0005	100	100	100	100

*All detergents were dissolved in E4 (v/v).
†All data are percentages.
‡Not determined.
§DOC (w/w) was dissolved in PBS, and all incubations were performed in PBS.
Note: All cells that did not stick to the collagen were nonviable. Thus, any inhibition observed was not due to an effect of the detergents on the mechanisms of attachment and spreading.

ineffective in inhibiting the processes of cellular attachment and spreading (TABLE 4). Thus, *de novo* protein synthesis is not involved in these processes.

Effect of Detergents

Nonlytic concentrations of detergents are effective solubilizers of several plasma membrane proteins.[15] Exposure of the cells to several detergents did not affect attachment or spreading. Therefore, it appears that no detergent-soluble protein was obligatory in cellular attachment or spreading on the collagen substratum (TABLE 5).

Role of Microfilaments and Microtubules

Microtubular and Microfilament Antagonists

The necessity of an intact microtubular system for spreading but not attachment was demonstrated by the use of colchicine and vinblastine sulfate (TABLE 6). At 10^{-4} M colchicine, attachment proceeded unimpeded, whereas spreading was completely inhibited. Vinblastine sulfate, within the concentration range 10^{-4}–10^{-6} M, also prevented spreading of cells without affecting attachment.

Cytochalasin B also influenced spreading, but not attachment, at a concentration as low as 25 μg/ml, thereby implicating microfilament involvement in CAF-mediated spreading on collagen (TABLE 6).

It was observed that as the doses of all of these agents were decreased, the size of the actual cell area attached to the collagen increased. These observations con-

TABLE 6

INHIBITION OF CELLULAR ADHESION AND SPREADING BY ANTAGONISTS OF MICROTUBULES AND MICROFILAMENTS

			BHK			PyBHK		
Group	Reagent	Concentration	Attach-ment	Polygonal Spreading	Spherical* Spreading	Attach-ment	Polygonal Spreading	Spherical Spreading
Microtubular and microfilament antagonists	colchicine	10^{-4} M	100†	0	100	100	12	88
		10^{-5} M	100	100	0	100	100	0
	vinblastine	10^{-4} M	83	0	0	100	0	0
		10^{-5} M	100	0	100	100	0	100
		10^{-6} M	100	0	100	100	0	100
	cytochalasin B‡	25 μg/ml	100	0	100	100	25	75
		1 μg/ml	100	100	0	100	100	0
Anesthetics§	dibucaine	1.0 mM	100	0	—	100	0	—
		0.1 mM	100	100	—	100	0	—
		100 mM	79	0	—	88	0	—
	procaine	10 mM	100	100	—	100	100	—
		1 mM	100	100	—	88	52	—
	lidocaine	12 mM	100	67	—	100	35	—
		1.2 mM	100	75	—	100	100	—

*Spherical shaped spreading was observed. The percentage reported above zero refers to normal polygon-shaped spreading, and all remaining cells are spherical.

†All data are percentages.

‡Cytochalasin B was dissolved in 0.5% dimethylsulfoxide in PBS, which by itself was not inhibitory or toxic. Results are not reported if non-adherence was due to cell death from toxic effects of the agent employed.

§All anesthetics were dissolved in 50% ethanol, which was not inhibitory or toxic in controls.

cur with those of Grinnell,[10] who also observed a spherical rather than polygonal type of spreading with colchicine and cytochalasin B at similar concentrations, using the same cell type but on an uncoated plastic surface in 10% fetal calf serum. No synergistic effect on attachment or spreading could be detected with the combination of one of the microtubular antagonists with cytochalasin B at various concentrations.

Local Anesthetics

Tertiary amine local anesthetics act on membrane-associated microtubules and microfilaments in a manner similar to a combination of microtubular and microfilament antagonists. The local anesthetics employed clearly inhibited spreading, but not attachment (TABLE 6), in agreement with the results obtained with the microtubular and microfilament antagonists. Because of the known effects of anesthetics on calcium displacement from cell membranes,[25,26] various amounts of calcium and magnesium were added during the incubation period. However, no amelioration of the anesthetic effect on spreading could be obtained with concentrations as high as 10 mM Ca^{2+} plus 5 mM Mg^{2+}. Dibucaine proved toxic to the cells at a dose of 10 mM. It was also noted that procaine at a concentration of 100 mM slightly affected attachment and that lidocaine caused a delay in attachment at both concentrations employed, until 100% of the cells attached 30 min later than did controls. The spherical type of spreading exhibited by cells under the influence of microtubular or microfilament antagonists was not noted with the local anesthetics.

DISCUSSION

Cell adhesion factor has been implicated as a mediator of cellular adhesion to collagen *in vitro*.[7] Direct evidence for this function has been provided by the observation that highly purified human plasma CAF will mediate attachment and spreading of fibroblasts, an effect that can be observed at 0.5 μg/ml. Fifty percent of cells will attach at 10 μg/ml. In addition, specific anti-CAF antibody, but not preimmune γ-globulin at the same concentration, inhibited both cellular attachment and spreading (TABLES 1 & 2). These results agree with those obtained by Yamada *et al.*,[16] who demonstrated that anti-CAF antibody prevented the reversion of transformed cellular morphology to normal in the presence of excess CAF. Allowing cells to attach in the presence of inhibitors of spreading permitted analysis of the effect of antibody on spreading independent of attachment. The results obtained (TABLE 2) clearly indicate a dose-dependent inhibition of spreading. Higher concentrations of antisera were required to completely eliminate spreading, and it therefore appears that attachment is more sensitive to inhibition by anti-CAF than is spreading. This difference may be a reflection of the increased number of sites generally considered to be associated with spreading as compared to those involved in attachment.

Increasing the pH of acid-soluble collagen in an NH_4OH atmosphere induces fibrillogenesis during gelation.[27,28] Although dehydration of the gel causes no breakdown of fibrils, treatment with 8 M urea will induce opening of the fibrillar structure.[8] Thus, nonurea-treated collagen is physiologically closer to *in vivo* forms. Although adhesion on untreated collagen is slower, and after 6 hr represents only 80% of the attachment on urea-washed collagen (FIGURE 5), adhesion to both types of collagen remains CAF dependent. This result, together with the knowledge of the coordinate distribution both *in vivo*[29] and *in vitro*[30] of CAF and collagen and the decrease in CAF[31,41] and collagen[32,41] synthesis after transformation, implies that CAF-mediated adhesion of fibroblasts to collagen has important *in vivo* applications.

Cell surface proteins and contractile systems have been shown to contribute functionally to cellular attachment and spreading on synthetic substrates in the presence and absence of serum.[10,24,34] In our CAF-mediated assay for cellular attachment to collagen, it was demonstrated that plasma membrane proteins soluble in nonlytic concentrations of nonionic detergents[15] or cells denuded of most external proteins with trypsin[31] were capable, even in the absence of *de novo* protein synthesis, to attach to CAF-collagen complexes.

In agreement with Klebe,[17] we have established that cellular energy is required for both attachment and spreading; however, spreading was more sensitive to the effects of metabolic inhibitors (TABLE 4). When metabolic inhibitors were tested in the presence of D-glucose, enough ATP could be generated via glycolysis to restore normal cellular attachment and spreading.

The previously noted cross-reactivity of CAF from different species[34] was also found in our heterologous system, because hamster cell (BHK) adherence to calf skin collagen, mediated by CAF from calf serum, was inhibited by rabbit antisera raised against human plasma CAF.

We have demonstrated (TABLE 6) that both normal and transformed cells require intact microtubules and microfilaments for spreading, but not attachment, during CAF-mediated spreading on a collagen substratum. The pattern of inhibition by antagonists of these cytoskeletal structures is identical for both cell types. The maintenance of apparent normality with respect to function of these structures is significant in light of studies by others who have observed both morphologic and functional changes in these systems after transformation.[35-40] It therefore appears that although there are morphologic and functional alterations in both microtubular and microfilament systems after transformation, these changes do not influence the ability of transformed cells to attach and spread in our assay. Furthermore, the pattern of inhibition by various agents (TABLE 5) is identical, regardless of the cell type tested. This finding can be expanded to include all substances tested for their effect on cellular adhesion and spreading. Therefore, although CAF cannot be detected on the surface of PyBHK cells, such cells are able to utilize this protein for cellular attachment and spreading in an identical manner as BHK cells within our experimental limits.

These results support the conclusion proposed by Grinnell[42] that cellular attachment and spreading are independent events that operate by different mechanisms and require different internal cellular structures.

ACKNOWLEDGMENTS

We thank Mr. Gary Feilich for excellent technical assistance and Ms. Marlene Chavis for expert secretarial help in the preparation of this manuscript.

REFERENCES

1. HYNES, R. O. 1973. Proc. Nat. Acad. Sci. USA **70:** 3170–3174.
2. VAHERI, A. & E. RUOSLAHTI. 1974. Int. J. Cancer **13:** 579–586.

3. Wickus, G. G., P. E. Branton & P. W. Robbins. 1974. *In* Control of Proliferation in Animal Cells. B. Clarkson & R. Baserga, Eds.: 541–546. Cold Spring Harbor Laboratory. Cold Spring Harbor, N.Y.
4. Stone, K. R., R. E. Smith & W. K. Joklik. 1974. Virology **58:** 86–100.
5. Pearlstein, E., R. O. Hynes, L. M. Franks & V. J. Hemmings. 1976. Cancer Res. **36:** 1475–1480.
6. Coll, J. M., S. W. Luborsky & P. T. Mora. 1977. Biochemistry **16:** 3169–3177.
7. Pearlstein, E. 1976. Nature (London) **262:** 497–500.
8. Harnes, M. J., J. L. Gordon & D. E. MacIntyre. 1976. Biochem. J. **160:** 647–651.
9. Engvall, E. & E. Ruoslahti. 1977. Int. J. Cancer **20:** 1–5.
10. Grinnell, F. 1974. Arch. Biochem. Biophys. **165:** 524–530.
11. Grinnell, F. & P. A. Srere. 1971. J. Cell. Physiol. **78:** 153–158.
12. Grinnell, F., M. Milam & P. A. Srere. 1972. Arch. Biochem. Biophys. **153:** 193–198.
13. Mosher, D. F. & E. R. Blout. 1973. J. Biol. Chem. **248:** 6896–6903.
14. Mosher, D. F. 1975. J. Biol. Chem. **250:** 6614–6621.
15. Pearlstein, E. & J. Seaver. 1976. Biochim. Biophys. Acta **426:**589–597.
16. Yamada, K. M., S. S. Yamada & I. Pastan. 1976. Proc. Nat. Acad. Sci. USA **73:** 1217–1221.
17. Klebe, R. J. 1975. J. Cell. Physiol. **86:** 231–236.
18. Lowry, O. H., N. J. Rosebrough, A. L. Farr & R. Randall. 1951. J. Biol. Chem. **193:** 265–275.
19. Grabar, P. & C. A. Williams. 1953. Biochim. Biophys. Acta **10:** 193.
20. Laemmli, U. K. 1970. Nature (London) **227:**680–685.
21. Studier, F. W. 1972. Science **176:** 367–376.
22. Maizel, J. 1971. *In* Methods in Virology. K. Maramorosch & H. Koprowski, Eds. Vol. 5: 179. Academic Press, Inc. New York, N.Y.
23. McConahey, P. J. & F. J. Dixon. 1966. Int. Arch. Allergy Appl. Immunol. **29:** 185–189.
24. Grinnell, F. 1974. Arch. Biochem. Biophys. **160:** 304–310.
25. Papahadjopoulos, D. 1972. Biochim. Biophys. Acta **265:** 169–186.
26. Seeman, P. 1974. Pharmacol. Rev. **24:** 583–655.
27. Hellauer, H. & R. Winkler. 1975. Connect. Tissue Res. **3:** 227–230.
28. Oegema, T. R., J. Laidlaw, V. C. Hascall & D. D. Dziewiatkowski. 1975. Arch. Biochem. Biophys. **170:** 698–709.
29. Linder, E., A. Vaheri, E. Ruoslahti & J. Wartiovaara. 1975. J. Exp. Med. **142:** 41–49.
30. Bornstein, P. & J. F. Ash. 1977. Proc. Nat. Acad. Sci. USA **74:** 2480–2484.
31. Pearlstein, E. & M. W. Waterfield. 1974. Biochim. Biophys. Acta **362:** 1–12.
32. Kamine, J. & H. Rubin. 1977. J. Cell. Physiol. **92:** 1–12.
33. Rabinovitch, M. & M. DeStefano. 1974. Exp. Cell Res. **88:** 153–162.
34. Kuusela, P., E. Ruoslahti, E. Engvall & A. Vaheri. 1976. Immunochemistry **13:** 639–642.
35. Pollack, R., M. Osborn & K. Weber. 1975. Proc. Nat. Acad. Sci. USA **72:** 994–998.
36. Inbar, M. & L. Sachs. 1969. Proc. Nat. Acad. Sci. USA **63:** 1418–1425.
37. Nicolson, G. L. 1971. Nature (London) [NB] **233:** 244–246.
38. Yin, H. H., T. E. Ukena & R. D. Berlin. 1972. Science **178:**867–868.
39. Nicolson, G. L. 1973. Nature (London) [NB] **243:** 218–220.
40. Poste, G., D. Papahadjopoulos & G. L. Nicolson. 1975. Proc. Nat. Acad. Sci. USA **72:** 4430–4434.
41. Adams, S. L., M. E. Sobel, B. H. Howard, K. Olden, K. M. Yamada, B. deCrombrugghe & I. Pastan. 1977. Proc. Nat. Acad. Sci. USA **74:** 3399–3403.
42. Grinnell, F. 1976. Exp. Cell Res. **97:** 265–274.

ARE HIGH-MOLECULAR-WEIGHT GLYCOPROTEINS REGULATORS OF CELLULAR GROWTH?*

John M. Buchanan, Lan Bo Chen,† and Bruce R. Zetter‡

Department of Biology
Massachusetts Institute of Technology
Cambridge, Massachusetts 02139

The central focus of this conference has been the exploration of the role of a high-molecular-weight glycoprotein in various cellular processes. This glycoprotein, which is approximately 250,000 daltons in size and readily iodinated, has been designated as LETS (large, external, transformation sensitive) protein, as fibronectin, or as cold-insoluble globulin. Since none of these names is in universal use, we will employ the first in the rest of this report.

Several early findings were suggestive of a role for LETS protein in modulating the proliferation of cultured fibroblasts. Among these findings was the loss of LETS from the surface of cells that had been treated with mitogenic concentrations of the serine protease trypsin.[1] It was considered that the removal of LETS might be the mechanism by which proteases induced cellular proliferation via an action at the cell surface. In addition, the loss of surface-associated LETS after transformation of cultured fibroblasts with viruses,[2] such as RSV, SV40, and polyoma, also seemed to suggest that the increased cellular proliferation characteristic of virally transformed cells might results from the disappearance of this protein. Finally, expression of the LETS protein was reported to be cell cycle dependent,[2,3] although there have been conflicting reports as to precisely which stages of the cycle are associated with increased or decreased amounts of LETS.

Our initial interest in the LETS protein grew out of a desire to understand the mechanism by which proteolytic enzymes could induce cell division in quiescent cultures of chick embryo fibroblasts. Although it had been previously shown that trypsin was mitogenic for chick embryo fibroblasts and could also remove LETS from cell surfaces, it was apparent that not every protease could elicit the same degree of mitogenic response. It was also likely that the specificity of any given protease for substrates at the cell surface would play a critical role in determining the cellular response to that enzyme. Therefore, if every mitogenic protease were able to cleave the LETS protein, and if nonmitogenic proteases were to leave it intact, a strong case could be made for a role for LETS in the regulation of cellular growth.

To investigate this possibility, we examined several proteases for their abilities to induce DNA synthesis and cellular proliferation and for their abilities to cleave cell surface proteins. Our earliest studies were with the clotting enzyme thrombin. The finding by Chen and Buchanan[4] that this protease was a potent mitogen for

*Supported by National Science Foundation Grant PCM-17669, American Cancer Society Grant BC-206A, and National Cancer Institute Grant CA-02015.

†Present address: Sidney Farber Cancer Research Institute, Harvard Medical School, Boston, Massachusetts 02115.

‡Present address: Cancer Research Institute, University of California Medical Center, San Francisco, California 94143.

0077-8923/78/0312-0293 $01.75/0

TABLE 1
EFFECT OF PROTEASES ON LETS PROTEIN AND DNA SYNTHESIS*

Additions to Basic Medium	Integrated Intensity of LETS Protein (% of control)	[^{3}H]Thymidine Incorporation (cpm)
None (control)	—	1915
Calf serum, 2%	100	53,253
Trypsin (1 μg/ml, 10 min)	0	11,265
Thrombin (50 μg/ml, 30 min)	100	45,314
Thrombin (1 μg/ml, 12 hr)	100	53,333

*From Chen *et al.*[7]

chick embryo fibroblasts provided a new tool for investigating the cell surface effects of mitogenic proteases, because, in comparison with trypsin, thrombin is far more limited in the number and kinds of peptide bonds that it can cleave[5,6] and could therefore be expected to exert a more limited and more easily interpretable effect on the cell surface.

As shown in TABLE 1, both trypsin and thrombin are able to induce DNA synthesis in resting chick embryo fibroblasts, but only trypsin removes LETS protein.[7,8] Two percent calf serum is mitogenic, but it, too, fails to remove surface-associated LETS. Cell surface iodination with ^{125}I or ^{131}I and slab gel electrophoresis and subsequent audiography were performed as described by Teng and Chen.[8,9] This result clearly demonstrates that the removal of LETS protein from the surface of chick embryo fibroblasts is not *necessary* for these cells to proliferate. The result does not, however, indicate whether under certain conditions the removal of LETS might be *sufficient* to induce cellular growth. However, in this connection, more recent investigations have demonstrated that chymotrypsin is capable of removing LETS protein but does not stimulate DNA synthesis.[10]

In addition to chymotrypsin, thrombin, and trypsin, we tested several other proteases for their abilities to induce DNA synthesis and cellular proliferation and for their abilities to remove the LETS protein and a slightly smaller protein (mol wt 205,000) that had been found by Teng and Chen[9] to be sensitive to removal by thrombin.[10] The results are shown in FIGURE 1 and TABLE 2.

Of the enzymes tested, the most potent mitogens are thrombin, bromelin, and trypsin, each of which is capable of causing a doubling in cell number within the first 36 hr of protease treatment. Significant but less marked stimulation of growth is caused by treatment with elastase. All of these enzymes give rise to at least a 10-fold increase in the rate of DNA synthesis after 12 hr of continuous incubation. It is important to note that all enzyme treatments are performed in serum-free media and that the observed increases in DNA synthesis and cell number occur under these conditions.

It is clear, however, that not every type of proteolysis is capable of inducing cellular growth, even when the protease is constantly present in culture. In our hands, papain, α-protease, and thermolysin alone cause no significant increase in DNA synthesis in resting cells and are unable to induce a significant increase in cell number after a 36-hr incubation.

Again, the presence or absence of LETS on the cell surface could not be correlated with the growth state of the protease-treated cells. Although it was removed by such mitogens as trypsin and elastase, LETS protein was also removed by the nonmitogenic proteases chymotrypsin, thermolysin, and papain. The removal of this protein by such nonmitogenic proteases and its presence after treatment by some mitogens are clear evidence that the absence of LETS is neither sufficient nor necessary for the initiation of cell division. The smaller, thrombin-sensitive cell surface protein of 205,000 daltons was found to be removed by every protease tested in our experiments.[10] Since this protein is sensitive to both mitogenic and nonmitogenic enzymes, its removal also cannot be sufficient to induce cell division.

While these results eliminate the possibility that either of these two cell surface proteins plays a primary role in the initiation of cell division, they leave unanswered the question of how a protease actually does stimulate cellular mitosis. Two seemingly conflicting results may shed some light on this problem. The first is our finding that thrombin can be internalized intact by chick embryo fibroblasts.[11] This observation suggests that thrombin might exert its mitogenic effects inside the cell; however, Carney and Cunningham[12] have demonstrated that trypsin and thrombin can be mitogenic if attached to beads that cannot penetrate the cell membrane. There is one model that might accommodate these two results. In this model, the critical event in mitogenesis would be the internalization and subsequent action of a cell surface receptor for thrombin. If thrombin is soluble, it will be internalized along with the receptor, whereas if thrombin were insolubilized on beads, the receptor might become internalized by itself and leave thrombin at the cell surface, perhaps after limited proteolysis of a thrombin-receptor complex.

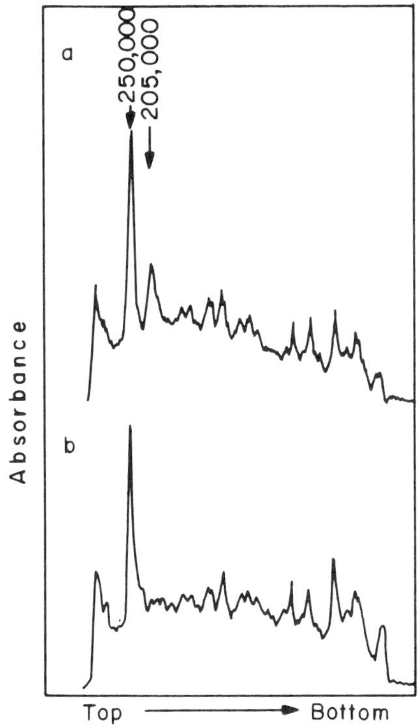

FIGURE 1. Densitometric scans of autoradiograms of gel strips. Autoradiograms of electropherograms of iodinated (^{131}I) cell surface proteins of cultures of chick embryo fibroblasts that were (a) untreated or (b) treated with thrombin at 20 μg/ml were scanned by a Canalco model G11 densitometer with a digital integrator (Autolab 6300) attachment.

TABLE 2

PROTEASE STIMULATION OF DNA SYNTHESIS AND CELLULAR PROLIFERATION IN CHICK EMBRYO FIBROBLASTS*

Treatment	Concentration (μg/ml)	Amount of ^{131}I-Labeled Protein (% of control) 250K	205K	Incorporation of† [^{3}H]Thymidine into DNA at 12 hr (cpm per plate)	Radioactivity, Experimental/Control	Cell Number* × 10^{-5} at 36 hr
None		100	100	1700	1.0	4.5
Serum	(5%)	100	100	33,000	19.4	10.6
Thrombin	1			32,000	18.8	9.2
	2.5	100	0	40,000	23.5	10.1
α-Protease	1			2000	1.2	4.4
	5	100	0	2200	1.3	4.5
	10			2800	1.6	4.8
	15			4200	2.5	5.1
Thermolysin	1	0	0	1800	1.1	4.8
	5	0	0	3800	2.2	4.9
	10			5200	3.0	5.1
Bromelin	1			9900	5.8	5.2
	2.5			17,900	10.5	9.0
	5.0	100	0	24,000	14.1	9.5
Trypsin	0.1	0	0	28,000	16.4	9.6
	0.2			41,000	24.1	10.3
Papain	0.5			1700	1.0	4.5
	1	0	0	2700	1.6	4.7
	5			3200	1.9	4.7
Elastase	1			14,000	8.2	5.8
	5	0	0	20,000	11.8	7.5
	10	0	0	20,000	11.8	7.6

*From Zetter *et al.*[10]

†Cells were incubated continuously with serum or protease for the duration of the experiment.

It remains to be determined whether all proteases might have the same receptor and whether they might share receptors with other mitogenic compounds, such as insulin.

If LETS protein were of interest only for its putative role in regulating cellular growth, there would be little reason to discuss it further. However, there is now considerable evidence to indicate that one of the major functions of this protein is to mediate the attachment of cells both to other cells and to the substratum on which they grow. By influencing the manner in which a cell attaches to a substratum, a protein can affect both the shape of a cell and its motility, and by influencing the manner in which cells attach to each other, it can affect both the positioning of cells relative to each other and the ability of cells to communicate with each other via intercellular junctions. Together, these two functions could allow a single protein to play a critical role in arranging the cells within a tissue into an organized three-dimensional structure.

Among the early indications that LETS might be involved in cell-cell interactions was the finding of Yamada *et al.*[13] that urea-extracted preparations of LETS could agglutinate formalinized sheep erythrocytes. This observation suggested that LETS might mediate the aggregation of other types of cells. In our studies on the effects of proteases on chick embryo fibroblasts, we noted that some, but not all, of the proteases tested caused the fibroblasts to migrate and form large aggregates of cells. When we compared the ability of a protease to induce aggregate formation to the distribution of cell surface proteins that remain after treatment, we found an intriguing correlation.[10] The proteases most likely to cause cellular aggregation were those that cleaved the 205K protein from the cell surface but left the LETS protein intact. Proteases that removed LETS protein were less likely to induce aggregation. These observations have led to speculation that LETS protein may be involved in the regulation of cell-cell recognition and cell-cell adhesion.

Nevertheless, the fortuitous correlation of the loss of LETS protein with initiation of cellular growth on application of certain proteases or during viral transformation has led to faulty speculation on the role of this protein as a signal of cellular growth. The loss of LETS protein during viral transformation may either result from reduced intracellular production[14] or may be a secondary consequence of the action of proteases produced by the transformed cell. In any event, its removal from the cell during viral transformation cannot at the present time be designated as the primary event from which other phenotypic characteristics are derived.

References

1. Sefton, B. M. & H. Rubin. 1970. Nature (London) **227:** 843–845.
2. Hynes, R. O. 1976. Biochim. Biophys. Acta **458:** 73–107.
3. Roblin, R., I.-N. Chou & P. H. Black. 1975. Adv. Cancer Res. **22:** 203–260.
4. Chen, L. B. & J. M. Buchanan. 1975. Proc. Nat. Acad. Sci. USA **72:** 131–135.
5. Magnusson, S. 1971. *In* The Enzymes. P. D. Boyer, Ed. 3rd edit. Vol. 3: 277–321. Academic Press, Inc. New York, N.Y.
6. Muszbek, L. & K. Laki. 1974. Proc. Nat. Acad. Sci. USA **71:** 2208–2211.
7. Chen, L. B., N. N. H. Teng & J. M. Buchanan. 1975. *In* Proteases and Biological Control. E. Reich, D. B. Rifkin & E. Shaw, Eds.: 957–965. Cold Spring Harbor Laboratory. Cold Spring Harbor, N.Y.
8. Teng, N. N. H. & L. B. Chen. 1975. Proc. Nat. Acad. Sci. USA **72:** 413–417.

9. Teng, N. N. H. & L. B. Chen. 1976. Nature (London) **259:** 575–580.
10. Zetter, B. R., L. B. Chen & J. M. Buchanan. 1976. Cell **7:** 407–412.
11. Zetter, B. R., L. B. Chen & J. M. Buchanan. 1977. Proc. Nat. Acad. Sci. USA **74:** 596–600.
12. Carney, D. H. & D. D. Cunningham. 1977. Nature (London) **268:** 602–606.
13. Yamada, K. M., S. S. Yamada & I. Pastan. 1975. Proc. Nat. Acad. Sci. USA **72:** 3158–3162.
14. Yamada, K. M., S. S. Yamada & I. Pastan. 1977. J. Cell Biol. **74:** 649–654.

ROLE OF THE HIGH-MOLECULAR-WEIGHT GLYCOPROTEIN IN CELLULAR MORPHOLOGY, ADHESION, AND DIFFERENTIATION*

Bruce R. Zetter, Gail R. Martin, Charles R. Birdwell,†
and Denis Gospodarowicz

Cancer Research Institute
University of California
School of Medicine
San Francisco, California 94143
and †Scripps Clinic and Research Foundation
La Jolla, California 92037

The high-molecular-weight cell surface glycoprotein that is the subject of this conference has attracted the attention of cell biologists for several reasons. Over the past few years, this protein has been implicated in processes as diverse as cellular growth control, transformation, tumorigenesis, myogenesis, cellular morphology, cell-cell aggregation, cell-substratum adhesion, hemostasis, and as one of the main components of extracellular limiting membranes (basement membranes). Several excellent reviews have documented the explosion of information concerning this unique glycoprotein.[1-3] In this paper, we will focus on the role of this protein in mediating and maintaining cell-cell and cell-substratum interactions. In particular, we will discuss its physiologic role in maintaining the geometric configuration of the cells in a given tissue, such as the contact-inhibited vascular and corneal endothelia. We will also discuss new evidence that this protein could play a role in the cell-cell interactions that occur in the developing mammalian embryo. Although several names have been proposed for this protein, none is yet in universal use, and for reasons of convenience only, we will use the term LETS (large, external, transformation-sensitive) protein.[1]

There is now considerable evidence to indicate that one of the major functions of the LETS protein is to mediate the attachment of cells both to other cells and to the substratum on which they grow. When purified LETS protein is added to cultures of transformed cells that adhere poorly to a substratum, the strength of their attachment to either plastic surfaces or collagen-coated surfaces is increased, and the morphology of the cells changes from round to flattened with a well-spread cytoplasm indicative of increased contact with the substratum.[4-6] In addition, LETS protein is found to be one of the components of a "substrate-attached material" (SAM) that remains in contact with the substratum, even after solubilization of cell membranes with detergent.[7] It should be noted, however, that several other glycoproteins have been suggested to play a role in the adhesion of cells to a substratum and that this adhesive process is a complex phenomenon that involves several types of surface macromolecule.[8,9]

Among the early indications that LETS might be involved in cell-cell interactions was the finding of Yamada *et al.*[10] that addition of LETS to suspensions of formalinized sheep erythrocytes resulted in their agglutination. This observation

*Supported by National Institutes of Health Grant HD 11082, American Cancer Society Grant VC 194, and National Science Foundation Grant PCM 77-05403.

0077-8923/78/0312-0299 $01.75/0

suggested that LETS protein might mediate the aggregation of other cell types; in later studies on the aggregation of protease-treated chick embryo fibroblasts, it was observed that the proteases most likely to induce cellular aggregation were those that failed to remove LETS protein from the cell surface.[11] The ability of the chick cells to aggregate was diminished, however, if a protease cleaved LETS protein from the cell surface.

In a more recent work, Chen *et al.*[12] demonstrated, with cultures of normal and transformed rat embryo fibroblasts, that certain cells in culture express LETS protein preferentially in areas of cell-cell contact. In sparse cultures, LETS was first observed on the intercellular processes between contacting cells. In dense cultures, the LETS protein distribution pattern was a reflection of the intercellular borders. These observations have led to speculation that the expression of LETS protein both influences and depends on the nature of cell surface contacts.

Vascular Endothelial Cells and LETS Protein

Endothelial cells constitute the inner lining of the blood vascular system. *In vivo,* these cells form a highly contact-inhibited monolayer of flattened cells that must adhere very tightly to the basement membranes on which they rest. Because of their location at the interface between blood and tissue, they are the chief element involved in the permeability of blood vessels. Thus, the continuity of the vascular endothelium is essential for the survival of the organism, and this continuity is, in turn, dependent on the molecules that mediate the adhesion of these

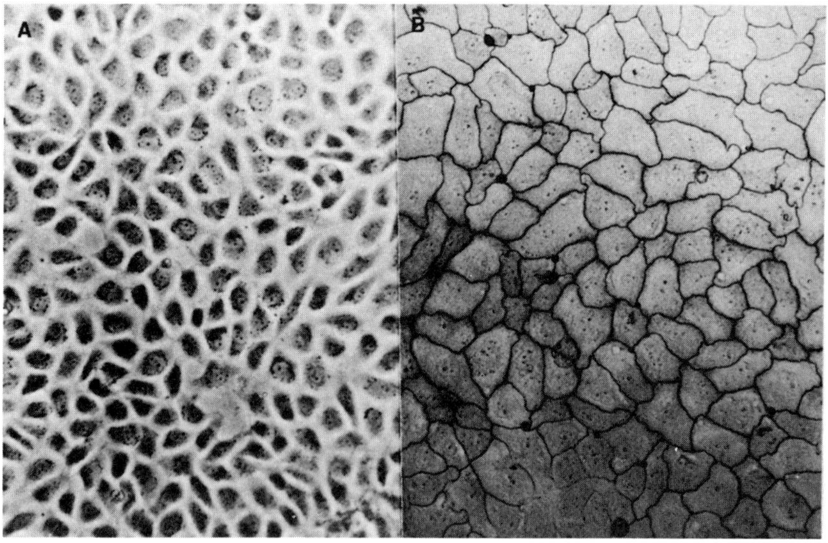

FIGURE 1. Morphologic features of a clonal cell line of adult aortic endothelial cells. A: Monolayer of cultured vascular endothelial cells photographed under phase-contrast microscopy. × 100. B: Monolayer of cultured vascular endothelial cells stained with silver nitrate. The sharp delineation of cell borders is similar to that observed in preparations of endothelium stained *in situ.*

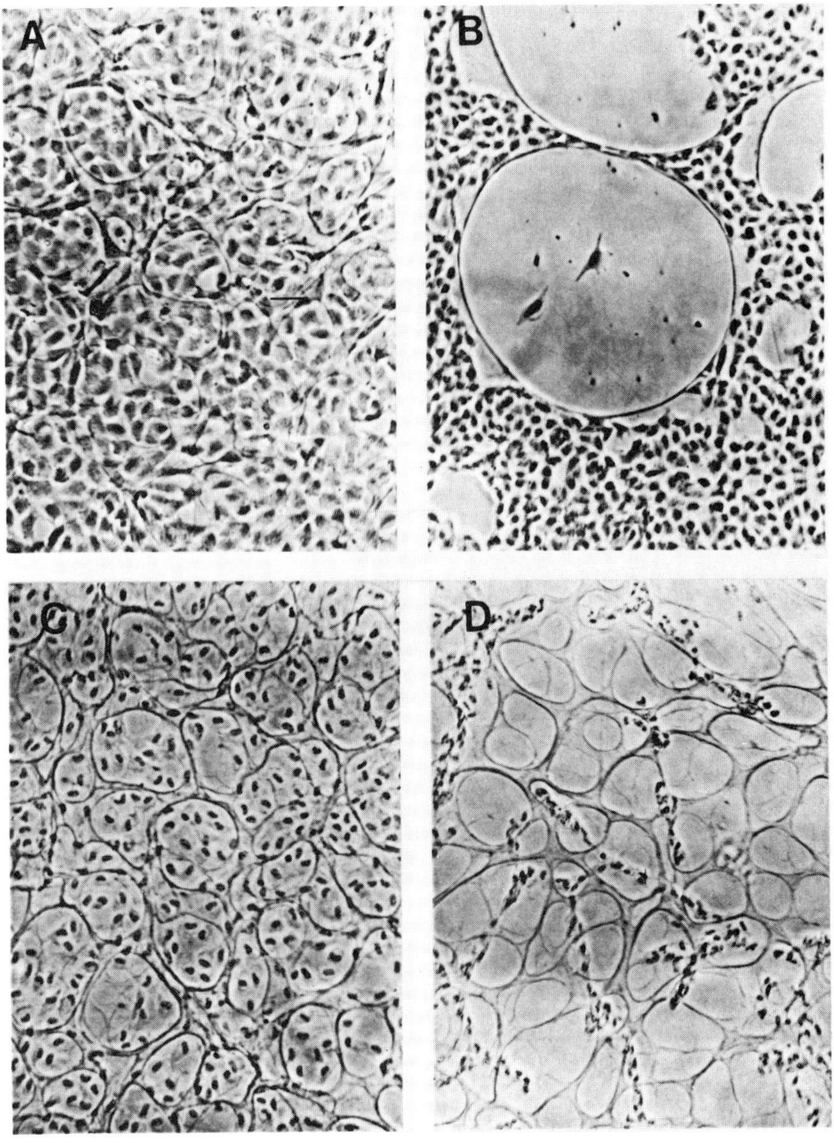

FIGURE 2. Appearance of an extracellular matrix produced by cultures of vascular endothelial cells. Bovine calf aortic endothelial cells were grown to confluence and maintained in medium that contained 10% calf serum and 10 ng/ml of FGF. After 2 additional weeks, an amorphous material appeared under the monolayer (A), and the cells began to retract, leaving circular areas devoid of cells (B). When treated with phosphate-buffered saline that contained 0.5% Triton X-100, the monolayer was dissolved, leaving behind only the nuclei and the extracellular matrix (C,D).

cells to their *in vivo* substratum and on the contacts between neighboring cells. We have therefore investigated whether the LETS protein might be involved in any of these processes. For these investigations, we have used a clonal line of cultured vascular endothelial cells derived from bovine aortic arch that has been described in detail elsewhere.[13,14] The morphologic characteristics of these cultured cells are shown in FIGURE 1. As is the case *in vivo*, the cells are flat and, at confluence, form a contact-inhibited monolayer.

One characteristic of cultured vascular endothelial cells is that they are capable of secreting enormous amounts of extracellular material. As shown in FIGURE 2, these cells are capable of covering the culture dish with a layer of an amorphous

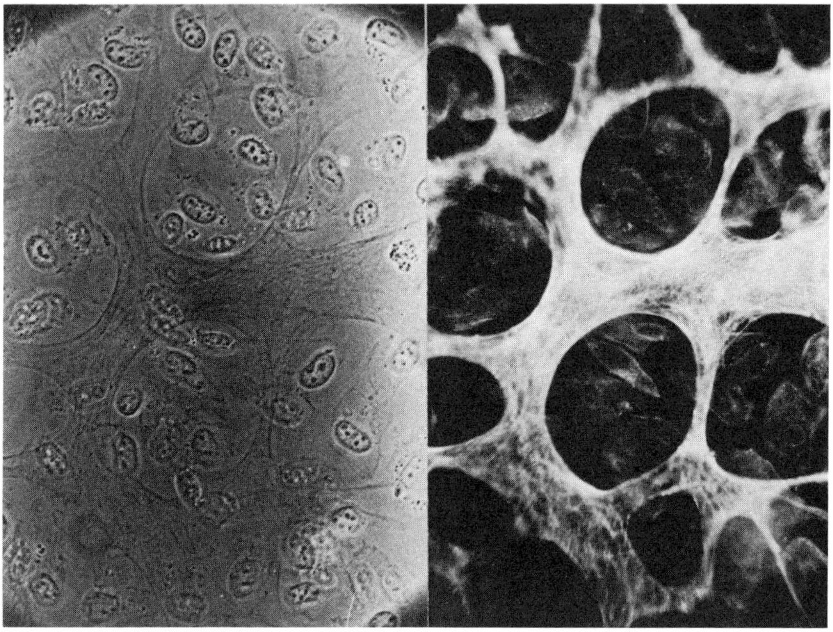

FIGURE 3. Codistribution of LETS protein with the extracellular matrix produced by vascular endothelial cells. Cultures of adult bovine aortic endothelial cells were solubilized with Triton X-100 as described in FIGURE 2 and assayed for the presence of LETS protein by use of indirect immunofluoresence as described in Reference 12. The LETS protein distribution is equivalent to the pattern in the extracellular matrix.

material that forms concentric circles of fibers and that resembles, in some respects, the pattern of the material in the basal lamina on which the endothelium rests *in vivo*. By use of indirect immunofluorescence (as described in Reference 12) as a probe, we have determined that the LETS protein is a major component of this amorphous material. FIGURE 3 demonstrates that the pattern of LETS protein deposition is equivalent to that observed when cultures of endothelial cells are treated with 0.5% Triton® X-100, a detergent that dissolves cellular membranes and reveals the underlying matrix. This result indicates that endothelial cells may produce and export a major component of the basement membranes on which they rest *in vivo*.

Since, as discussed above, LETS protein has been shown to be involved in morphology and cell-substratum adhesion, and because endothelial cells grow as a single layer of flattened, tightly adherent cells, it is probable that the LETS protein secreted by endothelial cells functions to maintain the proper morphologic features and positioning of these cells. It is likely that the endothelial cells that occupy the inner surface of arteries must secrete large quantities of LETS protein merely to remain attached to the basal lamina in the face of enormous shear forces. The production of LETS protein by these cells will therefore be a prerequisite for preservation of the integrity of the blood vessel.

Corneal Endothelial Cells and LETS Protein

The corneal endothelium consists of a single layer of flattened cells on the posterior surface of the cornea and serves to separate the corneal stroma from the aqueous humor.[15] Like the vascular endothelium, corneal cells grow as a contact-inhibited monolayer on top of a basement membrane (Descemet's membrane). We have cultured endothelial cells from the corneas of rabbits, cows, and humans,[16,17] and, again, the morphologic features seen *in vitro* parallel those observed *in vivo.* Colonies grow as monolayers of polygonal cells in close apposition (Figure 4). As the colonies differentiate, the cells synthesize collagenous material that is deposited extracellularly and that can be visualized in fixed cultures stained with alcian green or in electron micrographs of the cultured cells.[16] The staining properties of the material produced *in vitro* are very similar to those of Descemet's membrane observed in histologic preparations of whole corneas. These findings therefore suggest that corneal endothelial cells also secrete basement membrane materials and should therefore secrete LETS protein. When these cells are treated with Triton X-100 and assayed for the presence of LETS protein, a pattern very similar to that of the vascular endothelium is observed (Figure 5), in which LETS protein is secreted in a fine matrix on the floor of the culture dish and also in large deposits in the intercellular spaces. Our results suggest that LETS protein will be observed wherever cells are found that function as a barrier between tissue and fluid spaces because of the requirement in these situations for single layers of flattened, tightly connected cells.

LETS Protein and Contact Inhibition

The endothelial cells described above represent one of the few classes of cells that are organized *in vivo* into single layers of contact-inhibited cells. It is evident that if, for example, vascular endothelial cells were to overgrow the monolayer, they would rapidly fill in the vessel lumen and subvert their normal function. There is, then, great selective pressure for such cells to remain as a flattened monolayer. In tissue culture, and also *in vivo,* endothelial cells will form a monolayer of polygonal cells in close contact with one another. Unlike strains of fibroblast that have been selected for their ability to grow as a monolayer, contact-inhibited vascular endothelial cells will not overgrow the monolayer when treated with high concentrations of serum or mitogenic agents.[18]

We have investigated the expression of LETS protein on the surface of two endothelium-derived cell lines that have different morphologies at confluence. One is the clonal line of adult bovine aortic endothelial cells discussed above, and the other is a variant that was isolated after treatment of the normal endothelial cell

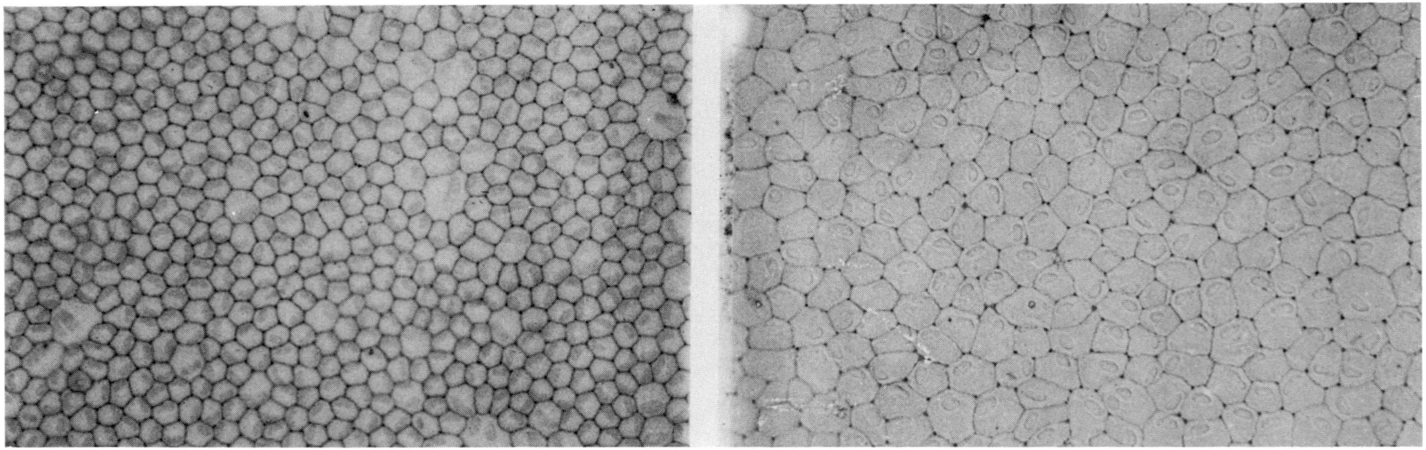

FIGURE 4. Organization of corneal endothelial cells *in vivo* and *in vitro*. *Left:* Segments of bovine cornea were removed from intact eyes, and the endothelial layer was stained with alizarin red. Intercellular boundary areas stain preferentially. *Right:* Cultures of corneal endothelial cells were grown to confluence in medium that contained 10% calf serum and 100 ng/ml of FGF and maintained for 14 days after reaching confluence. Intercellular boundary areas were stained *in vitro* with silver nitrate. The cells assume a polygonal shape both *in vitro* and *in vivo*.

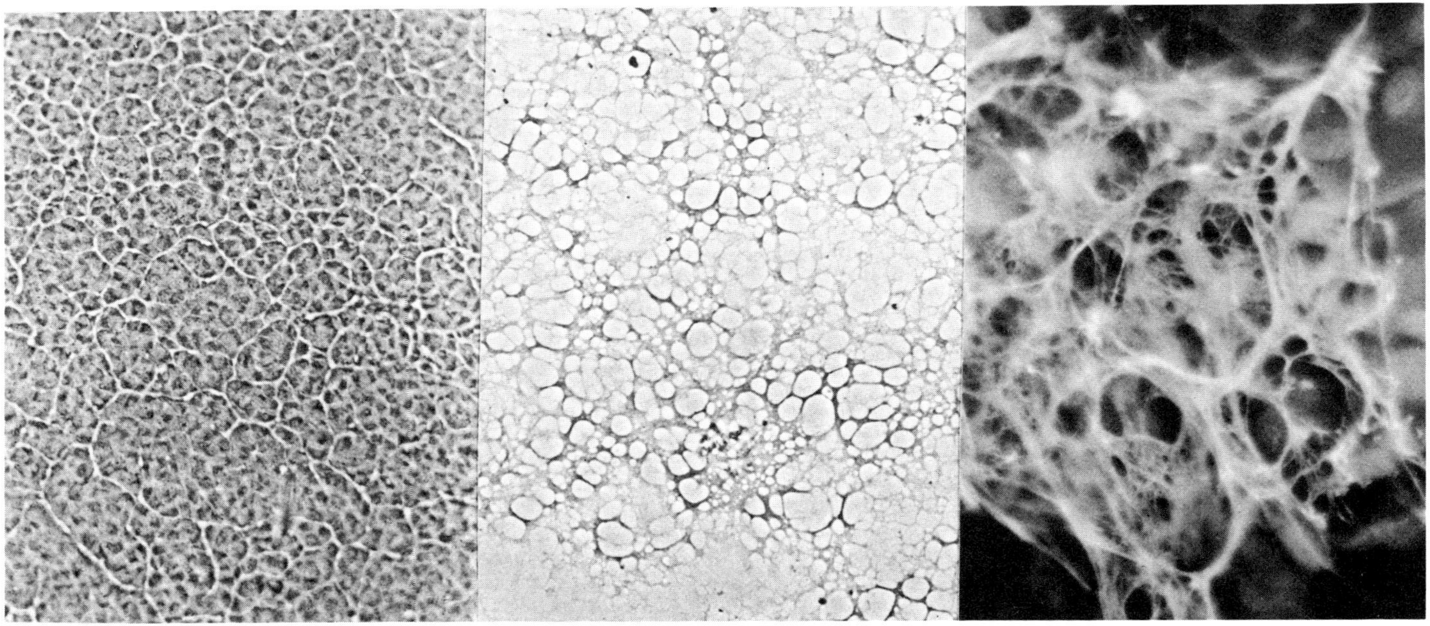

FIGURE 5. Production of LETS protein by cultured corneal endothelial cells. *Left:* Phase-contrast micrograph of a culture of corneal endothelial cells maintained at confluence for 14 days. × 150. *Center:* Amorphous basement membrane-like material visible after treatment of confluent monolayer with 0.5% Triton X-100. × 150. *Right:* Production of LETS protein by cultured bovine corneal endothelial cells. The distribution of LETS protein can be seen as a fine fibrillar network attached to the substratum and as a thicker accumulation that follows the polygonal outline of the inercellular spaces.

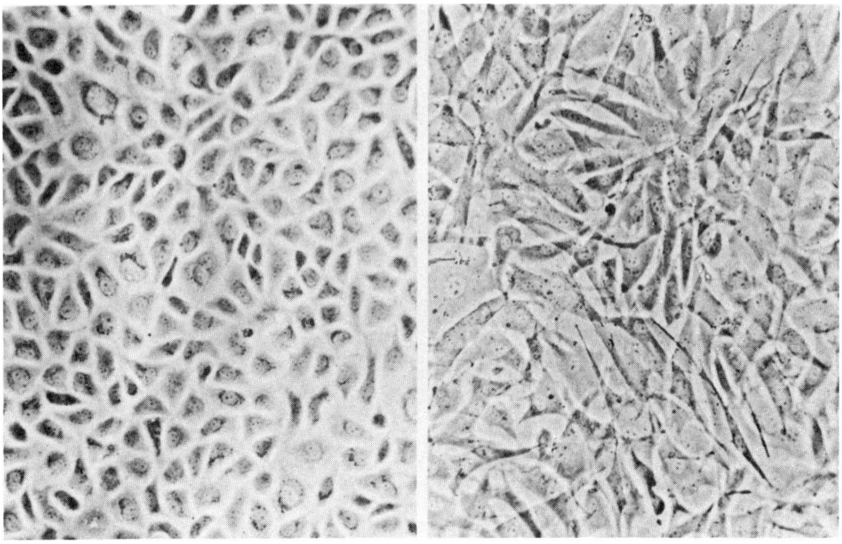

FIGURE 6. Morphologic features of normal and 2-chloroacetaldehyde-treated bovine endothelial cells. *Left:* Parental clonal cell line of normal bovine aortic endothelial cells. *Right:* Cell line isolated after treatment with 2-chloroacetaldehyde. The variant cells are elongated and overgrow the monolayer. × 150.

cultures with 2-chloroacetaldehyde, a metabolite of the carcinogen vinyl chloride.[19] As shown in FIGURE 6, the morphologic characteristics of the two cell types are strikingly different. Whereas normal cells are maintained at confluence as a highly ordered monolayer, the variant cells are larger and more elongated and display extensive overgrowth. When subconfluent cultures were examined for the presence of LETS protein, the two cell types revealed a striking difference in the distribution of LETS on their surfaces. Both cell types produce LETS protein, but contact-inhibited normal endothelial cells deposit it only onto the plastic dish and not onto the surface that is exposed to the medium. It is important here to note that the functions of the upper and lower surfaces of the vascular endothelial cell *in vivo* are totally different. One surface is responsible for maintaining attachment of the cells to the basal laminal, whereas the other must present an inert, nonthrombogenic surface to the bloodstream to prevent adhesion of circulating platelets and other blood cells. Thus, one surface of the endothelial cell *must* be sticky, and the other *must not* be sticky. It is therefore extremely interesting that in culture, endothelial cells deposit LETS protein only on the side attached to the substratum and not on the side exposed to the fluid environment.[20]

The importance of cell-cell contacts for LETS protein deposition is evident in subconfluent cultures of normal endothelial cells, which exhibit no evidence of LETS on isolated cells but only on cells in contact with others (FIGURE 7, *left*). In contrast, variant cells, which are capable of overgrowth, can anchor the protein to their surfaces in a pattern that corresponds to the distribution of stress fibers visible by phase microscopy. Here, totally isolated cells express LETS protein on their surfaces (FIGURE 7, *right*). These results demonstrate that a cell line selected for its ability to overgrow in culture has an altered distribution of LETS protein on its cell surfaces. It is possible that these differences are indicative of other cell surface alterations that are responsible for the altered morphologic features and growth pattern. Alternatively, the appearance of LETS protein on the upper surfaces of the overgrowing cells may provide a kind of guide or track that allows these cells to migrate over or under each other.

Our results indicate that endothelial cells may distribute their LETS protein in three distinct ways *in vivo* and *in vitro*. First, LETS protein can be directly associated with intercellular processes that connect growing normal endothelial cells. Second, normal vascular endothelial cells in culture deposit an enormous fibrillar network of LETS-containing material onto the floor of the culture dish, indicating that these cells will *in vivo* deposit a similar material onto or into the basement membrane and that this deposit could aid the cells in maintaining their adherence to the basal lamina. Finally, it has been observed that cold-insoluble globulin (CIg), a β-globulin present in plasma and serum,[21,22] is similar or identical to LETS protein.[23] Although CIg is known to precipitate with fibrinogen at 4°C, its actual role, if any, in hemostatis is still unknown. However, the presence of CIg in plasma at concentrations in excess of 300 μg/ml suggests that the protein may indeed play an important role in hemostasis. The question then arises as to which cell type in the body supplies CIg to the plasma; for this role, we suggest the vascular endothelium, because it is directly exposed to the bloodstream and can produce large quantities of LETS protein. It might therefore be concluded that vascular endothelial cells, by virtue of their shape (flattened), organization (monolayer), and localization (at the inner layer of blood vessels), will depend for their function and survival on the production of LETS protein.

It is important to note that the production of LETS protein by a given cell type is a regulatable phenomenon. It is now well known that viral transformation of certain types of LETS-producing cells results in a decrease in the expression of

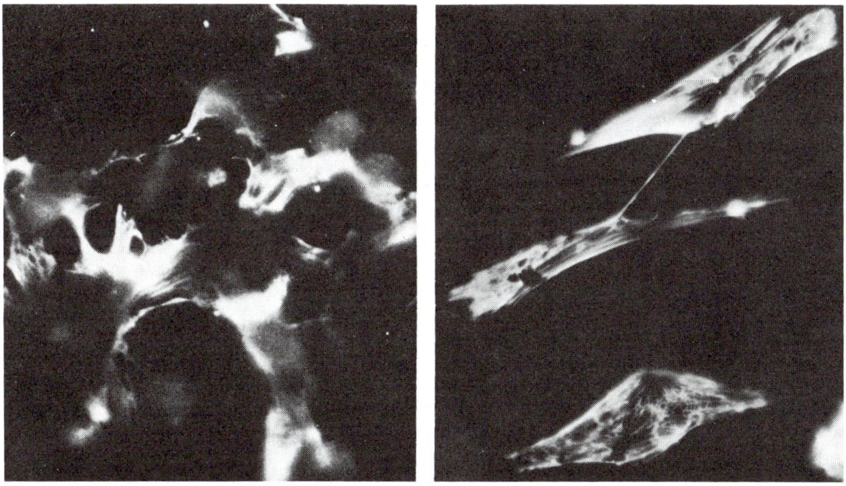

FIGURE 7. Production of LETS protein by normal and 2-chloroacetaldehyde-treated bovine endothelial cells. *Left:* Normal cells; LETS production is restricted to the intercellular spaces. × 630. *Right:* 2-Chloroacetaldehyde-treated cells; LETS is found on the surface of isolated cells. × 630.

LETS protein on the surface of these cells,[1-3] and this decrease has been found to be a consequence of reduced transcription of the specific mRNA for the protein.[24] As discussed above, the density of a cell population will influence LETS protein production, because cells in contact with other cells frequently synthesize more LETS than do isolated cells. Finally, certain hormones can influence LETS protein production, since it has been shown that cultures of Balb/c 3T3 cells produce more LETS when incubated with the epidermal growth factor (EGF)[25,26] than they do in its absence.[27] This result has been found to be true also for fibroblast growth factor (FGF),[28-30] and this finding[31] may be relevant to our earlier observation that FGF is required for the proliferation and survival of cultured normal bovine vascular endothelial cells.[13,14] When these cells are maintained in the presence of FGF, they proliferate rapidly if seeded at low cell density (32 cells/cm^2) and maintain their characteristic morphology and contact-inhibited organization after as many as 400 population doublings. On the other hand, when these cells are passaged in the absence of FGF, they fail to proliferate in sparse culture and, when plated at higher cell densities, will proliferate but become disorganized after only a few passages and fail to form a regular monolayer. It is possible, therefore, that one of the important functions of FGF is to ensure the production of LETS protein in adequate amounts and proper distribution. As discussed above, this ability to synthesize and properly distribute LETS could be crucial for maintaining the integrity of the vascular endothelium *in vivo.* This might also explain the finding that, *in vivo,* the vascular endothelium requires the presence of platelets for normal maintenance and survival.[32,33] Several investigators have found that factors isolated from platelets support the proliferation and survival of other cell types,[34-37] and FGF has been found to be similar to some of these platelet factors.[38] Thus, *in vivo,* FGF or similar factors from platelets may function to maintain the organization of the endothelial monolayer by stimulating the normal production of LETS protein by these cells.

LETS Protein and Preimplantation Mouse Embryo

If, as has been proposed, LETS were to play a major role in mediating cell-cell interactions *in vivo,* one likely place to exercise this function would be in the developing embryo. Although the adhesiveness of embryonic cells in the later stages of development have been investigated extensively,[39,40] relatively little is known about the nature of the cellular adhesions in the earliest (preimplantation) stages of mammalian development. Since LETS protein is probably not involved in the *specific* recognition of different cell types, it is conceivable that it would play a role in the cell-cell interactions of relatively undifferentiated cell types. We have therefore investigated the expression of LETS protein in developing embryos in two ways. First, we have used indirect immunofluorescence to test for the presence of LETS protein at each stage in the development of the preimplantation mouse embryo, from the two-cell stage through the formation of the blastocyst. In addition, we have employed previously described continuous cell lines of cultured teratocarcinoma stem cells, which are closely similar or possibly identical to early embryonic cells,[41] to study the expression of LETS protein *in vitro.* The advantage of using such cultures is that they can be more easily manipulated than can the normal embryo.

The results of the first series of experiments are summarized in Table 1. They reveal that no surface LETS protein is produced by the two-, four-, or eight-cell embryo or on the morula or the outer surface (trophectoderm) of the blastocyst.

When the trophectoderm is removed, however, by means of the immunosurgical technique of Solter and Knowles,[42] the isolated inner cell mass (ICM) of the late blastocyst (Day-4 embryo) exhibited large surface deposits of LETS protein (FIGURE 8). This result indicates that the expression of LETS protein is under temporal control during early development and that the appearance of this protein may serve as a cell surface marker for a particular stage of the developing embryo.

Of particular interest is our observation that the expression of LETS protein by ICM cells occurs only after a particular differentiative event has occurred in the ICM. Thus, little if any LETS protein is detected in ICMs isolated from the early (Day 3) blastocyst (FIGURE 8). However, 24 hr later, when a single layer of primitive endoderm has formed on the blastocelic surface of the ICM, a large matrix of LETS protein can be seen in the intercellular spaces and also in the form of large fibrils that crisscross the ICM surface. At present, it is not possible to determine whether the LETS protein observed is synthesized by the newly formed endodermal cells or whether the central core of embryonic ectodermal cells itself becomes capable of producing LETS only after the development of an outer layer

TABLE 1

EXPRESSION OF LETS PROTEIN ON PREIMPLANTATION MOUSE EMBRYOS

Developmental Stage	LETS Expression
Two cell	—
Four cell	—
Eight cell	—
Early morula	—
Late morula	—
Early Blastocyst (day 3)	
Trophectoderm	—
Inner cell mass	±
Late Blastocyst (day 4)	
Trophectoderm	—
Inner cell mass	+

of endoderm. It seems unlikely that the ability to synthesize LETS protein is limited solely to the primitive endoderm for two reasons. First, in the intact embryo, the endodermal cell layer does not completely surround the embryonic endoderm. If only these cells were producing LETS, the protein most likely would be found on only one side of the ICM. In our experience, LETS protein is frequently observed over the entire surface of the ICM. This observation suggests that the embryonic ectoderm is synthesizing the protein. Additional evidence to support this conclusion comes from the studies with teratocarcinoma stem cells described below.

To study the production of LETS protein by embryonic cells, there would be a distinct advantage in having a cell culture system that closely resembles some or all of the stages of mammalian development. The use of mouse teratocarcinoma stem cells now seems to satisfy these requirements (for reviews, see References 41 and 43). The fundamental similarity between the tumor cells and the cells of the early embryo is that both are multi- or totipotent and are thus capable of forming many or all of the tissues of the adult mouse.[44-46] Unlike the pluripotent cells of the early embryo, however, the teratocarcinoma stem cells (embryonal carcinoma

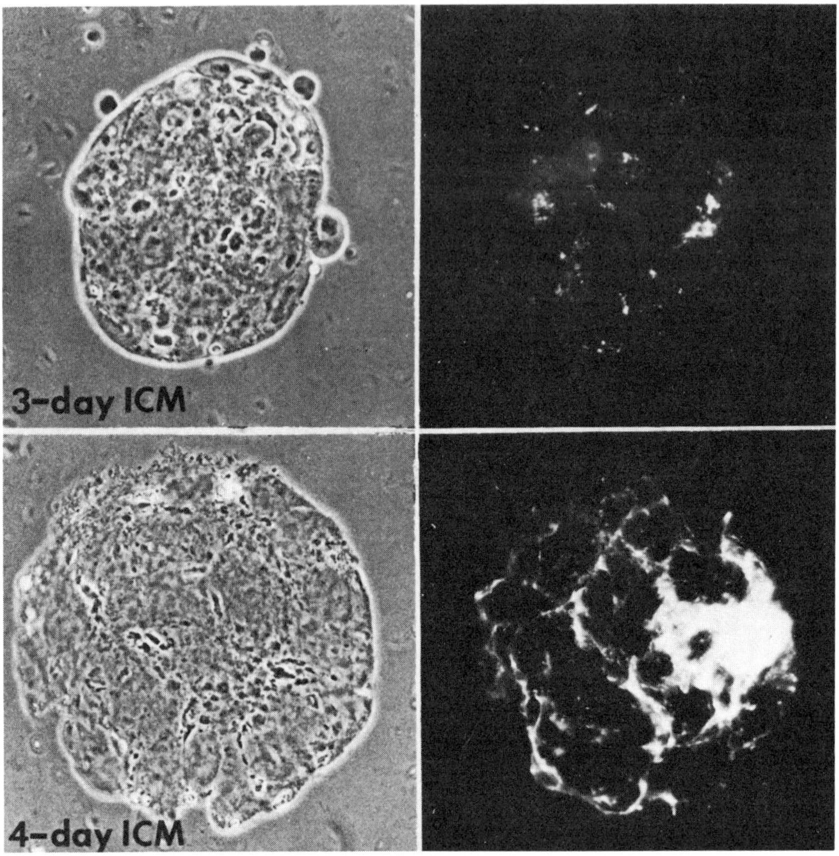

FIGURE 8. Distribution of LETS protein on isolated inner cell masses (ICMs). ICMs were isolated from Day-3 and Day-4 mouse blastocysts by an immunosurgical technique[42] and assayed for the presence of LETS by indirect immunofluorescence. Only the Day-4 ICM has significant quantities of LETS on its outer surface. × 820.

cells) can be cloned and maintained in the undifferentiated state *in vitro*. In addition, certain embryonal carcinoma cell lines are capable of differentiating *in vitro* in a manner that closely parallels the periimplantation development of the embryonic ICM.[47] Such cultures, therefore, afford certain practical advantages for experimental purposes over normal pluripotent embryonic cells.

FIGURE 9. Production of LETS protein by teratocarcinoma stem cells (embryonal carcinoma cells). Cultures of Nulli SCC 1, a nullipotent embryonal carcinoma cell line, were treated with 0.25% trypsin to detach the cells from the tissue culture dish and to remove LETS protein from the cell surfaces. Twelve hours after replating, no LETS protein is visible after immunofluorescent staining. After 24 hr, small amounts have been produced; by 48 hr, significant amounts of LETS protein can be seen in the intercellular spaces. × 630.

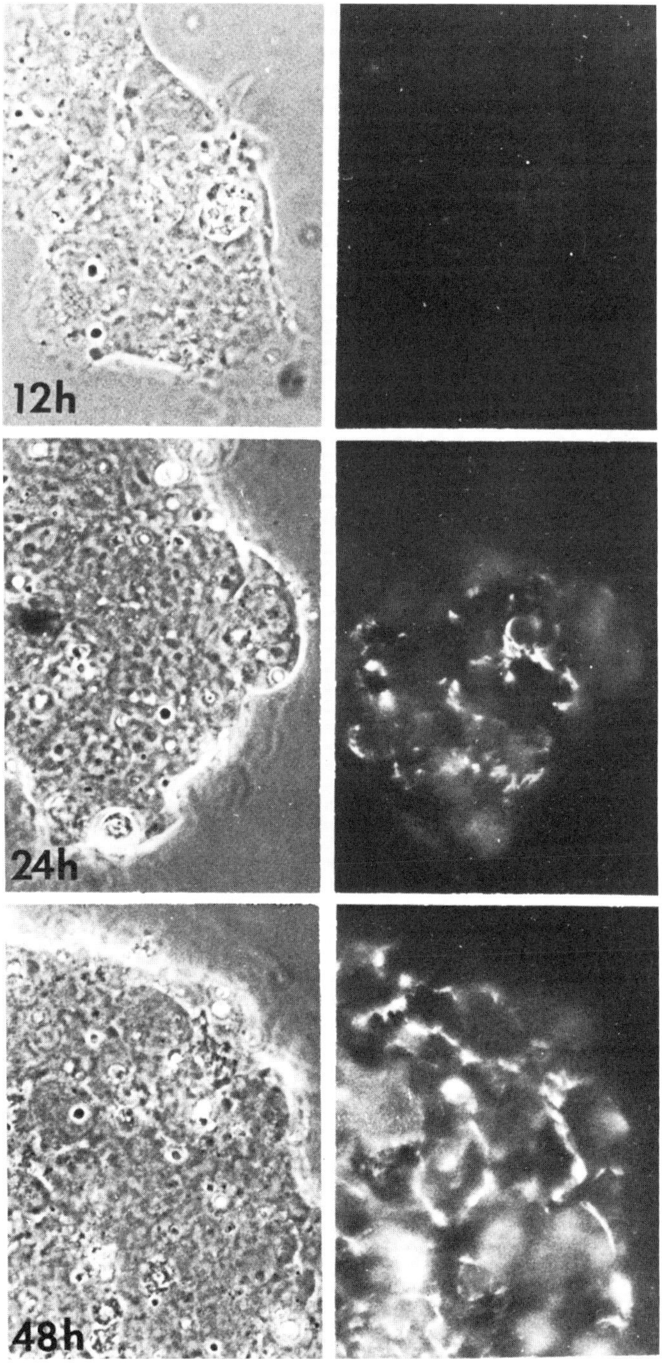
12h
24h
48h

Recent experiments that compared the distribution of proteins synthesized by embryonal carcinoma cells with those synthesized by mouse embryos at various stages of development have indicated that these cells are very similar to the embryonic ectoderm.[48] This pattern of synthesis common to the embryonal carcinoma cells and embryonic ectoderm has not been observed in any other cultured cell type. For this reason, we chose to examine cultured embryonal carcinoma cells to determine if, like the embryonic ectoderm, they, too, produce LETS protein.

Martin and Evans[49,50] have reported the isolation of two general classes of clonal mouse embryonal carcinoma cell line. One type is "nullipotent." These cells are morphologically and biochemically similar to pluripotent embryonal

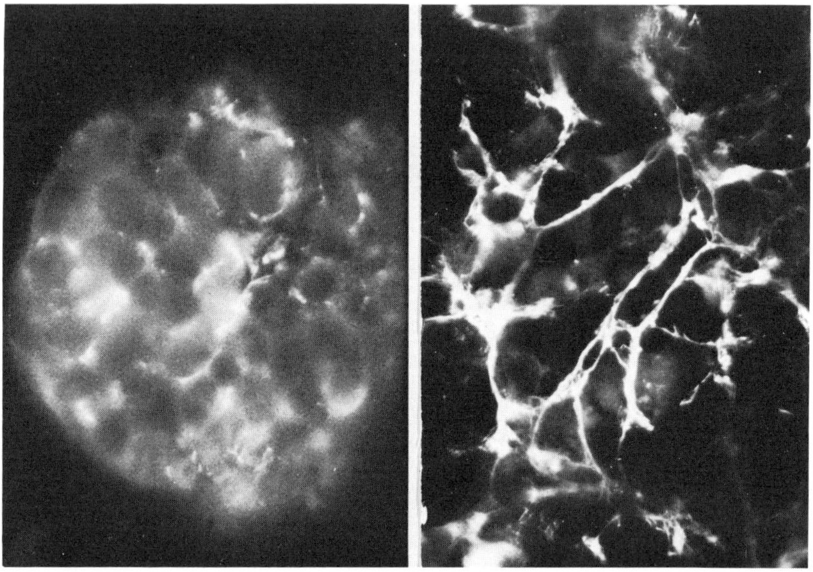

FIGURE 10. Comparison of the distribution of LETS protein on embryonal carcinoma cells and cultured endodermal cells. *Left:* PSA 1, a pluripotent embryonal carcinoma cell line, shown here before the differentiation of an outer layer of endoderm, deposits LETS protein in the intercellular spaces in the same manner as the nullipotent cell line shown in FIGURE 9. × 500. *Right:* PYS, a teratocarcinoma-derived endodermal cell line, produces and exports LETS protein in the form of long fibers that cross the surfaces of several cells. × 500.

carcinoma cells but are apparently incapable of differentiation *in vivo* or *in vitro.* The second class of embryonal carcinoma cell is pluripotent. These cells can be maintained in the undifferentiated state by frequent subculturing in the presence of feeder cells. By altering the culture conditions, the cells will differentiate *in vitro.* The first stage in their differentiation, like that of the ICM, is the appearance of endodermal cells on the outer layer of aggregates of embryonal carcinoma.

FIGURE 9 shows the appearance of LETS protein with time after disaggregation and replating of a nullipotential cell line (Nulli SCC 1). As with other cell types, trypsin treatment moves LETS protein from the surface of the embryonal carcinoma cells. Twelve hours after plating, there is no evidence of LETS protein, but by 24 hr, it begins to be apparent in the intercellular spaces; by 48 hr, there is a considerable deposition of the protein. The appearance of LETS protein in the intercellular spaces of colonies of embryonal carcinoma cells thus reinforces the similarity between these cells and the embryonic ectoderm and provides evidence that, as discussed above, the embryonic ectoderm may synthesize some or all of the LETS protein found on the Day-4 ICM.

Pluripotent embryonal carcinoma cells (PSA 1) that are maintained in the undifferentiated state by coculturing with feeder cells also express LETS protein in the spaces between the cells (FIGURE 10, *left*). When these cells are removed from the dish and maintained in suspension culture, they form "embryoid bodies," which consist of an inner core of embryonal carcinoma surrounded by an outer layer of endodermal cells. On these embryoid bodies, LETS protein can be observed on the central core of embryonal carcinoma cells and also in a basement membrane component that separates the central core from the outer endodermal cell layer. Little or no LETS protein is seen on the outer surfaces of the endodermal cells themselves.

To further investigate whether endodermal cells can produce LETS protein, we examined cultures of a teratocarcinoma-derived endodermal cell line (PYS)[51] kindly provided by Dr. John Lehman. These cells produce considerable quantities of LETS protein, but, as seen in FIGURE 10 (*right*), the distribution of the protein in these cultures is considerably different from that in cultures of embryonal carcinoma cells. Whereas with the embryonal carcinoma cells the majority of LETS protein is deposited in the spaces between cells (FIGURE 10, *right*), the endodermal cells deposit large fibers of LETS on the surface of the culture dish. This pattern is similar to that observed with cells, such as fibroblasts and endothelial cells, that secrete basement membrane-like material and implies that the endoderm may be the source of the LETS protein seen in Reichert's membrane, which separates the endoderm from the central core of embryonal carcinoma cells in the embryoid bodies. Thus, the conclusion of our studies with teratocarcinoma-derived cell cultures is that both embryonal carcinoma cells and endodermal cells produce LETS protein but deposit it in different locations, probably for different purposes. For the embryonal carcinoma cell, the intercellular deposits could play a role in maintaining adhesion and cell-cell positioning during critical phases of their embryo-like development. By analogy, the same may be true of the normal development of the embryo. For the endoderm, LETS protein is exported and employed as a component of a particular basement membrane.

ACKNOWLEDGMENT

We thank Dr. Lan Bo Chen for his kind gift of antibodies to human CIg that were used in the immunofluorescence studies described.

REFERENCES

1. HYNES, R. O. 1976. Cell surface proteins and malignant transformation. Biochim. Biophys. Acta **458:** 73–107.
2. VAHERI, A., E. RUOSLAHTI, E. LINDER, J. WARTIOVAARA, J. KESKI-OJA, P. KUUSELA & O. SAKSELA. 1976. Fibroblast surface antigen (SF): molecular properties, distribution *in vitro* and *in vivo,* and altered expression in transformed cells. J. Supramol. Struct. **4:** 63–70.

3. YAMADA, K. M. & I. PASTAN. 1976. Cell surface protein and neoplastic transformation. Trends Biochem. Sci. **1:** 222–224.
4. YAMADA, K. M., S. S. YAMADA & I. PASTAN. 1976. Cell surface protein partially restores morphology, adhesiveness, and contact inhibition of movement to transformed fibroblasts. Proc. Nat. Acad. Sci. USA **73:** 1217–1221.
5. PEARLSTEIN, E. 1976. Plasma membrane glycoprotein which mediates adhesion of fibroblasts to collagen. Nature (London) **262:** 497–500.
6. ALI, I. U., V. MAUTNER, R. LANZA & R. O. HYNES. 1977. Restoration of normal morphology, adhesion and cytoskeleton in transformed cells by addition of a transformation-sensitive surface protein. Cell **11:** 115–126.
7. CULP, L. A. 1976. Molecular composition and origin of substrate-attached material from normal and virus-transformed cells. J. Supramol. Struct. **5:** 239–255.
8. POUYSSEGUR, J. M. & I. PASTAN. 1976. Mutants of Balb/c 3T3 fibroblasts defective in adhesiveness to substratum: evidence for alteration in cell surface proteins. Proc. Nat. Acad. Sci. USA **73:** 544–548.
9. JULIANO, R. L. & E. GAGALANG. 1977. The adhesion of Chinese hamster cells: effects of temperature, metabolic inhibitors and proteolytic dissection of cell surface macromolecules. J. Cell. Physiol. **92:** 209–220.
10. YAMADA, K. M., S. S. YAMADA & I. PASTAN. 1975. The major cell surface glycoprotein of chick embryo fibroblasts is an agglutinin. Proc. Nat. Acad. Sci. USA **72:** 3158–3162.
11. ZETTER, B. R., L. B. CHEN & J. M. BUCHANAN. 1976. Effects of protease treatment on growth, morphology, adhesion, and cell surface proteins of secondary chick embryo fibroblasts. Cell **7:** 407–412.
12. CHEN, L. B., P. H. GALLIMORE & J. K. MCDOUGALL. 1976. Correlation between tumor induction and the large external transformation sensitive protein on the cell surface. Proc. Nat. Acad. Sci. USA **73:** 3570–3574.
13. GOSPODAROWICZ, D., J. MORAN, D. BRAUN & C. R. BIRDWELL. 1976. Clonal growth of bovine vascular endothelial cells in tissue culture: fibroblast growth factor as a survival agent. Proc. Nat. Acad. Sci. USA **73:** 4120–4124.
14. GOSPODAROWICZ, D., J. MORAN & D. BRAUN. 1977. Control of proliferation of bovine vascular endothelial cells. J. Cell. Physiol. **91:** 377–386.
15. HOGAN, M. J., J. A. ALVARADO & J. E. WEDDELL. 1971. Histology of the Human Eye. W. B. Saunders Co. Philadelphia, Pa.
16. GOSPODAROWICZ, D., A. L. MESCHER & C. R. BIRDWELL. 1977. Stimulation of corneal endothelial cell proliferation *in vitro* by fibroblast and epidermal growth factors. Exp. Eye Res. **25:** 75–89.
17. GOSPODAROWICZ, D., I. VLODAVSKY & P. F. FIELDING. 1978. The effects of the epidermal and fibroblast growth factors upon cell proliferation using vascular endothelial cells as a model. Excerpta Med. Int. Congr. Ser. In press.
18. HAUDENSCHILD, C., D. ZAHNISER, J. FOLKMAN & M. KLAGSBRUN. 1976. Human vascular endothelial cells in culture: lack of response to serum growth factors. Exp. Cell Res. **98:** 175–183.
19. ZETTER, B. R., L. K. JOHNSON, M. A. SHUMAN & D. GOSPODAROWICZ. 1978. The isolation of vascular endothelial cell lines with altered cell-surface and platelet binding properties. Cell. In press.
20. BIRDWELL, C. R., D. GOSPODAROWICZ & G. L. NICOLSON. 1978. Identification, localization and role of fibronectin in cultured bovine endothelial cells. Proc. Nat. Acad. Sci. USA. In press.
21. MOSESSON, M. W. & R. A. UMFLEET. 1970. The cold-insoluble globulin of human plasma: purification, primary characterization, and relationship to fibrinogen and other cold-insoluble fraction components. J. Biol. Chem. **245:** 5728–5736.
22. MOSESSON, M. W., A. B. CHEN & R. M. HUSEBY. 1975. The cold-insoluble globulin of human plasma: studies of its essential structural features. Biochim. Biophys. Acta **386:** 509–524.
23. RUOSLAHTI, E. & A. VAHERI. 1975. Interaction of soluble fibroblast surface antigen with fibrinogen and fibrin. J. Exp. Med. **141:** 497–501.
24. ADAMS, S. L., M. E. SOBEL, B. H. HOWARD, K. OLDEN, K. M. YAMADA, B. DE

CROMBRUGGHE & I. PASTAN. 1977. Levels of translatable mRNAs for cell surface protein, collagen precursors, and two membrane proteins are altered in Rous sarcoma virus-transformed chick embryo fibroblasts. Proc. Nat. Acad. Sci. USA **74:** 3399–3403.

25. COHEN, S. 1962. Isolation of a mouse submaxillary gland protein accelerating incisor eruption and eyelid opening in the new-born animal. J. Biol. Chem. **237:** 1555–1562.
26. CHEN, L. B., R. C. GUDOR, T. T. SUN, A. B. CHEN & M. W. MOSESSON. 1977. Control of a cell surface major glycoprotein by epidermal growth factor. Science **197:** 776–778.
27. COHEN, S. 1965. The stimulation of epidermal proliferation by a specific protein (EGF). Devel. Biol. **12:** 394–407.
28. GOSPODAROWICZ, D. 1975. Purification of a fibroblast growth factor from bovine pituitary. J. Biol. Chem. **250:** 2515–2520.
29. GOSPODAROWICZ, D., J. MORAN & H. BIALECKI. 1976. Mitogenic factors from the brain and pituitary. Excerpta Med. Int. Congr. Ser. **381:** 141–155.
30. GOSPODAROWICZ, D. 1975. Localization of a fibroblast growth factor and its effect alone and with hydrocortisone on 3T3 cell growth. Nature (London) **249:** 123–127.
31. CHEN, L. B. 1978. Personal communication.
32. GIMBRONE, M. A., JR., R. H. ASTER, R. S. COTRAN, J. CORKERY, J. H. JANDL & J. FOLKMAN. 1969. Preservation of vascular integrity in organs perfused *in vitro* with a platelet-rich medium. Nature (London) **222:** 33–36.
33. KITCHENS, C. S. & L. WEISS. 1975. Ultrastructural changes of endothelium associated with thrombocytopenia. Blood **46:** 567–574.
34. KOHLER, N. & A. LIPTON. 1974. Platelets as a source of fibroblast growth-promoting activity. Exp. Cell Res. **87:** 297–301.
35. ROSS, R., J. A. GLOMSET, B. KARIYA & L. HARKER. 1974. A platelet-dependent serum factor that stimulates the proliferation of arterial smooth muscle cells *in vitro.* Proc. Nat. Acad. Sci. USA **71:** 1207–1212.
36. RUTHERFORD, R. B. & R. ROSS. 1976. Platelet factors stimulate fibroblasts and smooth muscle cells quiescent in plasma serum to proliferate. J. Cell Biol. **69:** 196–203.
37. ANTONIADES, H. N. & C. D. SCHER. 1977. Radioimmunoassay of a human serum growth factor for Balb/c 3T3 cells: derivation from platelets. Proc. Nat. Acad. Sci. USA 74: 1973–1977.
38. GOSPODAROWICZ, D., G. GREENE & J. MORAN. 1975. Fibroblast growth factor can substitute for platelet factor to sustain the growth of Balb/c 3T3 cells in the presence of plasma. Biochem. Biophys. Res. Commun. **65:** 779–787.
39. MOSCONA, A. A. 1974. The Cell Surface in Development. John Wiley & Sons, Inc. New York, N.Y.
40. POSTE, G. & G. L. NICOLSON. 1976. The Cell Surface in Animal Embryogenesis and Development. North Holland Publishing Co. Amsterdam, The Netherlands.
41. MARTIN, G. R. 1975. Teratocarcinomas as a model system for the study of embryogenesis and neoplasia: review. Cell **5:** 229–243.
42. SOLTER, D. & B. KNOWLES. 1975. Immunosurgery of the mouse blastocyst. Proc. Nat. Acad. Sci. USA **72:** 5099–5102.
43. DAMJANOV, I. & D. SOLTER. 1974. Experimental teratoma. Curr. Topics Pathol. **59:** 69–130.
44. PAPAIOANNOU, V. E., M. W. MCBURNEY, R. L. GARDNER & M. J. EVANS. 1975. Fate of teratocarcinoma cells injected into early mouse embryos. Nature (London) **258:** 70–73.
45. MINTZ, B. & K. ILLMENSEE. 1975. Normal genetically mosaic mice produced from malignant teratocarcinoma cells. Proc. Nat. Acad. Sci. USA **72:** 3585–3589.
46. ILLMENSEE, K. & B. MINTZ. 1976. Totipotency and normal differentiation of single teratocarcinoma cells cloned by injection into blastocysts. Proc. Nat. Acad. Sci. USA **73:** 549–553.
47. MARTIN, G. R., L. M. WILEY & I. DAMJANOV. 1977. The development of cystic embryoid bodies *in vitro* from clonal teratocarcinoma stem cells. Devel. Biol. **61:** 230–244.

48. Martin, G. R., S. Smith & C. J. Epstein. 1978. Protein synthetic patterns in teratocarcinoma stem cells and mouse embryos at early stages of development. Devel. Biol. In press.
49. Martin, G. R. & M. J. Evans. 1975. The differentiation of clonal lines of teratocarcinoma cells: formation of embryoid bodies *in vitro*. Proc. Nat. Acad. Sci. USA **72:** 1441–1445.
50. Martin, G. R. & M. J. Evans. 1975. The formation of embryoid bodies *in vitro* by homogenous embryonal carcinoma cell cultures derived from isolated single cells. *In* Roche Symposium on Teratomas and Differentiation. M. Sherman & D. Solter, Eds.: 169–187. Academic Press, Inc., New York, N.Y.
51. Lehman, J. M., W. C. Speers, D. E. Swartzendruber & G. B. Pierce. 1974. Neoplastic differentiation: characteristics of cell lines derived from a murine teratocarcinoma. J. Cell. Physiol. **84:** 13–28.

A LARGE GLYCOPROTEIN LOST FROM THE SURFACES OF TRANSFORMED CELLS*

R. O. Hynes, I. U. Ali, A. T. Destree, V. Mautner, M. E. Perkins,†
D. R. Senger,‡ D. D. Wagner, and K. K. Smith§

Department of Biology
and Center for Cancer Research
Massachusetts Institute of Technology
Cambridge, Massachusetts 02139
and §The Rockefeller University
New York, New York 10021

About four years ago, we reported that several different types of transformed cell showed loss of or reduction in levels of a large external protein, as detected by lactoperoxidase-catalyzed iodination.[1] This protein was also observed to be extremely sensitive to proteolytic enzymes. Similar results have been reported by others.[2-5] Subsequently, it has become clear that this phenomenon is a fairly general observation, which has been described by many groups on a variety of cell types and with a variety of different detection procedures (see Reference 6 for a review of the literature). Since it became clear that the sensitivity to transformation was a general phenomenon, this protein has been referred to as LETS (large, external, transformation-sensitive) protein.

We also observed that the levels of LETS protein on normal cells varied, depending on growth rate and position in the cell cycle.[7,8] Growth-arrested cells exhibited more than did rapidly growing cells, and mitotic cells contained very little of this protein. This observation has also been confirmed by others in other systems (see Reference 6).

Cell fractionation experiments indicated that LETS protein could be separated from plasma membranes defined by their enzyme content[9] and that it is located in a nonmembranous fraction, suggesting that it is not an integral protein.[9a] It became clear that LETS protein was closely related to the antigen detected on fibroblastic cells by Ruoslahti and Vaheri.[10] This antigen is also transformation sensitive[11] and is found in fibrils at the cell surface.[12] Subsequent work has revealed the identity of these two proteins.[6,13,14] It has also been reported[15] that this protein is related to cold-insoluble globulin (CIg), a plasma protein.[16-18]

Another line of research also converged with that on LETS protein. Weston and Hendricks[19] had demonstrated that urea treatment of chick embryo fibroblasts leads to altered growth and culture morphology. Yamada and Weston showed that a major component of the urea extract is a large glycoprotein.[20] This protein, when purified and added to transformed cells, restored their morphology toward normal,[21] as did LETS protein purified from cells or culture medium.[22] It is clear that the two series of reports concern the same protein.

*Supported by Grants CA 14051 and CA 17007 from the National Institutes of Health and by Grant VC-258 from the American Cancer Society.

†Recipient of a Damon Runyon-Walter Winchell Postdoctoral Fellowship.

‡Recipient of a National Institutes of Health Postdoctoral Fellowship.

In this paper, we report some recent results on LETS protein and, for continuity, include relevant results from earlier work.

Materials and Methods

Cells and Culture

The cells employed for most of the experiments reported here were NIL8 hamster fibroblasts and their transformed derivatives, NIL8-HSV. The cells were cultured in Dulbecco's modified Eagle's medium (DME, Gibco, Grand Island, N.Y.) plus 5% fetal calf serum (Flow Laboratories, Rockville, Md.). For a few experiments, a line of rat cells (rat 1) transformed by a temperature-sensitive mutant of Rous sarcoma virus (tsRSV) were used. At 34°C, they are transformed, while at 39°C they are normal. They were cultured in DME plus 10% fetal calf serum. For immunofluorescence, cells were cultured on 12-mm^2 glass coverslips.

Radioactive Labeling and Acrylamide Gel Analysis

For labeling with [^{35}S]methionine, cells were cultured in DME with 10% of the usual level of methionine. Lactoperoxidase-catalyzed iodination was performed as previously reported.[1] Samples of cells, medium, and purified proteins were analyzed on slab gels by use of the sodium dodecyl sulfate (SDS) buffers described by Laemmli.[23,24]

Immunofluorescence

Cells on coverslips were fixed in formaldehyde and stained as described elsewhere.[25,26] For surface staining, cells were used live or fixed and were not permeabilized; for internal staining, cells were permeabilized with acetone. The antisera used were as follows: anti-LETS protein,[26] antiserum against the subunit of 100 Å filaments,[27] antiactin,[28] antitubulin,[29] and antimyosin.[30] The first three antisera were raised against SDS gel-purified proteins; antitubulin was affinity purified. The specificities of these antisera have been previously reported[26-30]; they all appear to be monospecific. Staining was indirect, with fluorescein-conjugated goat antirabbit IgG (Miles-Seravac, Elkhart, Ind. or Microbiological Associates, Bethesda, Md.), and slides were photographed, with Kodak Plus-X® film, on a Zeiss microscope equipped with epifluorescence illumination.

Immunoprecipitation

Samples were incubated with rabbit antiserum to LETS protein for 1 hr at 37°C, goat antirabbit IgG (Cappel Laboratories, Cochranville, Pa.) was added at a ratio to first antiserum previously determined to give maximal precipitation, and incubation was continued for 1 hr at 37°C and overnight at 4°C. Precipitates were washed and dissolved for SDS polyacrylamide gel electrophoresis.

Migration Assay

Migration of cells on coverslips coated with colloidal gold was assayed as described by Albrecht-Buehler.[31] Cells were plated at 500–1000 cells per 35-mm dish that contained one 18 × 18-mm coverslip.

Purification of LETS Protein

LETS protein was purified in several different ways: from cells by 1 M urea extraction according to the procedure of Yamada *et al.*,[21] from conditioned medium as described elsewhere,[22,26] and from conditioned medium by affinity chromatography on gelatin-Sepharose as described by Engvall and Ruoslahti.[32]

All three preparations contained different minor contaminants: the affinity-purified material was the purest (>95%). Cold-insoluble globulin (CIg) was purified from hamster and human plasma by the gelatin-Sepharose procedure.[32]

Platelet Binding to Collagen

Fresh human blood was drawn and mixed with acid-citrate-dextrose as the anticoagulant. Platelet-rich plasma was prepared by centrifugation at 150*g* for 10 min. Platelet-rich plasma was used directly or platelets were prepared from it by washing in buffers that contained chelating agents.[33] Washed platelets were resuspended in buffer, plasma, or plasma depleted of CIg (CIg^--plasma) prepared by affinity chromatography[32] from plasma from the same donor. Platelet-rich plasma or resuspended platelets were incubated for varying times up to 1 hr at 37°C on coverslips coated with collagen. After incubation, coverslips were rinsed, fixed, and processed for immunofluorescence as described above. The collagen layers were prepared from acid-soluble rat tail collagen by drying directly from acetic acid or by equilibration with ammonia vapor and drying. Before applying platelets, collagen-coated coverslips were rehydrated in Ca^{2+}-Mg^{2+}-free saline.

Peptide Fingerprinting of LETS Protein

Cells were labeled with methionine ([^{3}H]methionine, 200 μCi/ml, > 100 mCi/mmol or [^{35}S]methionine, 100 μCi/ml, 100–400 mCi/mmol; New England Nuclear, Boston, Mass.) for 24 hr and the labeled medium collected. The medium was separated on SDS polyacrylamide gels; the major labeled band in the case of normal cells is LETS protein. The gels were sliced, and slices were incubated with TPCK-trypsin (246 U/mg; Worthington Biochemicals, Freehold, N.J.) at 100 μg/ml for 16 hr at 37°C. Aliquots of the digests were counted to locate the peaks. The fractions that contained digested LETS protein were pooled, lyophilized, and oxidized with performic acid. After repeated lyophilization, the samples were applied to a column of Chromobeads® (Beckman Instruments, Wakefield, Mass.) and eluted with a pH gradient of pyridine acetate. Column fractions were dried and counted.

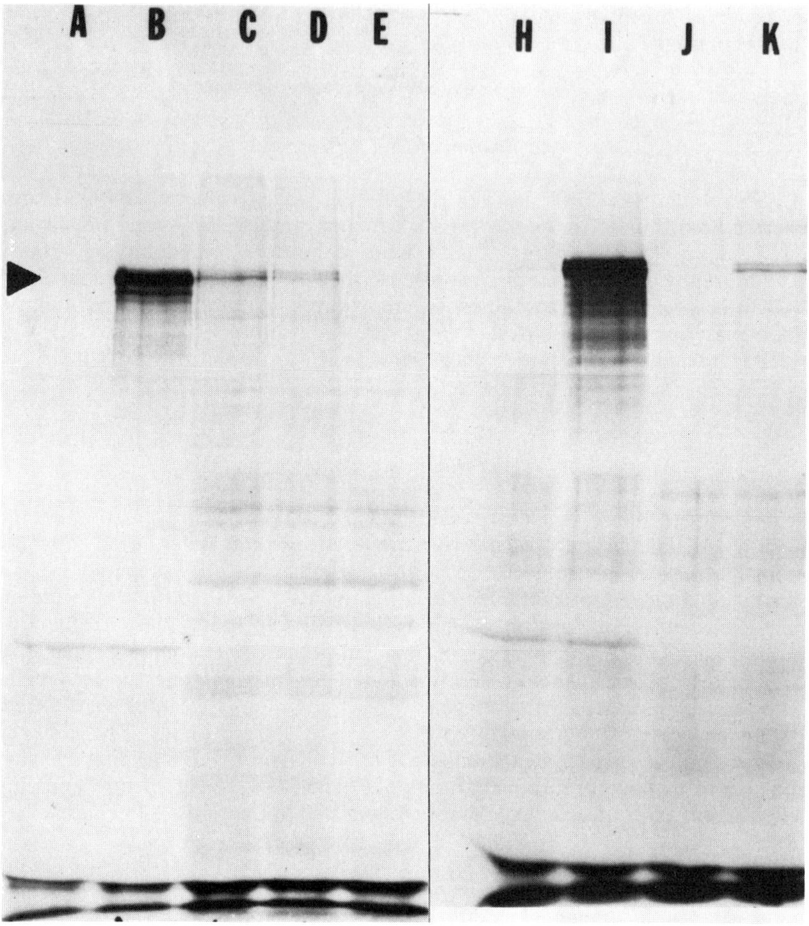

FIGURE 1. Biosynthesis of LETS protein in normal and transformed cells. Cells were pulse labeled with 200 μCi/ml of [^{35}S]methionine for 10 min. The cells were lysed in 2% deoxycholate, and aliquots of deoxycholate-soluble material were immune precipitated with anti-LETS protein or preimmune serum. Within each experiment (A–E and H–K), equal numbers of incorporated counts per minute were subjected to immune precipitation, and equal fractions were applied to SDS polyacrylamide slab gels. A & B, Confluent NIL8; A, preimmune; B, immune; C–E, NIL8 cells trypsinized and replated in 5% fetal calf serum; C, 22 hr after replating, immune serum; D, 41 hr after replating, immune serum; E, 41 hr after replating, preimmune serum; H & I, confluent NIL8; H, preimmune; I, immune; J & K, confluent NIL8-HSV; J, preimmune; K, immune. Arrows mark LETS protein. Note that growing NIL8 and NIL8-HSV both make less LETS protein than do confluent NIL8 cells.

RESULTS

Alterations in Surface Levels of LETS Protein

Since LETS protein is sensitive to a variety of proteolytic enzymes,[1-3,6,34-36] including plasmin,[35,36] it seemed possible that its absence from transformed cells might correlate with the increase in plasminogen activation that is commonly observed in transformed cells.[37,38] Consistent with this idea, an increase in turnover rate of LETS protein was observed on shift to the permissive temperature of cells transformed by tsRSV[36,39,40] and also when transformed cells were mixed with prelabeled normal cells.[36,41] However, addition of inhibitors of proteolytic enzymes[36,41] or depletion of plasminogen[36,42] was without effect on these phenomena, and the LETS protein released to the medium under all of these conditions appeared intact[42] and did not exhibit evidence of proteolytic cleavage. Furthermore, the expected correlation between fibrinolytic activity and loss of LETS protein could be broken.[43]

TABLE 1

BINDING OF ^{125}I-LABELED LETS PROTEIN TO NORMAL AND TRANSFORMED CELLS

Time After Addition of LETS Protein to Cells*	LETS Protein on Cells (cpm/dish)†	
	NIL8	NIL-HSV
0	230	210
2	2890	530
4	5170	1064
8	6460	1824
18	6130	2624
32		2736

*Confluent cultures of cells in 60-mm dishes were incubated with partially purified [^{125}I]LETS protein from conditional medium of iodinated cells. One hundred micrograms of protein (7.6×10^4 cpm) were added to 1.0 ml of medium.

†All bound radioactivity comigrated with LETS protein. A confluent dish of NIL8 cells contains 2–3×10^6 cells. A confluent dish of NIL-HSV contains 5–6×10^6 attached cells.

These experiments essentially ruled out plasminogen activation as an explanation for loss of LETS protein from transformed cells. Assays for the rate of biosynthesis of LETS protein by immunoprecipitation of pulse-labeled cells showed that the rate was diminished in NIL8-HSV-transformed cells as compared with normal NIL8 cells[44] (FIGURE 1). A more extensive analysis of chicken embryo fibroblasts by Olden and Yamada[45] showed a similar result on transformation by Rous sarcoma virus. They also confirmed the increase in turnover rate reported earlier. In neither system was the depression in biosynthetic rate sufficient to explain completely the drop in surface levels. It appears that some other factor also contributes to this phenomenon. This possibility is reflected in the increased turnover rates mentioned earlier.[39,40,42]

These results suggest an alteration in binding of LETS protein to transformed cells. To analyze the binding of LETS protein to cells, we have isolated radioactive LETS protein and added it to normal and transformed cells; transformed cells bind LETS protein significantly less efficiently (TABLE 1).

Levels of LETS protein at the surfaces of normal cells also vary, depending on growth rate and other culture conditions.[8,26] We have been unable to detect differences in turnover rates under varying conditions, for example, serum stimulation of resting cultures, although this treatment does lead to a reduction in surface LETS protein.[8] Assays for the rate of LETS biosynthesis do, however, indicate that the levels vary in normal cells (FIGURE 1). Dense NIL8 cultures that are arrested in G_1 (G_0, A state) show a high rate of synthesis. When arrested cultures are stimulated to grow by subculturing, the rate of synthesis of LETS protein drops fairly quickly (FIGURE 1). Since the normal level of turnover in NIL8 cells is quite high,[36,42] lowered biosynthetic rate leads eventually to reduced surface levels. These reduced levels do not occur immediately after growth stimulation in this system[8] or in others.[14,34]

Thus, there appear to be differences between the situation in transformed cells (lower rate of synthesis, higher rate of turnover) and growing normal cells (lower rate of synthesis, no evidence for increased turnover).

Readdition of LETS Protein to Transformed Cells

As reported elsewhere,[21,22] addition of purified LETS protein to cultures of transformed cells causes the cells to attach more firmly to the substratum, spread more extensively, elongate, and align with each other. The increased attachment can be used as a sensitive, quantitative bioassay.[22] LETS protein, isolated in several ways from established NIL8 hamster cells, secondary chick embryo fibroblasts, or the culture medium of either cell type, appears equally active on a mass basis by this assay. LETS protein from all of these sources also produces all of the other effects mentioned. Although LETS protein binds less well to transformed cells (TABLE 1), it does bind and can be detected by iodination or by immunofluorescence.[22] The latter technique reveals that the added LETS protein adopts a fibrillar array similar to that seen on normal cells.

Since CIg is similar to LETS protein, we also tested its activity in the readdition experiments. Both hamster and human CIg, even from freshly drawn plasma, were less active per milligram that was LETS protein isolated in the same way. TABLE 2 shows the effects on cellular attachment. Cold-insoluble globulin was similarly less effective in causing cellular flattening and elongation (data not shown).

We also reported that LETS protein causes reappearance of ordered actin microfilament bundles in treated transformed cells,[22] which typically lack them.[22,26,46–49] FIGURES 2a and 2b depict the results for four different cytoskeletal proteins. It can be seen that, along with the flattening of the cells on treatment with LETS protein (middle row), microfilament bundles detected by staining for actin or myosin reappear in patterns reminiscent of those seen in normal cells (bottom row). The arrangement of microtubules and 100-Å filaments is also observed more clearly in cells treated with LETS protein. However, the transformed cells studied do not lack the latter two filamentous structures, and the apparent difference may merely reflect the shape change. The same may be true for the microfilament bundles (see DISCUSSION).

Several aspects of cellular morphology are, therefore, altered toward normal by treatment with LETS protein. In contrast, no effect was observed on transport of sugar[22] or other nutrients[21] or on growth in agar.[49a] It appears that the predominant phenotypic changes brought about by this protein can probably be viewed as secondary effects of the increased adhesion (see DISCUSSION).

Effects of LETS Protein on Migration

Because LETS protein appeared to affect adhesion, and since adhesion and cellular motility are related on both a priori and experimental grounds,[50-52] we investigated the effects of LETS protein on migration of normal and transformed cells by use of the phagokinetic assay of Albrecht-Buehler.[31,52a] This assay exploits the fact that cells phagocytose colloidal gold particles during spreading and migration and therefore leave a track of cleared substratum as a record of their movement. We have observed that addition of LETS protein to several types of normal or transformed cell causes marked increases in migration as measured in this assay (FIGURE 3). The added LETS protein binds to the cells in a fibrillar network, and the cells leave trails of LETS protein behind them (FIGURE 4). The cells, in fact, bear much more LETS protein than they would normally at the low cell densities employed in this assay (see below). Under these conditions, the trans-

TABLE 2

EFFECTS OF LETS PROTEIN AND COLD-INSOLUBLE GLOBULIN ON ATTACHMENT OF NIL8-HSV CELLS

Source	Method of Preparation	Dose for 50% Reduction in Number of Floating Cells* (μg/ml)
LETS Protein		
NIL8 conditioned medium†	gelatin-Sepharose‡	10
NIL8 conditioned medium†	gelatin-Sepharose‡ dialysis vs CAPS buffer	8
Secondary chick embryo cells	extraction with urea§ dialysis vs CAPS buffer	8
Secondary chick embryo cell conditioned medium	gelatin-Sepharose‡	12
Secondary chick embryo cell conditioned medium	gelatin-Sepharose‡ dialysis vs CAPS buffer	13
Cold-Insoluble Globulin		
Hamster plasma (frozen, commercial)	gelatin-Sepharose‡	22
Hamster plasma (freshly drawn)¶	gelatin-Sepharose ‡ dialysis vs CAPS buffer	22
Human plasma (freshly drawn)¶	gelatin-Sepharose‡	31

*The attachment assay relies on the fact that NIL8-HSV cultures contain many non-attached cells that attach with increasing quantities of added LETS protein.[22] LETS protein was added either in 4 M urea or in pH 11 CAPS buffer.[21] Numbers of floating cells were compared with controls that received urea or buffer alone. All preparations were compared in a single experiment.

†Confluent NIL8 cells were switched to 5% fetal calf serum that had been depleted of CIg by ammonium sulfate precipitation.[26] After 3–4 days, the medium was harvested.

‡Medium or plasma was incubated with gelatin-Sepharose, which binds LETS protein and CIg.[32] After washing, the bound material was eluted with 4 M urea in pH 11 CAPS buffer.

§LETS protein was extracted from secondary chick embryo cell cultures 5 days after subculturing by use of 1 M urea.[21]

¶Plasma was drawn and acid-citrate dextrose was added as the anticoagulant. After removal of the cells, the plasma was immediately placed onto gelatin-Sepharose.

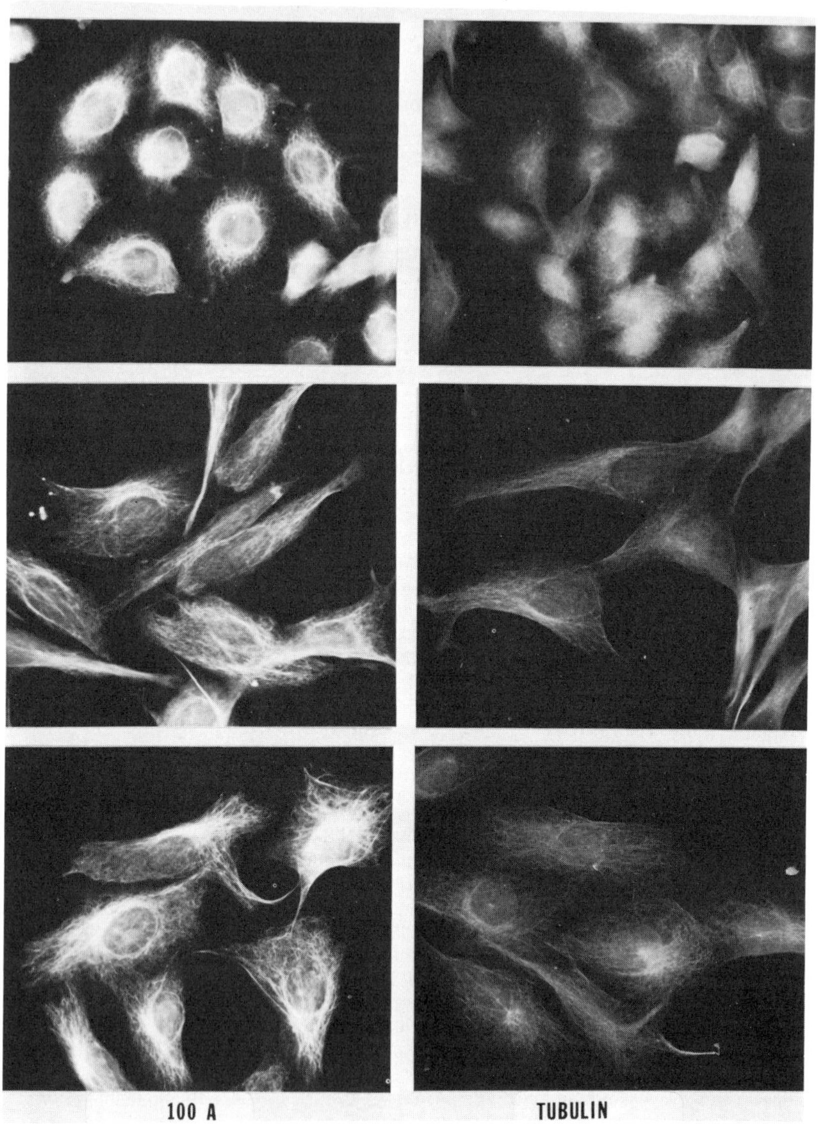

FIGURE 2a. Immunofluorescent staining of cytoskeletal fibers in normal cells (*bottom*), transformed cells (*top*), and transformed cells treated for 24 hr with 50 μg/ml of LETS protein (*middle*). Cells were fixed, permeabilized, and stained for 100 Å filaments and tubulin (FIGURE 2a) and for myosin and actin (FIGURE 2b). Note the readily visible cytoplasmic fibers and bundles of microfilaments in normal and LETS-treated cells but not in transformed cells.

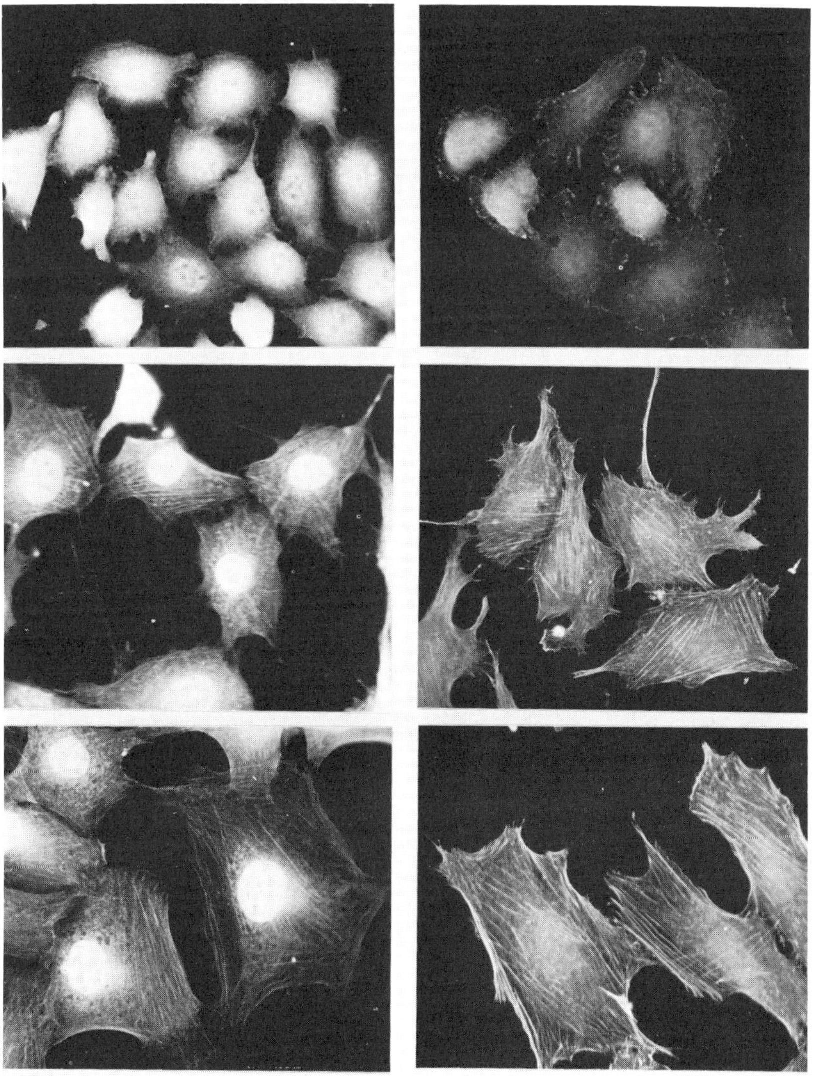

FIGURE 2b. See the legend to FIGURE 2a.

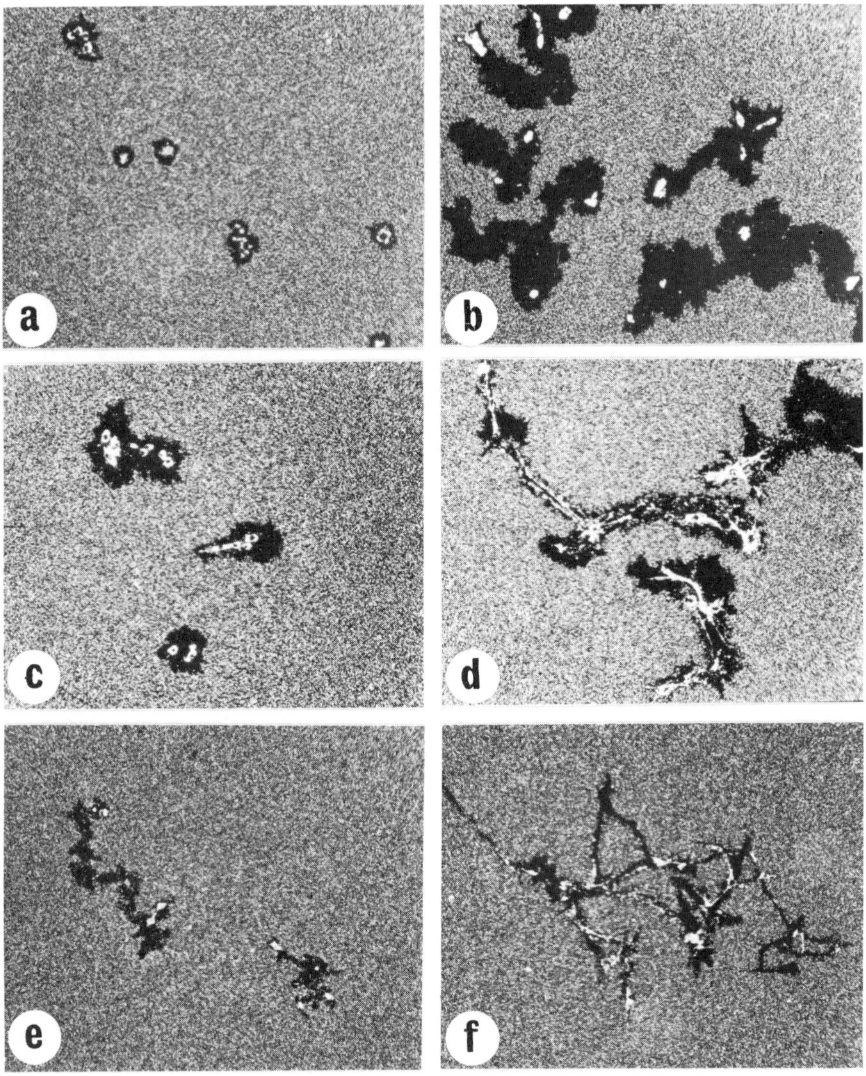

FIGURE 3. Migration of cells: effect of added LETS protein. Cells were seeded on gold-coated coverslips[31] (see MATERIALS AND METHODS) with or without LETS protein from chicken embryo cells. Coverslips were fixed 24 hr later. *Left*, controls, no LETS protein; *right*, plus 50 μg/ml of LETS protein. A & B, NIL8-HSV cells; C & D, NIL8 cells; E & F, tsRSV rat-1 cells at 39°C (nonpermissive temperature). Samples were photographed with dark-field illumination; the dark areas have been cleared of gold particles; the white spots and streaks within the tracks are, respectively, cells and deposits of gold particles plus LETS protein. In each case, note that addition of LETS protein markedly increases migration.

formed cells also bind less LETS protein than do normal ones, as is the case in dense cultures (TABLE 1). This was even true for a temperature-sensitive transformed cell line grown at the permissive versus nonpermissive temperature, indicating that reduced ability to bind LETS protein is a close correlate of transformation.

In cells that already displayed extensive migration (3T3 and Py3T3, for which the system was optimised[31]), the effect of LETS protein was small; with cells that migrate only little in the absence of added LETS protein (e.g., NIL8), the effect saturated at high doses of LETS protein (> 50 μg/ml). We have not observed significant inhibition of migration.

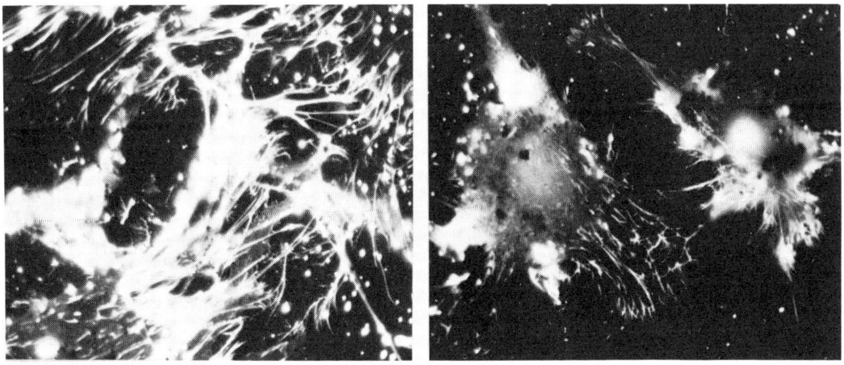

FIGURE 4. Immunofluorescent staining of LETS-treated cells after 2 days of migration. Procedures were the same as in FIGURE 3; coverslips were stained with anti-LETS protein. *Left*, NIL8 cells, several cells with strands of LETS protein around them; *right*, NIL8-HSV cells, two cells with surrounding LETS protein fibrils. Control, untreated cells contained very much less LETS protein; compare FIGURE 5 (b & c) for NIL8 cells. Also note that the transformed NIL8-HSV cells bind less LETS protein than do normal NIL8 cells.

Locations of LETS Protein in Cultured Fibroblasts

Since several results suggested a role for LETS protein in adhesion, it was of interest to examine the organization of LETS protein around cells. By means of immunofluorescence, we have analyzed the distribution of LETS protein under various conditions.[13,22,26,44,53] In agreement with others,[12,28,54-56] we find that LETS protein is arrayed in fibrillar structures. However, the nature of the patterns observed depends on the exact conditions employed. Consistent with a role in adhesion, LETS protein is commonly found in areas of cell-cell and cell-substratum contact.[13,26] Isolated NIL8 cells bear LETS protein almost exclusively beneath them (FIGURE 5). In spreading cells, LETS protein is found in short radial fibrils under the edge of the spreading lamellae (FIGURE 5,a). In migrating cells, it is found in short fibrils under the edges of and beneath lamellipodia (FIGURE 5, b & c). These fibrils are predominantly parallel to the major axes of the lamellipodia. In the very well-spread cells, which are characteristic of serum-starved cultures, a very extensive network of LETS fibrils occurs under the cells (FIGURE 5, d & e). The very well-spread and firmly attached cells are frequently so

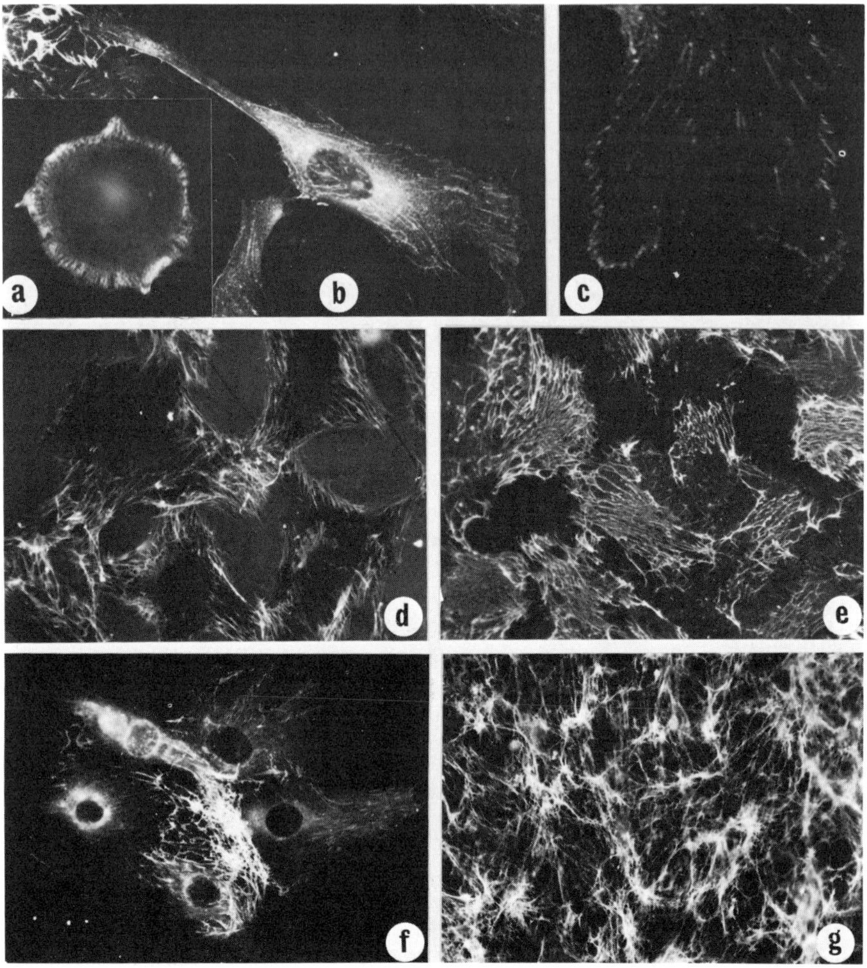

FIGURE 5. Immunofluorescent staining for LETS protein of NIL8 cells. A–E, Focused at level of substratum; F & G, focused on top of cells. A, Cell 3 hr after seeding, LETS protein concentrated in radial stitches under edge of spreading lamella; B & C, migrating NIL8 cells 24 hr after seeding, LETS protein arranged in short fibrils under lamellipodia, especially around edges but also back from leading edge, fibrils parallel to major axis; D & E, NIL8 cells cultured in 0.3% serum, cells very well spread and do not migrate; D, cells fixed and stained, cells located at positions of the dark shadows, LETS fibrils under cells but only partially stained, LETS fibrils also located between neighboring cells; E, cells removed with 2 mM EGTA and coverslips stained, no cells remaining but their former positions marked by areas of LETS fibrils left behind when cells detach; F, group of cells growing in 5% serum, LETS fibrils located beneath cells at periphery of group and between and on top of cells in areas of contact; G, confluent sheet of cells in 5% serum, LETS fibril network extending across entire culture.

strongly adherent that antibody cannot penetrate beneath them, and the "footprint" of LETS protein beneath them can be visualized only by permeabilizing or removing the cells[13,26] (FIGURE 5, d & e).

When cells make contact, LETS protein is also found between cells[26,54,55] (FIGURE 5, f); in confluent cultures, an elaborate network extends across the cell layer, and the contribution of one cell cannot be distinguished from that of its neighbor (FIGURE 5, g). In many situations, fibrils of LETS protein extend well away from cells[12,26] (FIGURES 4 & 5). As reported by others,[28,54] staining with antibody to human CIg yields identical pictures.

When live cells were stained, exactly similar results are obtained, and incubation at 37°C with a double layer of antibodies did not produce a major change in pattern.[13,26] These experiments would not detect small movements, such as those reported previously.[57] The most definitive analysis is that of Schlessinger *et al.*,[56] who showed by analysis of recovery from photobleaching that LETS protein is essentially immobile. These results taken together are most consistent with the interpretation that LETS protein is present in a fibrillar extracellular matrix and that little, if any, of it is freely mobile in the plane of the plasma membrane. This hypothesis is consistent with earlier cell fractionation studies[9] (see also Graham *et al.*[9a]). The solubility properties of LETS protein are also inconsistent with the belief that it is a typical membrane protein. It is insoluble in nonionic detergents and can be solubilized only by chaotropic agents.[13]

Involvement of Cold-Insoluble Globulin in Attachment of Platelets to Collagen

Given the role of LETS protein in cell-substratum adhesion of cultured fibroblasts, it would be of interest to examine whether similar proteins may function in the adhesion of other cell types to various substrata. A well-known adhesion reaction is that of platelets to collagen. When human platelets in citrated platelet-rich plasma were allowed to attach to collagen, rinsed, and stained with antisera to hamster LETS protein, it was observed that the attached platelets, whether single or in clumps, are surrounded by fibrils that stained for LETS protein (FIGURE 6). The collagen film stained a diffuse green, presumably reflecting LETS protein bound to collagen,[32] but fibrils were found only around the platelets. Neither the diffuse stain nor the fibrils were observed with preimmune sera or when untreated collagen films were stained. These controls confirm the specificity of staining observed and indicate that the antisera do not recognize collagen.[57a]

The LETS protein antigenicity present in these fibrils could originate from the platelets or from the CIg in the plasma. The latter appears to be the case for the following reasons. First, we were unable to detect LETS protein by iodination of washed platelets,[57a] in agreement with numerous reports in the literature (e.g., see Reference 33), none of which has described a protein of this size on platelet surfaces. Second, when washed platelets were allowed to attach to collagen in the presence of buffer or plasma depleted of CIg by affinity chromatography on gelatin-Sepharose,[32] the attached platelets were not surrounded by LETS-staining fibrils. When washed platelets were resuspended in complete plasma, attached platelets were surrounded by LETS-containing fibrils (FIGURE 6). Therefore, the platelets themselves appear not to have LETS protein but, rather, to recruit CIg from the plasma, concomitant with their adhesion to collagen. The bound CIg is found in the region of platelet-collagen contact and could be involved in this

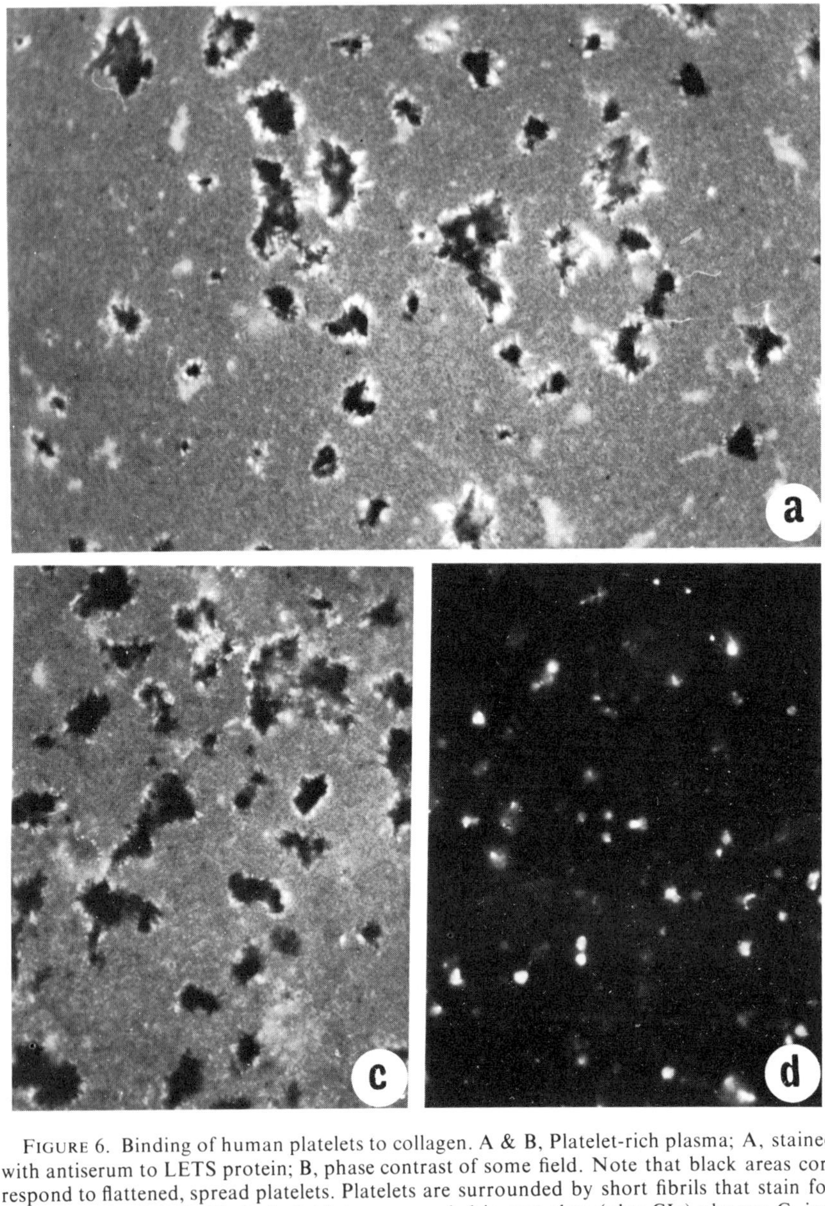

FIGURE 6. Binding of human platelets to collagen. A & B, Platelet-rich plasma; A, stained with antiserum to LETS protein; B, phase contrast of some field. Note that black areas correspond to flattened, spread platelets. Platelets are surrounded by short fibrils that stain for LETS protein. C & E, Washed platelets resuspended in complete (plus CIg) plasma; C, immunofluorescence, results very similar to A; E, phase contrast of different sample, platelets spread as in B. D & F, Washed platelets resuspended in CIg-deficient plasma; D, immunofluorescence, note absence of diffuse staining of background and of LETS-positive fibrils, preimmune staining of CIg-added samples yields identical pictures; that is, fluorescence on platelets is nonspecific; F, phase contrast shows that platelets do not spread in the absence of CIg.

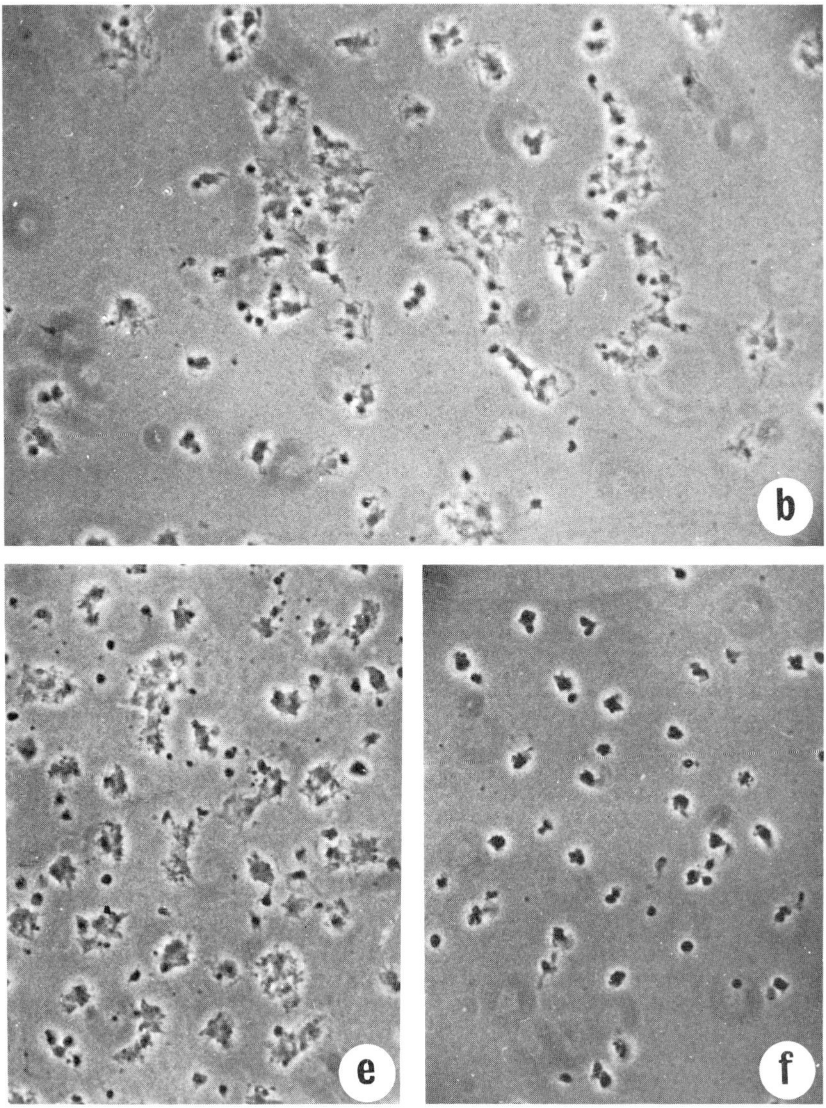

FIGURE 6 *continued.*

adhesion. Consistent with this interpretation, platelets in buffer or in CIg-depleted plasma remained predominantly single and poorly spread on the collagen films, whereas platelets in platelet-rich plasma or washed and resuspended in complete plasma were much more extensively clumped, and many were well spread. Examination revealed that spread platelets were surrounded by CIg-positive fibrils, whereas unspread phase-dense platelets were not. These results suggest that CIg is necessary for spreading of platelets on collagen films.

Disulfide Bonding at the Cell Surface

In analyzing the organization of a fibrillar network, such as that which contains LETS protein, one must consider the possibility of protein-protein interactions. These interactions are also suggested by a variety of experiments on mobility of membrane proteins and the possibility of interactions of cytoskeletal fibers with the surface membrane.[6,7,58,60] We have recently reported that there is extensive disulfide bonding within and among surface proteins of fibroblasts.[13,24] In particular, LETS protein is entirely present as disulfide-bonded complexes,[24] as is also true of CIg.[17,18] These disulfide bonds are necessary for the biologic activity of LETS protein,[53,61] and treatment of cells with reducing agents leads to rapid loss of LETS protein from the cells.[53,61] Disulfide bonds are thus essential for maintaining the structure of these cell surface complexes and the biologic activity of LETS protein.

Since the disulfide bonds in LETS protein appear important, we have analyzed their distribution in the molecule.[61a] Cleavage of LETS protein, with low concentrations of several proteases either on the cell surface or as a purified molecule in solution, leads to release of a large fragment with a molecular weight of 200,000 daltons. This fragment is released regardless of whether the disulfide bonds are reduced (FIGURE 7). Therefore, no interchain disulfide bond is present in this part of the LETS protein monomer, indicating that all interchain disulfide links occur near one or both ends of the molecule. The same result has been reported for CIg.[62,63]

To further investigate this question, purified LETS protein was digested with carboxypeptidase Y. This enzyme releases a large fragment (185,000 daltons) that contains no intermolecular disulfide bonds. Simultaneously, carboxypeptidase releases the majority (>80%) of the half-cystine residues of LETS protein.[61a] We have been unable to rule out completely the possibility that the carboxypeptidase contains minor endopeptidase contaminants, although the maximum possible contamination would be an unlikely explanation for our results. These results suggest, therefore, that the two chains of LETS protein are held together by interchain disulfide bonds near the C-terminal end. Amino acid analyses indicate the presence of 36–40 half-cystine residues per 230,000-dalton monomer,[61a] and similar values have been obtained by others for CIg.[17,18] We have so far been unable to detect free sulfhydryl groups in intact LETS protein by use of radioactive iodoacetic acid, although reduced LETS monomer is readily labeled.[61a] Thus, most of the half-cystines must be present as cystine. This, again, is also true for CIg.[62] Amino acid analysis and analysis of cystine-labeled LETS protein by tryptic cleavage indicate that there are from two to four disulfide loops present in each 200,000-dalton fragment, leaving 30–38 disulfide bonds in the 60,000-dalton (2 × 30,000) region removed by proteolysis. This cystine-rich region is removed in several fragments, including one of about

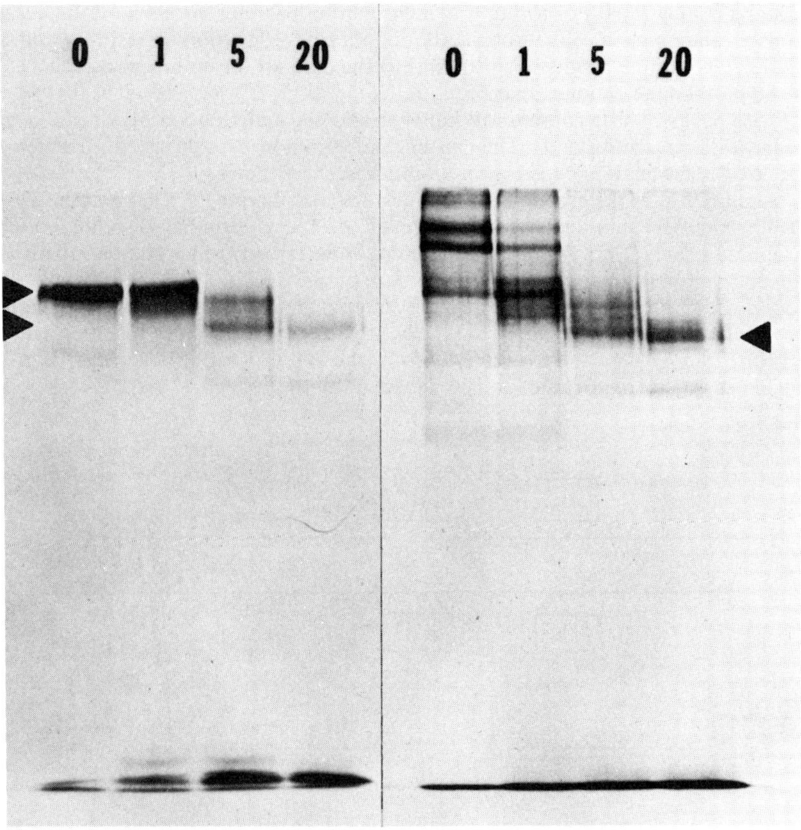

FIGURE 7. Partial digestion of LETS protein by trypsin. [^{35}S]Cystine-labeled LETS protein was treated with 1 μg/ml of trypsin for the time (minutes) indicated and run either reduced (*left*) or nonreduced (*right*). Note the appearance of 200,000-dalton band in both cases. Also note incomplete conservation of radioactivity in this band with cystine label. With glucosamine label, at least 95% of radioactivity was recovered in 200,000-dalton band. Arrows mark LETS protein monomer (230,000 daltons) and 200,000-dalton fragment.

25,000 daltons that contains intrachain, but not interchain, disulfides. Cleavage of glucosamine-labeled LETS protein demonstrates that all or most ($\geq$ 95%) of the carbohydrate is present on the 200,000-dalton fragment. The exact number of carbohydrate side chains is unknown but is greater than one.[61a] Thus, a picture emerges of LETS protein as a molecule with domains as depicted in FIGURE 8.

Peptide Fingerprinting

To analyze further the structure of LETS protein, we have turned to analysis on ion-exchange columns[64] of complete tryptic digests (longer treatments with higher concentrations will degrade the 200,000-dalton fragment). A typical analysis is illustrated in FIGURE 9, which compares LETS protein purified from hamster and chicken cell cultures. About 15% of the methionine residues and all of the glucosamine residues fail to bind to the column, in both species. The 85% of the methionine residues that bind are eluted as clearly separated peaks. Some of these peaks appear common to hamster and chicken LETS protein; others are clearly different (FIGURE 9). The existence of common peptides is consistent with the cross-reactions observed in immunofluorescence studies.

About 70% of the cysteic acid residues are also bound to the column. Thus, the chromatographic analysis covers much of the C-terminal disulfide-bonded region and excludes the carbohydrate-containing region, which is present on the 200,000-dalton fragment. An unknown part of the 200,000-dalton fragment that does not contain sugar may be present among the separated peptides. Alternatively, all of it may be excluded. Similarly, it is unclear how close together are the carbohydrate groups and disulfide loops in the 200,000-dalton fragment. Studies are under way in an attempt to answer these questions.

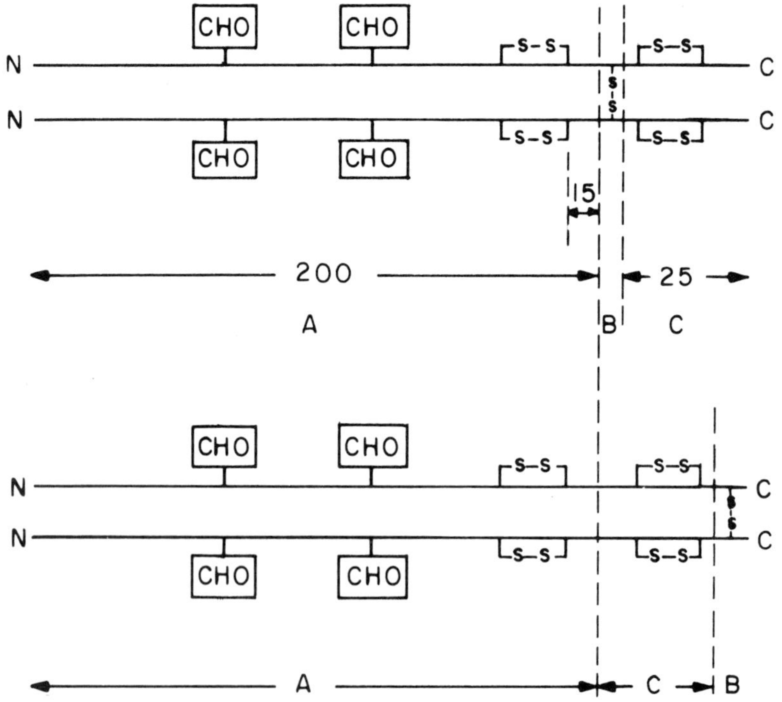

FIGURE 8. Diagramatic representation of possible arrangements of disulfide bonds in NIL8 hamster cell LETS protein. There are, in fact, 36–40 disulfide bonds; the diagrams indicate their positions but not their numbers. Most are located in the C-terminal region; only two to four disulfide linkages are present in the N-terminal 200,000-dalton fragment, along with the majority of the carbohydrate groups, whose exact positions are not known. A 25,000-dalton region near the C-terminal end has intrachain disulfide bonds but is free of interchain disulfides; its position in relation to the region of interchain disulfide linkages is currently uncertain.

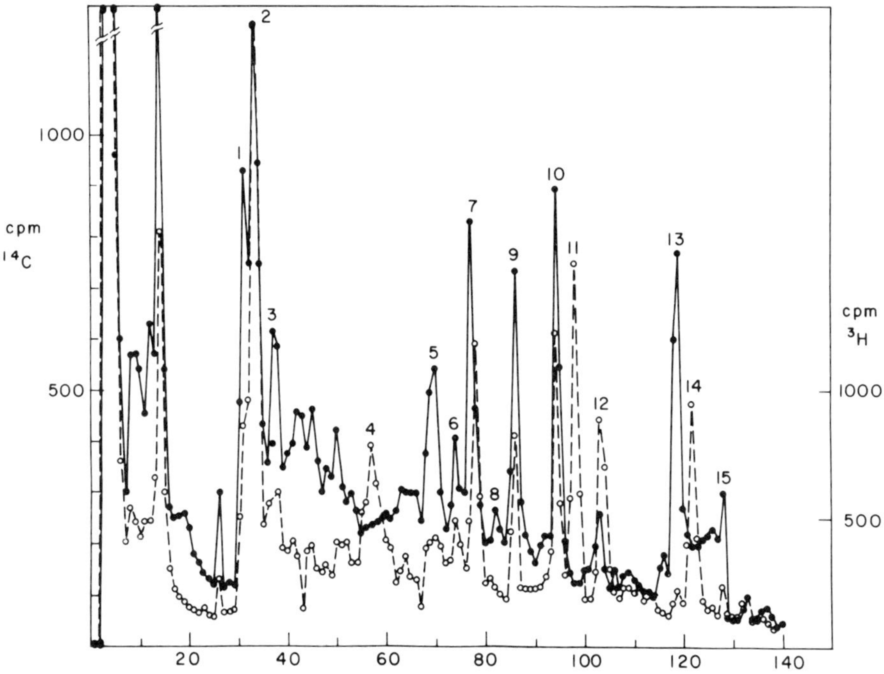

FIGURE 9. Ion-exchange chromatography[64] of tryptic digests of hamster ([^{3}H]-) and chicken [^{35}S]LETS protein. Methionine-labeled LETS protein was purified and digested as described in MATERIALS AND METHODS. Samples were pooled, retrypsinized, and oxidized together before analysis. Reproducible peaks are numbered. Peaks 2, 9, and 10 consistently exhibited coincidence between the two species; peak 7 consistently showed a small difference in elution, and peaks 4, 11, 13, and 14 were consistently unique to one species or the other as shown. Other numbered peaks (1, 3, 5, 6, 8, 12, and 15) usually demonstrated peaks with both isotopes, but their relative heights varied. The region between peaks 3 and 4 was variable in profile.

DISCUSSION

Altered Surface Levels of LETS Protein

Alterations in levels of LETS protein at the cell surface occur on transformation and, in normal cells, in response to growth stimulation and during the cell cycle. Clearly, a major contributory factor in the case of transformation is reduced biosynthesis (see References 44, 45, and 65 and FIGURE 1). It is also clear that normal cells can control their rate of biosynthesis of LETS protein (FIGURE 1), which could conceivably explain variations in surface levels on normal cells as a function of cell cycle[8] and, as recently reported, hormonal stimulation.[66] However, control of the rate of biosynthesis appears insufficient to explain completely the altered levels observed on transformed cells. Such a mechanism does not explain increased turnover rates on transformation[36,39,40,45] or on mixing of normal and transformed cells.[36,41,42] Nor does it explain the reduced binding of LETS protein to transformed cells (TABLE 1). These results suggest a further alteration in the molecules to which LETS protein binds.

Several groups have reported that LETS protein can be found in the culture medium of transformed cells.[36,44,67,68] We have recently quantitated the levels in NIL-HSV cells by metabolic labeling and by affinity chromatography on gelatin[32,44] and find levels that are 5–10% of those in the culture medium of normal NIL8 cells. This value is several-fold higher than the relative surface levels. The material accumulated in the culture medium can reattach to cells and presumably explains earlier results in which low surface levels of LETS protein could be restored to transformed cells in the absence of protein synthesis.[36,40,44] The reason for the reduced attachment of this LETS protein by transformed cells remains unclear but presumably reflects absence of, or alteration in, another protein to which LETS protein binds (e.g., collagen).

Role of LETS Protein in Adhesion

The results of readdition of LETS protein to transformed cells indicate that it causes increased attachment, stronger adhesion, and greater spreading and elongation of the cells.[21,22] The cells also show reductions in overlapping[21] and align with one another. These findings have been interpreted to represent increased "contact inhibition of movement," and it appears likely that contact inhibition of movement is in part a reflection of strong cell-substratum adhesion that prevents underlapping of cells.[50,51,69–71] Alterations in all of these culture parameters can, therefore, in the simplest interpretation, be viewed as pleiotropic effects of alterations in cell-substratum adhesion. Similarly, the elevated numbers of microvilli and ruffles characteristic of many transformed cells can be reduced by addition of LETS protein,[72] and microfilament bundles characteristic of normal cells can be restored[22,44,53,73] (FIGURE 2). It is quite possible that both of these sets of morphologic properties of transformed cells are also reflections of reduced adhesion and cellular spreading. Normal cells exhibit increased numbers of microvilli when they round up[74,75]; the best visualization of microfilament bundles is in well-spread normal cells,[26] and less well-spread cells demonstrate less well-developed bundles.[26,73,76] It thus seems likely that the cellular flattening produced by addition of LETS protein causes the changes observed in the arrangement of cytoskeletal structures inside the cell and in features of surface morphology as secondary effects. The corollary of this hypothesis would be that the reduced surface levels of LETS protein on transformed cells are the cause of the observed alterations in the properties discussed, all as indirect effects of a reduction in cell-substratum adhesion. While this model is likely to be an oversimplification, it provides a useful working hypothesis.

The effects of LETS protein on cell-substratum adhesion are also likely to be responsible for the increased cellular movement observed on addition of LETS protein to a variety of cell types, normal or transformed (FIGURE 3). It has long been known that increased adhesion leads to increased migration.[50–54,71] Very high levels of adhesion would be expected to inhibit migration. While we have not observed this phenomenon in readdition experiments, NIL8 cells cultures in low serum levels display very strong adhesion to the substratum and do not migrate. These cells also have a very high concentration of LETS protein beneath them[26] (FIGURE 5), consistent with (but not proving) the idea that very high levels of LETS protein in the area of cell-substratum contact are related to strong adhesion and inhibition of migration.

Also consistent with the postulation that LETS protein plays a role in cellular spreading and movement are the locations of LETS protein beneath spreading and migrating cells (FIGURE 5). In each case, LETS protein is found in small stitches parallel to the axis of cell extension; it occurs radially in spreading cells and perpendicular to the edge and parallel to the major axes of extending lamellipodia. One can suggest that the cells lay down fibrils of LETS protein as they extend on the substratum.

Finally, it is possible that LETS protein could be involved in cell-cell adhesion. It is found in regions of cell-cell contact[26,53-55] (FIGURE 5) and has been shown to promote red blood cell aggregation.[77] However, there is currently no direct evidence for a role in cell-cell adhesion.

Cold-insoluble globulin (CIg) also appears to be involved in adhesion of a completely different cell type, platelets. The evidence[57a] (FIGURE 6) suggests that CIg is recruited from the plasma concomitant with binding of platelets to collagen. The presence of CIg appears to promote spreading of the platelets and platelet aggregation. Thus, in an unrelated situation, CIg is also implicated in cell-substratum and cell-cell adhesion. An attractive hypothesis consistent with these data is that, on damage to the endothelial cell layer of capillaries or other blood vessels, the basal lamella, which contains collagen and perhaps also LETS protein, provides a substratum onto which platelets and CIg from the plasma adhere. The CIg could, then, mediate binding of fibrin by the platelet-basal lamella complex, thus providing a link between cellular and fibrin-mediated hemostasis.

Relationship between LETS Protein and CIg

LETS protein and CIg are clearly closely related both antigenically, as described here and elsewhere,[15,28,54] and structurally[16-18,61,62] (FIGURES 7 & 8). Both are dimers with subunits in excess of 200,000 daltons held together by terminal disulfide bonds. Both molecules bind to gelatin. However, LETS protein is an insoluble fibrillar protein, and CIg is a soluble plasma protein. In tests of the efficiency of the two proteins in promoting attachment of NIL-HSV cells to plastic dishes, CIg was consistently two- to threefold less active (TABLE 2), even when isolated from freshly drawn plasma and added to the cells within 4–6 hr of collection. This finding suggests that CIg as isolated is different from LETS protein isolated from cells or culture medium, both of which are equally active. This difference is not due to species variations, because LETS proteins isolated from cultures cells of different species (hamster and chicken) were about equally active, whereas CIg (hamster and human) was less active. We have also consistently observed that reduced CIg migrates as a closely spaced doublet, whereas LETS protein on the same gel runs as a single band. The higher-molecular-weight CIg band comigrates with LETS protein from the same species.[61a] Others have also noted heterogeneity in CIg preparations.[18,62] It is unclear whether this heterogeneity represents degradation and, if so, whether any such degradation occurs *in vivo* or *in vitro*. In our experiments, isolation of both LETS protein and CIg was very rapid, with affinity chromatography[32] in each case, and yet the difference remained. This difference in content of the 230,000-dalton monomer could explain the difference in biologic activity. However, it remains uncertain whether LETS protein on fibroblasts and CIg in plasma are identical *in vivo*. In any case, it is clear that although CIg appears less effective in mediating adhesion of transformed cells, it can perform similar functions regarding adhesion of platelets. It should be mentioned that platelet adhesion was assayed in the presence of collagen, whereas transformed cell attachment was measured in its absence. Several serum-derived cell attachment factors[78,79] have been described and are probably related to, or are identical with, CIg. The assays used in the different studies are significantly different; some use collagen, others do not. A careful comparison will be required to decide conclusively the question of exact identity or similarity of these various proteins.

Structure of LETS Protein

Another approach to the question of the relationships between LETS protein and CIg is to analyze the structure of the two proteins. At present, the analysis of the structure of CIg[16-18,62] and LETS protein (see RESULTS) indicate many parallels. In agreement with results on CIg, we find that LETS protein contains intersubunit disulfide bonds, all of which are near the ends of the chains. In the case of LETS protein, the interchain disulfide bonds are C-terminal, as are most of the intrachain disulfide bonds (FIGURE 8). A large N-terminal fragment (200,000 daltons) contains between two and four intrachain disulfide loops per chain and has no interchain disulfides. Another region (25,000 daltons) from near the C-terminal end contains intrachain but no interchain disulfides, leaving a stretch of about 5000–10,000 daltons that contains the interchain disulfide bonds. Nearly all, or all, of the carbohydrate residues are present on the 200,000-dalton fragment and could be responsible for its relative resistance to proteolytic cleavage. These features of the structure of LETS protein are incorporated in the models depicted in FIGURE 8. Exactly which of the numerous disulfide bonds are required for binding and activity of LETS protein remains to be analyzed. The approaches described in this paper should provide insights into the relationships between the structure and function of LETS protein and CIg.

ACKNOWLEDGMENTS

We thank Dr. John Wyke for the generous gift of tsRSV rat-1 cells. We are also grateful to Drs. Keith Burridge, Gerald Fuller, S. J. Singer, and Michael Mosesson for the antiactin, antitubulin, antimyosin, and antihuman CIg, respectively. Acid-soluble rat tail tendon collagen was kindly provided by Dr. Thomas Linsenmayer.

REFERENCES

1. HYNES, R. O. 1973. Alteration of cell-surface proteins by viral transformation and by proteolysis. Proc. Nat. Acad. Sci. USA **70:** 3170–3174.
2. WICKUS, G. G., P. E. BRANTON & P. W. ROBBINS. 1974. Rous sarcoma virus transformation of the chick cell surface. *In* Control of Proliferation in Animal Cells. B. Clarkson & R. Baserga, Eds.: 541–546. Cold Spring Harbor Laboratory. Cold Spring Harbor, N.Y.
3. HOGG, N. M. 1974. A comparison of membrane proteins of normal and transformed cells by lactoperoxidase labelling. Proc. Nat. Acad. Sci. USA **71:** 489–492.
4. GAHMBERG, C. G., D. KIEHN & S. HAKOMORI. 1974. Changes in a surface-labelled galactoprotein and in glycolipid concentrations in cells transformed by a temperature-sensitive polyoma virus mutant. Nature (London) **248:** 413–415.
5. STONE, K. R., R. E. SMITH & W. K. JOKLIK. 1974. Changes in membrane polypeptides that occur when chick embryo fibroblasts and NRK cells are transformed with avian sarcoma viruses. Virology **58:** 86–100.

6. HYNES, R. O. 1976. Cell surface proteins and malignant transformation. BBA Rev. Cancer **458:** 73–107.
7. HYNES, R. O. 1974. Role of surface alterations in cell transformation: the importance of proteases and surface proteins. Cell **1:** 147–156.
8. HYNES, R. O. & J. M. BYE. 1974. Density and cell cycle dependence of cell surface proteins in hamster fibroblasts. Cell **3:** 113–120.
9. GRAHAM, J. M., R. O. HYNES, E. A. DAVIDSON & D. F. BAINTON. 1975. The location of proteins labelled by the ^{125}I-lactoperoxidase system in the NIL8 hamster fibroblast. Cell **4:**353–365.
9a. GRAHAM, J. M., R. O. HYNES, C. ROWLATT & J. K. SANDALL. This monograph.
10. RUOSLAHTI, E. & A. VAHERI. 1974. Novel human serum protein from fibroblast plasma membrane. Nature (London) **248:** 789–791.
11. VAHERI, A. & E. RUOSLAHTI. 1974. Disappearance of a major cell-type specific surface glycoprotein antigen (SF) after transformation of fibroblasts by Rous sarcoma virus. Int. J. Cancer **13:** 579–586.
12. WARTIOVAARA, J., E. LINDER, E. RUOSLAHTI & A. VAHERI. 1975. Distribution of fibroblast surface antigen. Association with fibrillar structures of normal cells and loss upon viral transformation. J. Exp. Med. **140:** 1522–1533.
13. HYNES, R. O., A. T. DESTREE & V. M. MAUTNER. 1976. Spatial organization at the cell surface. *In* Membranes and Neoplasia: New Approaches and Strategies. V. T. Marchesi, Ed.: 189–201. Alan R. Liss, Inc. New York, N.Y.
14. KESKI-OJA, J., A. VAHERI & E. RUOSLAHTI. 1976. Fibroblast surface antigen (SF): the external glycoprotein lost in proteolytic stimulation and malignant transformation. Int. J. Cancer **17:** 261–269.
15. RUOSLAHTI, E. & A. VAHERI. 1975. Interaction of soluble fibroblast surface antigen with fibrinogen and fibrin. J. Exp. Med. **141:** 497–501.
16. MOSESSON, M. W. & R. A. UMFLEET. 1970. The cold-insoluble globulin of human plasma. J. Biol. Chem. **245:** 5728–5736.
17. MOSHER, D. F. 1975. Cross-linking of cold-insoluble globulin by fibrin-stabilizing factor. J. Biol. Chem. **250:** 6614–6621.
18. MOSESSON, M. W., A. B. CHEN & R. M. HUSEBY. 1975. The cold-insoluble globulin of human plasma: studies of its essential structural features. Biochim. Biophys. Acta **386:** 509–524.
19. WESTON, J. A. & K. L. HENDRICKS. 1972. Reversible transformation by urea of contact-inhibited fibroblasts. Proc. Nat. Acad. Sci. USA **69:** 3727–3731.
20. YAMADA, K. M. & J. A. WESTON. 1974. Isolation of a major cell surface glycoprotein from fibroblasts. Proc. Nat. Acad. Sci. USA **71:** 3492–3496.
21. YAMADA, K. M., S. S. YAMADA & I. PASTAN. 1976. Cell surface protein partially restores morphology, adhesiveness, and contact inhibition of movement to transformed fibroblasts. Proc. Nat. Acad. Sci. USA **73:** 1217–1221.
22. ALI, I. U., V. M. MAUTNER, R. P. LANZA & R. O. HYNES. 1977. Restoration of normal morphology, adhesion and cytoskeleton in transformed cells by addition of a transformation-sensitive surface protein. Cell **11:** 115–126.
23. LAEMMLI, U. K. 1970. Cleavage of structural proteins during the assembly of the head of bacteriophage T4. Nature (London) **227:** 680–685.
24. HYNES, R. O. & A. T. DESTREE. 1977. Extensive disulfide bonding at the mammalian cell surface. Proc. Nat. Acad. Sci. USA **74:** 2855–2859.
25. LAZARIDES, E. 1976. Actin, α-actinin and tropomyosin interaction in the structural organization of actin filaments in non-muscle cells. J. Cell Biol. **68:** 202–219.
26. MAUTNER, V. M. & R. O. HYNES. 1977. Surface distribution of LETS protein in relation to the cytoskeleton of normal and transformed fibroblasts. J. Cell Biol. **75:** 743–758.
27. HYNES, R. O. & A. T. DESTREE. 1978. 10nm Filaments in normal and transformed cells. Cell **13:** 151–163.
28. BURRIDGE, K. 1976. Changes in cellular glycoproteins after transformation: identification of specific glycoproteins and antigens in sodium dodecyl sulfate gels. Proc. Nat. Acad. Sci. USA **73:** 4457–4461.
29. BRINKLEY, B. R., G. M. FULLER & D. P. HIGHFIELD. 1975. Cytoplasmic microtubules

in normal and transformed cells in culture: analysis by tubulin antibody immunofluorescence. Proc. Nat. Acad. Sci. USA **72:** 4981–4985.

30. Sheetz, M., R. G. Painter & S. J. Singer. 1976. Relationships of the spectrin complex of human erythrocyte membranes to the actomyosins of muscle cells. Biochemistry **15:** 4486–4492.
31. Albrecht-Buehler, G. 1977. The phagokinetic tracks of 3T3 cells. Cell **11:** 395–404.
32. Engvall, E. & E. Ruoslahti. 1977. Binding of soluble form of fibroblast surface protein, fibronectin, to collagen. Int. J. Cancer **20:** 1–5.
33. Phillips, D. R. & P. P. Agin. 1977. Platelet plasma membrane glycoproteins. J. Biol. Chem. **252:** 2121–2126.
34. Teng, N. N. H. & L. B. Chen. 1975. The role of surface proteins in cell proliferation as studied with thrombin and other proteases. Proc. Nat. Acad. Sci. USA **72:** 413–417.
35. Blumberg, P. M. & P. W. Robbins. 1975. Effects of proteases on activation of resting chick embryo fibroblasts and on cell surface proteins. Cell **6:** 137–147.
36. Hynes, R. O., J. A. Wyke, J. M. Bye, K. C. Humphryes & E. S. Pearlstein. 1975. Are proteins involved in altering cell surface proteins during viral transformation? *In* Proteases and Biological Control. E. Reich, E. Shaw & D. B. Rifkin, Eds.: 931–944. Cold Spring Harbor Laboratory. Cold Spring Harbor, N. Y.
37. Unkeless, J. C., A. Tobia, L. Ossowski, J. P. Quigley, D. B. Rifkin & E. Reich. 1973. An enzymatic function associated with transformation of fibroblasts by oncogenic viruses. I. Chick embryo fibroblast cultures transformed by avian RNA tumor viruses. J. Exp. Med. **137:** 85–111.
38. Ossowski, L., J. P. Quigley, G. M. Kellerman & E. Reich. 1973. Fibrinolysis associated with oncogenic transformation. Requirement of plasminogen for correlated changes in cellular morphology, colony formation in agar, and cell migration. J. Exp. Med. **138:** 1056–1064.
39. Robbins, P. W., G. G. Wickus, P. E. Branton, B. J. Gaffney, C. B. Hirschberg, P. Fuchs & P. M. Blumberg. 1974. The chick fibroblast cell surface following transformation by Rous sarcoma virus. Cold Spring Harbor Symp. Quant. Biol. **39:** 1173–1180.
40. Hynes, R. O. & J. A. Wyke. 1975. Alterations in surface proteins in chicken cells transformed by temperature-sensitive mutants of Rous sarcoma virus. Virology **64:** 492–504.
41. Hynes, R. O. & E. S. Pearlstein. 1976. Investigations of the possible role of proteases in altering surface proteins of virally transformed hamster fibroblasts. J. Supramol. Struct. **4:** 1–14.
42. Mahdavi, V. & R. O. Hynes. 1978. Effects of cocultivation with transformed cells on surface proteins of normal cells. Biochim. Biophys. Acta. In press.
43. Pearlstein, E. S., R. O. Hynes, L. M. Franks & V. J. Hemmings. 1976. Surface proteins and fibrinolytic activity of cultured mammalian cells. Cancer Res. **36:** 1475–1480.
44. Hynes, R. O., A. T. Destree, V. M. Mautner & I. U. Ali. 1977. Synthesis, secretion, and attachment of LETS glycoprotein in normal and transformed cells. J. Supramol. Struct. **7:** 397–408.
45. Olden, K. & K. M. Yamada. 1977. Mechanism of the decrease in the major cell surface protein of chick embryo fibroblasts after transformation. Cell **11:** 957–969.
46. Pollack, R., M. Osborn & K. Weber. 1975. Patterns of organization of actin and myosin in normal and transformed cultured cells. Proc. Nat. Acad. Sci. USA **72:** 994–998.
47. Ash, J. F., P. K. Vogt & S. J. Singer. 1976. Reversion from transformed to normal phenotype by inhibition of protein synthesis in rat kidney cells infected with a temperature-sensitive mutant of Rous sarcoma virus. Proc. Nat. Acad. Sci. USA **73:** 3603–3607.
48. Edelman, G. M. & I. Yahara. 1976. Temperature-sensitive changes in surface modulating assemblies of fibroblasts transformed by mutants of Rous sarcoma virus. Proc. Nat. Acad. Sci. USA **73:** 2047–2051.

49. POLLACK, R. & D. RIFKIN. 1975. Actin-containing cables within anchorage-dependent rat embryo cells are dissociated by plasmin and trypsin. Cell **6:** 495–506.
49a. ALI, I. U. 1977. Unpublished results.
50. CARTER, S. B. 1967. Haptotaxis and the mechanism of cell motility. Nature (London) **213:** 256–260.
51. HARRIS, A. 1973. Behavior of cultured cells on substrata of variable adhesiveness. Exp. Cell Res. **77:** 285–297.
52. GAIL, M. H. & C. W. BOONE. 1972. Cell-substrate adhesivity. A determinant of cell motility. Exp. Cell Res. **70:** 33–40.
52a. ALI, I. U. & R. O. HYNES. 1978. Effects of LETS glycoprotein on cell motility. Cell. In press.
53. HYNES, R. O., I. U. ALI, V. M. MAUTNER & A. T. DESTREE. 1978. LETS glycoprotein: arrangement and function at the cell surface. *In* Molecular Basis of Cell-Cell Interactions: 139–153. Alan R. Liss, Inc. New York, N.Y.
54. CHEN, L. B., P. H. GALLIMORE & J. K. MCDOUGALL. 1976. Correlation between tumor induction and cell surface LETS protein. Proc. Nat. Acad. Sci. USA **73:** 3570–3574.
55. CHEN, L. B., N. MAITLAND, P. H. GALLIMORE & J. K. MCDOUGALL. 1977. Detection of the large external transformation-sensitive protein on some epithelial cells. Exp. Cell Res. **106:** 39–46.
56. SCHLESSINGER, J., L. S. BARAK, G. G. HAMMES, K. M. YAMADA, I. PASTAN, W. W. WEBB & E. L. ELSON. 1977. Proc. Nat. Acad. Sci. USA **74:** 2909–2913.
57. ALBRECHT-BUEHLER, G. & L. B. CHEN. 1977. Nature (London) **266:** 454–456.
57a. HYNES, R. O. & K. K. SMITH. 1977. Unpublished results.
58. NICOLSON, G. L. 1976. Transmembrane control of the receptors on normal and tumor cells. II. Surface changes associated with transformation and malignancy. Biochim. Biophys. Acta **458:** 1–72.
59. EDELMAN, G. M. 1976. Surface modulation in cell recognition and cell growth. Science **192:** 218–226.
60. BERLIN, R. D., J. M. OLIVER, T. E. UKENA & H. H. YIN. 1974. Control of cell surface topography. Nature (London) **247:** 45–46.
61. ALI, I. U. & R. O. HYNES. 1977. Effects of cytochalasin B and colchicine on attachment of a major surface protein of fibroblasts. Biochim. Biophys. Acta **471:** 16–24.
61a. WAGNER, D. D. & R. O. HYNES. In preparation.
62. CHEN, A. B., D. K. AMRANI & M. W. MOSESSON. 1977. Heterogeneity of the cold-insoluble globulin of human plasma (CIg), a circulating cell surface protein. Biochim. Biophys. Acta. **493:** 310–324.
63. JILEK, F. & H. HÖRMANN. 1977. Cold-insoluble globulin. II. Plasminolysis of cold-insoluble globulin. Hoppe-Seyler's Z. Physiol. Chem. **358:** 133–136.
64. EISENMAN, R., V. M. VOGT & H. DIGGELMAN. 1975. Synthesis of avian RNA tumor virus structural proteins. Cold Spring Harbor Symp. Quant. Biol. **39:** 1067–1075.
65. ADAMS, S. L., M. E. SOBEL, B. H. HOWARD, K. OLDEN, K. M. YAMADA, B. DE CROMBRUGGHE & I. PASTAN. 1977. Levels of translatable mRNAs for cell surface protein, collagen precursors and two membrane proteins in Rous transformed chick embryo fibroblasts. Proc. Nat. Acad. Sci. USA **74:** 3399–3403.
66. CHEN, L. B., R. C. GUDOR, T. T. SUN, A. B. CHEN & M. W. MOSESSON. 1977. Control of a cell surface major glycoprotein by epidermal growth factor. Science **197:** 776–778.
67. VAHERI, A. & E. RUOSLAHTI. 1975. Fibroblast surface antigen produced but not retained by virus transformed human cells. J. Exp. Med. **142:** 530–538.
68. VAHERI, A., E. RUOSLAHTI, B. WESTERMARK & J. PONTEN. 1976. A common cell-type specific surface antigen in cultured human glial cells and fibroblasts: loss in malignant cells. J. Exp. Med. **143:** 64–72.
69. ABERCROMBIE, M. 1970. Contact inhibition in tissue culture. In Vitro **6:** 128–142.
70. MARTZ, E., H. M. PHILLIPS & M. S. STEINBERG. 1974. Contact inhibition of overlapping and differential cell adhesion: a sufficient model for the control of certain cell culture morphologies. J. Cell Sci. **16:** 401–419.

71. Bell, P. B. 1977. Locomotory behavior, contact inhibition, and pattern formation of 3T3 and polyoma virus-transformed 3T3 cells in culture. J. Cell Biol. **74:** 963–982.
72. Yamada, K. M., S. H. Ohanian & I. Pastan. 1976. Cell surface protein decreases microvilli and ruffles on transformed mouse and chick cells. Cell **9:** 241–245.
73. Willingham, M. C., K. M. Yamada, S. S. Yamada, J. Pouyssegur & I. Pastan. 1977. Microfilament bundles and cell shape are related to adhesiveness to substratum and are dissociable from growth control in cultured fibroblasts. Cell **10:** 375–380.
74. Goldman, R. D., C. Chang & J. F. Williams. 1974. Properties and behavior of hamster embryo cells transformed by human adenovirus 5. Cold Spring Harbor Symp. Quant. Biol. **39:** 601–614.
75. Ukena, T. E. & M. J. Karnovsky. 1976. Patching, microvilli and the agglutination of normal and transformed cells. *In* Membranes and Neoplasia: New Approaches and Strategies. V. T. Marchesi, Ed.: 261–273. Alan R. Liss, Inc. New York, N.Y.
76. Tucker, R. W., K. K. Sanford & F. R. Frankel. 1977. Cell. In press.
77. Yamada, K. M., S. S. Yamada & I. Pastan. 1975. The major cell surface glycoprotein of chick embryo fibroblasts is an agglutinin. Proc. Nat. Acad. Sci. USA **72:** 3158–3162.
78. Klebe, R. J. 1974. Isolation of a collagen-dependent cell attachment factor. Nature (London) **250:** 248–251.
79. Grinnell, F. 1976. Cell spreading factor: occurrence and specificity of action. Exp. Cell Res. **102:** 51–62.

FIBRONECTIN AND THE PERICELLULAR MATRIX OF NORMAL AND TRANSFORMED ADHERENT CELLS*

Antti Vaheri,† Kari Alitalo,†‡ Klaus Hedman,†§
Jorma Keski-Oja,† Markku Kurkinen,† and Jorma Wartiovaara‡§

Departments of † Virology, ‡ Pathology, and § Electron Microsccopy
University of Helsinki
00290 Helsinki 29, Finland

Fibronectin (for review, see Reference 1) is a polymorphic glycoprotein found in vertebrates. The protein exists in soluble and insoluble forms both *in vivo* and *in vitro*. The soluble form of fibronectin *in vivo* (cold-insoluble globulin) has been known for nearly 30 years.[2] The concentration in normal human plasma is about 300 μg/ml.[3] In the insoluble form *in vivo*, fibronectin is found surrounding individual primitive mesenchymal cells and smooth muscle cells, in various basement membranes, including capillary walls and sarcolemmae of striated muscle fibers, in the stroma of lymphatic tissue, and as thin fibers in loose connective tissue.[4-6]

In cultured fibroblastic and astroglial cells, fibronectin is found in the cytoplasm, pericellularly in an insoluble form on and around cell surfaces, and in a soluble form in the culture medium.[7,8] Cultured transformed cells usually lack pericellular fibronectin.[1,9] Transformed cells secrete a fibronectin that closely resembles that secreted by normal cells, in amounts that vary with the cell type.[1,7,8,10] There is evidence that in cell cultures, fibronectin, either present in the serum used in the growth medium or synthesized by the cell, may mediate the adhesion of cells in the substratum.[11-14]

This report will discuss the organization and composition of insoluble fibronectin-containing structures found in layers of adherent cells. Fibronectin appears to be a major component of a pericellular matrix that can be regarded as an *in vitro* equivalent of the basal laminae and connective tissue matrix. The significance of the pericellular matrix for normal cells and of the loss of the matrix for transformed cells will be considered.

Properties of External Fibronectin in Cultures of Adherent Cells

External surface-labeled fibronectin in layers of cultured fibroblasts, when analyzed reduced in sodium dodecyl sulfate polyacrylamide gels, has an apparent molecular weight of 220,000 daltons.[1,9] When analyzed in nonreduced form, some of the labeled protein migrates at the dimeric position (comigrating with nonreduced plasma fibronectin), and some does not enter the 3% spacer gel (mol wt > 1,200,000 daltons). This finding indicates that pericellular fibronectin is largely disulfide bonded into complexes.[15] Two-dimensional polyacrylamide gel electrophoresis has revealed that other surface-associated proteins are also involved in disulfide bonding.[16] Plasma transglutaminase (blood coagulation factor XIII),

*Supported by National Cancer Institute Grant CA-17373 and by the Sigrid Jusélius Foundation.

0077-8923/78/0312-0343 $01.75/0 © 1978, The New York Academy of Sciences

activated by thrombin, also cross-links external fibronectin into a covalent, very-high-molecular-weight form.[17,42] A portion of radiolabeled cell surface fibronectin, iodinated 18 hr after subculture, is converted to a very-high-molecular-weight form under standard cell culture conditions.[18]

These results suggest that fibronectin is bound to itself and/or to other pericellular proteins by both disulfide and nondisulfide covalent linkages. Fibronectin, cross-linked either "spontaneously" or by added transglutaminase, is like the disulfide-bonded dimers and the complexed forms of fibronectin quite sensitive to treatment with proteolytic enzymes.[18]

Immunofluorescent and scanning electron microscopic studies have shown that pericellular fibronectin is located in fibrillar structures,[19,20] radially oriented striae,[7] or punctate striae,[8,21] depending on the cell type. Fibronectin is not retained in plasma membrane preparations unless the membranes are fixed with zinc chloride[22] and fractionates in a dense (ρ = 1.25–1.26) particulate fraction poor in phospholipid and plasma membrane markers.[23] Fibronectin is not solubilized by nonionic detergents but is solubilized by such agents as urea,[24,25] guanidine hydrochloride, and sodium iodide. The addition of a reducing agent is needed to fully solubilize surface-iodinated fibronectin.[26]

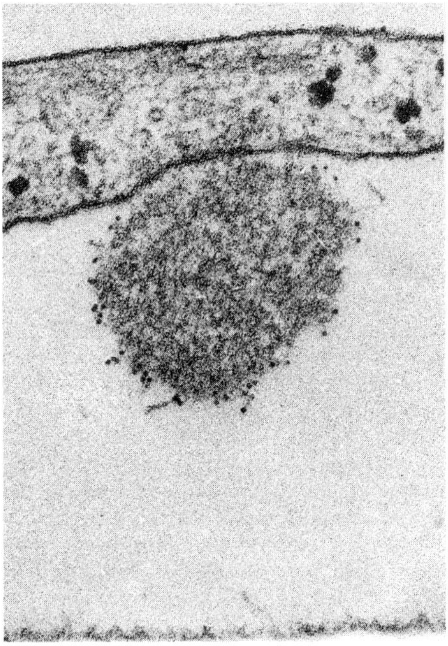

FIGURE 1. Human fibroblast stained for fibronectin by the immunoferritin method. The cell culture was fixed with a mixture of glutaraldehyde and paraformaldehyde and treated with antifibronectin serum and ferritin-conjugated sheep immunoglobulin against rabbit IgG. The culture was then refixed with glutaraldehyde and paraformaldehyde, postfixed with osmium tetroxide, and processed for sectioning. The samples were block-stained with uranyl acetate and lead citrate. The Figure shows ferritin-stained extracellular material. The material is seen as a 400 nm thick "pillow" beneath the cell surface in contact with the plasma membrane. Ferritin markers are attached to the material. ×130,000. (From Hedman *et al.*[27] By permission of *Journal of Cell Biology*.)

PERICELLULAR FIBRONECTIN IS A MATRIX PROTEIN

The above data suggested that cell surface fibronectin is not a conventional membrane protein. Immunoelectron microscopic studies[27] on the distribution of fibronectin have provided further proof of this conclusion and show that pericellular fibronectin in cultured cells is predominantly a matrix protein. Layers of cultured human fibroblasts were fixed with glutaraldehyde, and fibronectin was located with peroxidase- or ferritin-labeled antibodies. External fibronectin was visualized in structures that often mediate distant cell-cell and cell-substratum contacts, in some areas directly on the growth substratum, and to a lesser degree in close association with the upper and lower surface membranes of the cell. The pericellular fibronectin-containing structures consisted of amorphous or vaguely fibrillar material that formed strands or patches (50–500 nm in diameter) with no membranous elements visible (FIGURE 1). Both the absolute and the relative amounts of this form of fibronectin increased during incubation of the fibroblast cultures. This observation of pericellular structural fibronectin is compatible with the results of *in vivo* studies, in which fibronectin has been found in primitive mesenchymal and loose connective tissues, in vessel walls, and characteristically in basal laminae (see pp. 151–159).

Membrane-associated fibronectin had an uneven, patchy distribution and represented a minor proportion of the total external fibronectin, especially in layers of older dense cultures of human fibroblasts. An increased density of cytoplasmic submembranous filaments was often but not always seen beneath membrane-associated fibronectin. In perpendicularly sectioned membrane sites, the distance of ferritin from the outer edge of the membranous triple layer ranged between 20 and 50 nm. An electron-dense zone, which may have represented fibronectin molecules, was detected between ferritin and the membrane.[27] Whether the membrane-associated fibronectin molecules interact with the liquid bilayer itself of via membrane receptors is not known.

RELATIONSHIP OF FIBRONECTIN TO OTHER MATRIX COMPONENTS

After it was recognized that fibronectin is a matrix protein, we began to study its relationship to other matrix components. In addition to fibronectin, the connective tissue matrix *in vivo* is composed of collagen, proteoglycans, elastin, and other glycoproteins.[28,29] Development of antibodies specific for the distinct types of (pro)collagen[30] has made it possible to study the distribution of collagenous proteins in cell cultures. With such antibodies, it was previously shown that individual human fibroblasts synthesized simultaneously both types-I and -III collagen[31] and that (pro)collagen is present in layers of cultured fibroblasts in a fibrillar pattern,[31,32] which resembles that of pericellular fibronectin.[19,20] To compare the localization of fibronectin and (pro)collagen, fixed layers of cultured fibroblasts were first stained indirectly for (pro)collagen by use of rabbit antibodies[33] and fluorescein isothiocyanate-conjugated antirabbit IgG and then directly for fibronectin by use of tetramethylrhodamine isothiocyanate-conjugated goat IgG. In confluent cultures of human skin or lung fibroblasts, fibronectin and (pro)collagen, type I or III, demonstrated extensive codistribution in the pericellular fibrillar matrix (FIGURES 2 & 3). Fixation of cell layers with both paraformaldehyde and acetone to visualize intracellular antigen revealed cytoplasmic staining of fibronectin and procollagen in the same cells, suggesting simultaneous synthesis. Pericellular (pro)collagen was digested with collagenase without affecting the fibrillar distribution of matrix fibronectin. In contrast, brief treatment with trypsin removed both matrix proteins. In fibroblast cultures examined early (up to about 2 hr) after subculture of trypsinized cells, only pericellular fibronectin was seen and was observed by phase-contrast microscopy on the under surface of the cells. No (pro)collagen was detected in these prepara-

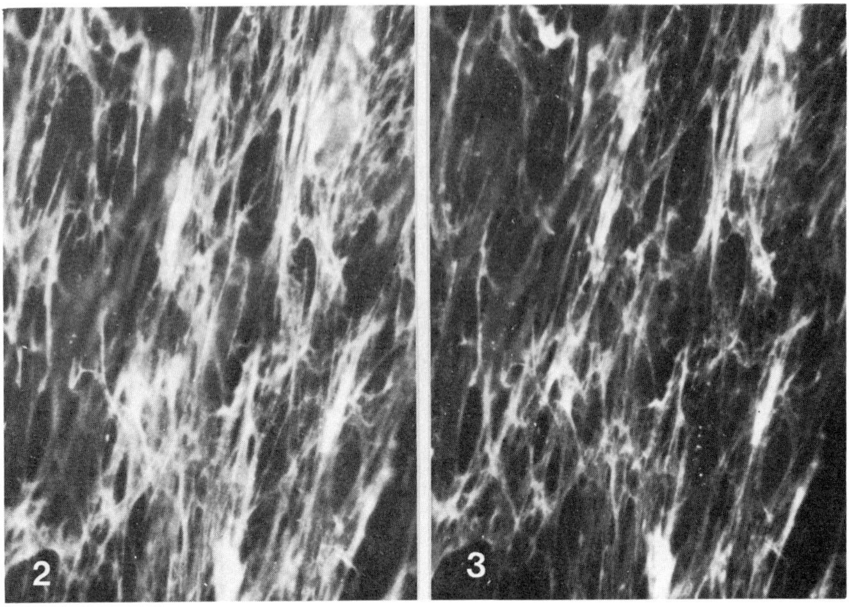

FIGURES 2 & 3. Human embryonic lung fibroblasts (WI-38) were cultured for 4 days, fixed with paraformaldehyde, and the same field was stained by immunofluorescence for both pericellular fibronectin (FIGURE 2) and procollagen type I (FIGURE 3). ×600.

tions, although in acetone-treated cells, both fibronectin and procollagen were visualized in the cytoplasm. The examination of cells at later times after subculture showed that when present pericellularly, (pro)collagen was detected exclusively at sites where matrix fibronectin was found.

The immunofluorescent results[33] should not be considered evidence of a molecular interaction between fibronectin and procollagen, because, at best, the resolution of the technique is only 200–500 nm. Such an interaction may, however, exist, because soluble fibronectin is known to bind to collagenous proteins, gelatin in particular.[34-36] Digestion experiments suggested that (pro)collagen is not essential for the stability of the pericellular matrix and supported the possibility that fibronectin may have an organizing role in matrix formation. The roles of proteoglycans and other components in matrix formation and structure are not known. Conceivably, such mechanisms as disulfide bonding, transglutaminase-mediated cross-linking, and polymerization of fibronectin may be among the factors that guide the maturation of the fibronectin-collagen matrix *in vivo*. Since most epithelial cells *in vivo* abut against basal laminae,[37] it seems likely that the matrix may not only have structural functions but may also contain information important for the orderly growth and positioning of cells. This hypothesis is also suggested by findings on the expression of fibronectin during early morphogenesis in differentiating teratocarcinoma cells in culture.[43] Since fibronectin, at least under cell culture conditions, appears to also interact directly with parenchymal cells, such as hepatocytes,[14] fibronectin may be one of the molecules that structurally links adherent cells to tissues.

LOSS OF THE FIBRONECTIN-COLLAGEN MATRIX IN TRANSFORMATION

Three human sarcoma cell lines were examined by immunofluorescence for the presence of fibronectin and (pro)collagen. The human tumor cell lines HT-1080 (fibrosarcoma) and RD (rhabdomyosarcoma) and SV40-transformed fibroblast lines (WI-38/VA13) are all known[8] to produce large amounts of fibronectin in culture. Layers of HT-1080 or RD cells, even in dense cultures, bear very little or no detectable pericellular fibronectin or (pro)collagen (FIGURES 4 & 5). Cytoplasmic fibronectin and procollagen were detected, however, in transformed cells, to a degree that varied with the type of procollagen and the cell line. The RD cells stained exceptionally strongly for procollagen type III. At high density, pericellular fibronectin that codistributed with procollagen type I.[33]

To study more closely the relationship of pericellular matrix protein to transformation, the Rous sarcoma virus (RSV)-chick embryo fibroblast system was used. Layers of normal chick embryo fibroblasts had a dense meshwork of fibrillar matrix in which both fibronectin and procollagen type I were present. In cultures of fibroblasts transformed by RSV, little or no pericellular fibronectin or procollagen was detected.[33] Chick fibroblasts infected with the NY68 mutant[38] of RSV contained both fibronectin and procollagen in the pericellular matrix at 41°C (the nonpermissive temperature for transformation). After a shift to 35°C,

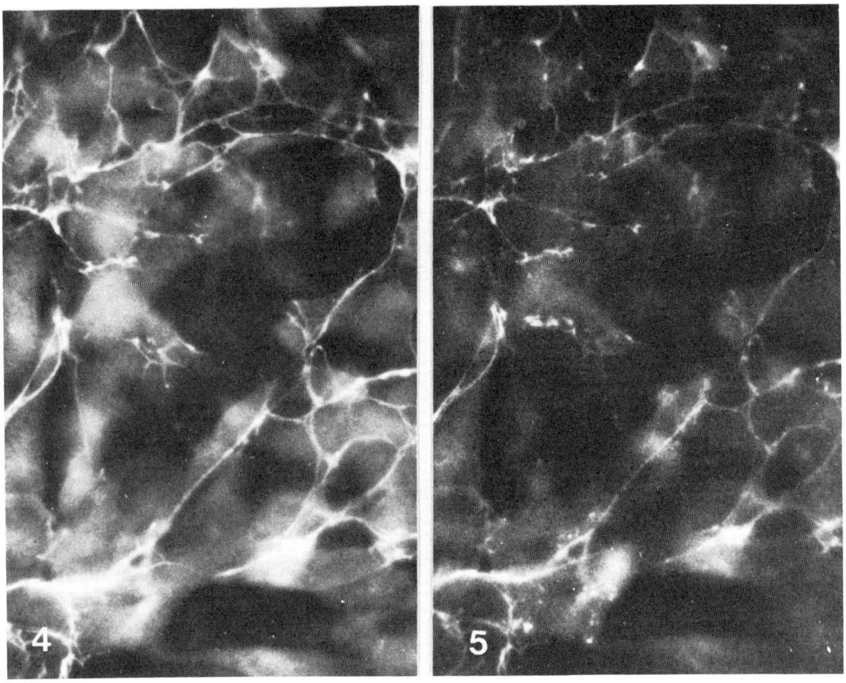

FIGURES 4 & 5. Human rhabdomyosarcoma cells (RD) were cultured for 2 days, fixed with paraformaldehyde and acetone, and the same field was stained for fibronectin (FIGURE 4) and procollagen type III (FIGURE 5). ×600.

which is permissive for transformation, they lost these pericellular proteins. Cytoplasmic fibronectin and procollagen were detected in cells of both normal (uninfected or NY68-infected cells at 41°C) and transformed phenotypes (RSV-infected or NY68-infected cells at 35°C.[33]

Two mechanisms, depending on cell type, seem to account for the loss of the pericellular fibronectin-collagen matrix in transformation (discussed in Reference 1). These mechanisms are decreased synthesis and/or decreased binding to the cell layers of the matrix components. The mechanism responsible for the decreased binding of fibronectin is not known; specifically, no change in the fibronectin molecule produced by the transformed cell has been reported. Loss of the pericellular matrix represents a major structural difference between normal and transformed cells. The loss may explain several of the phenotypic characteristics of transformed cultured cells.[39]

Decreased association with matrix components of malignant cells could be a factor involved in their unregulated growth behavior *in vivo*. Obviously, other factors, such as angiogenesis-stimulating factor (to assure nutrient supply), alterations in the pattern of cell-hormone interactions (for autonomous cell growth), and escape from immune surveillance, may be needed for oncogenicity. For tissue invasiveness, the tumor cell appears to be equipped for production of plasminogen activator, an enzyme that can activate other hydrolytic enzymes. This enzyme, either secreted[40] or membrane associated,[41] may be used to degrade fibronectin, collagenous proteins (through activation of procollagenase by plasmin), fibrin, and proteoglycans. The defective association of transformed cells with the fibronectin-collagen matrix may thus occur for two reasons. First, the transformed cell may fail to attach properly to the matrix components it produces; second, tumor cells in general may hydrolyze the fibronectin-collagen matrix in basement membranes and connective tissue stroma. The result of plasminogen activation may be degradation of the matrix. Loose association with the surrounding fibronectin-collagen matrix may be an essential feature of the malignant phenotype in adherent cell types.

Role of the Pericellular Matrix in Cellular Behavior

To study the matrix more directly, we have tried various approaches to isolate it from layers of cultured fibroblasts. A procedure[44] that involves sequential use of sodium deoxycholate and hypotonic buffers in the presence of protease inhibitors has been developed. This procedure, in essence, dissolves the cells and leaves the pericellular fibrillar matrix attached to the growth substratum. The prepared matrix, particularly when derived from confluent cultures, consists of a stable "clothlike" structure in which fibronectin (Figure 6) and (pro)collagen exhibit extensive codistribution in a manner similar to that seen in intact cell layers (Figures 2 & 3). The matrix prepared from confluent cultures represents approximately 20% of the total protein and less than 2% of total lipid and DNA in the fibroblast layer. Polypeptide analysis of the prepared matrix reveals the presence of fibronectin (220,000 daltons in reduced sodium dodecyl sulfate polyacrylamide gels) and procollagen (170,000 and 145,000 daltons) as the major components. In addition, about half of the total glycosaminoglycans are present.[44]

This new procedure has made it possible to prepare an *in vitro* connective tissue matrix that appears structurally intact and to study its role in cellular behavior. A profound effect of the matrix on cellular spreading is seen in Figures 7–10. The human sarcoma cells (HT-1080), when seeded on glass coverslips under standard cell culture conditions, attach, spread slowly (FIGURES 7 & 9), and retain a spindle-shaped, "transformed" morphology, even at high cell density. In contrast, the sarcoma cells, when seeded on matrix (prepared from cultures of normal confluent human fibroblasts), spread well and rapidly assume an extended flattened morphology (FIGURES 8 & 10) characteristic of normal well-spread fibroblasts, as described in more detail elsewhere.[45]

The above results reinforce the concept that the fibronectin-collagen matrix *in vitro* may provide a model system for studies on the functions on basal laminae and connective tissue matrix. It will be of interest to analyze the effect of preformed matrix on cellular differentiation, positioning and anchorage of epithelial cells, and cellular transformation.

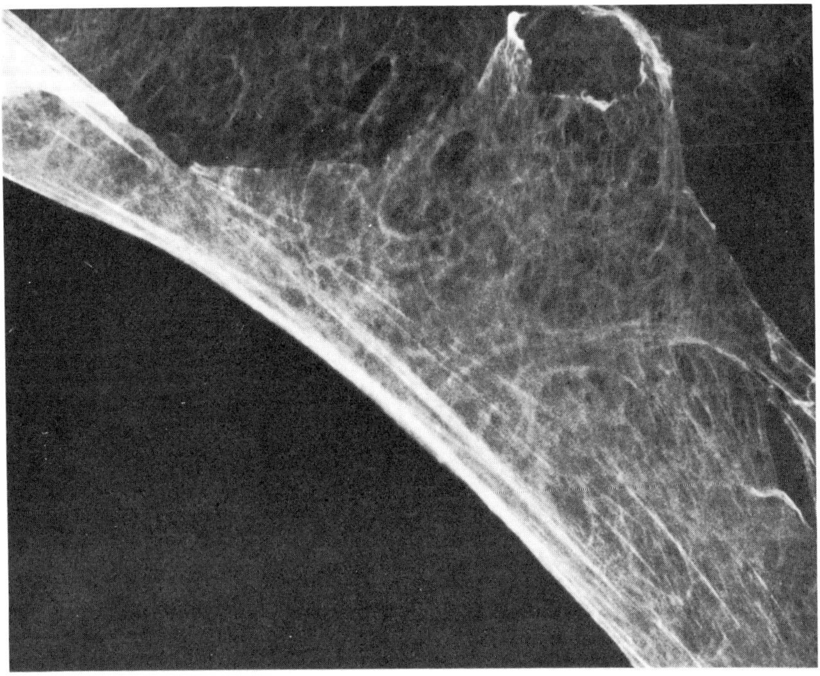

FIGURE 6. Fibronectin immunofluorescence of matrix prepared from a confluent culture of human embryonic skin fibroblasts by the deoxycholate-hypotonic buffer procedure. Note that the matrix forms a continuous "clothlike" structure and has here detached from the coverslip in the lower left-hand corner. ×240.

SUMMARY

Fibronectin is a major glycoprotein component of normal fibroblasts in culture. External fibronectin is predominantly present in a pericellular fibrillar matrix that mediates distant cell-cell and cell-substratum contacts. A small proportion of

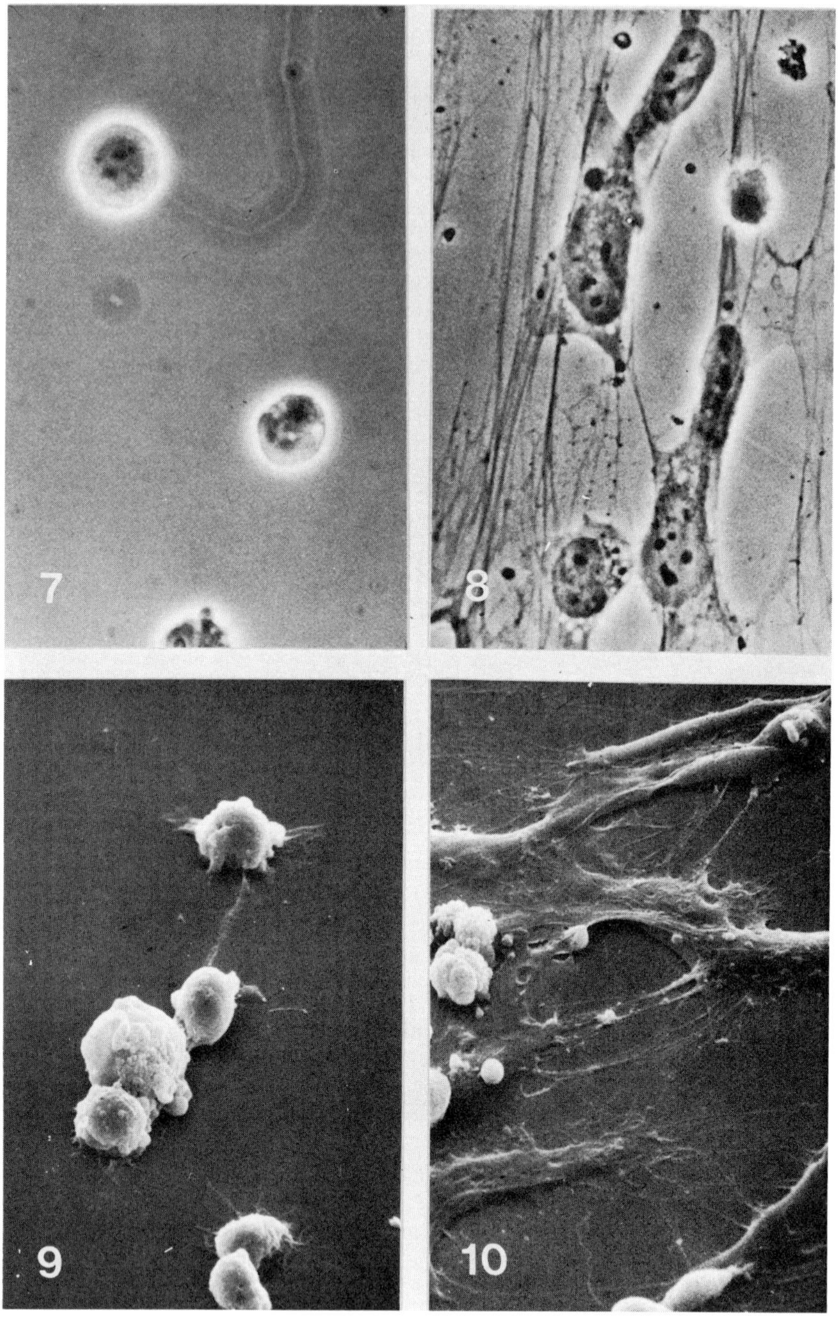

FIGURES 7–10. Effect of preformed matrix on spreading of transformed cells. Human fibrosarcoma cells (HT-1080) were subcultured in serum-free medium either under standard cell culture conditions (FIGURES 7 & 9) or on matrix-containing coverslips (FIGURES 8 & 10) and were fixed after 1 hr with paraformaldehyde and examined by phase-contrast microscopy (FIGURES 7 & 8) or were fixed after 9 hr with glutaraldehyde for scanning electron microscopy (FIGURES 9 & 10). Note that in the matrix-containing cultures, most of the transformed cells spread well and have relatively smooth surfaces. ×570 (FIGURES 7 & 8), ×840 (FIGURES 9 & 10).

external fibronectin is closely associated with the plasma membrane. In the matrix, fibronectin is partially disulfide bonded into complexes. Plasma transglutaminase, activated by thrombin, also cross-links external fibronectin into high-molecular-weight covalent complexes. In cultures of normal fibroblasts, pericellular matrix fibronectin displays extensive codistribution with (pro)collagens types I and III. Transformed adherent cells show decreased formation of the fibronectin-collagen matrix. The deficient synthesis of fibronectin and other matrix components and abnormal interactions with the matrix may account for several phenotypic characteristics of transformed cells.

The pericellular matrix structure has been prepared by use of deoxycholate and hypotonic medium to solubilize the cells. The matrix contains glycosaminoglycans, procollagens, and fibronectin. The fibronectin codistributes with the procollagens. The matrix may be considered to be an *in vitro* equivalent of the connective tissue matrix and basal laminae found *in vivo*. Human sarcoma cells spread rapidly on the prepared matrix and assume an elongated morphology characteristic of normal fibroblasts. The prepared matrix may provide a general tool to study the effects of matrix on cellular behavior and differentiation.

References

1. Vaheri, A. & D. F. Mosher. 1978. High molecular weight, cell surface-associated glycoprotein (fibronectin) lost in malignant transformation. Biochim. Biophys. Acta. In press.
2. Morrison, P. R., J. T. Edsall & S. G. Miller. 1948. Preparation and properties of serum and plasma proteins. XVIII. Separation of purified fibrinogen from fraction I of human plasma. J. Amer. Chem. Soc. **70:** 3103–3108.
3. Mosesson, M. W. & R. A. Umfleet. 1970. The cold insoluble globulin of human plasma. I. Purification, primary characterization and relationship to fibrinogen and other cold insoluble fraction components. J. Biol. Chem. **245:** 5728–5736.
4. Linder, E., A. Vaheri, E. Ruoslahti & J. Wartiovaara. 1975. Distribution of fibroblast surface antigen in the developing chick embryo. J. Exp. Med. **142:** 41–49.
5. Wartiovaara, J., S. Stenman & A. Vaheri. 1976. Changes in expression of fibroblast surface antigen (SFA) in induced cytodifferentiation and in heterokaryon formation. Differentiation **5:**85–89.
6. Stenman, S. & A. Vaheri. 1978. Distribution of a major connective tissue protein, fibronectin, in normal human tissues. J. Exp. Med. **147:** 1054–1064.
7. Vaheri, A., E. Ruoslahti, B. Westermark & J. Ponten. 1976. A common cell-type specific surface antigen (SF) in cultured human glial cells and fibroblasts: loss in malignant cells. J. Exp. Med. **143:** 64–72.
8. Mosher, D. F., O. Saksela, J. Keski-Oja & A. Vaheri. 1977. Distribution of a major surface-associated glycoprotein, fibronectin, in cultures of adherent cells. J. Supramol. Struct. **6:** 551–557.
9. Hynes, R. O. 1976. Cell surface proteins and malignant transformation. Biochim. Biophys. Acta **458:** 73–107.
10. Olden, K. & K. M. Yamada. 1977. Mechanism of the decrease in the major cell surface protein of chick embryo fibroblasts after transformation. Cell **11:** 957–969.
11. Klebe, R. J. 1974. Isolation of a collagen-dependent cell attachment factor. Nature (London) **250:** 248–251.
12. Pearlstein, E. 1976. Plasma membrane glycoprotein which mediates adhesion of fibroblasts to collagen. Nature (London) **262:** 497–500.
13. Grinnell, F. & D. G. Hay. 1976. Cell spreading factor—occurrence and specificity of action. Exp. Cell Res. **102:** 51–62.
14. Höök, M., K. Rubin, Å. Oldberg, B. Öbrink & A. Vaheri. 1977. Cold-insoluble globulin (fibronectin) mediates the adhesion of rat liver cells to plastic petri dishes. Biochem. Biophys. Res. Commun. **79:** 726–733.

15. Keski-Oja, J., D. Mosher & A. Vaheri. 1977. Dimeric character of fibronectin, a major cell surface-associated glycoprotein. Biochem. Biophys. Res. Commun. **74:** 699–706.
16. Hynes, R. O. & A. Destree. 1977. Extensive disulfide bonding at the mammalian cell surface. Proc. Nat. Acad. Sci. USA **74:** 2855–2859.
17. Keski-Oja, J., D. F. Mosher & A. Vaheri. 1976. Cross-linking of a major fibroblast surface-associated glycoprotein (fibronectin) catalyzed by blood coagulation factor XIII. Cell **9:** 29–35.
18. Keski-Oja, J. 1976. Polymerization of a major surface-associated glycoprotein, fibronectin, in cultured fibroblasts. FEBS Lett. **71:** 325–329.
19. Ruoslahti, E., A. Vaheri, P. Kuusela & E. Linder. 1973. Fibroblast surface antigen: a new serum protein. Biochim. Biophys. Acta **322:** 352–358.
20. Wartiovaara, J., E. Linder, E. Ruoslahti & A. Vaheri. 1974. Distribution of fibroblast surface antigen. Association with fibrillar structures of normal cells and loss upon viral transformation. J. Exp. Med. **140:** 1522–1533.
21. Chen, L. B., N. Maitland, P. H. Gallimore & J. K. McDougall. 1977. Detection of the large external transformation-sensitive protein on some epithelial cells. Exp. Cell Res. **106:** 39–46.
22. Marciani, D. J. & J. P. Bader. 1975. Polypeptide composition of cell membranes from chick embryo fibroblasts transformed by Rous sarcoma virus. Biochim. Biophys. Acta **401:** 386–398.
23. Graham, J. M., R. O. Hynes, E. A. Davidson & D. F. Bainton. 1975. The location of proteins labeled by the ^{125}I-lactoperoxidase system in the NIL 8 hamster fibroblasts. Cell **4:** 353–365.
24. Vaheri, A. & E. Ruoslahti. 1974. Disappearance of a major cell-type specific surface antigen (SF) after transformation of fibroblasts by Rous sarcoma virus. Int. J. Cancer **13:** 579–586.
25. Yamada, K. M. & J. A. Weston. 1974. Isolation of a major cell surface glycoprotein from fibroblasts. Proc. Nat. Acad. Sci. USA **71:** 3492–3496.
26. Hynes, R. O., A. T. Destree & V. Mautner. 1976. Spatial organization at the cell surface. *In* Membranes and Neoplasia: New Approaches and Strategies. V. Marchesi, Ed.: 189–201. Alan R. Liss, Inc. New York, N.Y.
27. Hedman, K., A. Vaheri & J. Wartiovaara. 1978. External fibronectin of cultured human fibroblasts is predominantly a matrix protein. J. Cell Biol. **76:** 748–760.
28. Anderson, J. C. 1976. Glycoproteins of connective-tissue matrix. Int. Rev. Connect. Tissue Res. **7:** 251–322.
29. Kefalides, N. 1973. Structure and biosynthesis of basement membranes. Int. Rev. Connect. Tissue Res. **6:** 63–104.
30. Timpl, R., G. Wick & S. Gay. 1977. Antibodies to distinct types of collagens and procollagens and their application in immunohistology. J. Immunol. Methods **18:** 165–182.
31. Gay, S., E. R. Martin, P. K. Muller, R. Timpl & K. Kuhn. 1976. Simultaneous synthesis of types I and III collagen by fibroblasts in culture. Proc. Nat. Acad. Sci. USA **73:** 4037–4040.
32. Bornstein, P. & J. F. Ash. 1977. Cell surface-associated structural proteins in connective tissue cells. Proc. Nat. Acad. Sci. USA **74:** 2480–2484.
33. Vaheri, A., M. Kurkinen, V.-P. Lehto, E. Linder & R. Timpl. 1978. Pericellular matrix proteins, fibronectin and (pro)collagen: codistribution in cultured fibroblasts and loss in transformation. Proc. Nat. Acad. Sci. USA. In press.
34. Kleinman, H. K., E. B. McGoodwin & R. J. Klebe. 1976. Localization of the cell attachment region in types I and II collagens. Biochem. Biophys. Res. Commun. **72:** 426–432.
35. Engvall, E. & E. Ruoslahti. 1977. Binding of soluble form of fibroblast surface protein, fibronectin, to collagen. Int. J. Cancer **20:** 1–5.
36. Dessau, W., B. C. Adelmann, R. Timpl & G. R. Martin. 1978. Identification of the sites in collagen α-chains binding serum anti-gelatin factor (cold insoluble globulin). Biochem. J. **169:** 55–59.

37. VRACKO, R. 1974. Basal lamina scaffold—anatomy and significance for maintenance of orderly tissue structure. Amer. J. Pathol. **77:** 313–346.
38. KAWAI, S. & H. HANAFUSA. 1971. The effects of reciprocal changes in temperature on the transformed state of cells infected with a Rous sarcoma virus mutant. Virology **46:** 470–479.
39. PONTÉN, J. 1976. The relationship between in vitro transformation and tumor formation in vivo. Biochim. Biophys. Acta **458:** 397–422.
40. REICH, E. 1975. Plasminogen activator: secretion by neoplastic cells and macrophages. *In* Cold Spring Harbor Conferences on Cell Proliferation. E. Reich, D. Rifkin & E. Shaw, Eds. Vol. 2: 333–341. Cold Spring Harbor Laboratory. Cold Spring Harbor, N.Y.
41. QUIGLEY, J. P. 1976. Association of a protease (plasminogen activator) with a specific membrane fraction isolated from transformed cells. J. Cell Biol. **71:** 472–486.
42. MOSHER, D. F. This monograph.
43. WARTIOVAARA, J., I. LEIVO, I. VIRTANEN, A. VAHERI & C. F. GRAHAM. 1978. Appearance of fibronectin during differentiation of mouse teratocarcinoma *in vitro*. Nature (London) **272:** 355–356.
44. HEDMAN, K., M. KURKINEN, K. ALITALO, A. VAHERI, S. JOHANSSON & M. HÖÖK. 1978. Isolation of fibronectin-collagen matrix of human fibroblast cultures. Submitted for publication.
45. ALITALO, K., M. KURKINEN, K. HEDMAN & A. VAHERI. In preparation.

CELLULAR TRANSGLUTAMINASE, GROWTH, AND TRANSFORMATION

Paul J. Birckbichler and M. K. Patterson, Jr.

Biomedical Division
The Samuel Roberts Noble Foundation, Inc.
Ardmore, Oklahoma 73401

Many properties are expressed at or mediated by the cell surface; among these properties appear to be those that distinguish cancerous from normal cells.[1] While a variety of normal and transformed cells have been studied in culture, few comparative studies have utilized cells of human origin. In such a study, we found that several membrane alterations noted in chick and rodent cells after transformation were absent from Simian virus 40 (SV40) transformed human cells and their normal parent counterparts.[2] While these findings suggested that the "transformed state" of human cells in culture did not conform rigidly to the "transformed state" observed in chick and rodent cell lines, such cells still possessed characteristics of the "transformed state."[2,3]

In a continuing effort to find differences between the membranes of normal and transformed human cells, we examined such cells for the presence of covalent isopeptide cross-links. Our studies have focused on the isopeptide ϵ-(γ-glutamyl)-lysine and the appropriate enzyme system that catalyzes isopeptide formation.

Transglutaminase (TGase) catalyzes a Ca^{2+}-dependent acyl transfer reaction between the γ-carboxamide group of peptide-bound glutamine and various primary amines to yield new γ-amide bonds of glutamic acid and ammonia.[4,5] If the amine is the ϵ-amino group of peptide-bound lysine, inter- or intramolecular ϵ-(γ-glutamyl)lysine isopeptide cross-links are formed. While stabilization of fibrin clots via ϵ-(γ-glutamyl)lysine formation is the physiologic role of plasma TGase[6-8] (fibrinoligase, fibrin-stabilizing factor, Laki-Lorand factor, activated factor XIII), the role of cellular TGase is still unclear. Cellular TGase has been suggested to be involved in isopeptide formation, leading to the production of cross-linked insoluble polymeric cellular proteins.[5,9,10] While ϵ-(γ-glutamyl)lysine has been localized in a variety of materials,[6-8,11-15] a direct correlation between the isopeptide product and TGase is still lacking in most systems.

Although our studies have utilized both *in vivo* and *in vitro* systems, we will confine this discussion primarily to studies with cells in culture. The cell system was that from human embryonic lung, strain WI-38, and an SV40-transformed counterpart, WI-38 VA13A.

Transglutaminase Activity in Normal and Transformed Cells

In paired cell systems that we have examined, transformed cells had significantly lower homogenate TGase activity than did their normal counterparts (Table 1). This finding was true *in vivo* for rat liver and malignant primary hepatoma induced with 3′-methyl-4-dimethylaminoazobenzene and for transplanted malignant Novikoff hepatoma.[16] Simian virus 40-transformed human and polyoma-transformed hamster cells, in addition to methylcholanthrene-transformed mouse cells, maintained in cell culture showed similar reductions in

0077-8923/78/0312-0354 $01.75/0 © 1978, The New York Academy of Sciences

TGase activity.[17] The differential in activity was independent of several homogenization procedures, not a result of incomplete cell rupture, and not the result of endogenous activators or inhibitors of TGase in the cells. Similarly, the possibility that the results were due to plasma TGase contamination of the preparations was minimal. Normal and transformed preparations exhibited similar Michaelis values for the amine donor putrescine, suggesting that the affinity of the enzyme for this substrate was unaffected by transformation.[16] Furthermore, these findings suggested that the reduction in activity was independent of the transforming agent, and there were differential levels of TGase-catalyzed cross-links in normal and transformed cells.

TABLE 1

TRANSGLUTAMINASE ACTIVITY IN NORMAL AND TRANSFORMED CELLS

Cell/Tissue	Source	Specific Activity* (U/mg of protein)	Normal / Transformed
Liver	rat	8.5 ± 0.4† (8)‡	—
Novikoff hepatoma	rat	2.0 ± 0.1 (6)	4.3
Primary hepatoma§	rat	5.4 ± 0.5 (10)	1.6
WI-38	human	7.0 ± 1.0 (7)	—
WI-38 VA13A	human	0.12 ± 0.02 (7)	58.3
WI-38 VA13-2RA	human	0.04	175.
BHK-21/C13	hamster	1.1 ± 0.2 (3)	—
PyBHK-21/C13	hamster	0.03 ± 0.02 (3)	36.7
C3H/10T1/2CL8	mouse	0.28 ± 0.10 (3)	—
C3H/10T1/2CL8MCATCL15	mouse	0.03 ± 0.01 (3)	9.3

*Homogenate activity except for C3H system, which is the 25,000 *g* supernatant.
†Means ± SE.
‡Numbers of preparations assayed.
§Induced with 3′-methyl-4-dimethylaminoazobenzene.

ε-(γ-GLUTAMYL)LYSINE IN NORMAL AND TRANSFORMED CELLS

To complement the enzyme studies, quantitation of enzyme product, ε-(γ-glutamyl)lysine, was performed. The isopeptide bond is resistant to proteolytic attack, so that after exhaustive enzyme digestion by pronase and leucine aminopeptidase, ε-(γ-glutamyl)lysine can be identified by amino acid chromatography.[18] In most cases, more than 80% of the protein is converted to free amino acids and isopeptide during enzymatic digestion. Analysis of pronase-digested total cellular protein revealed that normal WI-38 cultures contained 40–100 times more ε-(γ-glutamyl)lysine bonds per cell than did WI-38 VA13A cultures (TABLE 2). These values agree with the enzymatic differential shown in TABLE 1.

IMMUNOFLUORESCENT STUDIES

A differential between WI-38 and WI-38 VA13A cultures was also demonstrated when cells were stained by indirect immunofluorescence with rabbit anti-TGase antiserum. Sparse WI-38 cultures produced a weak intracellular fluor-

TABLE 2

ϵ-(γ-GLUTAMYL)LYSINE IN CULTURED HUMAN CELLS*

Cell	Days in Culture	Cells × 10^{-6}	pmoles of ϵ-(γ-Glutamyl)lysine / 10^6 Cells
WI-38			
1	6–7	10.1–24.6	515 ± 37† (4)‡
2	14	6.2 9.3	1498 ± 202 (4)
WI-38 VA13A			
1	7–9	27.3–44.1	14 ± 2 (5)

*Isolated from total cellular protein incubated with pronase for 168 hr.
†Means ± SE.
‡Numbers of preparations assayed.

escence (FIGURE 1, A). This fluorescence became more intense as cultures approached saturation density (FIGURE 1, B). Dense cultures also showed some surface-associated fluorescence when acetone fixation was omitted. WI-38 VA13A cultures appeared to display intracellular fluorescence, which did not increase as cultures became confluent (FIGURE 1, C). There was also no evidence of surface

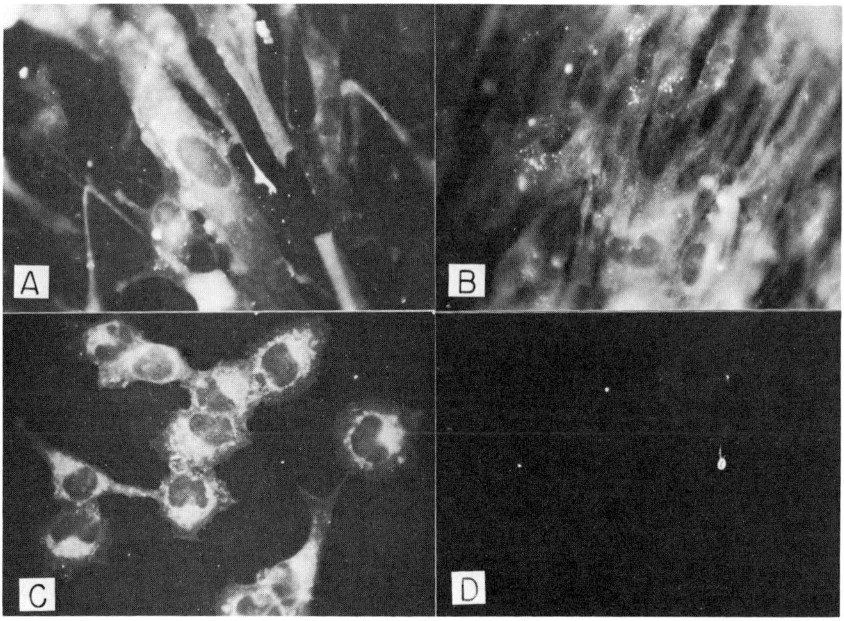

FIGURE 1. Transglutaminase immunofluorescence of normal and transformed human embryonic fibroblasts. Coverslip cultures were washed three times and incubated with phosphate-buffered saline. After 30 min, the liquid was removed, and the cells were washed, fixed with formaldehyde-acetone, and studied by indirect immunofluorescence. A, WI-38, preconfluent, rabbit anti-TGase; B, WI-38, confluent, rabbit anti-TGase; C, WI-38 VA13A, preconfluent, rabbit anti-TGase; D, WI-38 VA13A, preconfluent, rabbit serum.

fluorescence. These findings complement the enzymatic studies summarized in TABLE 1. Since the WI-38 VA13A cell showed fluorescence but had low TGase activity, the enzyme may be present in an inactive form in these cells.

TRANSGLUTAMINASE ACTIVITY AND CELLULAR PROLIFERATION

Examination of WI-38 homogenates at various times after subculturing (1:4 split ratio) for TGase activity revealed an apparent inverse relationship between growth of the cells and enzymatic activity (FIGURE 2). During the period of rapid proliferation, Days 1–5 after subculturing, enzymatic activity was low and relatively constant. As proliferation ceased, TGase activity increased and continued to rise for several days after confluence. Subsequently, enzymatic activity decreased to levels comparable to those of cultures 7 days after subculturing. In

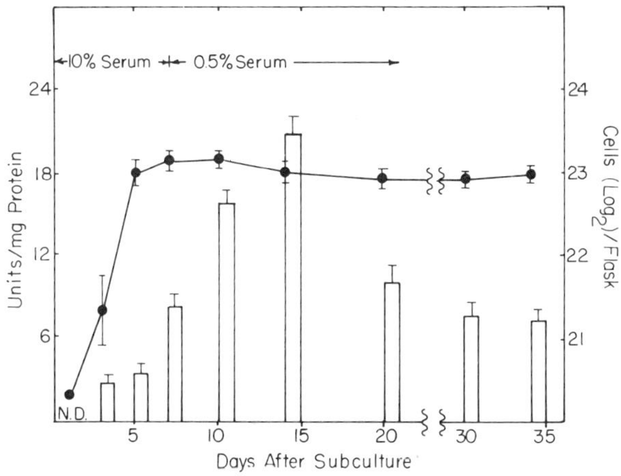

FIGURE 2. Transglutaminase activity in proliferating and nonproliferating WI-38 cells. Bar graph, TGase specific activity at various times after subculture; solid line, cell number. Values are the means ± SE for samples from three to five experiments. N.D., not determined.

FIGURE 2, the serum concentration in the medium was reduced to 0.5% after 7 days to ensure that essentially nonmitotic populations were maintained. However, cultures left in the presence of medium that contained 10% serum throughout the experiment exhibited a similar TGase pattern, although the maximal TGase activity attained was less than that seen in the low-serum cultures. That slowing of growth rate reflected TGase activity expression was further strengthened by the observation that preconfluent, rapidly growing WI-38 cultures treated with low-serum medium showed a reduced growth rate and an increase in TGase activity. Thus, it appears that the number of cells in an essentially nonmitotic state in normal cultures had an effect on the expression of TGase activity. Therefore, we postulated that the low enzymatic activity in transformed systems may reflect the presence of a rapidly proliferating population.[17]

TRANSGLUTAMINASE AND TRYPSIN TREATMENT

The regulating mechanism underlying the fluctuations of enzymatic activity under varying proliferative rates cannot be fully defined at this time. A model was suggested by the observation that treatment of intact cells with trypsin resulted in an increase in enzyme-specific activity.[17] The increase occurred between 30 and 60 min after trypsin administration and was dependent on the amount of trypsin added. Levels less than 1 μg of trypsin/ml did not increase TGase activity, whereas a two- to fourfold increase was evident at 5 μg of trypsin/ml. Higher concentrations of trypsin were not effective in increasing TGase activity further and caused the cells to detach from the flask. Increased enzymatic activity was not dependent on protein synthesis, because puromycin (100 μg/ml) failed to block the increase in enzymatic activity while effectively inhibiting [^{14}C]leucine incorporation into protein.[17] Thus, synthesis of new enzyme molecules probably did not occur; rather, existing TGase molecules were activated. Direct addition of trypsin to cell homogenates destroyed TGase activity, suggesting that trypsin acted at the membrane level.

In addition to activation of TGase, trypsin treatment of cell cultures resulted in solubilization of TGase activity. This phenomenon was demonstrated by a differential sedimentation of enzymatic activity when homogenates were centrifuged at 105,000*g* for 1 hr. Only small amounts of TGase activity were found in the cytosol fraction of untreated cells, whereas the majority of the activity was found in this fraction in trypsin-treated cells. The average cytosol distributions ($\pm$SE) for at least six determinations were 3.0 $\pm$ 0.6% for untreated and 79.2 $\pm$ 1.1% for trypsin-treated cultures. The qualitative nature of the enzyme appeared to be unaltered, because untreated and trypsin-treated preparations displayed the same affinity for the amine donor substrate, putrescine. The apparent K_m values, as calculated by the weighted least-squares method from triplicate experiments, were 0.37 $\pm$ 0.04 mM for untreated cells and 0.52 $\pm$ 0.06 mM for trypsin-treated cells. The increased enzymatic activity was a result of exposure of more catalytic sites, as evidenced by V_{max} values of 299 $\pm$ 28 and 1724 $\pm$ 178 pmol of putrescine incorporated into casein per minute per milligram of protein for untreated and trypsin-treated cultures, respectively.

Based on these trypsin findings, it could be speculated that nonproliferating normal cells have a higher TGase activity as a result of a cellular protease that effects "trypsin-like" activation of TGase. Several pieces of information, however, argue against this hypothesis. First, if this hypothesis were true, a higher percentage of the TGase activity should be cytoplasmic in confluent and nonproliferating cultures. Cytoplasmic TGase distributions similar to those observed after trypsin treatment were not evident in nonproliferating cells. The amount of cytoplasmic TGase activity was never greater than 6% in any culture examined. Second, the protease inhibitors *p*-tosyl-L-arginine methyl ester hydrochloride and L-1-tosylamide-2-phenylethylchloromethyl ketone, when added to proliferating cultures 48 hr after subculture, were ineffective in preventing the increase in TGase activity as the cultures progressed to confluence. In some instances, the cultures that contained the protease inhibitors had higher TGase activities. Third, proliferating fibroblasts reportedly demonstrate higher surface protease activity than do those that have stopped proliferating.[19]

While the trypsin concentrations used in these studies are known to be mitogenic, the increase in TGase activity that occurred after trypsin treatment did not appear to be a result of mitogenic stimulation per se (TABLE 3). Of several mitogenic stimuli (medium change, Ca^{2+} addition, proteases) examined, only proteases were effective in increasing TGase activity, with trypsin producing the largest stimulation. Note that thrombin was ineffective in activating TGase. Further evidence for a nonmitogenic role for trypsin activation is that high TGase activity appears to be contrary to the proliferative state in this system (FIGURE 2 & TABLE 1). Rather, the activation seemed to be related to some other consequence of trypsin administration.

TRANSGLUTAMINASE AND FIBRONECTIN

The rapidity with which trypsin activated TGase suggested that hydrolytic cleavage of a cell surface component might be involved, with cleavage of this component apparently leading to solubilization of TGase activity. Similarly, the differential effects of trypsin and pronase as compared to that of thrombin on activating TGase parallels their action in cleaving a cell surface glycoprotein found in several fibroblast cells. This glycoprotein has received many names,[20] including fibronectin,[21] which will be used in this discussion. Most of these surface proteins are thought to be the same protein and cross-react immunologically with the serum protein, cold-insoluble globulin.[22] These proteins are apparently transformation sensitive, since they are lost from the cell surface after transformation.[20,23] Trypsin and pronase catalyze removal of fibronectin from fibroblasts, whereas thrombin fails to do so under similar conditions.[24] To determine whether a relationship between TGase and fibronectin exists, a comparative study was undertaken.

TABLE 3

TRANSGLUTAMINASE ACTIVITY AFTER MITOGEN TREATMENT

Treatment*	Concentration	Treated Activity / Untreated Activity
Medium change†	—	0.9, 0.9
$CaCl_2$	1.8–5 mM	1.1 ± 0.2 (3)
Trypsin‡	5 μg/ml	2.7 ± 0.3 (5)
Pronase§	5 μg/ml	1.7, 1.8
Thrombin¶	5 μg/ml	1.1, 1.3

*Cultures were washed (except those that received the medium change) and incubated with the appropriate reagent in Earle's solution for 1 hr at 37°C prior to assay.

†Modified McCoy's medium 5a (2) plus 10% fetal bovine serum (Reheis Chemical Corporation).

‡Worthington Biochemical Corporation.

§Calbiochem.

¶Parke-Davis, topical.

Initially, surface fibronectin in WI-38 cells was labeled by the lactoperoxidase-catalyzed iodination procedure with ^{125}I.[25] Addition of trypsin to these labeled cells resulted in an enzyme-dependent loss of 60–75% of the radioactive label within 30 min. While the ^{125}I studies provided circumstantial evidence of fibronectin involvement, we desired to confirm these findings by another method.

WI-38 and WI-38 VA13A cells demonstrated a differential in fluorescence when stained by indirect immunofluorescence with goat antifibronectin antiserum (FIGURE 3). Preconfluent, growing WI-38 cells showed essentially an intracellular

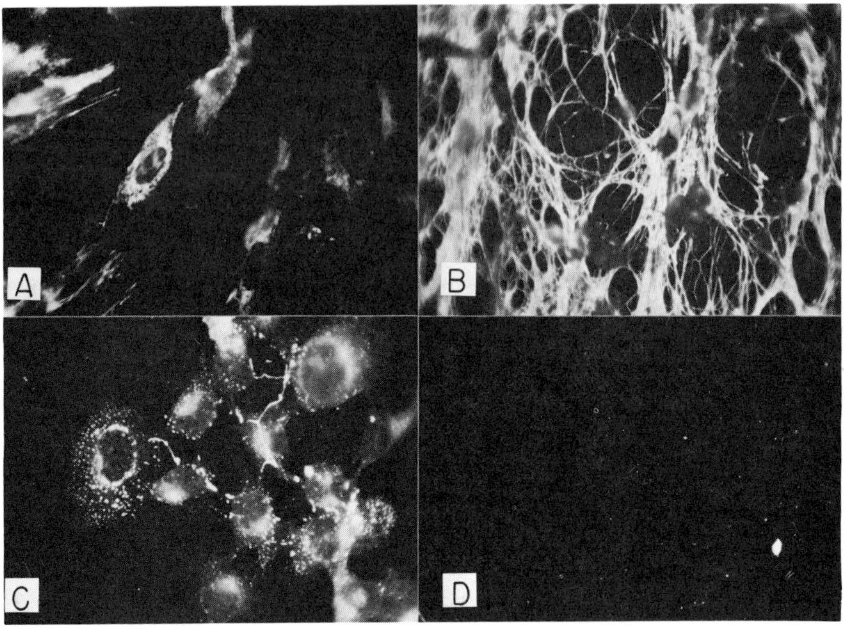

FIGURE 3. Fibronectin immunofluorescence of normal and transformed human embryonic fibroblasts. Coverslip cultures were washed three times and incubated with Earle's solution. After 1 hr, the liquid was removed, and the cells were washed, fixed with formaldehyde-acetone, and studied by indirect immunofluorescence. A, WI-38, preconfluent, goat antifibronectin; B, WI-38, confluent, goat antifibronectin; C, WI-38 VA13A, preconfluent, goat antifibronectin; D, WI-38, preconfluent, goat serum.

fibronectin pattern (FIGURE 3, A), while confluent cultures displayed an extensive fibrillar network (FIGURE 3, B). Transformed WI-38 VA13A cell cultures, however, did not show extensive fibronectin staining when either low-or high-density populations were examined (FIGURE 3, C). These observations are in agreement with those reported by other workers.[26,27] Recall that the normal and transformed human cells showed an analogous differential when stained for TGase (FIGURE 1).

When WI-38 cells were incubated with 5 μg of trypsin/ml for 1 hr, they demonstrated an altered staining pattern for both TGase and fibronectin (FIGURE 4). It should be noted that the staining pattern observed in FIGURE 4(A) differs from that seen in FIGURE 1(B). This difference has been attributed to the differential effect of incubation in Earle's solution or in phosphate-buffered saline. Most of the fibril-related fluorescence seen with anti-TGase antiserum prior to trypsin addition (FIGURE 4, A) was eliminated, leaving some intracellular fluorescence (FIGURE 4, C). The fibrillar pattern observed for fibronectin (FIGURE 4, B) was also destroyed by trypsin treatment, leaving a residual intracellular fluorescence (FIGURE 4, D).

Extraction of cells with nonionic detergent removed a majority of the cellular proteins but did not solubilize fibronectin.[28,29] The distinct pattern of fibronectin fibers was still visible after Triton® X-100 extraction (FIGURE 5, D). A fibrillar pattern was now also evident when the cells were stained with antiserum to TGase (FIGURE 5, C). It is presently not known if the TGase fibers observed after Triton X-100 extraction were associated with the cell cytoskeleton or with the extracellular fibronectin network. This finding raised the possibility that more than one class of TGase may be present in these cells. Homogenization of cells in Triton X-100 did not lead to an increase in TGase activity. However, enzymatic analysis revealed that both the Triton X-100 extract and the Triton X-100 residue contained TGase activity.

The parallel behavior of TGase and fibronectin in these studies raised the question as to whether fibronectin was a substrate for cellular TGase, since others have shown that cellular fibronectin is a substrate for plasma TGase.[21,30,31] Monomeric fibronectin purified from human plasma gave rise to a single band on sodium dodecyl sulfate polyacrylamide gel electrophoresis (FIGURE 6). The gel pattern remained essentially unchanged when fibronectin was incubated at 37°C for 1 hr with 20 μg of guinea pig liver TGase in the presence of 1 mM EDTA (FIGURE 6, gel B). However, replacement of EDTA with 5 mM $CaCl_2$ resulted in the appearance of material with a larger apparent molecular weight (FIGURE 6, gel A). It appeared that more than one species may be present after calcium-dependent polymerization.

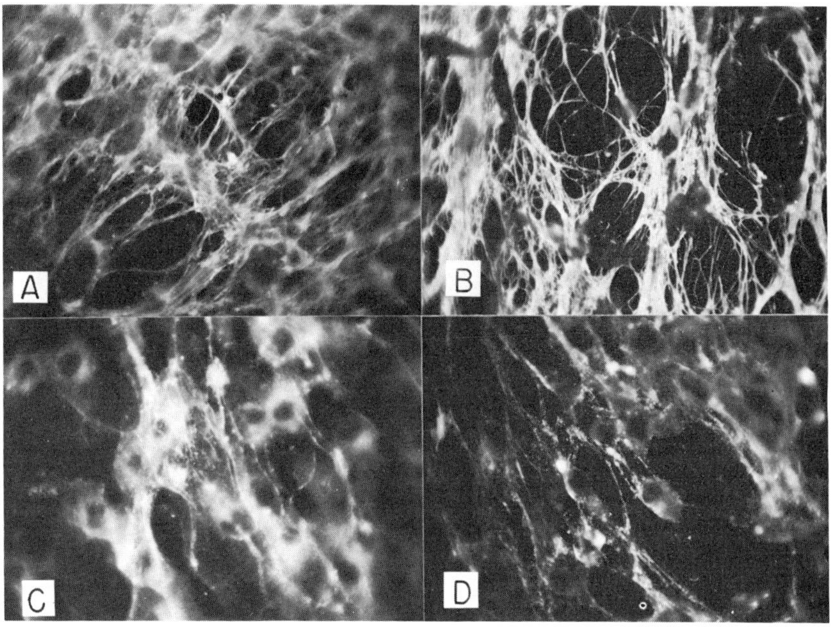

FIGURE 4. Transglutaminase and fibronectin immunofluorescence of trypsin-treated normal human embryonic fibroblasts. Coverslip cultures were washed three times and incubated with Earle's solution or Earle's solution that contained 5 μg of trypsin/ml. After 1 hr, the liquid was removed, and the cells were washed, fixed with formaldehyde-acetone, and studied by indirect immunofluorescence. A, WI-38, no trypsin, rabbit anti-TGase; B, WI-38, no trypsin, goat antifibronectin; C, WI-38, trypsin, rabbit anti-TGase; D, WI-38, trypsin, goat antifibronectin.

Incubation of fibronectin under catalytic conditions (5 mM $CaCl_2$, 1 hr, 37°C) in the absence of guinea pig TGase also induced a polymerization reaction (FIGURE 6, gel C). The degree of autopolymerization was considerably less than that seen when guinea pig TGase was present.

Appearance of polymeric material in gels A and C suggested the formation of intermolecular cross-links. Sufficient dithiothreitol (10 mM) was present in all samples to ensure that disulfide formation was not a factor. Isopeptide formation was confirmed by the isolation of 7.0 and 0.8 nmol of ϵ-(γ-glutamyl)lysine per milligram of protein from proteolytic digests of TGase-polymerized fibronectin (gel A) and autopolymerized fibronectin (gel C), respectively. The amount of cross-link in TGase-polymerized fibronectin (gel A) was equivalent to one to two cross-links per 210,000 daltons of protein. Monomeric fibronectin and fibronectin incubated with guinea pig TGase and EDTA contained no detectable isopeptide, in agreement with a previous report.[10] Some of the isopeptide (< 10%) in TGase-polymerized fibronectin (gel A) may have resulted from autopolymerization of the guinea pig TGase used to polymerize the fibronectin.[18] This phenomenon would not be a factor when guinea pig TGase was omitted (gel C).

The autopolymerization of fibronectin in the presence of calcium (gel C) suggested that fibronectin had catalytic activity. Some TGase activity was found when

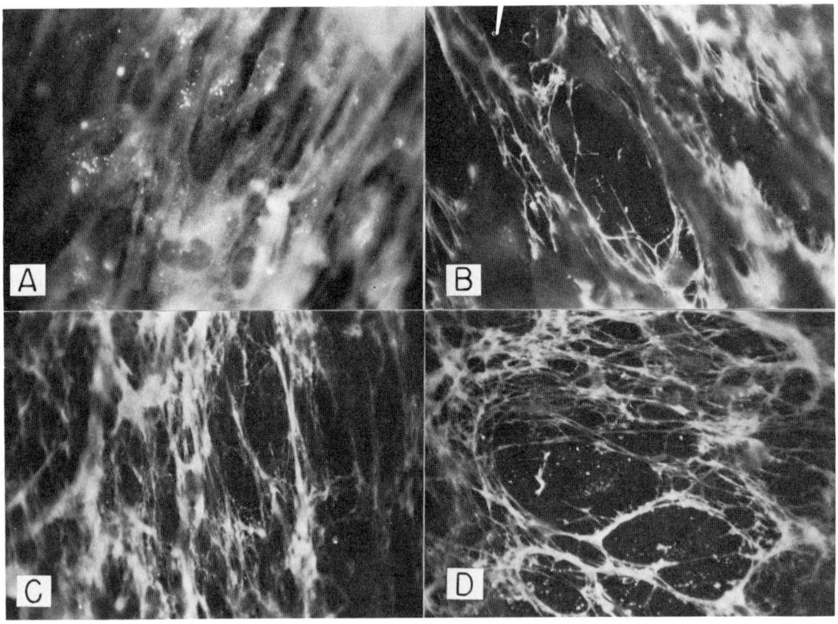

FIGURE 5. Transglutaminase and fibronectin immunofluorescence of Triton X-100-extracted normal human embryonic fibroblasts. Coverslip cultures were washed three times and incubated with phosphate-buffered saline or phosphate-buffered saline that contained 0.5% (v/v) Triton X-100. After 30 min, the liquid was removed, and the residue was washed, fixed with formaldehyde-acetone, and studied by indirect immunofluorescence. A, WI-38, no Triton X-100, rabbit anti-TGase; B, WI-38, no Triton X-100, goat antifibronectin; C, WI-38, Triton X-100, rabbit anti-TGase; D, WI-38, Triton X-100, goat antifibronectin.

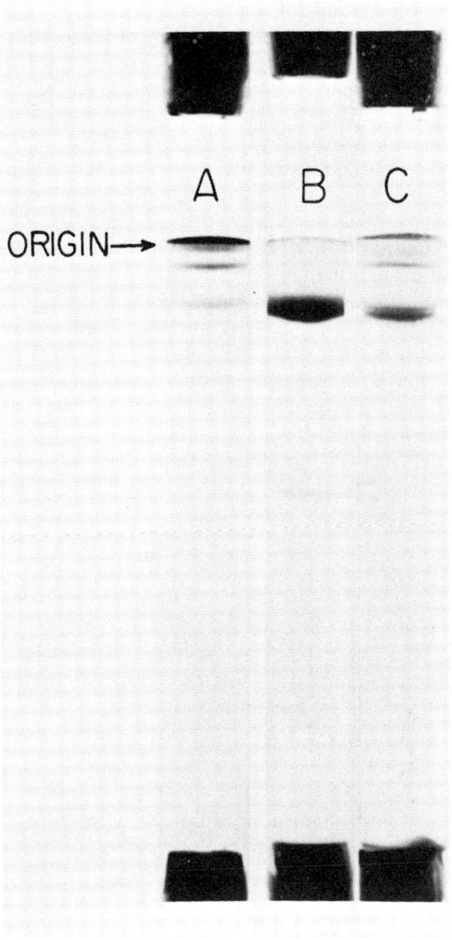

FIGURE 6. Sodium dodecyl sulfate polyacrylamide gel electrophoresis pattern of fibronectin. Each sample layered contained about 5 μg of protein. Electrophoresis (0.5 × 6 cm gels, 5.6% in acrylamide, 1% sodium dodecylsulfate) was run for 75 min at a current of 8 mA/gel.[18] A, 5 mM $CaCl_2$, no EDTA, 20 μg of guinea pig liver TGase, 37°C, 1 hr; B, no $CaCl_2$, 1 mM EDTA, 20 μg of guinea pig liver TGase, 37°C, 1 hr; C, 5 mM $CaCl_2$, no EDTA, no guinea pig TGase, 37°C, 1 hr. The faint band near the center in gel B is guinea pig liver TGase.

fibronectin was incubated with calcium, [^{14}C]putrescine, and casein in a standard TGase assay mixture. This activity, however, was probably the result of contamination of the fibronectin preparation with low levels of plasma TGase.[29]

Irrespective of this activity, the observation that fibronectin can be polymerized with a cellular TGase suggests that at least some of the fibronectin in WI-38 cultures might act as a substrate for the enzyme. The further significance of these findings is that they confirm the observation that the polymer formed was due to the catalytic formation of ϵ-(γ-glutamyl)lysine.[21,30,32]

CONCLUSIONS

Transglutaminase activity was higher in normal human embryonic lung cells than in its SV40-transformed counterparts. The same was true for several other paired cell systems. Similarly, the enzymatic reaction product, ε-(γ-glutamyl)-lysine, was present in greater amounts in normal lung cells. Both enzymatic activity and immunofluorescent staining with rabbit anti-TGase were greater in confluent, nongrowing normal cells, and TGase activity could be increased by incubation of cells with low levels of certain proteases.

The relationship between enzymatic activity and the proliferative state of the cells suggests a physiologic role for TGase in mammalian cells *in vitro*. Stability of the plasma membrane may be a necessary adjunct to the nonproliferative state. Transglutaminase-catalyzed cross-linking of cell membrane proteins may provide such stability. The low enzymatic activities and low ε-(γ-glutamyl)lysine levels observed in transformed cells may thus reflect their rapidly proliferating state.

Striking parallels between TGase and fibronectin in human lung fibroblasts in culture are evident: normal cells have higher TGase activity and bind fibronectin; TGase activity and cell surface fibronectin in the normal cell parallel one another (when one is low, the other is low; when one is high, the other is high); proteases that cleave fibronectin activate TGase: the time course for TGase activation by the protease parallels fibronectin removal from the cell surface by the same protease; fibronectin is a substrate for TGase. While TGase and fibronectin are probably not the same molecule, a close interrelationship between the two seems inescapable.

Acknowledgments

We thank Gerald R. Orr, Eugene Conway, Henry A. Carter, and Merle D. Maxwell for their technical assistance. We express our appreciation also to Dr. Deane Mosher for enlightening discussions on parts of this work and for supplying us with purified fibronectin and goat antifibronectin antiserum.

References

1. Patterson, M. K., Jr. 1974. Alterations in membrane structure and function associated with neoplastic transformation in vitro. J. Nat. Cancer Inst. **53:** 1493–1498.
2. Patterson, M. K., Jr., P. J. Birckbichler, E. Conway & G. R. Orr. 1976. Amino acid and hexose transport of normal and Simian virus 40-transformed human cells. Cancer Res. **36:** 394–397.
3. Girardi, A. J., D. Weinstein & P. S. Moorhead. 1966. SV_{40} transformation of human diploid cells. Ann. Med. Exp. Fenniae **44:** 242–254.
4. Clarke, D. D., M. J. Mycek, A. Neidle & H. Waelsch. 1959. The incorporation of amines into protein. Arch. Biochem. Biophys. **79:** 338–354.
5. Folk, J. E. & S. I. Chung. 1973. Molecular and catalytic properties of transglutaminases. *In* Advances in Enzymology. A. Meister, Ed. Vol. 38: 109–191. John Wiley & Sons, Inc. New York, N.Y.
6. Matačić, S. & A. G. Loewy. 1968. The identification of isopeptide cross-links in insoluble fibrin. Biochem. Biophys. Res. Commun. **30:** 356–362.
7. Lorand, L., N. G. Rule, H. H. Ong, R. Furlanetto, A. Jacobsen, J. Downey, N. Öner & J. Bruner-Lorand. 1968. Amine specificity in transpeptidation. Inhibition of fibrin cross-linking. Biochemistry **7:** 1214–1223.
8. Pisano, J. J., J. S. Finlayson & M. P. Peyton. 1968. Cross-link in fibrin polymerized by factor XIII:ε-(γ-glutamyl)lysine. Science **160:** 892–893.
9. Lorand, L. & P. Stenberg. 1976. Endo-γ-glutamine:ε-lysine transferases enzymes which cross-link proteins. *In* Handbook of Biochemistry and Molecular Biology. Proteins. G. D. Fasman, Ed. Vol. II: 669–685. CRC Press. Cleveland, Ohio.
10. Folk, J. E. & J. S. Finlayson. 1977. The ε-(γ-glutamyl)lysine cross-link and the catalytic role of transglutaminases. Adv. Protein Chem. **31:** 1–133.

11. HARDING, H. W. J. & G. E. ROGERS. 1971. ε(γ-Glutamyl)lysine cross-linkage in citrulline-containing protein fractions from hair. Biochemistry **10:** 624–630.
12. WILLIAMS-ASHMAN, H. G., A. C. NOTIDES, S. S. PABALAN & L. LORAND. 1972. Transamidase reactions involved in the enzymic coagulation of semen: isolation of γ-glutamyl-ε-lysine dipeptide from clotted secretion protein of guinea pig seminal vesicle. Proc. Nat. Acad. Sci. USA **69:** 2322–2325.
13. BIRCKBICHLER, P. J., R. M. DOWBEN, S. MATAČIĆ & A. G. LOEWY. 1973. Isopeptide bonds in membrane proteins from eukaryotic cells. Biochim. Biophys. Acta **291:** 149–155.
14. ABERNETHY, J. L., R. L. HILL & L. A. GOLDSMITH. 1977. ε-(γ-Glutamyl)lysine cross-links in human stratum corneum. J. Biol. Chem. **252:** 1837–1839.
15. RICE, R. H. & H. GREEN. 1977. The cornified envelope of terminally differentiated human epidermal keratinocytes consists of cross-linked protein. Cell **11:** 417–422.
16. BIRCKBICHLER, P. J., G. R. ORR & M. K. PATTERSON, JR. 1976. Differential transglutaminase distribution in normal rat liver and rat hepatoma. Cancer Res. **36:** 2911–2914.
17. BIRCKBICHLER, P. J., G. R. ORR, E. CONWAY & M. K. PATTERSON, JR. 1977. Transglutaminase activity in normal and transformed cells. Cancer Res **37:** 1340–1344.
18. BIRCKBICHLER, P. J., G. R. ORR, H. A. CARTER & M. K. PATTERSON, JR. 1977. Catalytic formation of ε-(γ-glutamyl)lysine in guinea pig liver transglutaminase. Biochem. Biophys. Res. Commun. **78:** 1–7.
19. HATCHER, V. B., M. S. OBERMAN, M. S. WERTHEIM, C. Y. RHEE, G. TSIEN & P. G. BURK. 1977. The relationship between surface protease activity and the rate of cell proliferation in normal and transformed cells. Biochem. Biophys. Res. Commun. **76:** 602–608.
20. HYNES, R. O. 1976. Cell surface proteins and malignant transformation. Biochim. Biophys. Acta **458:** 73–107.
21. KESKI-OJA, J., D. F. MOSHER & A. VAHERI. 1976. Cross-linking of a major fibroblast surface-associated glycoprotein (fibronectin) catalyzed by blood coagulation factor XIII. Cell **9:** 29–35.
22. RUOSLAHTI, E. & A. VAHERI. 1975. Interaction of soluble fibroblast surface (SF) antigen with fibrinogen and fibrin. Identity with cold insoluble globulin of human plasma. J. Exp. Med. **141:** 497–501.
23. VAHERI, A. & E. RUOSLAHTI. 1974. Disappearance of a major cell-type-specific surface glycoprotein antigen (SF) after transformation of fibroblasts by Rous sarcoma virus. Int. J. Cancer **13:** 579–586.
24. TENG, N. N. H. & L. B. CHEN. 1975. The role of surface proteins in cell proliferation as studied with thrombin and other proteases. Proc. Nat. Acad. Sci. USA **72:** 413–417.
25. HYNES, R. O. & J. M. BYE. 1974. Density and cell cycle dependence of cell surface proteins in hamster fibroblasts. Cell **3:** 113–120.
26. VAHERI, A. & E. RUOSLAHTI. 1975. Fibroblast surface antigen produced but not retained by virus-transformed human cells. J. Exp. Med. **142:** 530–535.
27. STENMAN, S., J. WARTIOVAARA & A. VAHERI. 1977. Changes in the distribution of a major fibroblast protein, fibronectin, during mitosis and interphase. J. Cell Biol. **74:** 453–467.
28. HYNES, R. O., A. T. DESTREE & V. MAUTNER. 1976. Spatial organization at the cell surface. *In* Progress in Clinical and Biological Research: Membranes and Neoplasia: New Approaches and Strategies. V. T. Marchesi, Ed. Vol. 9: 189–201. Alan R. Liss, Inc. New York, N.Y.
29. MOSHER, D. 1977. Personal communication.
30. KESKI-OJA, J. 1976. Polymerization of a major surface-associated glycoprotein, fibronectin, in cultured fibroblasts. FEBS Lett. **71:** 325–329.
31. MOSHER, D. F. 1977. Labeling of a major fibroblast protein (fibronectin) catalyzed by blood coagulation factor XIII. Biochim. Biophys. Acta **491:** 205–210.
32. MOSHER, D. F. 1975. Cross-linking of cold-insoluble globulin by fibrin-stabilizing factor. J. Biol. Chem. **250:** 6614–6621.

MODULATION OF CELL SURFACE GLYCOCALYX: STUDIES ON LARGE, EXTERNAL, TRANSFORMATION-SENSITIVE PROTEIN*

Lan Bo Chen,† Keith Burridge, Andrew Murray, Marcia L. Walsh, Christine D. Copple, Anne Bushnell, James K. McDougall, and Phillip H. Gallimore‡

Cold Spring Harbor Laboratory
Cold Spring Harbor, New York 11724

One of the distinctive features of cancer cells *in vivo* is their antisocial behavior in a sea of well-organized normal cells. The symptoms of such abnormality include cell division at the wrong time and in the wrong place, penetration to neighboring tissues, travel in the circulation, which should normally be confined to only a few cell types, and adhesion to forbidden tissues. Numerous factors contribute to the antisocial behavior of cancer cells. One of them agreed upon by many investigators, is the alteration in the cell surface. Among the alterations on the cell surface that can potentially contribute to the antisocial behavior of cancer cells are changes in composition of the lipid bilayer, in glycolipids, in glycocalyceal components, in topographic structure, and in the motile properties of the cell surface. Each of these aspects is being intensively investigated. The current status of this field has recently been reviewed.[1-4]

Among numerous alterations in cancer cells, a reduction in one of the cell surface components seems to warrant further investigation, because it may be relevant to the antisocial behavior of cancerous cells. This component is the large, external, transformation-sensitive (LETS) protein. LETS protein was first detected in fibroblasts by a variety of techniques and has since been termed as band 1,[5] galactoprotein a,[6] fibronectin,[7] zeta,[8] and cell surface protein, CSP.[9] It is a glycoprotein with a molecular weight of 220–250,000 daltons. It is significantly reduced in many virally transformed fibroblasts (for review, see References 3 and 4). LETS protein was first partially purified by Yamada and Weston[9] from chick embryo fibroblasts by mild urea extraction. The addition of such purified LETS protein restores the normal morphology of transformed cells.[10,11] It has also been suggested by Ruoslahti and Vaheri[12] that LETS protein is closely related to a plasma protein, cold-insoluble globulin.[13,14] It is likely that cold-insoluble globulin in the blood circulation is the modified shedding product of cell surface LETS protein.

Around the spring of 1974 in Dr. J. M. Buchanan's laboratory at Massachusetts Institute of Technology (MIT), we found that thrombin, a highly specific protease responsible for blood clotting, is a potent mitogen for cultured

*Supported by grants from the National Cancer Institute, the Muscular Dystrophy Association, the American Heart Association (Nassau County, New York Chapter), the American Cancer Society, and the Helen Hay Whitney Foundation.

†Present address: Sidney Farber Cancer Institute, Harvard Medical School, Boston, Mass. 02115.

‡Present address: Department of Cancer Studies, University of Birmingham, The Medical School, Birmingham B15 2TJ, England.

0077-8923/78/0312-0366 $01.75/0 © 1978, The New York Academy of Sciences

chick embryo fibroblasts.[15] Dr. Nelson Teng, who was using lactoperoxidase-catalyzed iodination to analyze alterations in cell surface protein during myogenesis, proposed that we should check whether thrombin can remove the cell surface LETS protein that was being intensively studied at the time in Dr. P. W. Robbin's laboratory. We found that thrombin does not remove this protein from the cell surface of chick embryo fibroblasts, even though it stimulates DNA synthesis.[16-18] Because of the apparent uninvolvement of LETS protein in growth control per se, we did not pursue further investigations on this protein; later, however, we did identify a thrombin-sensitive, cell surface 205K protein.[19-21]

In the summer of 1975, we observed a peculiar feature of fibrin, in that large bundles of this component can be detected by phase-contrast microscopy.[22] We thought that such bundles may have some relationship to cold-insoluble globulin (CIg), because it is known that CIg can be cross-linked to fibrinogen by factor XIIIa,[14] and it is possible that the large bundles of fibrin could result from further cross-linking between polymerized fibrin and CIg. To test such a possibility, we obtained a small amount of antiserum against CIg from Dr. M. W. Mosesson (Downstate Medical Center, State University of New York, Brooklyn) to perform immunofluorescent staining on our fibrin preparation. The result turned out negative; we therefore did not conduct further studies on CIg, even though by this time an intriguing paper by Ruoslahti and Vaheri[12] on the similarity between CIg and LETS protein had already appeared. In retrospect, it seems that we missed two chances in a row to work on LETS protein at MIT.

In early 1976, Dr. J. D. Watson suggested that we study the cell biology of a series of adenovirus-transformed rat fibroblasts, in an attempt to determine why some transformed lines are tumorigenic and others are not. Overwhelmed by the power of the indirect immunofluorescence technique for localization of cellular protein, I was eager to try any antiserum for a potential surprise. I grew many lines of adenovirus-transformed cells and also tried any antiserum available. Among the numerous sera tested, two kinds of serum gave interesting results. One was the CIg antiserum mentioned earlier, which had already been in a freezer for 8 months at MIT. The other was fibrinogen antiserum provided by Dr. John Fenton (Health Department, State of New York, Albany). These two antisera gave identical results. (It turned out that fibrinogen antiserum always contains CIg antibody.)

Figure 1 shows a typical example of the distribution of antigen as detected by CIg antibody in nontumorigenic adenovirus-transformed rat cells. As seen, fluorescence always localizes in areas of cell-cell contact. Subsequent experiments demonstrated that cell-cell contact is responsible for the expression of cell surface CIg-cross-reacting material in these cells. This observation immediately intrigued us. We have already reported on most of these experiments, conducted at Cold Spring Harbor.[23-29] Some of the highlights are further discussed here.

Relationship between LETS Protein and CIg

Although it had been demonstrated previously that CIg and LETS protein are immunologically related, it is important to confirm this result by other techniques. Soon after realizing that immunofluorescence with CIg antibody may reveal the cell surface structure of fundamental interest, Keith Burridge at Cold Spring Harbor Laboratory undertook a series of experiments to determine what

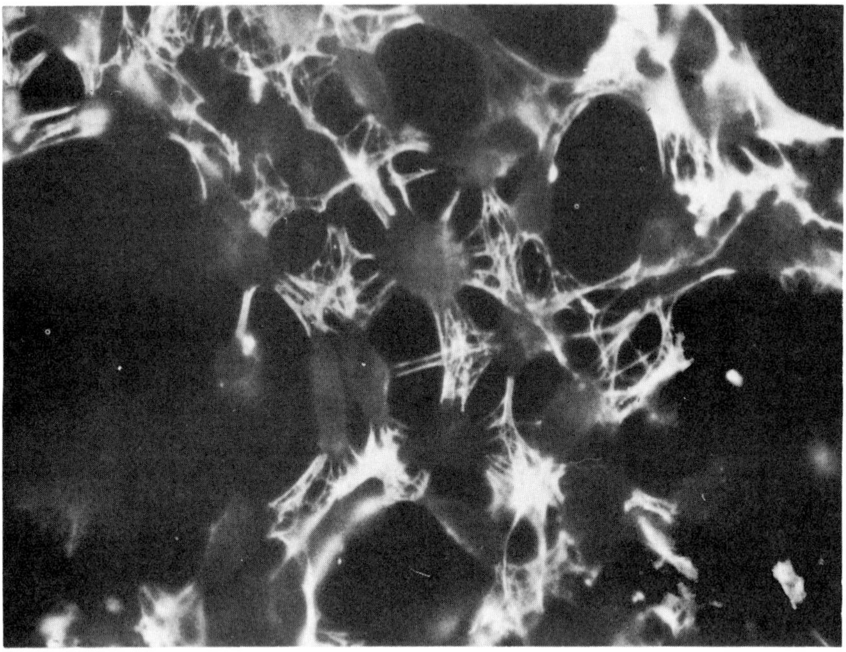

FIGURE 1. Immunofluorescent staining with LETS protein antibody of adenovirus-transformed rat fibroblast line F18. ×2250.

protein CIg antibody actually recognizes. He developed an ingenious technique that enables identification of the cellular antigens recognized by antibody on SDS polyacrylamide gels.[30] By use of such techniques, it was established that the antiserum that we originally obtained from Dr. Mosesson does "stain" LETS protein on SDS gels. With this knowledge, we have a firm basis for studying the immunofluorescent distribution of LETS protein on the cell surface with such antiserum.

Although LETS protein and CIg are antigenically very similar, they may not be identical molecules. The amino acid compositions of these two molecules are very similar.[31] The following differences, however, have been reported. The N-terminal amino acids are Glu-Ala for human CIg[32] but are blocked (possibly acetylated) in chick LETS protein.[31] The sialic acid contents are 1.2% for human CIg and 0.6% for chick LETS protein.[31,32] We noted on SDS polyacrylamide gels that CIg always migrates slightly faster than LETS protein. Recently, we found that purified human CIg does not restore normal morphologic features in Rous sarcoma virus-transformed chick cells; however, urea-extracted chick LETS protein does reverse the morphology to normal, a finding similar to those reported by Yamada *et al.*[10] and Ali *et al.*[11] Finally, the most puzzling difference between CIg and LETS protein is their solubilities. Cold-insoluble globulin is soluble in isotonic buffer at pH 7; however, LETS protein readily precipitates under such conditions.[31] Nonetheless, it is still possible that CIg in the bloodstream may be a modified LETS protein shed by a nonfibroblastic cell type, such as endothelial cells.

LETS Protein and Oncogenicity

Among numerous virally transformed cell lines tested for the expression of LETS protein, we found a correlation between the increase in tumorigenicity *in vivo* and the decrease in the percentage of cells that express LETS protein on cell-cell contact in a series of adenovirus-transformed rat cells.[23] This correlation was further confirmed in various clones of herpes simplex virus-transformed hamster cell line 333-8-9. Table 1 shows that the increase in oncogenicity is accompanied by a decrease in cell surface LETS protein. The molecular mechanism responsible for such a correlation is presently unknown. Several possibilities may be considered. First, the loss of cell surface LETS protein may be directly related to the antisocial behavior of transformed cells, which leads to the increase in oncogenicity. To prove this hypothesis, we must know the function of LETS protein and be able to elucidate the relationship between cell-cell interaction and growth control. It is encouraging that addition of LETS protein does "correct" abnormal cellular morphologic features in transformed cells.[10,11] If cell shape is indeed the prime factor involved in the social behavior of cells in animals, Yamada *et al.*'s and Hynes *et al.*'s observations could lead to an exciting breakthrough regarding the role of LETS protein in oncogenicity. Attempts to demonstrate that LETS protein is directly involved in growth control in culture systems have unfortunately yielded negative results.

With thrombin, we previously demonstrated the lack of a relationship between LETS protein and cellular growth.[16,17,20] In experiments in which LETS protein was resupplied to transformed cells, both Yamada *et al.* and Hynes *et al.* showed

Table 1

Correlation between Increase in Oncogenicity and Decrease in Cell Surface LETS Protein among Various Clones of Herpes Simplex Virus-Transformed Hamster Cell Line 333-8-9

Cells (2 × 10^5/animal)	Latency (days)	100% with Tumor (days)	LETS Protein* (cpm/10^5 cells)
Primary BHK	0	0	4300
333-8-9	17	23	1200
C1	14	22	380
C3	6	9	770
C19	5	7	600
C18	9	14	
└ C18-1	0	0	2150
└ C18-4	9	50	890
Hamster			
└ Tu35	7	7	430
Hamster			
└ 2Tu1	5	5	280

*Radioimmunoassay for cell surface LETS protein. One hundred percent LETS-negative cells give 200 cpm/10^5 cells.

that LETS protein does not influence the growth rate, although Hynes *et al.* did report a decrease in final cell density in LETS protein-replenished cultures.

The second possibility is to consider LETS protein as a differentiated product and to consider the loss of LETS protein as a reflection of the dedifferentiation observed in transformed cells. Dedifferentiation suggests the possibility of reprogramming in gene expression. In the course of reprogramming, some "forbidden" genes may be expressed, leading to the loss of growth control. Therefore, the correlation between loss of LETS protein and tumorigenicity occurs because LETS protein serves as an excellent index for alteration in gene expression. If this is the case, the relationship between LETS protein and tumorigenicity is completely fortuitous. Following such reasoning, even the morphologic changes, the hallmark of transformed cells, may turn out to be fortuitous with regard to oncogenicity. If this remote possibility can be substantiated, all pathologists should feel embarrassed, because they have long considered alterations in morphologic characteristics as the sole means for identification of cancer cells in tissue sections.

The third possibility is that LETS protein is not directly involved in the loss of growth control directly but, rather, in other oncogenetic phenomena, such as invasion and metastasis. In other words, the correlation should be between the decrease in cell surface levels of LETS protein and the increase in invasive and metastatic potentials. Recent data from our laboratory support such a possibility. The initial phase of the experimental approach toward resolution of this problem involves collection of cell lines that induce benign tumors in animals and of cell lines that induce metastatic tumors to compare LETS protein expression both *in vitro* and *in vivo.* We have already noted some intriguing results. For example, all SV40-transformed mouse fibroblast cell lines tested induce apparently benign tumors in their syngeneic hosts, but most of the SV40-transformed hamster cell lines induce visible massive metastases in the hamster. When we investigated expression of LETS protein by these cell lines *in vitro,* all SV40-transformed hamster cell lines that induce visible metastases were found to contain very little surface LETS protein; all SV40 mouse fibroblasts express significant amounts of LETS protein. Although these results are encouraging, the molecular basis of such a correlation is confounding. The metastasis observed is hopelessly complex and involves so many steps, that it is no easier to explain than is growth control. More work is needed on the molecular biology of metastasis before the specific pathologic phenomenon due to the loss of LETS protein can be elucidated.

Control of Expression of LETS Protein in Normal Cells

We shall first summarize our results on the distribution of LETS protein in "normal" cells obtained after screening more than 100 cell lines.

Type I. LETS protein is confined in the cell-substratum contact area, and its distribution is not affected by cell-cell contact. This type includes all epithelial cell groups that express LETS protein, such as liver epithelial cells from 3-week-old rats and kidney epithelial cells.[25]

Type II. LETS protein is present in the cell-substratum and cell-cell contact areas. On extensive contact, the three-dimensional LETS protein network that covers the whole culture will be formed. Type II includes 3T3, endothelial cells, WI-38 cells, and most of the embryonic fibroblasts and myoblasts from a variety of species, including snake, frog, mouse, hamster, rat, chick, and human.

Type III. LETS protein is found only in the cell-cell contact area, not in the cell-substratum contact region. After an extensive cell-cell contact, the three-dimensional LETS protein network is still present. Among these cell types are non-tumorigenic adenovirus-transformed rat cells, Yaffe's myogenic cell lines, and smooth muscle cells.

Type IV. LETS protein occurs only in some of the cell-cell contact areas, and no three-dimensional LETS protein network is formed, even after extensive cell-cell contact and a prolonged culture period. Included in type IV are some of the tumorigenic adenovirus-transformed rat cells, Hu Tu 80 (a cell line derived from human stomach carcinoma), and some Rous sarcoma virus-transformed rat and mouse cells.[23]

In view of the above differences, studies of the control of LETS protein expression should be undertaken on a category by category basis. It is possible that the controlling factors involved in the expression of LETS protein in cell-substratum contact areas may be different from those involved in cell-cell contact regions. In the initial phase of such studies, we focus our attention on factors that may influence the formation of three-dimensional networks of LETS protein.

Cell-Cell Contact

Without exception, cellular contact appears to precede the formation of LETS protein networks. Because LETS protein networks appear to be interlinked throughout the confluent monolayer, cellular contact may be a physical prerequisite for the formation of such networks. Interestingly, the timing of the formation of the LETS protein network after confluence is reached may be relevant. Since the total amount of LETS protein is significantly increased after confluence, cellular contact may potentiate the further production of LETS protein. The existence of a lag time between acquisition of extensive cell-cell contact and the appearance of LETS protein networks may indicate that the cells require this time period to "process" such a signal of potentiation.

We found that 3T3 cells newly seeded at a very high density express the initial form of the LETS protein matrix about 30 hr after plating. The internal pool of LETS protein synthesized prior to trypsinization is clearly not adequate for the formation of the protein network, because cycloheximide inhibits its appearance. Our results demonstrate that continuous protein synthesis is required for the formation of the LETS protein matrix at 30 hr in confluent density. At present, the requirement for RNA synthesis in such a system has yet to be confirmed. However, when we shifted confluent NRK cells transformed by Rous sarcoma virus with a temperature-sensitive lesion in the *src* gene (tsRSV)[33] from the permissive temperature (no LETS protein) to the nonpermissive temperature to observe the reappearance of the LETS protein matrix, we found that both protein synthesis and RNA synthesis are required.[39] In the same experiment, if the tsRSV-transformed cells were maintained at the permissive temperature under conditions of low cell density, where very little cellular contact occurred, and then shifted to the nonpermissive temperature, no LETS protein was visualized.

That cell-cell contact is a prerequisite for the formation of the extracellular protein matrix may be manifest from a different viewpoint. The observation may be interpreted that fibroblasts in culture are trying to express their differentiated function, such as the formation of a stromal matrix or basement membrane. When cells are isolated individually, they are found to bear low levels of this protein, possibly because they "sense" that if a single cell attempts to build a stromal matrix or basement membrane, the effort is most unlikely to succeed. When two cells touch each other, they may "tell" each other, "Perhaps we can start to try to make the extracellular matrix." This is probably the reason why we frequently detect LETS protein in cell-cell contact areas. However, two or three cells are not enought to build a true matrix. We found that, depending on cell type, from five to 20 cells in close proximity within an "island" are needed to form a protein network.

Hormonal Effects

In the course of experiments undertaken to study the effect of thrombin on LETS protein of mammalian cells, we noticed that LETS protein in the serum-free cultures is invariably absent. We therefore began to investigate the effect of serum on the expression and maintenance of LETS protein. We have found that Cohen's epidermal growth factor (EGF) is particularly interesting in this respect. When a confluent culture of 3T3 cells is incubated for 2 days in medium that contains 0.7% calf serum, LETS protein is significantly reduced. Addition of 1 ng/ml of EGF to such cultures restores the normal three-dimensional network of LETS protein.[26] Among 13 hormones tested, the effect of EGF appears to be specific. These observations suggest that studies on the effect of EGF on LETS protein expression may serve as a useful model for an evaluation of control of the expression of differentiated products by hormones. We have found that this process requires continuous protein synthesis.

Retraction Fibers

The sequential events involved in the expression of fibrillar LETS protein in cell-cell contact areas may be explained by the following: Collision of cells leads to the formation of retraction fibers. This phenomenon triggers the expression of LETS protein on the retraction fibers. We attempted to prove this hypothesis by determining whether a cell that is not in contact with other cells and that has no LETS protein on its surface can be forced to form such retraction fibers and to express LETS protein. Our studies revealed that this goal can be accomplished. By use of Mn^{2+}, formation of retraction fibers can be induced in an individually isolated cell, and LETS protein is visualized (FIGURE 2). With this information, we believe that a further understanding of the molecular mechanisms involved in the expression of LETS protein may be achieved in the near future. Again, in this process, protein synthesis is required for the appearance of LETS protein on retraction fibers.

ELECTRON MICROSCOPIC STUDIES ON LETS PROTEIN MATRIX

Since the fibrillar structure that contains LETS protein can be readily detected by immunofluorescence, the question arises as to whether such a structure can be detected by scanning electron microscopy. We found that although 3T3 cells contain a massive fibrillar structure, as detected by LETS protein antibody, a similar structure cannot be demonstrated by the latter technique. However, transmission electron microscopy does reveal fibrillar structures in between cells; these

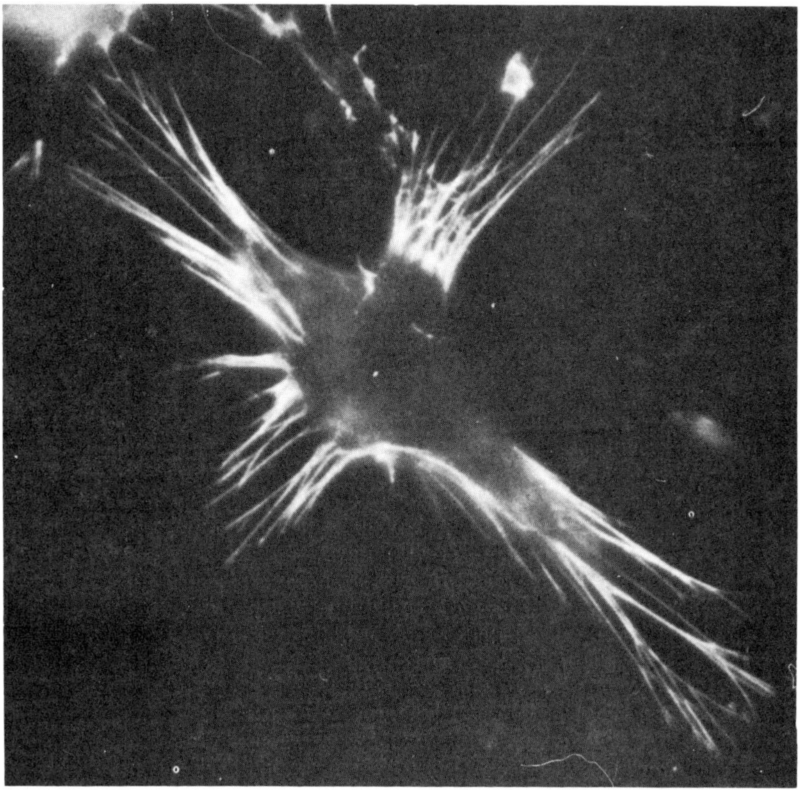

FIGURE 2. Immunofluorescent staining with LETS protein antibody of sparse 3T3 cells treated with 1 mM Mn^{2+}. ×5500.

structures are about 40–60 Å in diameter. Since a single fiber with such a diameter can be detected only with difficulty by scanning electron microscopy, and because any fiber with a diameter less than 200 Å will be revealed as 200 Å in size by immunofluorescence, if detectable at all, we believe that most fibers that contain LETS protein in 3T3 cells are probably about 50 Å in diameter and are therefore too small for resolution by scanning electron microscopy. However, in some cells, these 50-Å fibers may be clustered together, analogous to the clustering of actin-containing microfilaments into so-called stress fibers (or actin cables), so that some of the fibers may, then, be detectable by scanning electron microscopy. Indeed, we found that the fibrillar structures of chick embryo fibroblasts that secrete massive quantities of LETS protein in confluent culture can be detected by scanning electron microscopy, as illustrated in FIGURE 3. By means of transmission electron microscopy, these structures are found to be composed of numerous fine fibers (40–60 Å), and many of them are seen to be clustered together, very similar to the formation of stress fibers (FIGURE 4). Scanning electron microscopy probably detects the large bundles, such as those shown in FIGURE 4. That LETS protein is one of the components of these fine fibers is demonstrated in FIGURE 5, where the immunoferritin antibody technique was employed to localize LETS

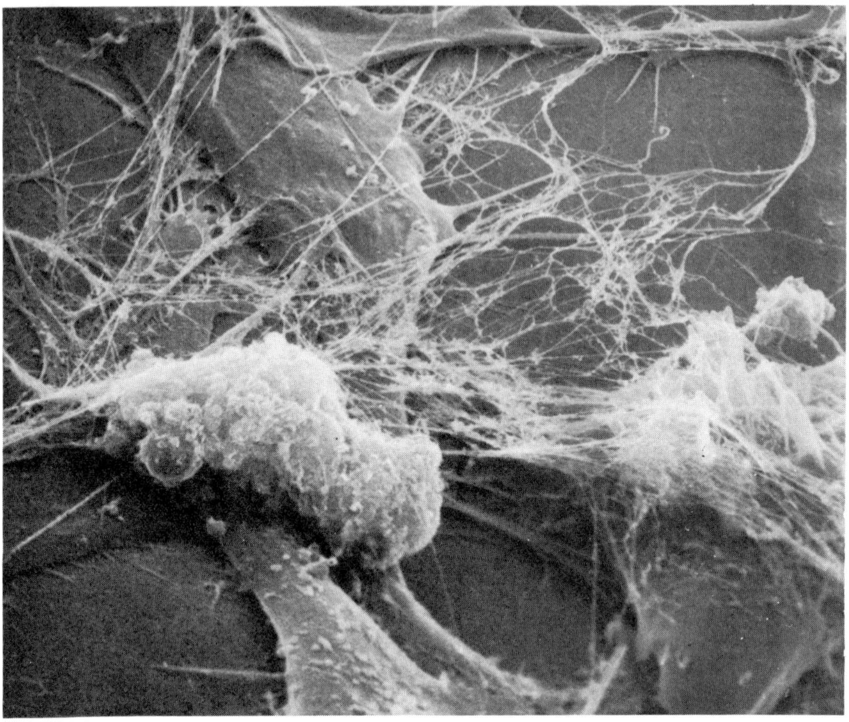

FIGURE 3. Scanning electron micrograph of the extracellular matrix secreted by secondary chick embryo fibroblasts. ×8000.

protein. We shall later substantiate further that LETS protein is a component of these fibers.

What are these matrix structures? What is the function of these matrices? We believe that they play a role in the formation of either the basement membrane or the stromal matrix in culture. To confirm this hypothesis, we have stained frozen sections of various tissues with CIg antibody, and the results indicate that LETS protein is localized preferentially in the stromal matrix and the basement membrane in various tissues, in agreement with a previous report by Linder *et al.*[34] Since the biologic functions of the stromal matrix and basement membranes are still unknown, except for their obvious role in maintaining the structural integrity of tissues, we will be investigating the possible biologic function of this extracellular matrix during the next few years. One step toward such a goal is preparation of a cell-free matrix system. FIGURE 6 shows that by use of 1% Nonidet P-40 at pH 9.5 and subsequent washing with 0.3 M KCl, a modification of the procedure of Hynes *et al.*,[35] a cell-free matrix can be obtained from a confluent culture of chick embryo fibroblasts. That LETS protein is one of the major components of the matrix system is shown by analysis of total proteins of the cell-free matrix by sodium dodecyl sulfate polyacrylamide gel electrophoresis and positive immunofluorescent staining with LETS protein antibody in this cell-free matrix. In addition to LETS protein, actin, myosin, and other minor components are present, as demonstrated by use of such gels. We have stained the fibrillar structure with actin and myosin antibody but failed to detect them on the fibrillar matrix; instead, they distribute as speckles in between the fibers on the glass substratum. We conclude that actin and myosin are the major contaminants of the detergent-preserved cell-free matrix. It is important to note that a substantial amount of protein did not enter the gel. These materials may represent LETS protein that has become cross-linked during detergent treatment, possibly by the action of cellular transglutaminase.[36] It should be emphasized that all of the above experiments were performed in chick embryo fibroblasts. We must stress that such observations may not have general application to all fibroblasts in culture.

By use of such cell-free matrix systems, we are currently studying their possible biologic effects. For example, we are exploring whether the matrix exerts effects on cell growth, rate of glucose uptake, protease production, and cellular morphologic features. When a variant clone of Chinese hamster ovary cells seeded in the cell-free matrix was compared with clones seeded in culture dishes for the above cellular phenotypes, there was no difference, except in cellular morphologic characteristics. Eighty percent of the cells that are normally round in culture dishes assume a fibroblastic shape in the environment of the cell-free matrix.

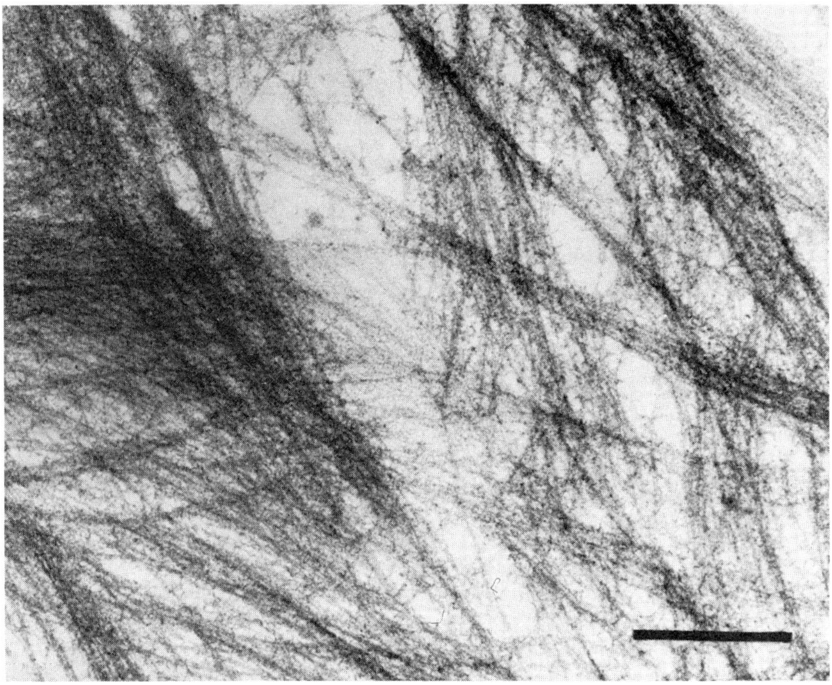

FIGURE 4. Transmission electron micrograph of a thin section of the extracellular matrix secreted by chick embryo fibroblasts. Bar = 0.5 μm.

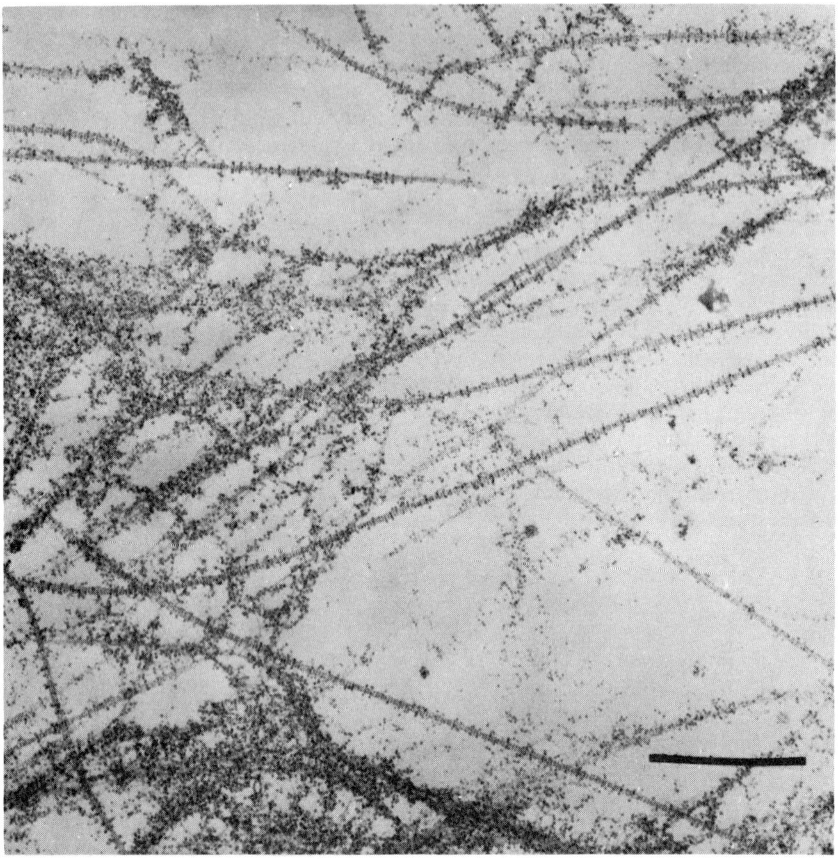

FIGURE 5. Transmission electron micrograph of extracellular matrix labeled with ferritin-conjugated LETS protein antibody. Bar = 0.5 μm.

RELATIONSHIP BETWEEN LETS PROTEIN AND COLLAGEN

Since LETS protein distributes as a three-dimensional matrix in confluent culture, and because collagen is the major matrix protein *in vivo,* their relationship inevitably attracts attention. Earlier, Hynes rigorously proved that LETS protein is not procollagen.[3] Subsequently, in light of Pearlstein's experiment, the possibility arose that LETS protein may bind to collagen in a significant way.[37] Indeed, more recent reports tend to support the notion that LETS protein may have a high affinity for collagen; in particular, Engvall and Ruoslahti showed that collagen can be employed in affinity column chromatography to purify LETS protein.[38]

FIGURE 7 shows that on the glycocalyx of 3T3 cells in 2-day culture, collagen (type I) distributes as patches or spotty speckles, whereas LETS protein forms fibrillar patterns. However, in some cultures, such as 10-day confluent cultures of chick embryo fibroblasts, collagen and LETS protein distribute very similarly as a three-dimensional extracellular matrix. Our results suggest therefore that the secretion and early fibrillogenesis of these two molecules are distinct but that in the later stages of matrix formation, LETS protein and collagen may interact in certain cell types. Our electron microscopic studies suggest strongly that in the later stages, when LETS protein and collagen codistribute, collagen (type I) probably binds to LETS protein-containing fibers rather than the converse of such a binding phenomenon. The following data support such a hypothesis. (i) The three-dimensional matrix in confluent culture is insensitive to collagenase treatment (both animal and bacterial sources) but is sensitive to trypsin and a variety of other proteases, as is LETS protein but not collagen.[16,20] (ii) Alkaline treatment, which sometimes preserves the collagen matrix, fails to preserve the matrix seen in culture, and LETS protein is known to be soluble at pH 11.[31] (iii) There are several fibroblastic cell lines that form a three-dimensional matrix that contains LETS protein but not collagen, even though these cells do secrete collagen (type I) (TABLE 2). It is not clear why collagen fails to bind to the matrix in these cultures, in contrast to chick embryo fibroblasts. (iv) In numerous electron microscopic studies on thin sections of chick embryo fibroblasts in culture, we have never encountered fibers with typical collagen periodic patterns. We have also screened a large number of cell lines to compare the distributions of collagen and LETS protein by immunofluorescence. TABLE 1 shows that in sparse culture, there is no evidence of a similar distribution of both LETS protein and collagen. In confluent 1-week cultures, some cell lines contain a fibrillar collagen matrix similar to the LETS protein matrix, some cell lines express very little matrix

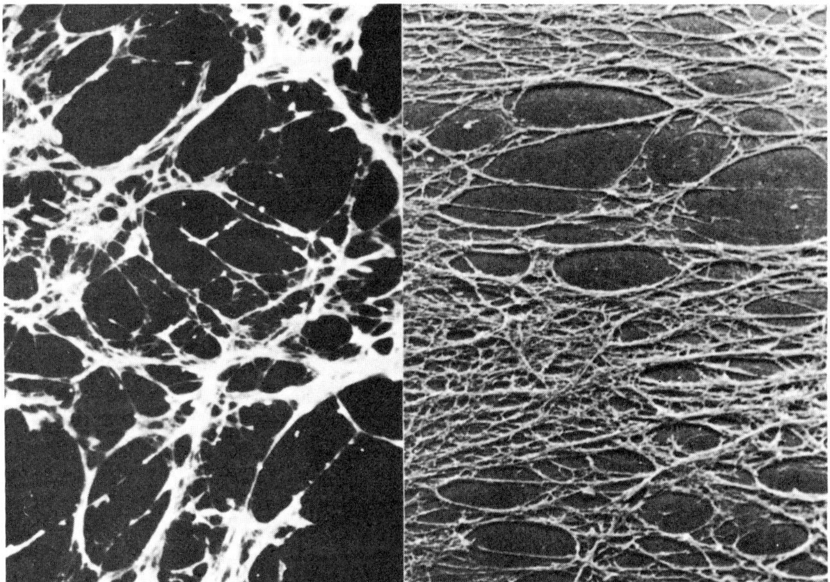

FIGURE 6. *Right:* Scanning electron micrograph of cell-free extracellular matrix preserved by detergent treatment; *left:* immunofluorescent staining with LETS protein antibody.

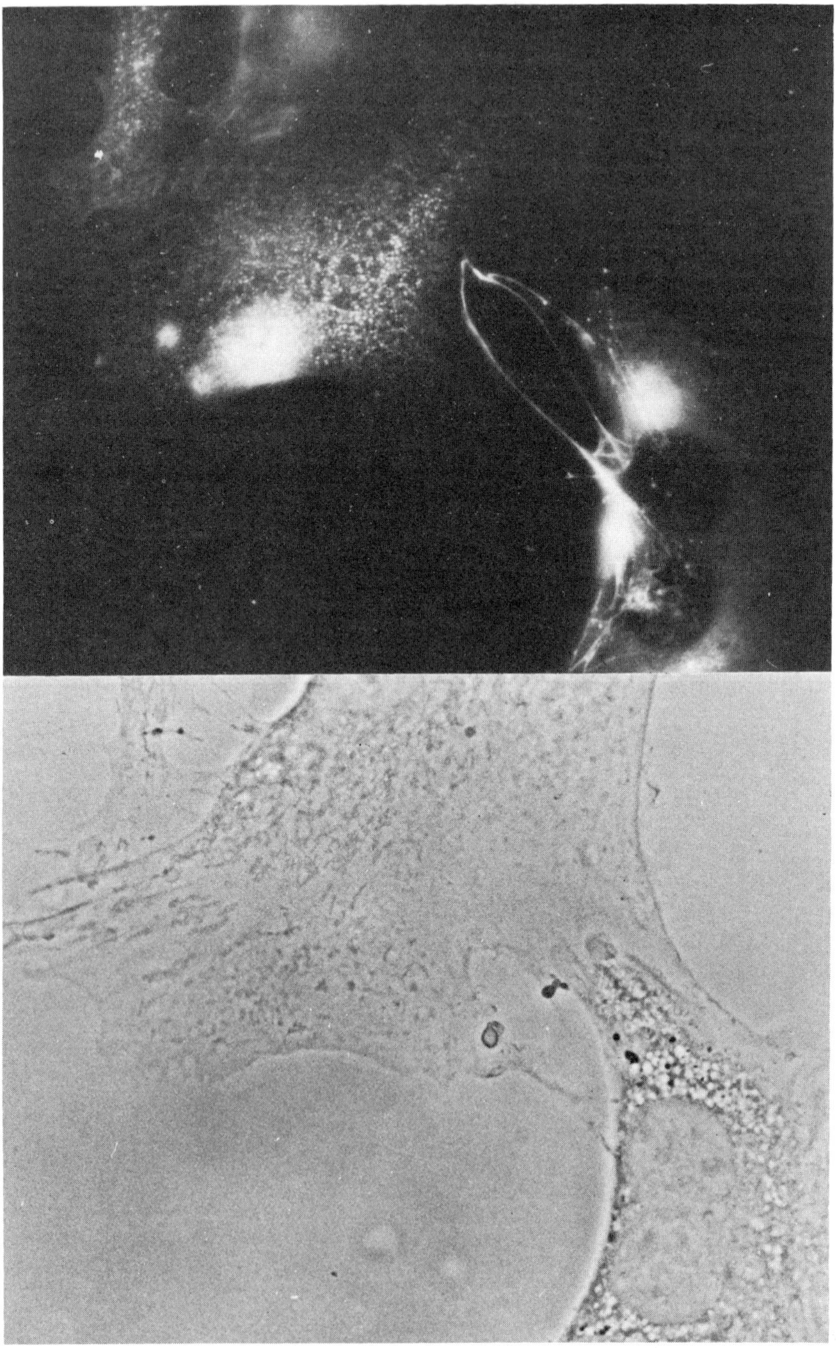

FIGURE 7. *Top:* Immunofluorescent staining of 3T3 cells with collagen antibody; *bottom:* phase-contrast micrograph of the same field.

TABLE 2

COMPARISON OF IMMUNOFLUORESCENCE WITH LETS PROTEIN ANTIBODY AND WITH COLLAGEN ANTIBODY ON CULTURED CELLS

	Sparse		Confluent	
Cell Line	LETS Protein	Collagen (type I)	LETS Protein	Collagen (type I)
CEF	fibrillar	patchy	matrix	matrix
3T3	fibrillar	patchy fibrillar	matrix	matrix
F18	fibrillar	patchy	matrix	patchy
NRK	fibrillar	none	matrix	none
L8 myoblast	fibrillar	none	matrix	patchy fibrillar
Rat liver epithelial	fibrillar	none	fibrillar	none
Sertoli cells	none	none	none	none

collagen, and a few cell lines bear no collagen matrix, even though LETS protein matrices are detected.

Our studies therefore suggest that although collagen (type I) and LETS protein exhibit a high affinity for each other, the appearance of fibrillar LETS protein in early culture does not "guarantee" the codistribution of collagen (type I), even though collagen (type I) is secreted. Such observations indicate that LETS protein and collagen (type I) are most likely two separate matrix systems. Since the LETS protein matrix appears much earlier than does the collagen matrix, and because the collagen matrix seems to align with the LETS protein matrix in later stages, it is tempting to assume that the LETS protein matrix may function as the initial backbone for the fibrillogenesis of collagen.

CONCLUSIONS

Abnormal intercellular behavior is always assumed to be the hallmark of cancerous cells. Despite many years of research, this commonly accepted concept still largely remains without scientific verification. To this end, cell culturists assume that some of the abnormalities in social behavior *in vivo* may also be manifest *in vitro*, and a comparison between normal and transformed cells at this level should result in some meaningful observations. In the last 20 years, a wealth of such comparisons have been made. Although there are exceptions, it is encouraging that normal cells always grow in a well-organized fashion in culture, whereas transformed cells always proliferate in a disorganized manner. Thus, at least some of the abnormalities in intercellular behavior of transformed cells can be "exposed" in the cell culture system. No worker has yet been able to correlate the abnormal growth patterns observed in culture to tumor behavior *in vivo*. Nonetheless, we should feel fortunate that there is a place to start: at least we can analyze the abnormalities that have already been visualized *in vitro*. It is possible that some of the abnormalities seen in culture may never occur *in vivo*. However, the fact that the normal cell and the transformed cell react differently to the plastic or glass substratum, granted an artifact, is still a useful bit of knowledge to use to begin to probe the essence of abnormality in the social behavior of transformed cells. Why do normal and transformed cells react differently to the plastic dish? It is reasonable to expect that the cellular machinery that controls the reaction to the plastic dish may be the same cellular apparatus that determines the social behavior of cells *in vivo*.

So far, LETS protein is the only cell surface component that has been shown unambiguously to be involved in abnormal morphologic characteristics, adhesion, and contact inhibition of movement of transformed cells.[10,11] That the addition of LETS protein in LETS-deficient transformed cells restores the abnormal intercellular behavior of transformed cells to normal in culture is evidence of a role for this molecule in cell-cell interaction, recognition, and communication. Indeed, this cell surface protein has been shown to originate from cell-cell contact, to be localized preferentially in cell-cell contact regions in sparse culture, and to be interconnected to itself in confluent cultures.[23] Moreover, this protein appears to inhibit cell surface movement, such as the transport of gold particles on the cell surface by centripetal flow.[28] Finally, this protein seems to play an important role in the fusion of myoblasts, which strengthens the notion that LETS protein is involved in cell-cell interaction.[24] Therefore, it appears that more thorough studies on LETS protein will be fruitful and may help to provide a greater understanding of the abnormal social behavior of transformed cells.

Finally, we should emphasize that since there is an apparent correlation between the loss of LETS protein and the increase in tumorigenicity,[23] studies on this protein may lead to a better understanding of the complex tumorigenesis *in vivo.* LETS protein may still turn out to be another red herring, but at the moment it represents one of our better hopes in pursuing the essence of tumor formation. If LETS protein is not *the* molecule, other culprit proteins may be found while we rigorously study LETS protein.

Acknowledgments

We are grateful for the stimulating advice and continuing support of our Director, Dr. J. D. Watson. Without his encouragement, the studies reported herein would not have been possible. The senior author acknowledges the continuous moral support of his thesis adviser, Dr. John M. Buchanan. We also appreciate valuable discussions with our colleagues Drs. Guenter Albrecht-Buehler, William Gordon, Michael W. Mosesson, John Fenton, Nelson Teng, and Bruce Zetter.

References

1. Nicolson, G. L. 1976a. Biochim. Biophys. Acta **457:** 57–108.
2. Nicolson, G. L. 1976b. Biochim. Biophys. Acta **458:** 1–72.
3. Hynes, R. O. 1976. Biochim. Biophys. Acta **458:** 73–107.
4. Vaheri, A. 1976. *In* Virus-Transformed Cell Membranes. C. Nicolan, Ed. Academic Press, Inc. New York, N.Y. In press.
5. Hynes, R. O. 1973. Proc. Nat. Acad. Sci. USA **70:** 3170–3174.
6. Gahmberg, C. G. & S. Hakomori. 1973. Proc. Nat. Acad. Sci. USA **70:** 3329–3333.
7. Ruoslahti, E., A. Vaheri, P. Kuusela & W. Linder. 1973. Biochim. Biophys. Acta **322:** 352–358.
8. Wickus, G. G., P. E. Branton & R. W. Robbins. 1974. *In* Control of Proliferation in Animal Cells. B. Clarkson & R. Baserga, Eds.: 541–546. Cold Spring Harbor Laboratory. Cold Spring Harbor, N.Y.
9. Yamada, K. M. & J. A. Weston. 1974. Proc. Nat. Acad. Sci. USA **71:** 3492–3496.
10. Yamada, K. M., S. S. Yamada & I. Pastan. 1976. Proc. Nat. Acad. Sci. USA **73:** 1217–1221.
11. Ali, I. U., V. Mautner, R. Lanza & R. O. Hynes. 1977. Cell **11:** 115–126.
12. Ruoslahti, E. & A. Vaheri. 1975. J. Exp. Med. **141:** 497–501.

13. Mosesson, M. W. & R. A. Umfleet. 1970. J. Biol. Chem. **245:** 5728–5736.
14. Mosher, D. F. 1975. J. Biol. Chem. **250:** 6614–6621.
15. Chen, L. B. & J. M. Buchanan. 1975. Proc. Nat. Acad. Sci. USA **72:** 131–135.
16. Teng, N. N. H. & L. B. Chen. 1975. Proc. Nat. Acad. Sci. USA **72:** 413–417.
17. Chen, L. C., N. N. H. Teng & J. M. Buchanan. 1975. *In* Proteases and Biological Control. E. Reich, D. Rifkin & E. Shaw, Eds.: 957–965. Cold Spring Harbor Laboratory. Cold Spring Harbor, N.Y.
18. Blumberg, P. M. & P. W. Robbins. 1975. Cell **6:** 137–147.
19. Teng, N. N. H. & L. B. Chen. 1976. Nature (London) **259:** 578–580.
20. Zetter, B. R., L. B. Chen & J. M. Buchanan. 1976. Cell **7:** 407–412.
21. Zetter, B. R., L. B. Chen & J. M. Buchanan. 1977. Proc. Nat. Acad. Sci. USA **74:** 595–600.
22. Buchanan, J. M., L. B. Chen, T. Hamazaki, E. Lenk & D. F. Waugh. 1977. *In* Chemistry and Biology of Thrombin. R. L. Lundblad, J. W. Fenton & K. G. Mann, Eds. Ann Arbor Science Publishers. Ann Arbor, Mich. In press.
23. Chen, L. B., P. H. Gallimore & J. K. McDougall. 1976. Proc. Nat. Acad. Sci. USA **73:** 3570–3574.
24. Chen, L. B. 1977. Cell **10:** 393–400.
25. Chen, L. B., N. Maitland, P. H. Gallimore & J. K. McDougall. 1977. Exp. Cell Res. **106:** 39–46.
26. Chen, L. B., R. C. Gudor, T. T. Sun, A. B. Chen & M. W. Mosesson. 1977. Science **197:** 776–778.
27. Chen, L. B., D. F. Mosher, A. B. Chen & M. W. Mosesson. 1977. Exp. Cell Res **108:** 375–383.
28. Albrecht-Buehler, G. & L. B. Chen. 1977. Nature (London) **266:** 454–456.
29. Gallimore, P. H., J. K. McDougall & L. B. Chen. 1977. Cell **10:** 669–678.
30. Burridge, K. 1976. Proc. Nat. Acad. Sci. USA **73:** 4457–4461.
31. Yamada, K. M., D. H. Schlesinger, D. W. Kennedy & I. Pastan. 1977. Biochemistry. In press.
32. Mosesson, M. W., A. B. Chen & R. M. Huseby. 1975. Biochim. Biophys. Acta **386:** 509–524.
33. Chen, Y. C., M. J. Hayman & P. K. Vogt. 1977. Cell **11:** 513–521.
34. Linder, E., A. Vaheri, E. Ruoslahti & J. Wartiovaara. 1975. J. Exp. Med. **142:** 41–49.
35. Hynes, R. O., A. T. Destree & V. Mautner. 1976. J. Supramol. Struct. Suppl. **1:** 189–201.
36. Keski-Oja, J., D. F. Mosher & A. Vaheri. 1976. Cell **9:** 29–35.
37. Pearlstein, E. 1976. Nature (London) **262:** 497–500.
38. Engvall, E. & E. Ruoslahti. 1977. Int. J. Cancer **80:** 1–5.
39. Walsh, M. L. & L. B. Chen. In preparation.

GLYCOPROTEIN-CONTAINING FACTOR THAT MEDIATES CONTACT INHIBITION OF GROWTH*

G. Lipkin, M. E. Knecht, and M. Rosenberg

Department of Dermatology
New York University School of Medicine
New York, New York 10016

Normal cells *in vitro* characteristically exhibit a form of growth control that has been termed "contact inhibition of growth,"[1] "density-dependent inhibition of growth,"[2] or "topoinhibition."[3] Such benign cells limit their growth by cellular interactions that depend on population density. On growing together to form confluent (and in some cases well-oriented) monolayers, they stop proliferating, even though the culture medium is still capable of supporting growth at lower cell densities. The population level at which such cells stop dividing under defined culture conditions has been termed the saturation density. Such arrest of growth occurs in the G_1 (G_0) phase of the cell cycle and is associated with a marked decrease in the rate of synthesis of DNA, RNA, and protein, increased turnover of certain cell surface proteins, and early diminution of transport of hexose and phosphate. By contrast, malignant cells do not exhibit this type of growth control; under identical culture conditions, on reaching confluence, they continue to multiply to form multilayered, disoriented heaps of cells, with significantly higher saturation densities than are found in cultures of their benign counterparts. Even in such crowded cultures, considerable DNA synthesis may continue, and there is little decline in transport of either hexose or phosphate. The capacity for normal contact inhibition of growth, although by no means the only component of normal growth control *in vitro,* as shown by work on characterization of serum and density revertants,[4] nevertheless is significantly related to the malignant process *in vivo.*

The relevance of contact inhibition of growth as a valid experimental model for the study of growth control rests on the observation that it is a property that occurs *in vitro* whose loss is closely correlated with malignancy *in vivo.*[5] For example, among different strains of a malignant cell line, there is a direct correlation between the degree of loss of contact inhibition of growth (measured as increasing saturation densities of cultures) and ease of transplantability of tumors *in vivo.*[6] In contrast, increased contact inhibition of cell division in cultures is associated with decreased efficiency of tumor initiation *in vivo.*[7,8]

We recently found that a contact-inhibited line of hamster melanocytes (FF) produces a diffusible factor that restores contact inhibition of growth to hamster malignant melanocytes.[9] The possible presence of a growth inhibitor in the contact-inhibited (FF) cultures was suggested by crossover experiments in which mixtures of conditioned media (CM) from contact-inhibited FF cultures and non-contact-inhibited RPMI 1846 hamster melanoma cultures significantly suppressed growth of the RPMI 1846 but not FF cells. Qualitative comparison of

*Supported by Grant CA 18034 from the National Cancer Institute through the National Large Bowel Cancer Project, Grant CA 16013 from the National Cancer Institute, Postdoctoral Research Training Program Grant AM 07190 from the National Institute of Arthritis, Metabolism, and Digestive Diseases, and a grant from the Rudolf L. Baer Foundation for Skin Diseases, Inc.

0077-8923/78/0312-0382 $01.75/0

proteins from CM of contact-inhibited FF and noncontact-inhibited RPMI 1846 cultures by polyacrylamide gel electrophoresis revealed the presence of a distinctive high-molecular-weight protein in the former cultures that was absent from the latter (FIGURE 1). To study the possible relationship between the presence of this unusual high-molecular-weight protein and the observed growth-inhibitory effects of CM from FF cultures, a partially purified fraction that contained the high-molecular-weight protein was prepared from 2-day-old serum-free CM of

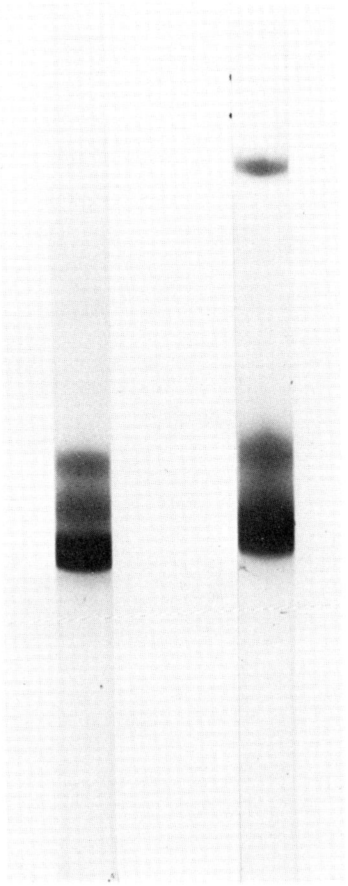

FIGURE 1. Polyacrylamide gels show electrophoretic separation of proteins in serum-free conditioned medium from cultures of FF cells (*right*) and RPMI 1846 cells (*left*). Electrophoresis was performed at 10 mA per tube for 4 hr. Gels were stained with 0.25% Coomassie brilliant blue in H_2O-methanol-acetic acid (17:1:2).

confluent FF cultures by column chromatography on Sephadex® G-200.[9] Two peaks were obtained (FIGURE 2), and these materials were subsequently tested for possible effects on *in vitro* growth of malignant melanocytes by addition to subconfluent cultures of RPMI 1846 cells. At a concentration of 50 μg/ml, the second peak material failed to alter the usual growth pattern. However, first-peak (void volume) material produced the obligatory acquisition, at confluence, of the contact-inhibited state, with a characteristic morphologic change from disoriented overgrowth of pleomorphic forms to well-oriented monolayers of fibroblast-like

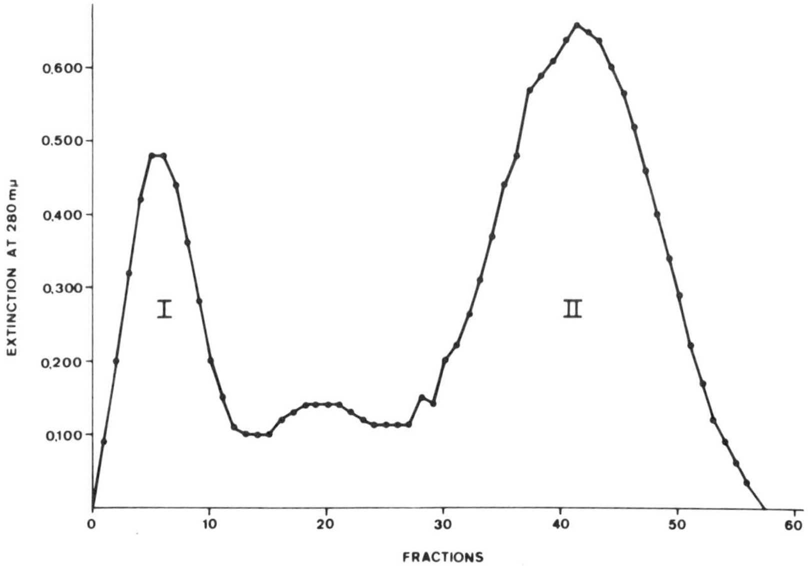

FIGURE 2. Sephadex G-200 column chromatography of lyophilized supernatant from contact-inhibited FF cell cultures.

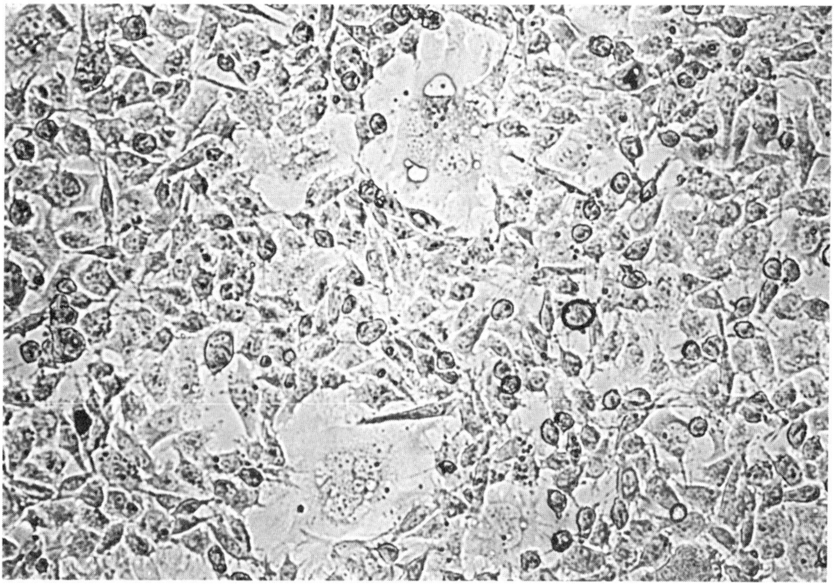

FIGURE 3. Hamster malignant melanoma RPMI 1846 culture at confluence, after 10 days in growth medium without added MCIF. ×115.

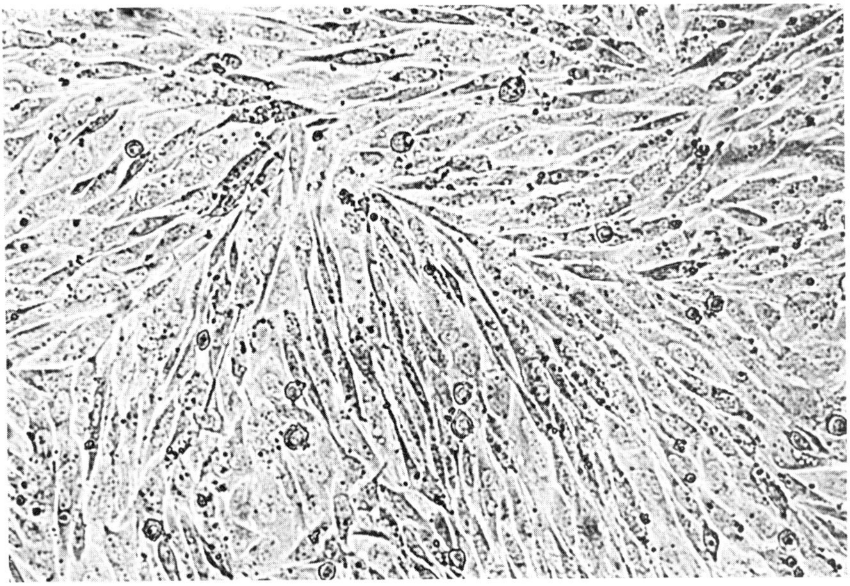

FIGURE 4. RPMI 1846 culture at confluence, grown for 10 days in culture medium with added MCIF (50 μg/ml). ×115.

cells (FIGURES 3 & 4). The arrest of growth occurred only at confluence, so that cells plated initially at lower densities proliferated longer than did those plated at higher densities. Saturation densities of cultures treated with the first-peak material were reduced by an average of 55%; inhibition was nontoxic, rapidly reversible (<24 hr) on removal of the first-peak fraction (FIGURE 5), and prevented by pretreatment of the latter with pronase or repeated freeze-thawing (three times). In addition to the high-molecular-weight protein, the first peak also included DNA and a number of lower-molecular-weight proteins; DNA was not required for the observed biologic activity, because pretreatment with DNase failed to abolish

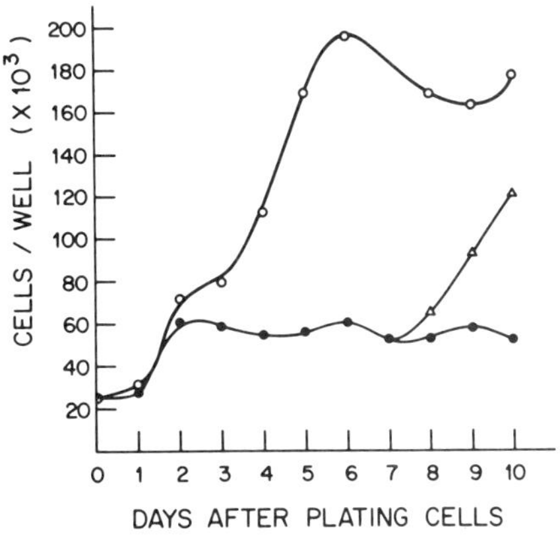

FIGURE 5. Effect of MCIF (60 μg/ml) on growth of hamster malignant cell line RPMI 1846. Upper curve (○) represents growth in complete medium without added MCIF. Lower curve (●) represents growth in complete medium with added MCIF. Removal of MCIF at Day 7 (△) leads to prompt resumption of growth to control levels. Cell cultures were plated originally in Falcon Micro Test II plates at density of 25,000 cells per well; MCIF was added at 6 hr to half of cultures, and the cells were refed at Days 4 and 7. Each point represents the average of duplicate determinations.

activity,[9] nor did the separated DNA exhibit activity.[10] Thymidine pulse labeling experiments following release of RPMI 1846 cells from the contact-inhibited state provided evidence that growth arrest was in the G_1 (G_0) phase of the cell cycle, with a synchronous wave of DNA synthesis occurring at 10–16 hr after removal of the first-peak material (maximum at 14 hr), followed 7 hr later by near-doubling of the cell population.[11] Other studies indicated that cessation of growth was not a result of depletion of essential component(s) of the nutrient medium by the added fraction.[11]

The species and tissue specificities, in addition to some mechanisms underlying the contact-inhibitory effects of melanocyte contact-inhibitory factor (MCIF) were investigated. Addition of partially purified MCIF (first-peak material) to subconfluent cultures of both murine (B16 and S91) and human melanomas produced identical morphologic and growth-inhibitory effects (FIGURES 6 & 7).[11,12] Furthermore,

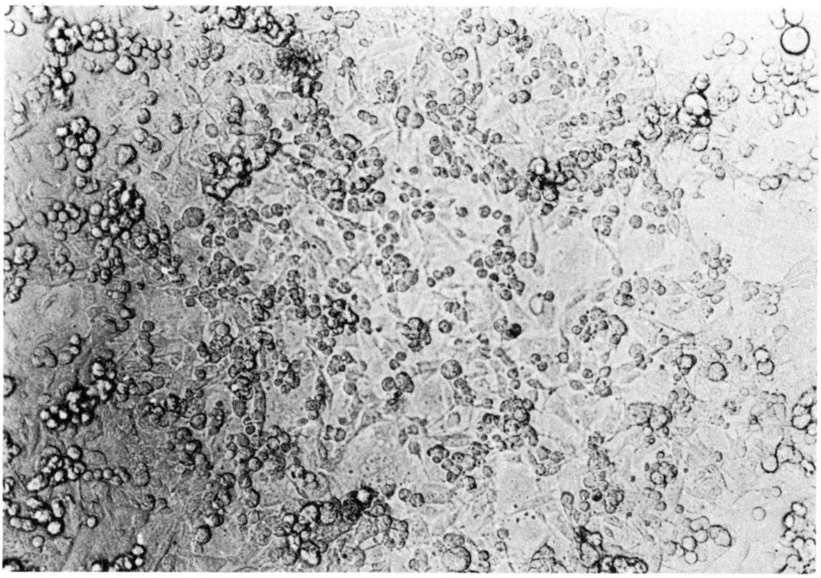

FIGURE 6. Human malignant melanoma cell line (H.F.) grown to confluence in complete medium without added MCIF. ×100.

the growth-inhibitory effects transcended tissue barriers, extending to a broad spectrum of transformed and benign (contact-inhibited) cell types of ectodermal, mesodermal, and endodermal origins[13] (TABLE 1). Maintenance of the contact-inhibited morphologic features of MCIF-treated melanoma cultures required intact cytoskeletal structures: inhibition of microtubules by addition of Colcemid® or vinblastine (5 μg/ml), or of microfilaments by addition of cytochalasin B (5 μg/ml), rapidly abrogated the oriented, fibroblast-like appearance, within 4 hr and 15 min, respectively. Inhibition of *de novo* RNA or protein synthesis gave similar effects by 24 hr.

The effect of MCIF on cyclic nucleotide levels was also determined in cultures of hamster melanoma cells (RPMI 1846).[10] Within 15 min after addition of 50

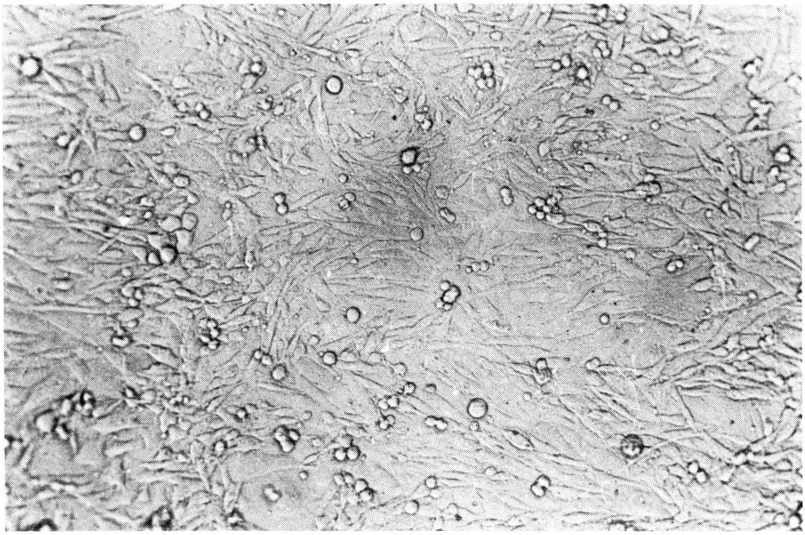

FIGURE 7. Human malignant melanoma cell line (H.F.) grown to confluence in complete medium that contained added MCIF (60 μg/ml). ×100.

μg/ml of first-peak material to subconfluent cultures, cGMP levels decreased by 67%; within 60 min, cAMP increased by 86%. These findings agree with those of other studies in which contact-inhibited cells showed higher cAMP and lower cGMP values than did their transformed counterparts[14,15] and indicate that the morphologic and growth-inhibitory effects of MCIF are preceded and paralleled by biochemical events known to be associated with *in vitro* regulation of growth.

TABLE 1

GROWTH *in Vitro* OF CELL LINES IN THE PRESENCE OF MCIF (50 μg/ml)

Cell Type	Species	Number of Cultures	Decrease in Saturation Density (%)
Transformed			
Melanoma	hamster	25	54.6 ± 17.7*
	mouse	21	62.4 ± 19.9
	human	24	52.2 ± 20.4
Neuroblastoma	mouse	8	60.6 ± 16.7
Neurinoma	rat	5	70.7 ± 27.0
Glioma	rat	1	72.0
Mammary carcinoma	rat	4	77.4 ± 1.0
	human	4	36.0 ± 5.5
Colonic carcinoma	human	25	50.5 ± 20.7
Rhabdomyosarcoma	human	9	53.7 ± 25.1
Contact Inhibited			
Epidermal	human	10	12.1 ± 7.3
Fibroblastic	human	13	27.3 ± 27.8

*Means ± SD.

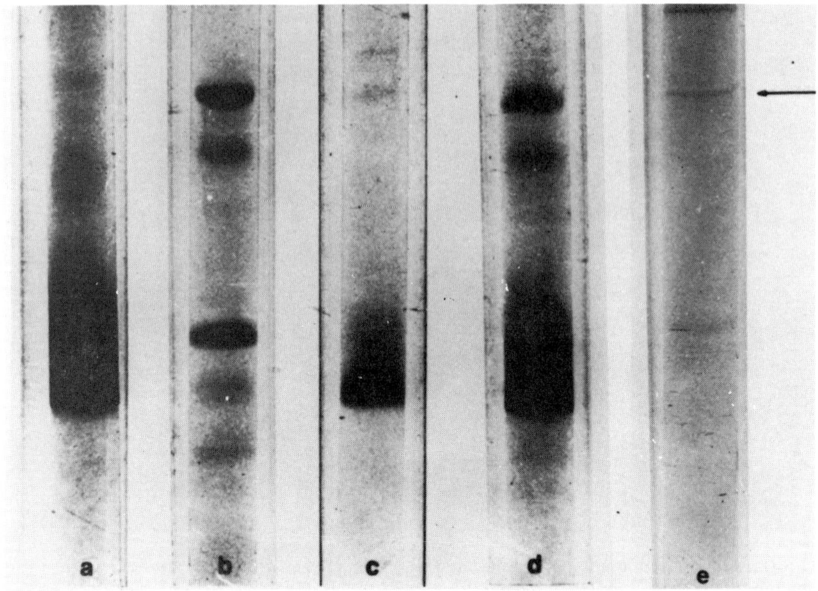

FIGURE 8. Sodium dodecyl sulfate polyacrylamide gels loaded with material from 2-day-old serum-free conditioned media of cultures of several cell lines. Arrow denotes level of major glycoprotein band of FF cultures. a, Human fibroblasts (ST); b, mouse fibroblasts (3T3); c, SV40-transformed 3T3; d, FF line (reference standard); e, human epidermal cells (BE).

Despite these MCIF-induced changes in cyclic nucleotides, perturbations of the latter cannot account for all of the observed effects of MCIF, because addition of exogenous cAMP or its analogs (10^{-8}–10^{-3} M) to RPMI 1846 cultures, while growth inhibitory, nevertheless fails to duplicate the characteristic morphologic conversion regularly induced by MCIF, causing, instead, a marked increase in number and elongation of dendritic processes. It thus appears that altered cyclic nucleotide levels are just one feature of a pleotypic response of melanoma cells to MCIF.

The nature and sources of the distinctive high-molecular-weight protein found in contact-inhibited (FF) cultures, but lacking from noncontact-inhibited RPMI 1846 cultures, were studied. It was found to be an acidic glycoprotein with an approximate molecular weight of 160,000 daltons. Subcellular fractionation studies of both FF and RPMI 1846 cells revealed that this glycoprotein was present in a fraction of both cell types that included endoplasmic reticulum and plasma membrane.[10] Although present in both cell types, it apparently was released in readily detectable quantities only into culture medium of the contact-inhibited FF cell line. It therefore was of interest to determine whether any correlation exists between the presence of this high-molecular-weight glycoprotein in culture medium and the capacity for density-dependent regulation of growth. Polyacrylamide gel electrophoresis of identically prepared CM from several cell types revealed an electrophoretically identical protein in media from every contact-inhibited cell type examined, including 3T3 mouse fibroblasts, a human fibroblast line (ST), and a line derived from human epidermal cells (BE); much reduced amounts were detected in SV40-transformed 3T3 cultures (FIGURE 8). In contrast, this band could not be detected by the gel technique in media from a variety of noncontact-inhibited (malignant) cell types, including B16 and S91 mouse melanomas, human melanomas, HeLa cells, and human colonic carcinomas (FIGURE 9).[16] Thus, the presence of a protein band with the same electrophoretic mobility as that of the high-molecular-weight glycoprotein first observed in MCIF does appear to be correlated with the capacity for density-dependent growth regulation; whether this protein is actually required or sufficient for such capacity cannot be determined until final purification of its contact-inhibitory activity.

It was further found that partially purified MCIF also contained at least three protease inhibitors. The latter could be demonstrated by a sensitive gel inhibition technique;[17,18] inhibition zones from FF gels appeared to coincide with identical zones in gels prepared from fetal calf serum. Thus, although serum had been routinely omitted for 48 hr prior to harvesting CM, it seemed very likely that serum protease inhibitors previously adsorbed onto, or incorporated into, FF cells had subsequently been released into the 48-hr serum-free medium. It therefore was interesting to find that protease-inhibitory gels similarly prepared from noncontact-inhibited RPMI 1846 melanoma cultures contained much smaller (trace) quantities of inhibitors compared to gels from the contact-inhibited FF cultures. These observations suggested the possibility that contact-inhibited cells either are much better able to take up and bind serum protease inhibitors or are less able to degrade such materials after adsorption.

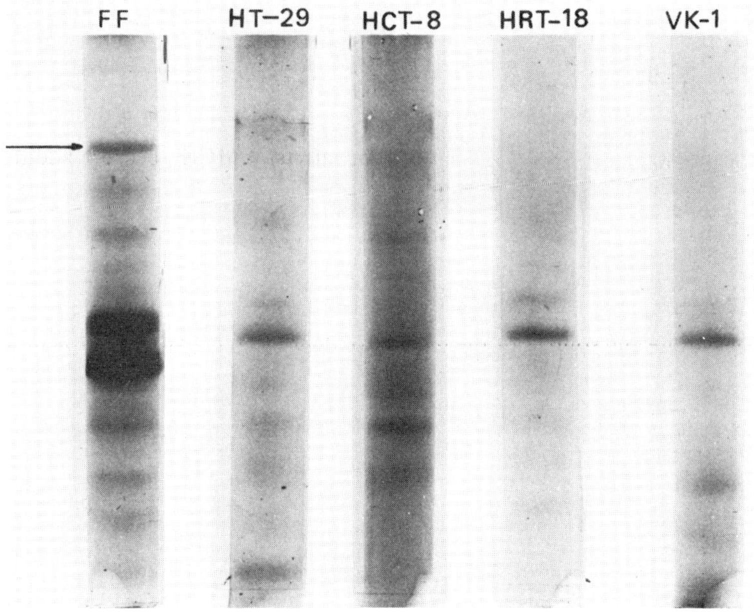

FIGURE 9. Sodium dodecyl sulfate polyacrylamide gels loaded with material from culture media of contact-inhibited FF cells and from four noncontact-inhibited lines of human colonic carcinoma (HT-29, HCT-8, HRT-18, VK-1).

The presence of protease inhibitors in these materials may imply a relationship between these inhibitors and the observed biologic effects. Restoration by antiproteases of features of the contact-inhibited state in transformed cells has been reported.[19] It is possible that in our system, protease inhibitors act synergistically with one or more other components to provide a more complete restoration of normal growth control than could be provided by any of the separate components alone.[20] Preliminary experiments have shown that when CM from FF cultures are further fractionated, somc of the fractions contain both antiprotease and contact-inhibitory activity, while others contain only contact-inhibitory activity. Further purification studies will be required to elucidate the precise function of each component of this complex mixture produced by contact-inhibited cultures.

Acknowledgments

We thank Ms. Vera Klaus and Ms. Ann Powell for their expert technical assistance.

References

1. Todaro, G. J., G. K. Lazar & H. Green. 1965. Initiation of cell division in a contact inhibited mammalian cell line. J. Cell. Comp. Physiol. **66:** 325–334.
2. Stoker, M. G. P. & H. Rubin. 1967. Density dependent inhibition of cell growth in culture. Nature (London) **215:** 171–172.
3. Dulbecco, R. 1970. Topoinhibition and serum requirement of transformed and untransformed cells. Nature (London) **227:** 802–806.
4. Oey, J., A. Vogel & R. Pollack. 1974. Intracellular cyclic AMP concentration responds specifically to growth regulation by serum. Proc. Nat. Acad. Sci. USA **71:** 694–698.
5. Aaronson, S. A. & G. J. Todaro. 1968. Basis for the acquisition of malignant potential by mouse cells cultivated *in vitro.* Science **162:** 1024–1026.
6. Pollack, R. E., H. Green & G. Todaro. 1968. Growth control in cultured cells: selection of sublines with increased sensitivity to contact inhibition and decreased tumor-producing activity. Proc. Nat. Acad. Sci. USA **60:** 126–133.
7. Stoker, M. G. P. 1967. Short range factors affecting cell growth and movement. UICC Monogr. **6:** 193–203.
8. Pollack, R. E. & G. W. Teebor. 1969. Relationship of contact inhibition to tumor transplantability, morphology and growth rate. Cancer Res. **29:** 1770–1772.
9. Lipkin, G. & M. Knecht. 1974. A diffusible factor restoring contact inhibition of growth to malignant melanocytes. Proc. Nat. Acad. Sci. USA **71:** 849–853.
10. Knecht, M. E. & G. Lipkin. 1977. Biochemical studies of a protein which restores contact inhibition of growth to malignant melanocytes. Exp. Cell Res. **108:** 15–22.
11. Lipkin, G. & M. E. Knecht. 1976. Contact inhibition of growth is restored to malignant melanocytes of man and mouse by a hamster protein. Exp. Cell Res. **102:** 341–348.
12. Lipkin, G. & M. E. Knecht. 1975. Restoring contact inhibition of growth to malignant melanocytes of man, mouse and hamster. Schweiz. Med. Wochschr. **105:** 1360–1364.
13. Lipkin, G., M. E. Knecht & M. Rosenberg. 1978. A potent inhibitor of normal and transformed cell growth derived from contact inhibited cells. Cancer Res. **38:** 635–643.
14. Holley, R. W. & J. A. Kiernan. 1968. Contact inhibition of cell division in 3T3 cell lines. Proc. Nat. Acad. Sci. USA **60:** 300–303.
15. Heidrick, M. L. & W. L. Ryan. 1971. Adenosine 3′,5′-cyclic monophosphate and contact inhibition. Cancer Res **31:** 1313–1315.

16. LIPKIN, G., M. E. KNECHT & M. ROSENBERG. 1977. Restoration of in vitro growth control to malignant cells. Cancer **40:** 2699–2705.
17. ROSENBERG, M., V. E. ROEGNER & F. F. BECKER. 1975. The quantitation of rat serum esterases by densitometry of acrylamide gels stained for enzyme activity. Anal. Biochem. **66:** 206–212.
18. ROSENBERG, M., V. E. ROEGNER & F. F. BECKER. 1976. Isolation and characterization of two α_1-protease inhibitors in rat serum. Amer. Rev. Respir. Dis. **113:** 779–785.
19. SCHNEBLI, H. P. & M. M. BURGER. 1972. Selective inhibition of growth of transformed cells by protease inhibitors. Proc. Nat. Acad. Sci. USA **69:** 3825–3827.
20. LIPKIN, G., M. ROSENBERG & M. E. KNECHT. 1976. Factors affecting growth of normal and malignant cells in vitro. Biochem. Pharmacol. **25:** 1333–1337.

SYNTHESIS AND FUNCTION OF CARBOHYDRATE UNITS OF CELL SURFACE GLYCOPROTEINS*

S. J. Turco† and P. W. Robbins

Department of Biology
and Center for Cancer Research
Massachusetts Institute of Technology
Cambridge, Massachusetts 02139

As illustrated by the two examples in FIGURE 1, the carbohydrate units of glycoproteins are attached to the polypeptide backbone of the protein through either the amide group of asparagine ("N-linked") or the hydroxyl group of serine, threonine, or hydroxylysine ("O-linked"). Few if any generalizations can be made about the O-linked glycopeptides. O-linked structures vary from a single galactose residue to the complex "megalosaccharide" structures of the blood group substances (see Reference 1 for a review). Little is known about the sites of assembly of O-linked structures, and their functions are at present obscure.

In contrast, the N-linked glycopeptides form a homogeneous group of related structures. Although in most instances the structural chemistry is incomplete, considerable progress has been made recently in understanding their synthesis and processing. Further, inhibitor studies are beginning to bring the subject of N-linked glycopeptide function out of the realm of pure speculation. For these reasons, the present brief review will be limited to a discussion of the synthesis and function of asparagine-linked oligosaccharides.

POSSIBLE FUNCTIONS OF ASPARAGINE-LINKED GLYCOPEPTIDES

Role in Protein Processing

It seems likely that glycosylation and processing of glycoproteins are closely coupled. Evidence that this notion is true comes from studies of the effects of 2-deoxyglucose, glucosamine, and tunicamycin on glycoprotein synthesis. These compounds, at appropriate concentrations, block protein glycosylation while affecting protein synthesis to a relatively lesser extent. With the aid of these inhibitors, the glycosylation of protein has been shown to be an essential feature of the synthesis and secretion of some proteins but not of others. For example, the lack of glycosylation inhibited the secretion of immunoglobulin molecules from several myeloma tumors.[2,3] Conversely, 2-deoxyglucose did not inhibit the secretion of a normally glycosylated k-type immunoglobulin light chain from another myeloma line,[4] and tunicamycin minimally affected the synthesis and secretion of procollagen from fibroblasts and bone.[5]

Concurrent with the reported effects of glycosylation inhibitors on protein synthesis and processing have been findings in mutants. Weitzman and Scharff have described mouse myeloma mutant cell lines defective in assembly, glycosyla-

*Supported by Grants CA 14142 and CA 14051 from the National Cancer Institute.
†Fellow of the Cystic Fibrosis Foundation and of the Leukemia Society of America, Inc.

0077-8923/78/0312-0392 $01.75/0 © 1978, The New York Academy of Sciences

tion, and secretion of immunoglobulins.[6] They conclude that glycosylation is not essential for the synthesis and export of immunoglobulins by these cells. In another mutant system,[7] a glycosylation deficiency apparently leads to the induction of specific cell surface proteins.

Role in Determining Turnover Rates

For proteins and cells of the circulatory system, it has been suggested that the carbohydrate of glycoproteins plays an important role in determining the rate of turnover.[8] A carbohydrate-binding protein exists on the surface of liver cells that specifically recognizes exposed sugars in circulating glycoproteins and removes

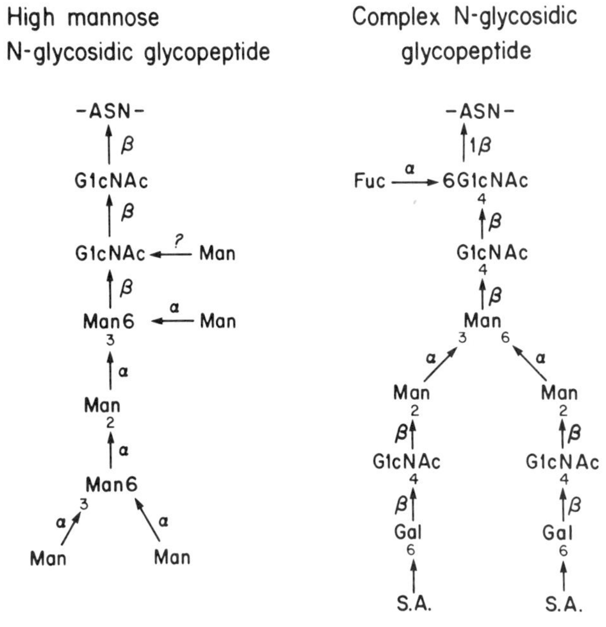

FIGURE 1. Typical examples of serine-threonine-linked and asparagine-linked glycopeptides.

glycoprotein from the circulation. Pioneering work by Ashwell and Morell has shown that a liver-binding protein in the rabbit recognizes terminal galactose residues on the complex N-glycosidically linked oligosaccharides;[9] in avian cells, a binding protein plays a similar role by recognizing terminal *N*-acetylglucosamine residues.[10] In the rat, there exists a mechanism for clearing glycoproteins based on specific recognition of terminal[11] and nonterminal mannose residues.[12] Thus, the structures of oligosaccharide chains serve as regulatory determinants in the catabolism of circulating glycoproteins.

Role in Intercellular and Intracellular Recognition

This proposed function for surface carbohydrates is one of the most exciting and has therefore received the largest amount of attention. However, thus far, most of the experimental evidence to support this proposed function is indirect. Specific recognition processes at the cell surface level have important functions in development and normal growth control, while altered recognition plays an important role in many disease processes, including cancer. Much effort has therefore gone into understanding the molecular basis of these specific recognition processes. One hypothesis involves specific recognition of complex carbohydrates of cell surfaces by specific binding proteins on other cells. Roth *et al.* have suggested that the binding proteins might be surface glycosyl transferases, which recognize substrates on the surface of an interacting cell.[13] Shur and Roth have recently reviewed the status of this hypothesis.[14] While the hypothesis remains attractive, there are many unanswered questions that prevent an evaluation of the significance of such a mechanism in cell-cell interactions.

Structure of Asparagine-Linked Glycopeptides

Structural studies of asparagine-linked oligosaccharide chains of glycoproteins have demonstrated the common structural feature of an inner core region,

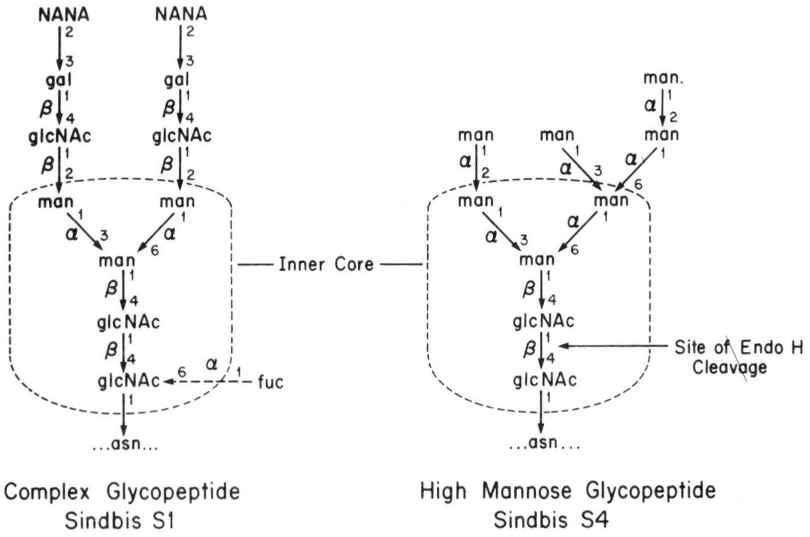

FIGURE 2. Structures of Sindbis virus glycopeptides S1 and S4. The structures are those suggested by Burke.[15] They are typical members of the "complex" and "high-mannose" classes of oligosaccharides. As pointed out by Burke and Keegstra,[16] "S4" is heterogeneous and may contain components with fewer or more mannose residues than the structure shown here. The structure with seven mannose molecules illustrated here is considered by Burke to be a major component of the S4 mixture. Two other complex glycopeptides of Sindbis virus are S2 and S3, which differ from S1 only in that they contain one and no residues of sialic acid (NANA), respectively.

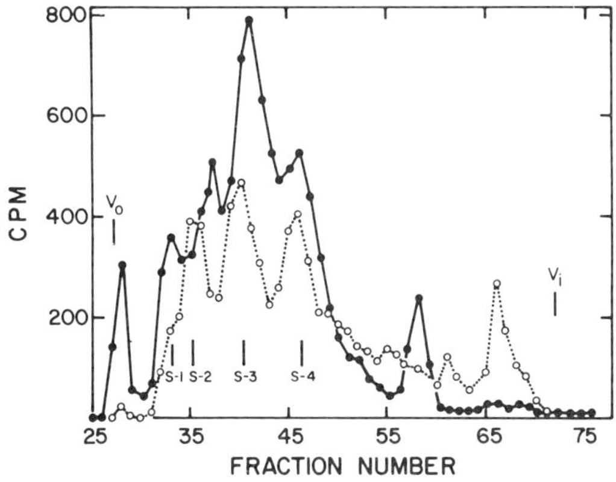

FIGURE 3. BioGel P-6 profile of large, external, transformation-sensitive (LETS) glycopeptides. Normal chicken cells were labeled with [1-^{3}H]glucosamine (●, 50 μCi/ml) for 24 hr. LETS was extracted from these cells by the method of Yamada *et al.*[17] and purified further by acrylamide gel electrophoresis. After electrophoresis, the LETS band was extracted, mixed with [^{14}C]glucosamine-labeled Sindbis virus (○), and the glycopeptides were separated on BioGel P-6, as previously described, after pronase digestion.[18]

The inner core is composed of three mannose and two *N*-acetylglucosamine residues and is attached to the polypeptide by an N-glycosidic linkage between *N*-acetylglucosamine and the amide nitrogen of an asparagine residue of the polypeptide backbone. Glycopeptides that contain additional mannose residues linked to the inner core are classified as "high-mannose" units, whereas glycopeptides that contain other sugars, namely, fucose, sialic acid, galactose, and *N*-acetylglucosamine residues, are classified as "complex" units (FIGURE 2). There are many glycoproteins that contain both types of oligosaccharide chains. For example, as shown in FIGURE 3, BioGel® P6 gel filtration of the glucosamine-labeled glycopeptides from the two Sindbis virus glycoproteins (E_1 and E_2) reveals the presence of a high-mannose unit, S4, and three complex units, S1, S2, and S3, that differ in their amounts of sialic acid. On the other hand, there are other glycoproteins, such as the G-protein of vesicular stomatitis virus, that contain only complex units.

LIPID INTERMEDIATES

During the last several years, evidence has accumulated indicating that the biosynthesis of the region of the oligosaccharide chains that contains mannose and *N*-acetylglucosamine residues involves a cyclic sequence (see Reference 19 for a review). As shown in FIGURE 4, the cycle consists of the preassembly of mannose and *N*-acetylglucosamine, in addition to glucose residues, in an activated state bound to the isoprenoid lipid dolichol pyrophosphate. The cyclic sequence is initiated by the reversible transfer of *N*-acetylglucosamine phosphate from UDP GlcNAc to dolichol phosphate to form GlcNAc-P-P-dolichol. A second *N*-acetylglucosamine residue and then a mannose residue are added directly from the respective nucleotide sugars. Additional mannose and glucose residues are believed to be donated by two other lipid intermediates. Mannose is transferred from GDP mannose to dolichol phosphate to form Man-P-dolichol. Similarly, glucose is transferred from UDP glucose to dolichol phosphate to form Glc-P-dolichol.[20] These lipid-linked monosaccharides then serve as the mannosyl or glucosyl donors, respectively, of the outer mannose or glucose residues, completing the synthesis of an oligosaccharide lipid common to many cultured cells[21]

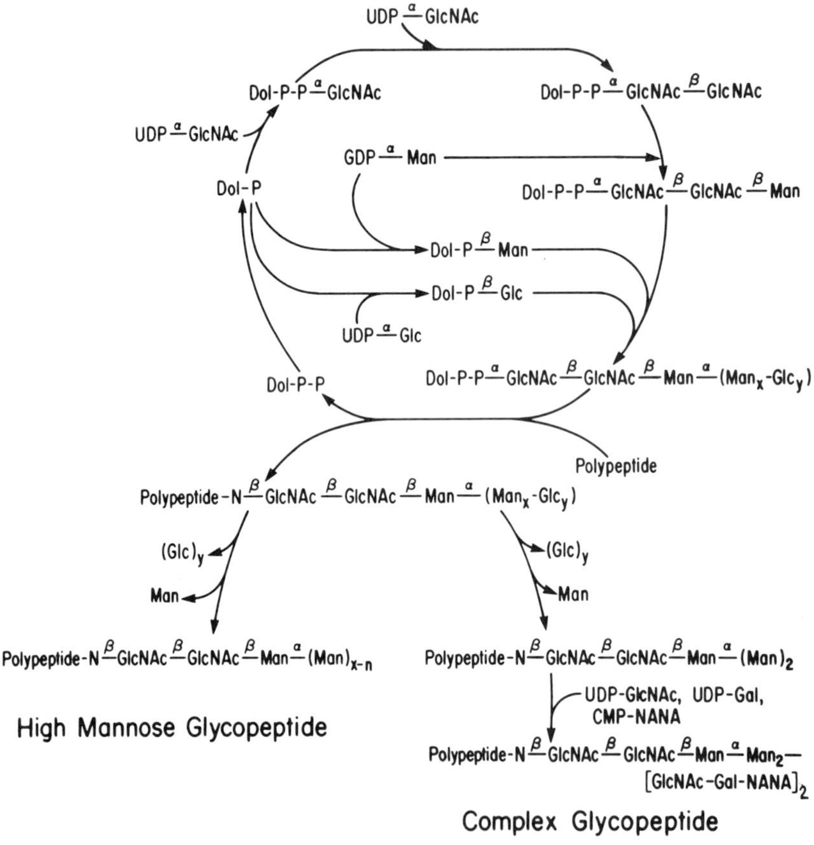

FIGURE 4. Schematic pathway of biosynthesis of asparagine-linked glycopeptides.

and tissues.[22] The composition and structure of the lipid-linked oligosaccharide are under intensive investigation, and at present it is believed to contain two *N*-acetylglucosamine, from eight to 11 mannose, and two to four glucose residues. The transfer of the oligosaccharide to protein generates dolichol pyrophosphate, which, presumably, is hydrolyzed to the monophosphoryl derivative to reinitiate the cycle.

Processing of Oligosaccharide Units

Evidence that the glucose-containing oligosaccharide lipid is an obligate intermediate in the biosynthesis of the core region of glycoproteins is not yet conclusive. However, evidence exists that the glucose-containing oligosaccharide lipid is a preferred substrate in comparative transfer reactions *in vitro*.[23] If further studies confirm the importance of glucose-containing structures in the initial transfer of mannosyl oligosaccharide units to protein, it will be clear that processing occurs before these glycoprotein molecules appear at the cell surface. While several soluble glycoproteins possess glucose as a constituent,[24-26] mature asparagine-linked oligosaccharides of membrane glycoproteins probably do not,[1,27] strongly suggesting that processing is a selective and important event.

As depicted in Figure 4, processing of oligosaccharide after transfer to protein may result in the formation of either high-mannose or complex types. The mechanisms of these two pathways remain to be elucidated. It is possible that the high-mannose units result from removal of all of the glucose residues and, possibly, a few mannose residues. The complex type probably arises from removal of all of the glucose residues and most of the mannose residues, resulting in the appearance of the inner core structure, followed by the addition of the terminal sugar residues. One approach to this problem involves virus-infected cells as model systems. Pulse-chase experiments in Sindbis- and vesicular stomatitis virus-infected cells[21] have shown the processing of a protein-bound oligosaccharide to form high-mannose and complex carbohydrate chains.

Nascent Polypeptide Chains as Acceptors

There is evidence to suggest that transfer of the oligosaccharide from the lipid carrier to appropriate asparagine residues occurs either during or immediately after translation of the polypeptide chain. Schimke *et al.* have recently demonstrated the incorporation of radioactive glucosamine and mannose into nascent ovalbumin polypeptides.[28] Furthermore, Sefton reported that the two Sindbis virus glycoproteins are glycosylated before release from the polysome and that glycosylation involves the transfer of a large oligosaccharide from a lipid to polypeptide chains.[29] Recently, the membrane glycoprotein of vesicular stomatitis virus, synthesized *in vitro* in the presence of pancreatic microsomes, has been shown to be glycosylated in two distinct steps while its polypeptide chain is nascent.[30]

References

1. Kornfeld, R. & S. Kornfeld. 1976. Annu. Rev. Biochem. **45:** 217–237.
2. Melchers, F. 1973. Biochemistry **12:** 1471–1476.
3. Hickman, S., A. Kulczychi, R. G. Lynch & S. Kornfeld. 1977. J. Biol. Chem. **252:** 4402–4408.
4. Eagon, P. K. & E. C. Heath. 1977. J. Biol. Chem. **252:** 2372–2383.
5. Duksin, D. & P. Bornstein. 1977. J. Biol. Chem. **252:** 955–962.
6. Weitzman, S. & M. D. Scharff. 1976. J. Mol. Biol. **102:** 237–252.
7. Pouyssegur, J., R. P. C. Shiu & I. Pastan. 1977. Cell **11:** 941–947.
8. Lunney, J. & G. Ashwell. 1976. Proc. Nat. Acad. Sci. USA **73:** 341–350.

9. Ashwell, G. & A. G. Morell. 1974. *In* Advances in Enzymology. A. Meister, Ed. Vol. 41: 99–128. John Wiley & Sons, Inc. New York, N.Y.
10. Lunney, J. & G. Ashwell. 1976. Proc. Nat. Acad. Sci. USA **73:** 341–343.
11. Baynes, J. W. & F. Wold. 1976. J. Biol. Chem. **251:** 6016–6024.
12. Achord, D. T., F. E. Brot & W. S. Sly. 1977. Biochem. Biophys. Res. Commun. **77:** 409–415.
13. Roth, S., E. J. McGuire & S. Roseman. 1971. J. Cell Biol. **51:** 536–547.
14. Shur, B. D. & S. Roth. 1975. Biochim. Biophys. Acta **415:** 473–512.
15. Burke, D. 1976. Ph.D. Thesis. State University of New York. Stony Brook, N.Y.
16. Burke, D. & K. Keegstra. 1976. J. Virol. **20:** 676–684.
17. Yamada, K. M., S. S. Yamada & I. Pastan. 1976. Proc. Nat. Acad. Sci. USA **73:** 1217–1221.
18. Robbins, P. W., S. S. Krag & T. Liu. 1977. J. Biol. Chem. **252:** 1780–1785.
19. Waechter, C. J. & W. J. Lennarz. 1976. Annu. Rev. Biochem. **45:** 95–112.
20. Herscovics, A., B. Bugge & R. W. Jeanloz. 1977. J. Biol. Chem. **252:** 2271–2277.
21. Robbins, P. W., S. C. Hubbard, S. J. Turco & D. F. Wirth. 1977. Cell **12:** 893–900.
22. Spiro, M. J., R. G. Spiro & V. Bhoyroo. 1976. J. Biol. Chem. **251:** 6420–6425.
23. Turco, S. J., B. Stetson & P. W. Robbins. 1977. Proc. Nat. Acad. Sci. USA **74:** 4411–4414.
24. Tulsiani, D. R. P., R. K. Keller & O. Touster. 1975. J. Biol. Chem. **250:** 4770–4776.
25. Tomino, S., K. Paigen, D. R. P. Tulsiani & O. Touster. 1975. J. Biol. Chem. **250:** 8503–8509.
26. Opheim, D. J. & O. Touster. 1977. J. Biol. Chem. **252:** 739–743.
27. Ginsberg, V. 1964. *In* Advances in Enzymology. A. Meister, Ed. Vol. 26: 35–38. John Wiley & Sons, Inc. New York, N.Y.
28. Kiely, M. L., G. S. McKnight & R. T. Schimke. 1976. J. Biol. Chem. **251:** 5490–5495.
29. Sefton, B. M. 1977. Cell **10:** 659–668.
30. Rothman, J. E. & H. F. Lodish. 1977. Nature (London) **269:** 775–780.

COLD-INSOLUBLE GLOBULIN: TRANSAMIDATION-SENSITIVE SITE AND AFFINITY BEHAVIOR*

Helmut Hörmann and Franz Jilek

Max Planck Institute for Biochemistry
8033 Martinsried, Munich, Federal Republic of Germany

Cold-insoluble globulin (CIg) consists of two polypeptide chains that have approximate molecular weights of 220,000 daltons;[1,2] they are connected by disulfide bonds that, as revealed by plasminolysis of CIg,[3,4] are located near the subunit ends. In this paper, we suggest that at least two forms of CIg may exist. Cold-insoluble globulin specimens prepared from two plasma samples of single donors differed in their ability to agglutinate erythrocytes of blood group A. One sample was highly active, whereas the other did not react. Erythrocytes of blood groups B and AB were less effectively agglutinated by the active sample, and cells of blood group O were not reactive.

As shown by Mosher,[2] CIg is a substrate of plasma transglutaminase. In our investigations,[5] 0.76 mol of [^{14}C]putrescine/220,000 g could be coupled by this enzymatic reaction to CIg, suggesting the presence of one transamidation-sensitive site per subunit. Both prolonged plasminolysis of [^{14}C]putrescine-CIg and cyanogen bromide degradation each yielded one labeled fragment that had an approximate molecular weight of 27,000 daltons. By use of cyanogen bromide degradation, a labeled fragment with a higher molecular weight was also formed; when reduced, it converted into the former 27,000-dalton peptide. Further studies suggested that the transamidation-sensitive site is located in a folded domain of the peptide chain that, in the native molecule, is resistant to further cleavage.

An important feature of CIg is its affinity for collagen. Obviously, CIg is identical to a collagen-dependent fibroblast attachment factor[5] of serum detected by Klebe[6] and to an antigelatin factor.[7] This paper reports that in affinity chromatography, CIg was retained at low temperatures from plasma by denatured collagen of various types immobilized by coupling with Sepharose 4B. It was only weakly adsorbed by immobilized native collagen (type I or II), while native collagen (type III) showed some affinity. The retained CIg was easily desorbed by increasing the temperature to 37°C.

More detailed results were obtained if [^{125}I]labeled soluble collagen preparations were complexed with varying amounts of CIg and if the aggregates formed were precipitated with anti-CIg antiserum. The binding curves shown in FIGURE 1 demonstrate the following sequence of affinity: type III denatured > type I denatured = type II denatured > type III native. Affinity between CIg and soluble native collagen type I or II could not be detected under these conditions.

*Supported by a grant from the Deutsche Forschungsgemeinschaft.

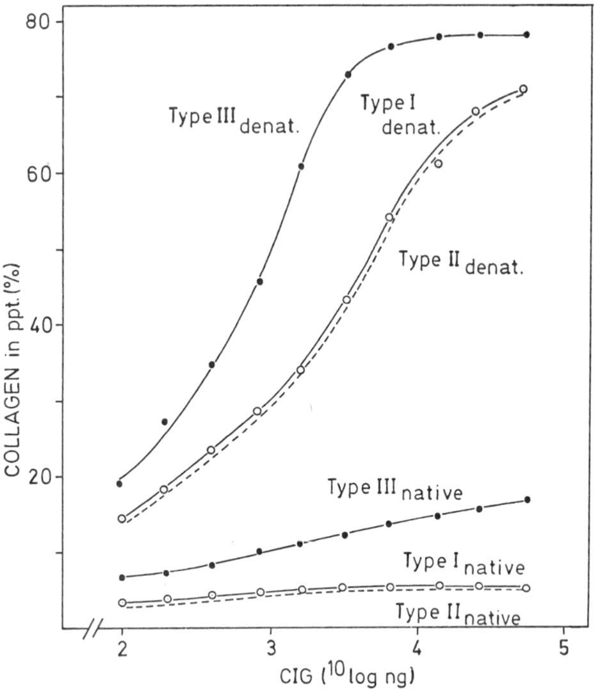

FIGURE 1. Binding of soluble [^{125}I]collagen by CIg. Complexes formed are precipitated by anti-CIg antiserum.

REFERENCES

1. MOSESSON, M. W., A. B. CHEN & R. M. HUSEBY. 1975. The cold-insoluble globulin of human plasma: studies of its essential structural features. Biochim. Biophys. Acta **386:** 509–524.
2. MOSHER, D. F. 1975. Cross-linking of cold-insoluble globulin by fibrin-stabilizing factor. J. Biol. Chem. **250:** 6614–6621.
3. JILEK, F. & H. HÖRMANN. 1977. Cold-insoluble globulin, II. Plasminolysis of cold-insoluble globulin. Z. Physiol. Chem. **358:** 133–136.
4. CHEN, A. B., D. L. AMRANI & M. W. MOSESSON. 1977. Heterogeneity of the cold-insoluble globulin of human plasma (CIg), a circulating cell surface protein. Biochim. Biophys. Acta **493:** 310–322.
5. JILEK, F. & H. HÖRMANN. 1977. Cold-insoluble globulin, III. Cyanogen bromide and plasminolysis fragments containing a label introduced by transamidation. Z. Physiol. Chem. **358:** 1165–1168.
6. KLEBE, R. J. 1974. Isolation of a collagen dependent cell attachment factor. Nature (London) **250:** 248–251.
7. DESSAU, W., F. JILEK, B. C. ADELMANN & H. HÖRMANN. 1978. Similarities between antigelatin factor and cold-insoluble globulin. Biochim. Biophys. Acta **533:** 227–237.

PROPERTIES OF FIBRONECTIN ISOLATED FROM CULTURE MEDIUM OF HUMAN FIBROBLASTS AND FROM HUMAN PLASMA*

Matti Vuento†

Division of Immunology
City of Hope National Medical Center
Duarte, California 91010

Fibronectin[1] is a major fibroblast surface glycoprotein that has recently been studied by several laboratories under various names (SF antigen, LETS protein, cell surface protein, galactoprotein a).[2] Antibodies to fibronectin recognize in plasma a similar protein[1] that has been identified as cold-insoluble globulin of human plasma.[3,4]

Sedimentation velocity measurements and analysis by sodium dodecyl sulfate polyacrylamide gel electrophoresis (SDS-PAGE) have revealed that cold-insoluble globulin of human plasma consists of two subunits. These subunits have an identical or nearly identical size. Molecular weights of 200,000–220,000 daltons have been obtained for the subunits of cold-insoluble globulin.[4,5] Fibronectin present on fibroblastic surfaces has not been characterized in detail. Analysis of labeled surface proteins from fibroblasts by SDS-PAGE has identified fibronectin as a polypeptide that has a molecular weight of 210,000–250,000 daltons and that is lacking from transformed cells.[6,7] By SDS-PAGE performed under nonreducing conditions, labeled surface proteins that migrate like a dimer of this polypeptide have been observed.[8]

We have isolated and analyzed fibronectin secreted into the medium by cultured human fibroblasts.[9] We used an immobilized antibody, which was specific for human fibronectin and did not react with fetal calf serum, to adsorb fibronectin from spent fibroblast culture medium. Fibronectin was eluted with urea and further purified by isoelectric focusing in 9 M urea. A similar procedure was employed for isolation of fibronectin from human plasma.[9]

The amino acid compositions of fibronectins isolated from spent fibroblast culture medium and from plasma are given in TABLE 1. Amino acid analysis was performed after hydrolysis with *p*-toluenesulfonic acid.[10] The amino acid compositions are very similar. The carbohydrate compositions, obtained by gas chromatography after methanolysis and trifluoroacetylation,[11] are listed in TABLE 2. Again, a close similarity between medium and plasma fibronectins is obvious. The amino acid and carbohydrate compositions obtained in the present experiments agree closely with those of cold-insoluble globulin.[4,5]

The fibronectins isolated by immunoaffinity chromatography from culture medium and from plasma focused in a pH range of 5.5–6.2 in 9 M urea, as detected by radioimmunoassay. A main peak that represented more than 70% of the total fibronectin activity focused at pH 6.0. Medium and plasma fibronectins thus both showed heterogeneity, but the major fractions had an identical isoelectric point of 6.0.

*Supported by Grants CA 22108 and CA 16434 from the National Cancer Institute.

†Present address: Department of Biochemistry, University of Helsinki, 00170 Helsinki 17, Finland.

By SDS-PAGE, the reduced medium and plasma fibronectins were shown to have identical migration rates. We observed two closely migrating bands. These bands had molecular weights of about 250,000 daltons. A weak doublet band with a molecular weight of 220,000 daltons was consistently present. The latter band may represent degraded fibronectin. The molecular weights obtained in these experiments are in accord with those obtained by other laboratories for cellular and plasma fibronectins.[4-8] Our results indicate that, within the limits of the detection system used, medium and plasma fibronectins have an identical subunit size.

After cyanogen bromide fragmentation of culture medium and plasma fibronectins (100-fold molar excess of cyanogen bromide to methionine, reaction in 70% formic acid), we obtained identical polypeptide patterns by SDS-PAGE.[9] Considerable portions of the reduced cleavage products had migration rates consistent with molecular weights greater than 18,000 daltons. Two strong bands that had molecular weights of 52,000 and 29,000 daltons were observed. This finding indicates that some of the methionine residues of fibronectin are relatively resistant to cleavage by cyanogen bromide. Resistance of plasma cold-insoluble globulin to cyanogen bromide cleavage has been reported by other workers.[12] We also obtained identical polypeptide patterns by SDS-PAGE after limited proteolysis of culture medium and plasma fibronectins.[9] Although in these experiments the cleavages of the fibronectin polypeptide chains were incomplete, our results suggest an extensive similarity between the polypeptide structures of medium and plasma fibronectins.

Our results suggest that fibronectin secreted by cultured fibroblasts into the medium is identical with fibronectin present in plasma. Recent results from our laboratory show that both medium and plasma fibronectins have collagen-binding activity.[13] This observation further supports the contention of identity of the two

TABLE 1

AMINO ACID COMPOSITIONS OF FIBRONECTINS ISOLATED FROM SPENT FIBROBLAST CULTURE MEDIUM AND FROM PLASMA

Residue	Fibroblast Culture Medium (type I)	Plasma (type II)
Tyrosine	4.2*	4.3
Phenylalanine	2.5	2.6
Tryptophan	2.6	2.7
Lysine	5.1	4.4
Histidine	2.0	1.8
Arginine	4.5	4.5
Aspartic acid	9.5	10.1
Threonine	8.4	10.4
Serine	8.3	8.6
Glutamic acid	11.6	11.8
Proline	8.5	7.0
Glycine	8.9	8.3
Alanine	4.9	5.1
Half-cysteine	2.8	2.1
Valine	6.2	6.0
Methionine	0.9	1.1
Isoleucine	3.6	3.1
Leucine	5.5	6.2

*All determinations are in mole percent.

TABLE 2

CARBOHYDRATE COMPOSITIONS OF FIBRONECTINS ISOLATED FROM SPENT FIBROBLAST CULTURE MEDIUM AND FROM PLASMA

Residue	Fibroblast Culture Medium (type I)	Plasma (type II)
Galactose	1.2*	0.9
Mannose	1.4	1.2
Glucosamine	1.4	1.5
Sialic acid	0.5	0.8
Total	4.5	4.4

*All determinations are in percent of total weight.

fibronectins. The possibility remains, however, that fibronectin present on cell surfaces differs from the secreted form. Further studies are needed to clarify this point.

ACKNOWLEDGMENTS

I thank Drs. E. Ruoslahti and M. Wrann for their collaborative efforts and Ms. Aulikki Pekkala and Mr. David Bills for technical assistance.

REFERENCES

1. RUOSLAHTI, E., A. VAHERI, P. KUUSELA & E. LINDER. 1973. Fibroblast surface antigen: a new serum protein. Biochim. Biophys. Acta **322:** 352–358.
2. HYNES, R. O. 1976. Cell surface proteins and malignant transformation. Biochim. Biophys. Acta **458:** 73–107.
3. RUOSLAHTI, E. & A. VAHERI. 1975. Interaction of soluble fibroblast surface antigen with fibrinogen and fibrin. Identity with cold insoluble globulin of human plasma. J. Exp. Med. **141:** 497–501.
4. MOSESSON, M. W., A. B. CHEN & R. M. HUSEBY. 1975. The cold insoluble globulin of human plasma: studies on its essential structural features. Biochim. Biophys. Acta **386:** 509–524.
5. MOSHER, D. F. 1975. Cross-linking of cold-insoluble globulin by fibrin-stabilizing factor. J. Biol. Chem. **250:** 6614–6621.
6. HYNES, R. O. 1973. Alteration of cell surface proteins by viral transformation and by proteolysis. Proc. Nat. Acad. Sci. USA **70:** 3170–3174.
7. HOGG, N. M. 1974. A comparison of membrane proteins of normal and transformed cells by lactoperoxidase labeling. Proc. Nat. Acad. Sci. USA **71:** 489–492.
8. KESKI-OJA, J., D. F. MOSHER & A. VAHERI. 1977. Dimeric character of fibronectin, a major cell surface associated glycoprotein. Biochem. Biophys. Res. Commun. **74:** 699–706.
9. VUENTO, M., M. WRANN & E. RUOSLAHTI. 1977. Similarity of fibronectins isolated from human plasma and spent fibroblast culture medium. FEBS Lett. **82:** 227–231.
10. LIU, T.-Y. & Y. H. CHANG. 1971. Hydrolysis of proteins with p-toluenesulfonic acid. J. Biol. Chem. **246:** 2842–2848.
11. WRANN, M. & C. W. Todd. 1978. Sensitive determination of sugars utilizing packed capillary columns and electron capture detection. J. Chromatogr. **147:** 309–316.
12. JILEK, F. & H. HÖRMANN. 1977. Cold insoluble globulin, III. Cyanogen bromide and plasminolysis fragments containing a label introduced by transamidation. Hoppe-Seyler's Z. Physiol. Chem. **358:** 1165–1168.
13. ENGVALL, E. & E. RUOSLAHTI. 1977. Binding of soluble form of fibroblast surface protein, fibronectin, to collagen. Int. J. Cancer **20:** 1–5.

ROLE OF FIBRONECTIN AND COLLAGEN TYPES I AND II IN CHONDROCYTIC DIFFERENTIATION *IN VITRO*

W. Dessau, J. Sasse, R. Timpl, and K. von der Mark

Max Planck Institute for Biochemistry
8033 Martinsried, Munich, Federal Republic of Germany

By use of immunofluorescence and lactoperoxidase-catalyzed iodination of cell surfaces, it is possible to identify fibronectin, a major cell surface glycoprotein of fibroblasts,[1,2] on the surface of chondrocytes maintained in suspension or monolayer cultures. The synthesis and secretion of this protein were followed by pulse labeling of chondrocytes and analysis of the labeled proteins secreted into the medium or retained on the cells by sodium dodecyl sulfate polyacrylamide gel electrophoresis and fluorography. Our data therefore show that fibronectin, although not present in the differentiated cartilaginous matrix,[3] can be produced by chondrocytes *in vitro* and thus may have a distinct function during chondrogenesis *in vivo.*

When chondrocytes are isolated from embryonic chick sternal cartilage by collagenase and trypsin treatment, no membrane-associated fibronectin or type-II collagen can be detected by immunofluorescence. After 15 min of incubation in culture medium (Ham's F12 plus 10% fetal calf serum), type-II collagen appears first apparently in membrane-associated form. Fibronectin was detected on the cells only after 2 hr in culture. The immunofluorescence findings were confirmed by a time course experiment in which the synthesis and secretion of [^{3}H]glucosamine or [^{35}S]methionine labeled proteins were studied by biochemical and immunologic techniques. While type-II collagen could already be detected after 30 min in the culture medium and in the cell layer as the precursor form procollagen, fibronectin began to appear only after 2 hr.

After 12–24 hr in culture, chondrocytes attach to the bottom of the plastic dishes and become flattened. They develop a dense pericellular filamentous matrix that can be intensely stained with antibodies to type-II collagen. Antifibronectin antibodies were detected only as distinct patches on the cell surface. These data indicate that the cells attach directly to type-II collagenous filaments. Fibronectin is apparently not involved in cellular attachment, because it was not found between cells and substrate and because addition of exogenous fibronectin did not accelerate or increase the rate of chondrocytic attachment and flattening. When cells started to divide in culture, fibronectin was located mostly at cell borders of postmitotic cells. Cells formed in clones of polygonal chondrocytes were interconnected by short fibronectin strands. In such clones, intracellular, but not extracellular, type-II collagen was found. However, after approximately 1 week in culture, when cells in the center of epithelioid clones rounded up and began to form a cartilagenous matrix, this matrix could be strongly stained with antitype-II collagen antibodies. Differentiated chondrocytes in such cartilagenous colonies exhibited less surface-bound fibronectin than did flattened chondrocytes in the logarithmic growth phase.

Chondrocytes that dedifferentiated in monolayer cultures, as indicated by a switchover to type-I collagen synthesis,[4,5] were covered by a dense network of long fibronectin strands that were in most regions intermingled with extracellular type-I collagen fibers, as revealed by double-antibody staining. This extracellular collagen-fibronectin network was similar to that observed in cultures of chick tendon fibroblasts.[6,7]

From these observations, we conclude that fibronectin, although not detectable in differentiated cartilagenous tissue, plays an important role in the *in vitro* expression of the differentiated functions of chondrocytes. Presumably, it provides intercellular contacts and thus enables the synthesis of the cartilagenous matrix. In contrast to fibroblasts, chondrocytes apparently do not require fibronectin for cellular attachment, because they attach directly to type-II collagenous filaments. The pattern of extracellular fibronectin deposition becomes fibroblast-like when single chondrocytes switch over to type-I collagen synthesis.

References

1. Ruoslahti, E., A. Vaheri, P. Kuusela & E. Linder. 1973. Biochim. Biophys. Acta **322:** 352–358.
2. Hynes, R. O. & J. M. Bye. 1974. Cell **3:** 113–120.
3. Linder, E., A. Vaheri, E. Ruoslahti & J. Wartiovaara. 1975. J. Exp. Med. **142:** 41–49.
4. Müller, P., C. Lemmen, S. Gay, K. von der Mark & K. Kühn. 1975. *In* Extracellular Matrix Influences on Gene Expression. H. C. Slavkin & R. C. Greulich, Eds. :293–302. Academic Press. San Francisco, Calif.
5. Mayne, R., M. S. Vail, P. M. Mayne & E. J. Miller. 1976. Proc. Nat. Acad. Sci. USA **73:** 1674–1678.
6. Dessau, W., J. Sasse, R. Timpl, F. Jilek & K. von der Mark. 1978. Submitted for publication.
7. Vaheri, A., M. Kurkinen, V. Lehto, E. Linder & R. Timpl. 1978. Submitted for publication.

CHANGES IN CELL SURFACE PROTEINS DURING NORMAL AND VITAMIN A-INHIBITED CHONDROGENESIS *IN VITRO*

J. R. Hassell, J. P. Pennypacker, K. M. Yamada,* and R. M. Pratt

Laboratory of Developmental Biology and Anomalies
National Institute of Dental Research
*and *Laboratory of Molecular Biology*
National Cancer Institute
Bethesda, Maryland 20014

Changes in cell surface proteins have been correlated with altered cellular adhesiveness and shape in culture. For example, the major cell surface protein of chick embryo fibroblasts is a 220,000-dalton cell surface glycoprotein (LETS, fibronectin, or CSP) that is synthesized in greatly reduced amounts after malignant transformation. After transformation, these cells lose their adhesiveness, cell-cell contacts, and become rounded. Yamada *et al.*[1-3] have suggested that the 220,000-dalton cell surface glycoprotein plays a role in maintaining cellular adhesion and shape.

Several studies[4,5] have shown that changes in cellular morphology and proteoglycan synthesis occur during the differentiation of limb mesenchymal cells into chondrocytes. During this differentiation, the mesenchymal cells establish transitory cell-cell contacts that are lost as cartilage matrix deposition occurs. Other studies[6-8] have shown that vitamin A enhances the adhesiveness of mesenchymal cells and inhibits chondrogenesis. Vitamin A also causes mature chondrocytes in culture to assume a more flattened, fibroblastic morphology.[9] This study was undertaken to determine whether changes in cell surface proteins occur during normal and vitamin A-inhibited chondrogenesis.

Mesenchymal cells were derived from prechondrogenic limb buds of stage-24 chick embryos and cultured at high density as previously described.[4-6] The plating day was designated as Day 0, and vitamin A (retinoic acid, 3 μg/ml of media) was added on Day 1. Cells in control cultures differentiated into chondrocytes as judged by staining of the cartilaginous matrix with alcian blue (FIGURE 1). Vitamin A-treated cultures remained mesenchymal in appearance and did not stain with alcian blue.

Mature chondrocytes were isolated from sterna of Day-14 chick embryos, as previously described,[9] and plated out at 3×10^6 cells/35 mm culture dish. The plating day was designated as Day 0, and vitamin A (retinoic acid, 3 μg/ml of media) was added 2 hr after plating. By Day 1, cells in control cultures appeared rounded, while cells in vitamin A-treated cultures assumed a flattened morphology (FIGURE 2).

Cell surface proteins were iodinated with ^{125}I by means of the lactoperoxidase-glucose oxidase technique.[10] The cell layers were harvested and electrophoresed in 5% polyacrylamide gels that contained sodium dodecyl sulfate according to previously described procedures.[1] The gels were dried, and autoradiograms were made by exposure to Kodak X Omat® film for 2–4 days.

Autoradiograms of the polyacrylamide gels obtained from the iodinated limb bud cell cultures revealed that the Day-1 cultures contained a 220,000-dalton protein as the major cell surface protein (FIGURE 3,A). This protein was lost, and

0077-8923/78/0312-0406 $01.75/0 © 1978, The New York Academy of Sciences

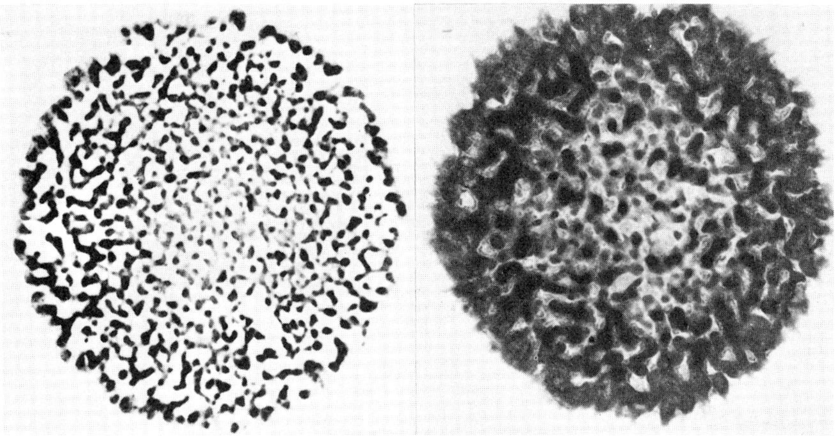

FIGURE 1. Limb bud cultures stained with alcian blue. *Left*, Control Day 4, distinct cartilaginous nodules have developed; *right*, control Day 6, cartilaginous nodules become continuous. Control Day-1 and vitamin A-treated cultures demonstrate no uptake of stain. ×10.

an 80,000-dalton cell surface protein appeared after the cells differentiated into chondrocytes (FIGURE 3,B). Vitamin A-treated cultures retained the 220,000-dalton cell surface protein and, compared to control Day-6 cultures, had less of the 80,000-dalton cell surface protein (FIGURE 3,C).

The major iodinated cell surface protein of sternal chondrocytes was the 80,000-dalton chain seen in autoradiograms of control Day-4 cultures (FIGURE 4,A). In Day-4 vitamin A-treated cultures, this 80,000-dalton cell surface protein was substantially reduced, and the 220,000-dalton cell surface protein was present (FIGURE 4,B).

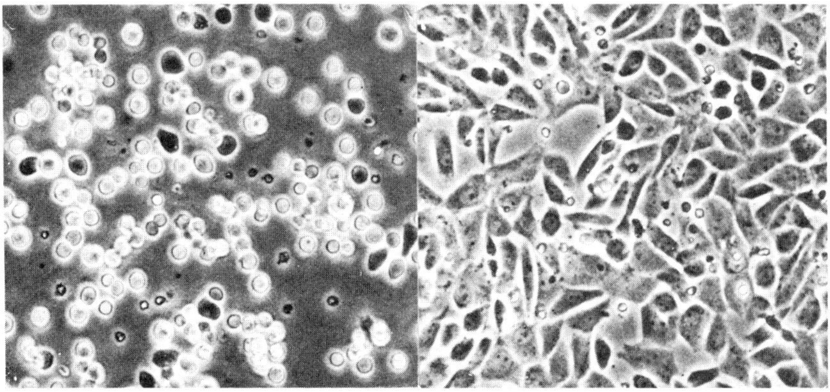

FIGURE 2. Phase-contrast photomicrographs of chick sternal chondrocytes in culture. *Left*, Control Day 1; *right*, vitamin A Day 1. ×97.

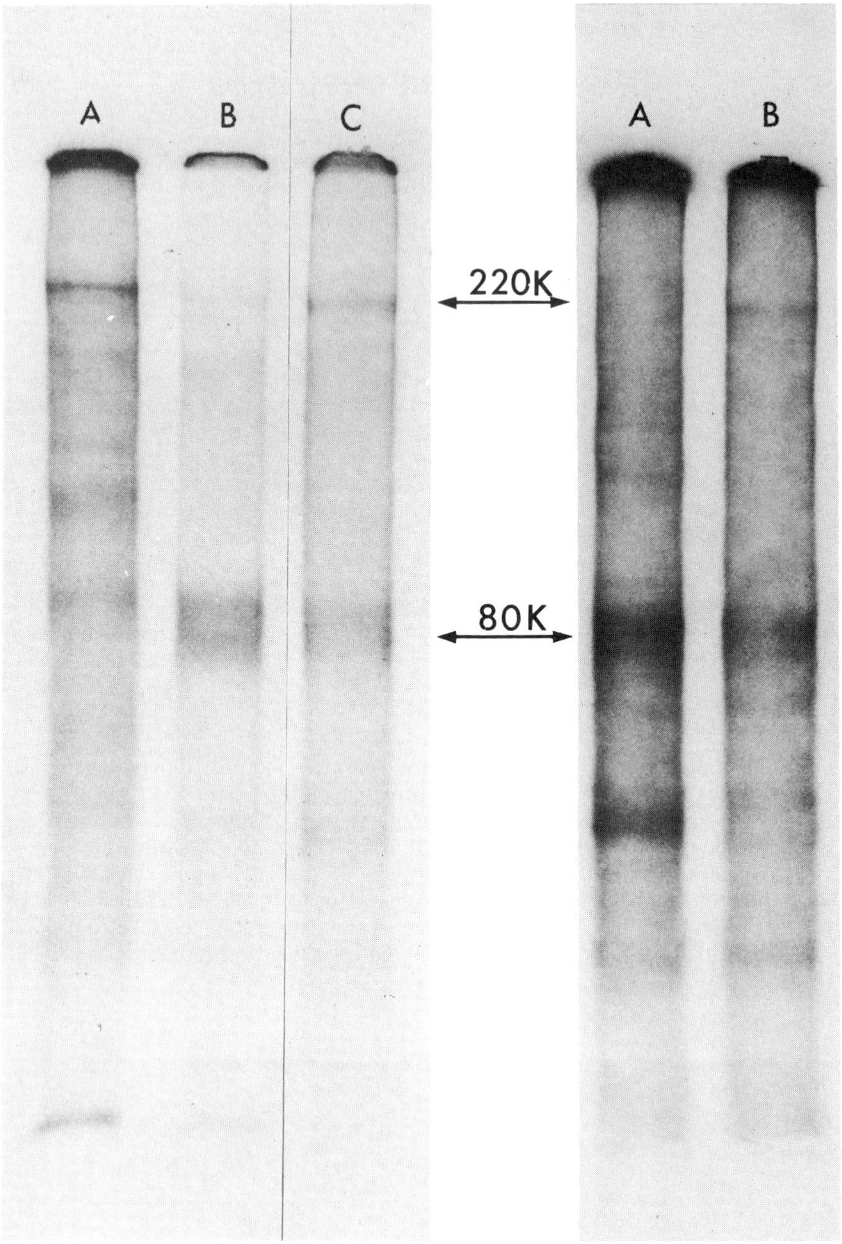

FIGURE 3 (*left*). Autoradiograph of ^{125}I-labeled limb bud cultures after electrophoresis in 5% polyacrylamide gels that contained sodium dodecyl sulfate. A, Control Day 1; B, control Day 6; C, vitamin A Day 6.

FIGURE 4 (*right*). Autoradiograph of ^{125}I-labeled sternal chondrocytic cultures after electrophoresis in 5% polyacrylamide gels that contained sodium dodecyl sulfate. A, Control Day 4; B, vitamin A Day 4.

These results indicate that the differentiation of mesenchymal cells into chondrocytes involves, in part, the loss of the 220,000-dalton cell surface protein and the appearance of an 80,000-dalton cell surface chain. It has been shown that close cellular contact is necessary for initiation of chondrogenesis in the limb bud culture system.[11] Since it has been suggested that the 220,000-dalton cell surface protein plays a role in cellular adhesion,[1-3] it is possible that this cell surface protein may be involved in initiating chondrogenesis in mesenchymal cells. However, the loss of the 220,000-dalton cell surface protein may be necessary for maintenance of the chondrocytic phenotype.

Vitamin A treatment had no effect on the 220,000-dalton cell surface protein in the developing limb bud cultures and caused it to appear in mature sternal chondrocytes. Although the mechanism of action of vitamin A is not completely understood, evidence from several laboratories[12,13] indicates that it may act as a lipid intermediate in the glycosylation of glycoproteins. Furthermore, high levels of vitamin A have been found to increase the incorporation of radiolabeled sugars into glycoproteins.[7,14] Thus, vitamin A's glycosylation function may play a role in regulating the appearance of the 220,000-dalton cell surface glycoprotein in these cells.

References

1. Yamada, K. M. & J. A. Weston. 1974. Proc. Nat. Acad. Sci. USA **71:** 3492.
2. Yamada, K. M., S. S. Yamada & I. Pastan. 1976. Proc. Nat. Acad. Sci. USA **73:** 1217.
3. Yamada, K. M., K. Olden & I. Pastan. This monograph.
4. Levitt, D., P. L. Ho & A. Dorfman. 1975. Dev. Biol. **43:** 75.
5. Royal, P. D. & P. F. Goetinck. 1977. J. Embryol. Exp. Morphol. **39:** 79.
6. Pennypacker, J. P., C. L. Lewis & J. R. Hassell. 1976. J. Cell Biol. **70:** 109(Abst.).
7. Pennypacker, J. P., C. L. Lewis & J. R. Hassell. 1978. Arch. Biochem. Biophys. **186:** 351.
8. Lewis, C. L., R. M. Pratt, J. P. Pennypacker & J. R. Hassell. 1978. Dev. Biol. In press.
9. Solursh, M. & S. Meier. 1973. Calif. Tissue Res. **13:** 131.
10. Hynes, R. O. & J. M. Bye. 1974. Cell **3:** 113.
11. Umansky, R. 1966. Dev. Biol. **13:** 31.
12. Rosso, G. C., S. Masushige, H. Quill & A. Wolf. 1977. Proc. Nat. Acad. Sci. USA **74:** 3762.
13. De Luca, L., C. S. Silverman-Jones & R. M. Barr. 1975. Biochim. Biophys. Acta **409:** 342.
14. Hassell, J. R., C. S. Silverman-Jones & L. M. De Luca. 1978. J. Biol. Chem. **235:** 1627.

AMNIOTIC FLUID FIBRONECTIN AND THE SYNTHESIS OF FIBRONECTIN BY AMNIOTIC FLUID CELLS IN CULTURE

E. Crouch, G. Balian, K. Holbrook, H. Hoehn, and P. Bornstein

Departments of Biochemistry, Biological Structure, and Pathology
University of Washington
Seattle, Washington 98195

Chen *et al.* have described the presence, in human amniotic fluid, of a high-molecular-weight protein recognized by antibodies to plasma cold-insoluble globulin (CIg), which they identified as CIg.[1] The observed enrichment of CIg relative to other proteins of fetal and maternal plasma (1.2% as compared with 0.2 and 0.3%, respectively), together with the high molecular weight of this protein, led the authors to conclude that the CIg could not be derived from plasma by passive ultrafiltration. Instead, it was suggested that the CIg was synthesized locally by fetal amniotic tissues, possibly by amniotic fluid cells.

The observations and suggestions of Chen *et al.* are of interest inasmuch as the origin of plasma CIg and its precise relationship to immunologically related molecules synthesized by a variety of cell types in culture are unknown. Furthermore, only a few amniotic fluid proteins have been convincingly demonstrated to be synthesized within the amniotic compartment.[2] We therefore wished to determine whether amniotic fluid cells synthesize a CIg-like protein in culture and to compare such a protein with the amniotic fluid protein and plasma CIg.

There are at least three phenotypically distinct cell types that can be cloned from second-trimester amniotic fluid and subsequently propagated in culture.[3] These cell types have been designated as E (true epithelial cells), AF (the predominant amniotic fluid cell type), and F (fibroblastic, the least frequently isolated type).[3] These cells have been distinguished on the basis of their clonal and cellular morphologic features, their growth potential under specified conditions of culture,[3] and by their patterns of cellular and secreted proteins on sodium dodecyl sulfate polyacrylamide gel electrophoresis (SDS-PAGE) and DEAE cellulose. We have studied these cell types, and cells explanted from human amnion, with particular attention given to the possible synthesis and accumulation of a CIg-like protein in culture.

In this paper, we describe data that confirm the presence of a high-molecular-weight amniotic fluid protein (referred to here as fibronectin) recognized by antibodies to plasma CIg and describe its quantitation, purification, and preliminary characterization by biochemical and immunochemical means. In addition, we discuss experiments that demonstrate that amniotic fluid cells synthesize and accumulate fibronectin in culture. Furthermore, we discuss data that suggest that the amniotic fluid protein is distinct from CIg but that it is similar or identical to the fibronectin that accumulates in the medium of cultured amniotic fluid cells.

AMNIOTIC FLUID FIBRONECTIN

We have confirmed the presence of a CIg-like protein in second-trimester amniotic fluids. Radial immunodiffusion of whole and ammonium sulfate-

0077-8923/78/0312-0410 $01.75/0

precipitated amniotic fluid [15 or 20% (w/v), 4°C] against monospecific rabbit antihuman plasma CIg gave a single line of fusion with plasma CIg and with concentrated amniotic fluid cell media. The immunoreactive material also comigrated with CIg in immune electrophoresis against anti-CIg.

By use of SDS-PAGE of amniotic fluid, we observed a somewhat heterogeneous, reducible component (1–2% of the total Coomassie blue-stained protein) with an apparent molecular weight of 220,000–250,000 daltons after reduction. This band could be completely and specifically precipitated by affinity-purified anti-CIg.

The amount of material that cross-reacted with antibodies to CIg has been quantitated by radial immunodiffusion by use of a standard curve calibrated with purified CIg or, alternatively, by radioimmune inhibition assays with [^{125}I]CIg, rabbit antihuman CIg, and goat antirabbit IgG. The result of approximately 70 μg/ml is in good agreement with that published by Chen *et al.*[1] and with independent estimates obtained by densitometry of stained gels.

The protein has been purified from amniotic fluid by affinity chromatography on antibody-conjugated Sepharose after elution with 3 M sodium thiocyanate. The eluted protein was examined by SDS-PAGE and found to be essentially pure after a single passage through the column. The migration of the eluted protein on SDS-PAGE was compared with that of CIg under oxidizing and reducing conditions. Like CIg, most of the amniotic fluid protein behaved as a dimer. However, unlike CIg, which migrates as a relatively narrow band or as a closely spaced doublet with a maximum apparent molecular weight, after reduction, of 220,000 daltons, the fluid protein migrated as a broad band with a maximum apparent molecular weight, after reduction, of 240,000–250,000 daltons. The two proteins could be completely resolved with 5 or 6% slab gels.

The purified protein shows complete immunologic cross-reactivity with CIg by use of antibodies to CIg and gives maximal inhibition in radioimmune inhibition assays at concentrations equal to that of purified CIg. Its amino acid composition is essentially identical to those of CIg and human skin fibroblast fibronectin. Although the protein exhibits a high degree of structural homology with CIg and fibroblast fibronectin after limited enzymatic or chemical cleavage, reproducible differences from CIg peptide maps have been observed.

Synthesis of Fibronectin by Amniotic Fluid Cells

By use of SDS-PAGE of radiolabeled medium proteins of E and AF cells, we observed a major, broad, reducible protein that migrated more slowly than CIg but that comigrated with the amniotic fluid protein and with human skin fibroblast fibronectin. In dialyzed medium from cultures of E or AF cells labeled for 12–24 hr with [^{3}H]tyrosine or [^{3}H]amino acids, this protein constituted 45–60% of the radioactivity, as determined by densitometry of fluorescent autoradiograms or by immune titration of the medium with antibodies to CIg or fibroblast fibronectin. The protein was completely and specifically precipitated by these antibodies, as assessed by SDS-PAGE of the immune precipitates. This protein was also observed to be a major medium component when proteins were stained with Coomassie blue after electrophoresis. This protein was, however, a relatively minor component of the F-cell medium (only 10–20% of the radioactivity in [^{3}H]tyrosine-labeled cultures). The CIg-like protein in E, AF, and F media has been shown to be resistant to bacterial collagenase and to contain significant carbohydrate. On the basis of the above information and the observed differences from CIg, the protein that accumulates in the medium of amniotic fluid cells has been called amniotic fluid cell fibronectin.

E and AF cells appear to synthesize fibronectin at approximately twice the rate of F cells. This difference was revealed by quantitation of the immune precipitable band in the cells pulsed for 15 min in the presence of [^{3}H]amino acids. These results are consistent with radioimmune inhibition assay findings, which indicated that approximately twice as much fibronectin accumulates in the media of E and AF cultures in a 24-hr period compared with results with F-cell medium. Significant degradation of fibronectin was not observed in cultures under these conditions.

The synthesis of fibronectin was also observed in amniotic fluid cell mass cultures and has been demonstrated in whole amniotic fluid labeled with [^{3}H]amino acids shortly after amniocentesis. In addition, a fibronectin-like protein can be immune precipitated by anti-CIg from the medium of epithelioid cells explanted from human amnion. This protein migrated somewhat more slowly on SDS-PAGE than did the previously described fibronectins, and its relationship to these other proteins is uncertain. Amniotic fluid E cells are morphologically distinguishable from epithelioid cells derived from amnion under comparable culture conditions.

Fibronectin has been demonstrated in the cell layers of cultured amniotic fluid cells by immune precipitation of biosynthetically labeled cell layer proteins. The corresponding band constituted 3–5% of the Coomassie blue-stained proteins in the cell layer and was trypsin sensitive. Cell layer fibronectin was also identified by lactoperoxidase-catalyzed surface iodination and by indirect labeling with affinity-purified rabbit anti-CIg and fluorescein isothiocyanate-conjugated goat antirabbit IgG. Fibronectin was identified in the extracellular matrix and in association with the cell surface by electron microscopy by use of affinity-purified rabbit anti-CIg and ferritin-conjugated goat antirabbit IgG. By these techniques, E and AF cells, again, appear to accumulate more fibronectin in the cell layer than do F cells.

In pulse-chase experiments with AF cells, more than 50% of the newly synthesized fibronectin appeared in the medium after 2 hr of chase and over 70% after 6–8 hr. The amount of fibronectin in the medium remained relatively constant for at least 48 hr. The fibronectin that remained in the cell layer was localized at trypsin-sensitive sites and had an apparent half-life of 2 days. We have no evidence for significant redistribution of fibronectin beyond the first few hours after synthesis. We also have no evidence for time-dependent conversion of fibronectin to a protein with the migration rate of CIg. The transit time of fibronectin from its site of synthesis to trypsin-sensitive sites was estimated to be approximately 60 min by comparing the amount of radiolabeled fibronectin in trypsinized to nontrypsinized cell layers at short intervals of chase after a 15-min pulse with labeled amino acid. Maximum synthesis and secretion time was estimated to be approximately 90 min by quantitating the amount of immune precipitable fibronectin in the medium after 2, 3, and 4 hr of incorporation and extrapolating to zero incorporation. On the basis of these experiments, we conclude that the majority of newly synthesized AF-cell fibronectin has, at most, a short half-life on the cell surface. Thus, medium fibronectin behaves more like a secreted protein than a "sloughed" membrane protein. This behavior would be consistent with its observed presence and presumed role in the extracellular matrix *in vivo* and *in vitro*.

In summary, amniotic fluid contains a protein that is similar to CIg but that appears to be more closely related to the fibronectin found to accumulate in the media of cultured amniotic fluid cells. Whether the observed differences may be explained by differences in the level or type of glycosylation, by differences in the primary amino acid sequence, or by a combination of both is under active investigation. Fibronectin may be a precursor of CIg. However, we have little data to indicate that fibronectin may be converted to a molecule with the mobility on SDS-PAGE of CIg. Our data are consistent with the synthesis of fibronectin by fetal cells within the amniotic fluid compartment.

References

1. Chen, A. B., M. W. Mosesson & G. I. Solish. 1976. Identification of the cold-insoluble globulin of plasma in amniotic fluid. Amer. J. Obstet. Gynecol. **125:** 958–961.
2. Sutcliffe, R. G. 1975. The nature and origin of the soluble protein in human amniotic fluid. Biol. Rev. **50:** 1–33.
3. Hoehn, H., E. M. Bryant, L. E. Karp & G. M. Martin. 1974. Cultivated cells from diagnostic amniocentesis in second trimester pregnancies. I. Clonal morphology and growth potential. Pediatr. Res. **8:** 746–754.

TRANSMEMBRANE LINKAGE OF FIBRONECTIN TO INTRACELLULAR ACTIN-CONTAINING FILAMENTS IN CULTURED HUMAN FIBROBLASTS*

Michael H. Heggeness, J. F. Ash, and S. J. Singer†

Department of Biology
University of California at San Diego
La Jolla, California 92093

Fibronectin was shown by immunofluorescence studies[1] to be present in streak-like patterns on the surfaces of normal chicken fibroblasts, and the suggestion was made that these surface streaks were associated with underlying fibrous structures within the cells. In this paper, we provide strong evidence for such an association by means of double-fluorescent staining of fibronectin and of intracellular actin with normal human fibroblasts. We have also examined the time course of appearance of fibronectin and of actin-containing filaments on plating out freshly trypsinized fibroblasts. The actin-containing filaments reappear before the surface streaks of fibronectin, suggesting therefore that the filaments are involved directly in the determination of the fibronectin surface distribution.

WI38 cells were grown in Dulbecco's modified minimal essential medium supplemented with 10% fetal calf serum and antibiotics in an atmosphere of 90% air, 10% CO_2. Cells were treated with 0.25% trypsin (1:250; Gibco, Grand Island, N.Y.) for 15 min, replated onto coverslips in fresh medium, and allowed to spread.

At varying times after replating, cells were fixed with 3% formaldehyde, washed with phosphate-buffered saline (PBS), and incubated with dilute antiserum (kindly provided by Dr. R. Doolittle) against cold-insoluble globulin (which cross-reacts antigenically with fibronectin protein). The cells were then washed in PBS and treated with rhodamine-conjugated goat antirabbit IgG.[2] After another PBS wash and a 2-min treatment with 0.1% Triton® X-100 to render the interior of the cell accessible to staining reagents, the intracellular actin was localized by use of biotinyl-conjugated rabbit heavy meromyosin and fluorescein-labeled avidin.[3] As a control for the specificity of the antibody reagents, it was found that absorption of the rabbit serum with purified cold-insoluble globulin eliminated the fluorescent staining of the cell surfaces. This general scheme of two-color fluorescent staining has been applied successfully to several other systems.[4,5]

As many investigators have shown (reviewed in Reference 6), trypsin treatment of cultured WI38 normal human fibroblasts removes essentially all surface fibronectin, as can also be demonstrated by immunofluorescence (FIGURE 1,A). In the process, the cells round up, and the organized actin-containing filaments (stress fibers) inside the cells appear to break down. One hour after replating, the actin distribution appears essentially entirely diffuse (FIGURE 1,C). Fibronectin staining begins to appear at this time (FIGURE 1,D) as a punctate ring at the bottom surface of the cell near the cell periphery, but not elsewhere on the surface. After 2 hr, the actin-containing fibers and fiber bundles become clearly visible (FIGURE 1,F), while the fibronectin is still confined to a peripheral ring (FIGURE 1,G). The positions of these fibronectin rings are within the leading edges of the cells (FIGURE

*Supported by United States Public Health Service Grants GM-15971 and CA-22031.
†American Cancer Society Research Professor.

0077-8923/78/0312-0414 $01.75/0

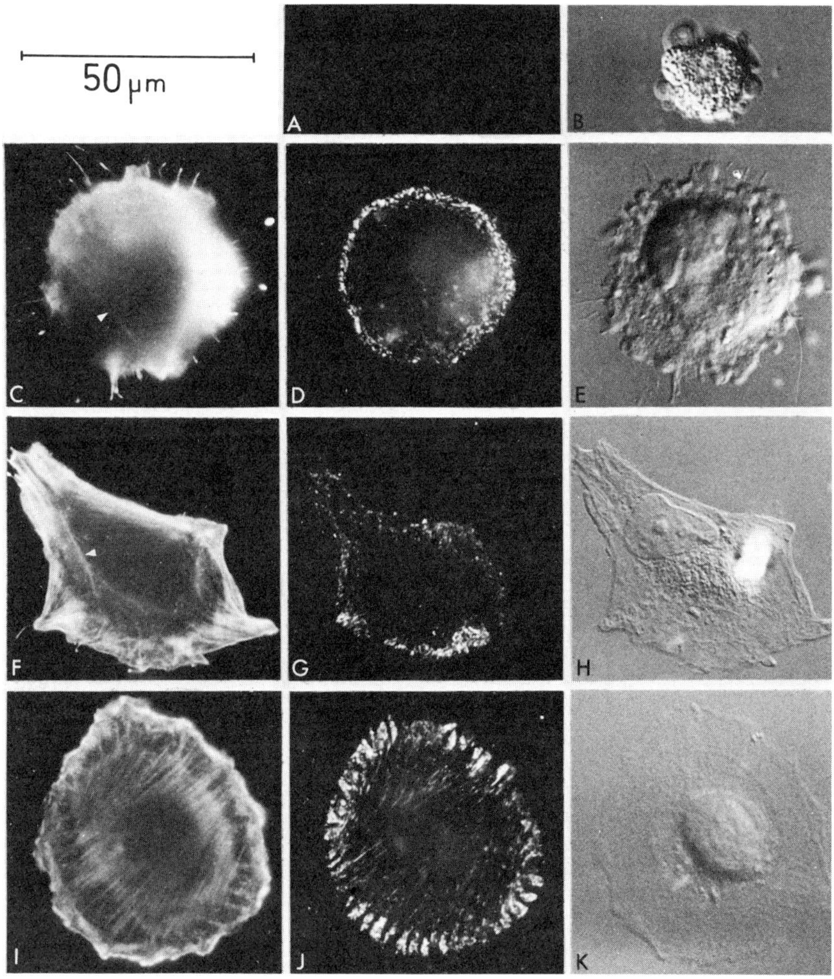

FIGURE 1. Trypsinized WI38 cells. A & B, freshly replated; C–E, 1 hr after plating; F–H, 2 hr after plating; I–K, 6 hr after plating. Specific staining for intracellular F actin is shown in C, F, and I; for surface fibronectin, it is demonstrated in A, D, G, and J; the cellular appearance by Nomarski optics is illustrated in B, E, H, and K. In all cases, the focus is at the bottom surface of the cell. Immediately after trypsinization, there is no detectable staining for surface fibronectin (A). The small white arrows in C and F point to actin-containing fiber bundles.

1,E&H), probably where the lower surfaces of the cells adhere to the substrate. (This finding supports the contention that fibronectin serves in cellular adhesion to the substrate.[7]) After 6 hr, the actin-containing stress fibers are more prominent and are visualized as parallel arrays running across the width of the spreading cell (FIGURE 1,I, in focus near the bottom surface of the cell). At this time, fibronectin is present not only in the peripheral ring but also, to a lesser extent, in dots and

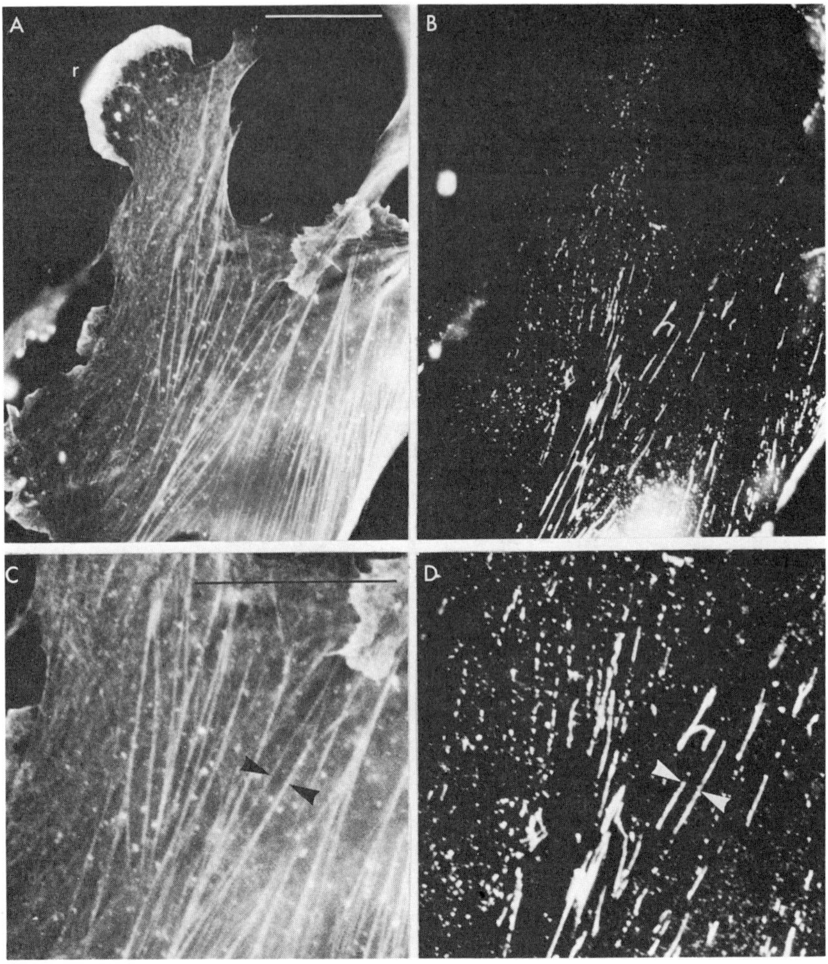

FIGURE 2. Trypsinized WI38 cells 24 hr after replating. C and D are higher magnifications of the sections shown in A and B, respectively. A and B show intracellular F actin and surface fibronectin staining, respectively. The focus is at the bottom surface of the cell. In A, r designates a ruffle, which is heavily stained for actin but shows no surface fibronectin staining. In C and D, the arrows call attention to surface streaks of fibronectin that are particularly clearly superimposed on the actin-containing stress fibers under the cell membrane. Bars in A and C represent 25 μm.

streaks on the lower surface of the cell (FIGURE 1,J). (No fibronectin is present on the upper surface at this time.) Superimposition of negatives reveals that the fibronectin streaks are coincident with actin-containing stress fibers (FIGURE 1,I&J). After 24 hr, more fibronectin appears in dots and streaks on the bottom surface, and these dots and streaks are clearly superimposed as a discontinuous assembly over the continuous stress fibers (FIGURE 2). Where a ruffle (r) is present

(FIGURE 2,A) that contains actin not in the form of stress fibers, no fibronectin is found (FIGURE 2,B).

This superimposition implies that a molecular linkage has formed that connects the fibronectin on the cell surface through the plasma membrane to the actin-containing stress fibers underneath the membrane. The relative times of appearance of the fibronectin and stress fibers after trypsinization suggest that the stress fibers determine the ultimate surface distribution of fibronectin. (What factors determine the organization of the fibronectin that initially appears on the cell surface as a ring around the cellular periphery is an interesting subject for further study.) Whether surface components other than fibronectin, such as collagen,[8] are similarly organized can be investigated by the same double-labeling techniques.

The molecular mechanism responsible for the transmembrane linkages of fibronectin to the stress fibers is not understood. It may, however, be similar to that proposed recently[5,9] for the antibody- or lectin-induced transmembrane linkage of membrane receptors generally to actin- and myosin-containing structures inside cells. Fibronectin aggregated into filaments (which may also contain other surface components) may serve as multivalent ligands (analogous to antibodies or lectins) for one or more specific membrane receptors, thereby inducing a clustering of those receptors in the plane of the membrane. The clustered receptors, it is proposed,[5,9] then become attached to a unique integral protein X in the membrane; protein X provides the binding site for actin attachment to the cytoplasmic face of the membrane.

REFERENCES

1. WARTIOVAARA, J., E. LINDER, E. RUOSLAHTI & A. VAHERI. 1974. J. Exp. Med. **140:** 1522–1533.
2. BRANDTZAEG, P. 1973. Scand. J. Immunol. **2:** 273–290.
3. HEGGENESS, M. & J. F. ASH. 1977. J. Cell Biol. **73:** 783–788.
4. HEGGENESS, M., K. WANG & S. J. SINGER. 1977. Proc. Nat. Acad. Sci. USA **74:** 3883–3887.
5. ASH, J. F., D. LOUVARD & S. J. SINGER. 1977. Proc. Nat. Acad. Sci. USA **74:** 5584–5588.
6. HYNES, R. O. 1976. Biochim. Biophys. Acta **458:** 73–107.
7. YAMADA, K. M., S. S. YAMADA & I. PASTAN. 1976. Proc. Nat. Acad. Sci. USA **73:** 1217–1221.
8. BORNSTEIN, P. & J. F. ASH. 1977. Proc. Nat. Acad. Sci. USA **74:** 2480–2484.
9. BOURGUIGNON, L. Y. W. & S. J. SINGER. 1977. Proc. Nat. Acad. Sci. USA **74:** 5031–5035.

SUBUNIT STRUCTURE OF SURFACE AND SHED LARGE, EXTERNAL, TRANSFORMATION-SENSITIVE PROTEIN OF CHICK EMBRYO FIBROBLASTS*

Michael R. McConnell† and Peter M. Blumberg

Department of Pharmacology
Harvard Medical School
Boston, Massachusetts 02115

Several laboratories have recently reported that large, external, transformation-sensitive (LETS) glycoprotein exists as disulfide-linked polymers on the surface of chick embryo fibroblasts[1,2] and Nil hamster cells.[3] For the chick fibroblast system, quantitation after lactoperoxidase-catalyzed iodination indicated that 21.3 ± 1.3% of total LETS protein was present as the dimer and 77.4 ± 1.9% as the oligomer (mean values ± SEM; n = 20). The term oligomer refers to polymeric LETS protein that was too large to enter a 3.5% sodium dodecyl sulfate (SDS) polyacrylamide gel. These results do not appear to be an artifact of the lactoperoxidase labeling procedure. Alkylation of cells with iodoacetamide either before or after iodination did not affect the proportion of LETS protein dimer; quantitation of the LETS protein dimer by protein staining with Coomassie brilliant blue yielded values similar to those obtained by iodination. Trapping of LETS protein dimer likewise cannot explain the apparent presence of the oligomer. The quantitation of LETS protein dimer by protein staining was linear with respect to the amount of protein added, thus ruling out concentration-dependent aggregation. When purified radioactive LETS protein dimer was added back to the nonradioactive cell lysate, less than 6% of the radioactivity was retained at the top of the gel under the conditions routinely employed for measuring the proportion of LETS protein dimer.

The subunit structure of the LETS protein on the cell surface exhibited little sensitivity to growth conditions. Chick embryo fibroblasts cultured for 1, 2, or 3 days to equal final cell densities showed no difference in the proportions of the LETS protein dimer and oligomer. Growth of cells for 2 days to final cell densities that ranged from 1 to 20 × 10^4 cells/cm^2 caused the proportions of LETS protein dimer to vary by no more than 4%.

LETS protein is known to be shed by fibroblasts into the medium. We have examined the subunit structure of this shed material. Chick embryo fibroblasts were labeled by lactoperoxidase-catalyzed iodination, then incubated for 20 hr with conditioned culture medium. Only 23.5 ± 3.3% (mean value ± SEM; n = 7) of the LETS protein released into the medium was present as the dimeric form. The remainder was primarily oligomer. Comparable results were obtained if incubation occurred in fresh rather than in conditioned medium. Controls ruled out the possibility of LETS protein dimer trapping due to high serum protein concentration in the medium samples. Preliminary experiments indicate, moreover, that the subunit structure of the shed LETS protein was not altered if the incubation was prolonged for an additional day. It would thus appear that the subunit

*Supported by Grant CA-18294-2 from the National Institutes of Health.

†Formerly a National Institutes of Health Viral Oncology trainee and presently a post-doctoral fellow of the National Cancer Institute.

structure of the shed LETS protein closely resembles that of surface LETS protein. Its structure differs from that of human cold-insoluble globulin, which is reported to be entirely dimeric.[4]

The relationship between the dimer and the oligomer on the cell surface was further investigated by measuring changes in the proportions of the two components as a function of time. Cells were pulse labeled with [^{35}S]methionine for 1 hr, and conditioned medium was added back at 39°C. The total amount of LETS protein and the proportion of the LETS protein dimer were then monitored with time. At the beginning of the chase period, 70% of the LETS protein was present as the dimeric form. This proportion rapidly decreased for the first 3 hr ($t\frac{1}{2}$ = 1.9 hr). The rate then slowed substantially ($t\frac{1}{2}$ = 14.5 hr), so that the overall kinetics were biphasic. An increase in LETS protein surface oligomer levels occurred concomitantly with loss of the dimer, indicating the conversion of dimer into oligomer. If cells were labeled by lactoperoxidase-catalyzed iodination rather than by metabolic pulse labeling with [^{35}S]methionine, the decrease in the proportion of the LETS protein dimer showed simple first-order kinetics, with a half-life of 11.7 ± 1.7 hr (mean value ± SEM; n = 7). This decrease probably corresponds to the second phase of the [^{35}S]methionine pulse labeling curve. The failure to observe the first rapid phase of the decrease in the dimer was expected, because the iodinated LETS protein would be predominantly protein that is older than 3 hr. These results indicate that the LETS protein dimer is not in equilibrium with the LETS protein oligomer but is, instead, converted into the oligomeric form by an irreversible mechanism. Moreover, the shedding data indicate that the LETS protein dimer is as stable in its attachment to the cell surface as is the oligomeric form, because both the dimer and the oligomer appear to be released from the cell surface into the medium at similar rates.

Cross-linking data provide a possible explanation for the failure to observe differential shedding of the LETS protein dimer and oligomer. Treatment of ^{125}I-labeled cells with the bifunctional cross-linking reagent dithiobispropionimidate resulted in quantitative retention of the LETS protein dimer at the top of the 3.5% SDS polyacrylamide gel. The LETS protein dimer would thus appear to be interacting, albeit noncovalently, with other surface proteins, possibly molecules of the LETS protein oligomer.

References

1. Rossow, P. W., M. R. McConnell & P. M. Blumberg. 1977. Fed. Proc. **36:** 358.
2. Keski-Oja, J., D. F. Mosher & A. Vaheri. 1977. Biochem. Biophys. Res. Commun. **74:** 699–706.
3. Hynes, R. O. & A. Destree. 1977. Proc. Nat. Acad. Sci. USA **74:** 2855–2859.
4. Mosesson, M. W., A. B. Chen & R. M. Huseby. 1975. Biochim. Biophys. Acta **386:** 509–524.

ATTACHMENT OF CHICKEN MYOBLASTS TO A GELATINIZED SUBSTRATUM IS MEDIATED BY CELL ATTACHMENT PROTEIN FROM HORSE SERUM AND BY CHICKEN FIBROBLAST CELL SURFACE PROTEIN

Matthias Chiquet, Emilio C. Puri, and David C. Turner

Institute of Cell Biology
Swiss Federal Institute of Technology
CH-8093 Zurich, Switzerland

When cells from embryonic chicken muscle are cultured in serum-free "LH" medium (see legend to FIGURE 1), fibroblasts (Fb) remain attached to the plate, and myoblasts (Mb) are found in suspension.[1] The suspended Mb attach to, and spread upon, a gelatinized substratum in LH if the medium is supplemented with horse serum (HS) or if either substratum or Mb is preincubated with HS, demonstrating that one or more components of HS mediate Mb attachment (FIGURE 1).

Cell attachment protein (CAP) was purified from HS by a modification of a procedure described for fetal calf serum[2] and was shown to mediate Mb attachment if added to LH or if used to pretreat gelatinized dishes (FIGURE 1,c). Human cold-insoluble globulin (CIg) was purified according to the procedure of Mosesson and Umfleet[3]; it showed only weak attachment-promoting activity. Antisera to the purified CIg were raised in rabbits, and a crude IgG fraction was obtained (see legend to FIGURE 1). In immunodiffusion and immunoelectrophoresis tests, anti-CIg reacted with CAP and with no other component of HS. Moreover, CAP appeared serologically identical to CIg, to a single component of chicken serum, and to cell surface protein (CSP) extracted from chicken fibroblasts according to the method of Yamada *et al.*[4] Cell surface protein promoted Mb attachment in our assay in the same range of concentrations as did CAP. The three proteins (CAP, CIg, and CSP) comigrated when, after reduction in mercaptoethanol, they were electrophoresed in polyacrylamide gels that contained sodium dodecyl sulfate.

Experiments with anti-CIg demonstrated that CAP is not merely one of several components that promote Mb attachment but that CAP alone is necessary and sufficient: addition of anti-CIg IgG to medium that contained CAP or HS blocked Mb attachment almost completely; preincubation of HS with anti-CIg IgG removed almost all attachment-promoting activity; and Mb failed to attach to gelatinized plates treated first with HS and then with anti-CIg IgG (FIGURE 1,b).

Cell surface proteins capable of reacting with anti-CIg were localized by indirect immunofluorescence after formaldehyde fixation (FIGURE 2). Fibroblasts displayed intense fibrillar fluorescence when cultured in LH supplemented with HS (FIGURE 2,a). In LH supplemented with CAP-free HS, the fluorescent networks were less intense but still prominent (FIGURE 2,b). This observation demonstrates that Fb produce CSP and suggests they may also adsorb CAP from HS, although the difference in intensity could be due to enhanced CSP synthesis in complete compared to CAP-free HS. Fibroblasts cultured in LH without supplements showed, at best, weak fluorescence. Labeling of Fb with [^{35}S]methionine in HS-containing medium, followed by urea extraction, immunoprecipitation with anti-CIg, gel electrophoresis, and autoradiography, confirmed that CSP is synthesized by Fb and that adsorption of CAP from HS cannot account for all of the Fb surface protein capable of reacting with anti-CIg.

Myoblasts suspended in LH did not stain with anti-CIg, but neither did attached myotubes and Mb, even after 5 days of subculturing in medium that contained HS, although the adjacent substratum was weakly fluorescent (FIGURE 2,c). The serum-derived CAP that mediates Mb attachment is thus not detectable by our immunofluorescence procedure, either because too little is bound or because the undersides of the cells are inaccessible to antibody. Metabolic labeling of Mb

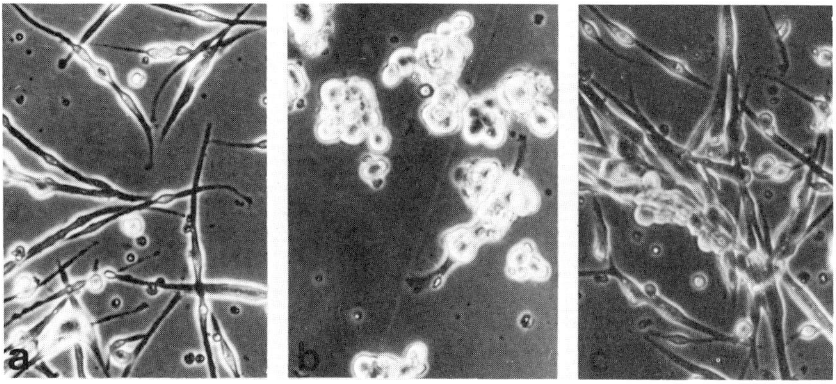

FIGURE 1. Cell-substratum attachment in horse serum is prevented by antiserum against human cold-insoluble globulin (CIg), and purified cell attachment protein (CAP) from horse serum promotes cell-substratum attachment.

Freshly trypsinized cells from breast muscle of 11-day chick embryos were plated at a density of 2 × 10^5/ml in "LH" medium,[1] which consists of 82% L-15 basal medium, 16% 150 mM $NaHCO_3$, and 2% "H" fraction of chick embryo extract,[7] supplemented with glutamine, antibiotics, and bovine insulin (200 ng/ml). After 2 days, nonattached cells were collected, pooled, and subcultured in the same medium (4 ml/dish) in 60-mm tissue culture plates, all of which had been pretreated first with 0.1% gelatin and then as follows: a, HS for 1 hr at 37°C, three washes with buffered salt solution (BSS), crude IgG fraction (0–40% ammonium sulfate cut, redissolved in BSS at original volume, dialyzed against BSS) of rabbit control serum for 1 hr at 37°C, three washes with BSS; b, as in a, except that the second incubation was with the crude IgG fraction of rabbit anti-CIg serum; c, purified CAP (0.5 mg/ml) for 1 hr at 37°C. Cultures were photographed 20 hr after subculturing. Note the typical bipolar morphology of attached and spread Mb in a and c. The rounded cells in b are not attached and can be decanted with the medium. With gelatinization as the sole pretreatment, almost no cells attached (as in b); cells attached to gelatinized plates pretreated with HS (no second incubation) to about the same extent as in a. × 120.

revealed no synthesis of protein precipitable with anti-CIg. In serum-containing cultures of rat L8 cells (in apparent contrast to our results with chick myotubes), iodinatable large, external, transformation-sensitive protein was found to increase along with myoblast fusion,[5] and foci of a surface protein capable of reacting with anti-CIg were detected on myotubes by immunofluorescence.[6]

Our results indicate that chicken Mb and myotubes neither synthesize nor accumulate CSP. They are, however, dependent on fibroblast-derived CSP or a similar protein for attachment and spreading on a gelatinized substratum *in vitro.*

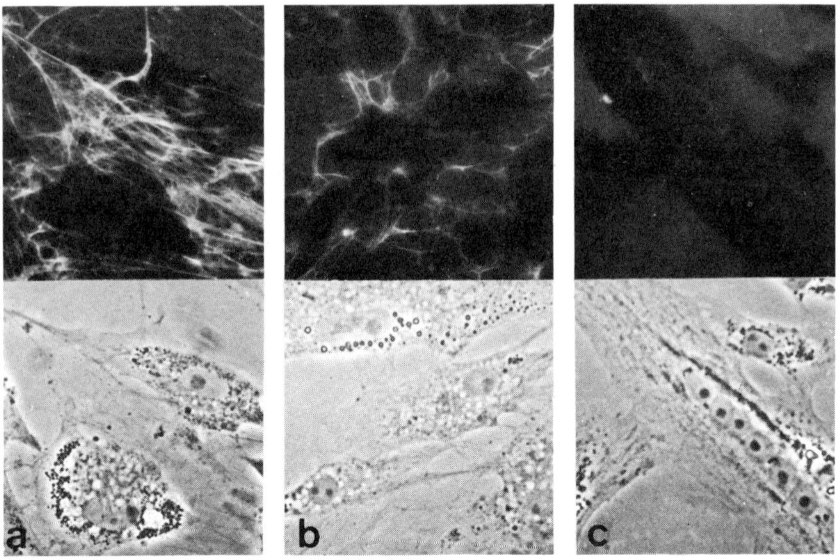

FIGURE 2. Detection by immunofluorescence of surface antigen capable of reacting with anti-CIg antiserum.

Cells were cultured for 2 days in LH medium. Nonattached Mb were then decanted into new gelatinized dishes, and the medium was supplemented with HS. Fibroblasts that had remained attached were given fresh LH medium that contained either HS or "CAP-free HS," a fraction obtained by chromatography on Sephadex® G-150 that contained no proteins with a molecular weight greater than 150,000 daltons and that did not cross-react with anti-CIg. Cells were fixed with formaldehyde (3.5% in BSS) after 7 days in culture, washed three times with BSS, incubated (30 min at room temperature) first with a 1:80 dilution of anti-CIg and then with a 1:100 dilution of fluorescein-labeled goat antirabbit IgG. Other regions of the same plates served as controls; they were reacted with nonimmune serum (1:80) in the first incubation. For details of photography, see Turner *et al.*[8] a, Fibroblasts in 10% HS; b, Fb in 10% CAP-free HS (prominent CSP networks in the absence of exogenous CAP); c, Mb and myotubes in 10% HB (uniform extracellular fluorescence due to CAP attached to substratum). Fluorescence (*top*) and phase-contrast (*bottom*) micrographs are shown for each region. ×700.

We suggest that a similar mechanism might function *in vivo* to anchor muscle cells to the extracellular matrix, of which collagen is a component.

REFERENCES

1. PURI, E. C. & D. C. TURNER. 1978. Exp. Cell Res. In press.
2. KLEBE, R. J. 1974. Nature (London) **250:** 248.
3. MOSESSON, M. W. & R. A. UMFLEET. 1970. J. Biol. Chem. **245:** 5736.
4. YAMADA, K. M., S. S. YAMADA & I. PASTAN. 1975. Proc. Nat. Acad. Sci. USA **72:** 3158.
5. HYNES, R. O., G. S. MARTIN, M. SHEARER, D. R. CRITCHLEY & C. J. EPSTEIN. 1976. Dev. Biol. **48:** 35.
6. CHEN, L. B. 1977. Cell **10:** 393.
7. COON, H. & R. CAHN. 1965. Science **153:** 116.
8. TURNER, D. C., R. GMÜR, H. G. LEBHERZ, M. SIEGRIST, T. WALLIMANN & H. M. EPPENBERGER. 1976. Dev. Biol. **48:** 284.

BIOAUTOGRAPHY OF SERUM- AND CELL-DERIVED CELL ATTACHMENT PROTEINS*

W. Daryl Dickey, James R. Hall,
Cary M. Seals, and Robert J. Klebe

Division of Human Genetics
University of Texas Medical Branch
Galveston, Texas 77550

Several proteins that mediate cellular adhesion have been purified and characterized in the past few years.[1-3] In this paper, the existence of two different collagen-dependent cell attachment proteins (c-CAP) is described. It has previously been demonstrated that a c-CAP can be isolated from bovine serum that has the following properties. First, the serum c-CAP has an isoelectric point (p*I*) of 4.8, is of high molecular weight, and binds to the CB-7 cyanogen bromide peptide of the α_1(I) chain of collagen I.[1,4] Second, divalent cations and cellular metabolic energy[9] are required for the cell to respond to a c-CAP-collagen complex.[1,5] It has previously been shown that BHK baby hamster kidney cells can elaborate a c-CAP, and this material has been interpreted to be related to large, external, transformation-sensitive protein, cold-insoluble globulin, and cell surface protein.[6] In this report, a cultured rat hepatoma cell has been shown to synthesize a c-CAP that differs in its molecular properties from the c-CAP present in serum.

A new c-CAP, termed hepatoma c-CAP, was purified from the conditioned medium of 6TG-11 cells, a derivative of the HTC rat hepatoma cell line.[7] Hepatoma c-CAP has been purified 2345-fold by batch preparation procedures. The purified protein, at 0.7 μg/60-mm collagen-coated petri plate, will permit the attachment of 2.5×10^5 Chinese hamster ovary (CHO) cells to collagen. The molecular and cellular characteristics of the hepatoma c-CAP are compared to those of the serum c-CAP below.

The molecular characteristics of bovine serum c-CAP and rat hepatoma c-CAP are as follows. Both c-CAPs appear to be high-molecular-weight proteins, because both are inactivated by treatment for 60 min at 37°C with 2.5 U/ml of trypsin covalently bound to polyacrylamide. Both c-CAPs are unaffected by DNase or RNase. Under reducing conditions, both c-CAPs yield high-molecular-weight peptides on sodium dodecyl sulfate polyacrylamide gel electrophoresis. The molecular properties of the serum and hepatoma c-CAPs differ in the following respects. First, bovine serum c-CAP can be 50% inactivated by heating at 65°C for 10 min, while hepatoma c-CAP, at comparable protein concentrations and cellular attachment activities, is 50% inactivated by heating at 55°C for 10 min. Second, bovine serum c-CAP and hepatoma c-CAP have isoelectric points of 4.8 and 4.1, respectively. When the two cell attachment proteins are mixed and isoelectrically focused, two peaks of cellular attachment activity are seen at p*I* 4.1 and 4.8.[8]

In addition to their molecular characteristics, the cellular response characteristics of bovine serum c-CAP and hepatoma c-CAP have been studied. The cellular

*Supported in part by Grants CA 19017, GM 21433, and GM 07204 from the National Institutes of Health.

attachment response of several differentiated cell lines to bovine serum c-CAP and hepatoma c-CAP have been examined. Differentiated cell types, such as lymphocytes and mouse neuroblastoma cells, do not respond to either c-CAP. Of those cell lines tested, several cell lines would respond to both serum c-CAP and hepatoma c-CAP. The following list gives the order in which the cell lines studied respond to both CAPs: 3T3 mouse fibroblasts > CHO cells > RJK-36 Chinese hamster embryonic lung fibroblasts > SV3T3-transformed mouse fibroblasts ≫ B-16 mouse melanoma cells. Thus, both serum and hepatoma c-CAPs appear to have a broad specificity for several fibroblasts. It is noteworthy that BALB/c 3T3 fibroblasts respond about three times as well to both both serum c-CAP and hepatoma c-CAP as does their transformed counterpart, SV3T3. Kinetic studies undertaken with a saturating amount of bovine serum c-CAP (10 μg/plate) indicate that some cells display maximum attachment within 30 min, while other cell lines require as long as 120 min. In other studies, the ability of cultured cells to elaborate c-CAPs into their conditioned medium was ascertained. The following cell lines have been shown to elaborated c-CAPs: 6TG-11 rat hepatoma cells, SV3T3-transformed mouse fibroblasts, RJK-36 Chinese hamster lung cells, and WI38-Va-transformed human fibroblasts. During the course of these studies, SV3T3-transformed mouse fibroblasts were found to be capable of liberating a c-CAP into their culture medium, while the normal conterpart, 3T3, did not liberate a c-CAP.

Lastly, a new technique, termed bioautography, has been used to investigate the electrophoretic mobilities of bovine serum c-CAP and hepatoma c-CAP.[5] The principle of bioautography involves the "staining" of electrophoretically separated bands of c-CAP by live mammalian cells. The methodology employed in performing the bioautographic technique is summarized below. Bioautography can be divided into three steps, namely, electrophoresis of c-CAPs, replication of the electrophoretic pattern, and visualization of c-CAP bands by staining with live mammalian cells. Briefly, bioautography was performed as follows: c-CAP samples were applied to a Mylar-backed cellulose acetate gel, and electrophoresis was carried out; after electrophoresis was completed, gels were placed face down on a collagen surface to permit replication of the electrophoretic pattern; freshly trypsinized cells were than planted on the treated collagen surface and incubated for 1.5 hr at 37°C; unattached cells were washed off with normal saline, and attached cells were then fixed with 1% glutaraldehyde for 10 min, followed by staining with toluidine blue O to visualize sites of active c-CAP bands.[5]

Bioautography, as described above, has been employed to study the electrophoretic mobility of serum and hepatoma c-CAPs. Bioautography indicates that hepatoma c-CAP has approximately half of the electrophoretic mobility of either bovine or rat serum c-CAP.[8]

In summary, a new c-CAP has been isolated and characterized from the conditioned medium of rat hepatoma cells. Evidence gained from isoelectric focusing, heat inactivation, and bioautographic determination of electrophoretic mobility indicate the existence of a second c-CAP different from the previously described bovine serum c-CAP.[1]

References

1. Klebe, R. J. 1974. Nature (London) **250:** 248–251.
2. Hausman, R. E. & A. A. Moscona. 1974. Proc. Nat. Acad. Sci. USA **72:** 916–920.

3. OPPENHEIMER, S. B. & T. HUMPHREYS. 1972. Nature (London) **232:** 125–127.
4. KLEINMAN, H. K., E. B. MCGOODWIN & R. J. KLEBE. 1976. Biochem. Biophys. Res. Commun. **72:** 426–432.
5. KLEBE, R. J., J. R. HALL, S. L. NAYLOR & W. D. DICKEY. 1978. Exp. Cell Res. In press.
6. PEARLSTEIN, E. 1976. Nature (London) **262:** 497–500.
7. THOMPSON, E. B., G. M. TOMKINS & J. F. CURRAN. 1966. Proc. Nat. Acad. Sci. USA **56:** 296–303.
8. DICKEY, W. D., C. M. SEALS & R. J. KLEBE. 1978. Exp. Cell Res. In press.
9. KLEBE, R. J. 1975. J. Cell. Physiol. **86:** 231–236.

DIFFERENCES IN THE ULTRASTRUCTURAL LOCALIZATION OF FIBRONECTIN ON NORMAL AND TRANSFORMED CELLS*

Leo T. Furcht,†‡ Deane F. Mosher,§
and Gwen Wendelschafer-Crabb‡

†St. Paul Regional Red Cross Blood Center
‡Department of Laboratory Medicine and Pathology
University of Minnesota Medical School
Minneapolis, Minnesota 55455
and §Department of Medicine
University of Wisconsin Medical School
Madison, Wisconsin 53706

Cell surfaces and associated structures of normal and transformed fibroblasts may differ in many ways, including content of glycosylated proteins and lipids, lectin interactions with the cell surface, and transport of small molecules. Recently, much attention has been paid to the large, external, transformation-sensitive (LETS) glycoprotein fibronectin (FN) [also known as cell surface protein (CSP) and cold-insoluble globulin (CIg)] that decreases on transformation (reviewed in References 1 and 2). Indirect immunofluorescence studies have localized FN to pericellular structures that are sensitive to trypsin and other proteases. Add-back studies suggest a role for FN in cellular morphology and adhesion.[3-5] Clearly, techniques are needed that afford the resolution necessary to make critical examinations of the organization of FN and its relationship to cytoskeletal components and other extracellular proteins. In studies now in progress, we are using the highly sensitive peroxidase-antiperoxidase method to determine FN organization on normal and transformed human fibroblasts.

Materials and Methods

Low-passage human foreskin fibroblasts and SV40-transformed human fibroblasts are grown in Falcon dishes with Dulbecco's minimal essential medium. The preparation and characterization of rabbit antihuman fibronectin (RaFN) and immunocytochemical localization of FN by the peroxidase-antiperoxidase technique on glutaraldehyde-fixed cells have been described.[6] In the standard experiment, cells are reacted with a 1:100 dilution of RaFN, followed by goat antirabbit IgG in excess, then rabbit peroxidase-antiperoxidase complex. Controls include the following: nonimmune rabbit serum, RaFN absorbed with bovine globulins (since bovine FN in growth media may bind to cells), RaFN absorbed with hu-

*Supported by a Basil O'Connor starter research grant, by the National Foundation March of Dimes, National Institutes of Health Grants 1R01-CA21463-01 and 5P01-CA16228-03, the Leukemia Task Force (to L.T.F.), the University of Wisconsin Graduate School, and an institutional grant to the University of Wisconsin from the American Cancer Society (to D.F.M.).

0077-8923/78/0312-0426 $01.75/0

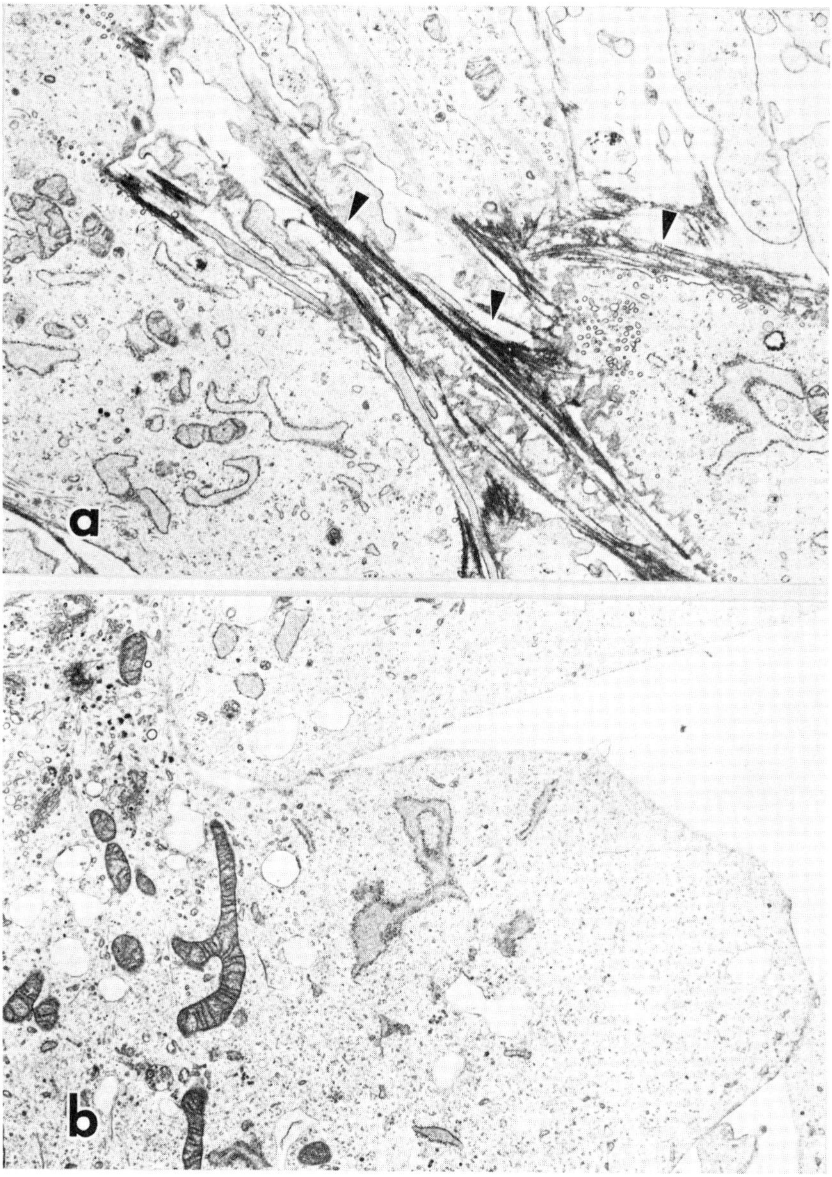

FIGURE 1. Thin-section micrographs of cells cut parallel to the monolayer and counter-stained with heavy metals to enhance contrast. a, High-density normal human fibroblasts are surrounded by a dense matrix of EFM (arrowheads, ×6300); b, high-density transformed human fibroblasts lack EFM (×6300).

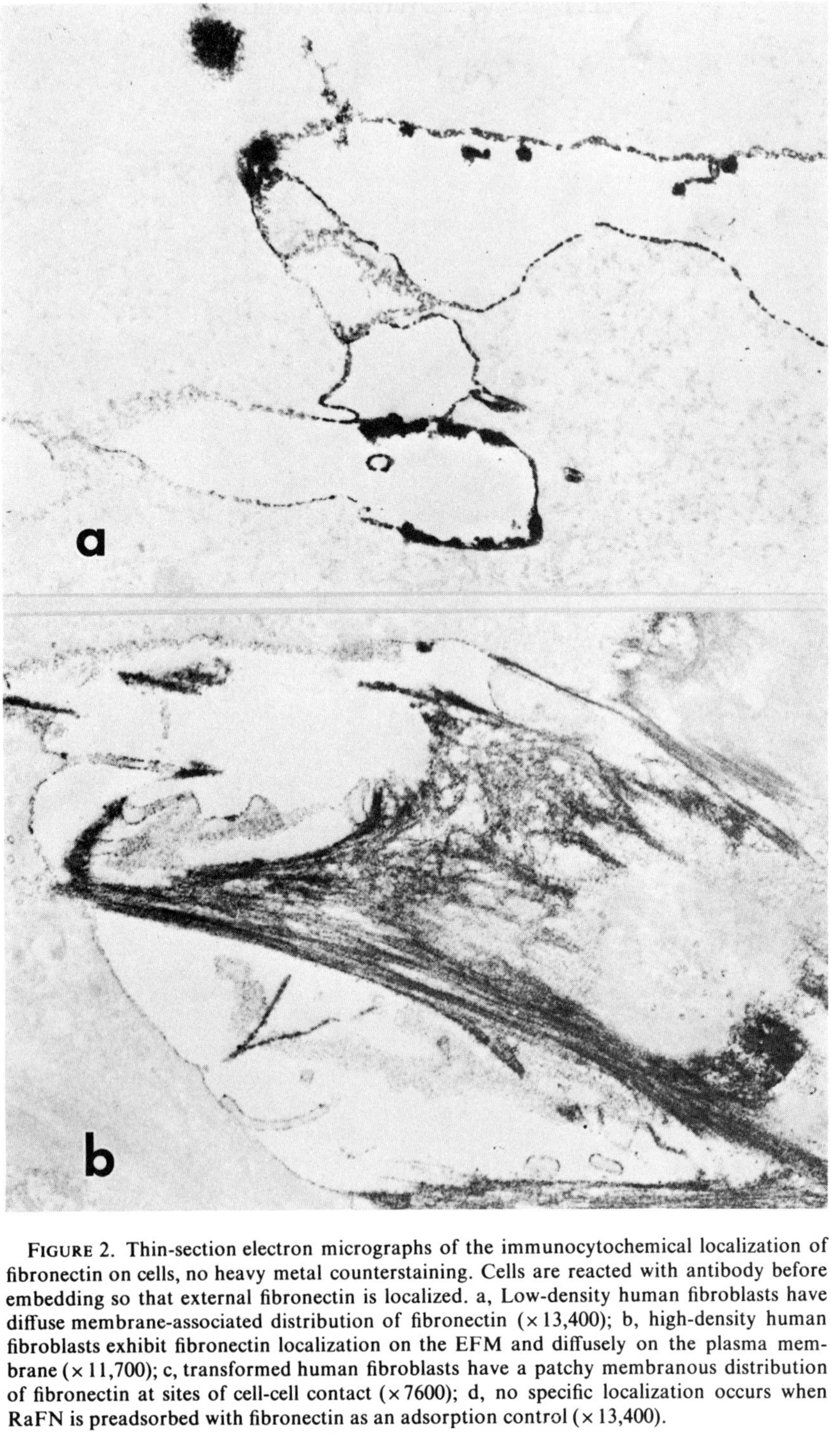

FIGURE 2. Thin-section electron micrographs of the immunocytochemical localization of fibronectin on cells, no heavy metal counterstaining. Cells are reacted with antibody before embedding so that external fibronectin is localized. a, Low-density human fibroblasts have diffuse membrane-associated distribution of fibronectin (× 13,400); b, high-density human fibroblasts exhibit fibronectin localization on the EFM and diffusely on the plasma membrane (× 11,700); c, transformed human fibroblasts have a patchy membranous distribution of fibronectin at sites of cell-cell contact (× 7600); d, no specific localization occurs when RaFN is preadsorbed with fibronectin as an adsorption control (× 13,400).

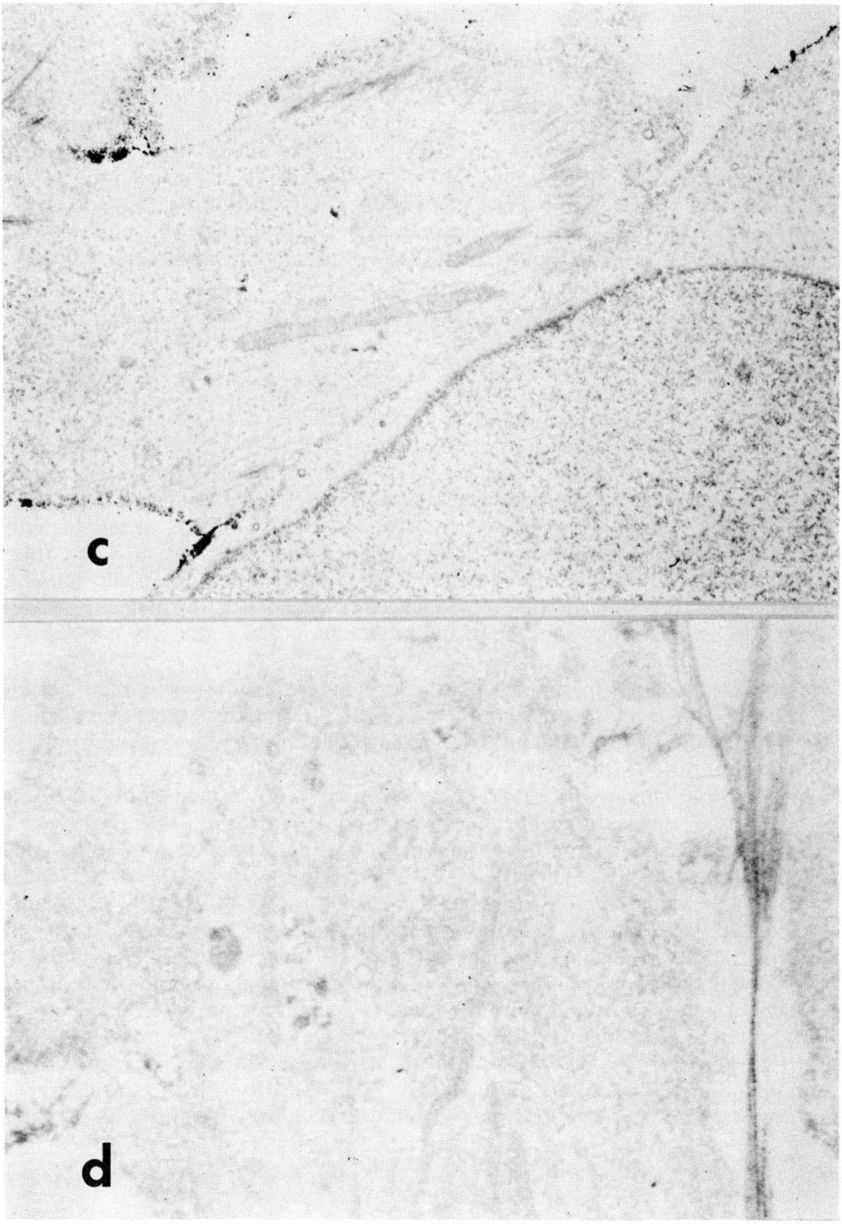

FIGURE 2 *Continued.*

man FN, and substitution of normal sheep serum for antirabbit IgG for the linking antibody.

RESULTS AND DISCUSSION

Thin-section electron microscopy reveals that low-density (noncontacted) human fibroblasts lack any significant amount of extracellular components. Associated with increasing density and cellular contacts, there is the development of an extensive extracellular filamentous matrix (EFM), most pronounced at cell-cell and cell-substratum contact sites (FIGURE 1, a). SV40-transformed human fibroblasts fail to develop the EFM, even when overlapping and multilayered (FIGURE 1, b). Two different patterns of RaFN localization are observed in contacted human fibroblasts: an intense reaction to the EFM and a diffuse nonstructured binding to the plasmalemma. Low-density, noncontacted, normal fibroblasts exhibit binding only in the diffuse membranous form (FIGURE 2, a). High-density normal cells display both patterns (FIGURE 2, b). SV40-transformed human fibroblasts clearly and specifically show reaction with RaFN in a patchy diffuse membrane-associated form that is decreased in intensity relative to the diffuse form seen in contact-inhibited normal human fibroblasts (FIGURE 2, c). An identical reaction is found when using RaFN absorbed with bovine globulins, and a complete absence of staining to the EFM or the membrane occurs when normal rabbit serum is substituted for RaFN, when RaFN is absorbed with human FN, or when normal sheep serum is substituted for goat antirabbit IgG. FIGURE 2(d) illustrates the lack of reaction when RaFN is absorbed with human FN. These controls indicate that FN localized to both the EFM and the plasma membrane is human FN produced by the cultured cells and not bovine FN derived from growth media.

Our work supports the numerous studies that suggest the pericellular nature of FN in high-density contact-inhibited cells.[5,7-9] In addition, we have demonstrated that a diffuse form of RaFN occurs on membranes and is distinct from the EFM. The diffuse form is observed in noncontacted low-density human fibroblasts, on SV40 human fibroblasts at high density, and in concert with EFM on high-density normal human fibroblasts. Our preliminary studies show that the EFM can be disrupted by mild trypsin digestion while remaining substantially intact in collagenase-treated cultures.

Although decreased production and increased turnover of FN (CSP) by transformed chick embryo cells can account for the lack of FN associated with cell surfaces,[10] SV40-transformed human cells produce much more FN than is found in cell layers of normal cells.[11,12] Thus, the fundamental perturbation that occurs on transformation of human fibroblasts may be a failure to organize FN into an EFM. It seems likely that the presence or absence of an EFM composed of FN may be intimately involved in some of the varied phenotypic characteristics of normal and transformed cells. Whether an EFM is formed in transformed cells and then degraded or never formed is currently under investigation.

ACKNOWLEDGMENTS

We thank S. Gentry, R. Scott, J. Ihrke, and S. Palm for technical assistance and C. Furcht for assistance in preparation of this manuscript.

References

1. Hynes, R. O. 1976. Biochim. Biophys. Acta **458:** 73–107.
2. Vaheri, A. 1977. *In* Virus-Transformed Cell Membranes. C. Nicholau, Ed. Academic Press. New York and London. In press.
3. Yamada, K. M., S. H. Ohanian & I. Pastan. 1976. Cell **9:** 241–245.
4. Yamada, K. M., S. S. Yamada & I. Pastan. 1976. Proc. Nat. Acad. Sci. USA **73:** 1217–1221.
5. Ali, I. U., V. Mautner, R. Lanza & R. O. Hynes. 1977. Cell **11:** 115–126.
6. Furcht, L. T., D. F. Mosher & G. Wendelschafer-Crabb. 1978. Cell **13:** 263–271.
7. Wartiovaara, J., E. Linder, E. Ruoslahti & A. Vaheri. 1974. J. Exp. Med. **140:** 1522–1533.
8. Chen, L. B., P. H. Gallimore & J. K. McDougall. 1976. Proc. Nat. Acad. Sci. USA **73:** 3570–3574.
9. Yamada, K. M., S. S. Yamada & I. Pastan. 1977. J. Cell Biol. **74:** 649–654.
10. Olden, K. & K. M. Yamada. 1977. Cell **11:** 957–969.
11. Vaheri, A., E. Ruoslahti, B. Westermark & J. Pontén. 1976. J. Exp. Med. **143:** 64–72.
12. Mosher, D. F., O. Saksela, J. Keski-Oja & A. Vaheri. 1977. J. Supramol. Struct. **6:** 551–557.

THROMBIN-INDUCED LECTIN ACTIVITY OF HUMAN PLATELETS: A POSSIBLE DETERMINANT OF PLATELET AGGREGATION

T. Kent Gartner and Danny C. Williams

Department of Biology
Memphis State University
Memphis, Tennessee 38152

David R. Phillips

Department of Biochemistry
St. Jude Children's Research Hospital
Memphis, Tennessee 38103

Lectin-lectin receptor binding has been implicated as the mechanism responsible for membrane-membrane interactions in at least two different cell types.[1,2] According to this model, the lectin (hemagglutinating factor[3]) on the membrane surface of one cell binds to its receptor on an adjacent cell, thereby facilitating cellular interactions.[4,5] Platelet aggregation is an example of a process that occurs as a consequence of the interaction of membrane surfaces. We have undertaken a study to test the hypothesis that platelet aggregation is mediated by lectin-lectin receptor interactions.

The lectin activity of washed human platelets was measured by determining their ability to agglutinate trypsinized, formalinized erythrocytes from various species.[2] We found that erythrocytes from cow, rabbit, sheep, catfish, and electric eel were agglutinated readily by the washed platelets, whereas human, goat, and hog erythrocytes reacted weakly. All efforts to inhibit this lectin activity by common mono- and disaccharides were unsuccessful. Therefore, it is not known if these hemagglutination activities of unstimulated platelets are due to nonspecific interactions or to specific interactions with a complex receptor.

Thrombin, a proteolytic enzyme, is a potent inducer of platelet aggregation. We have found that thrombin also causes a marked enhancement of the platelet lectin activity. This enhanced activity was most active against bovine erythrocytes; no enhanced activity was detected when goat, hog, or rabbit erythrocytes were used as a source of lectin receptors. In contrast to the lectin activity of unstimulated platelets, the thrombin-enhanced activity appeared to be directed against specific sugar residues. For example, we found that 30 mM glucosamine, galactosamine, or mannosamine inhibited this activity. In contrast, *N*-acetylated derivatives or other neutral sugars, such as glucose, galactose, mannose, fructose, fucose, xylose, ribose, lactose, and sucrose, were without effect on the enhanced lectin activity.

The thrombin-enhanced, amino sugar-sensitive lectin activity was shown to reside in proteins within the plasma membrane. The plasma membrane location of the lectin was demonstrated by comparing the lectin activity of isolated membranes to that of intact platelets. This comparison revealed that the isolated membranes had a sixfold greater lectin-specific activity than did intact platelets. The lectin activity of isolated membranes was destroyed by trypsin but was resistant to phospholipase C and was stable to freezing and thawing. We found that the same inhibitors that inhibited the thrombin-enhanced lectin activity also inhibited all of the lectin activity of isolated membranes. Accordingly, the hexosamines inhibited, while *N*-acetylated derivatives and other neutral sugars were without effect. Thus, platelet plasma membranes offer a useful source of the thrombin-enhanced lectin activity.

In an effort to correlate the thrombin-enhanced lectin activity with thrombin-induced platelet aggregation, we evaluated the effect of sugars on platelet aggregation. A perfect correlation was found between sugars that inhibited the thrombin-enhanced lectin activity and those that inhibited thrombin-induced platelet aggregation. Therefore, it appears that thrombin-induced aggregation is mediated by a membrane surface lectin.

References

1. Rosen, S. D., J. A. Kafka, D. L. Simpson & S. H. Barondes. 1973. Proc. Nat. Acad. Sci. USA **70:** 2554.
2. Gartner, T. K. & T. R. Podleski. 1975. Biochem. Biophys. Res. Commun. **67:** 972.
3. Boyd, W. C. 1971. Ann. N.Y. Acad. Sci. **169:** 168.
4. Rosen, S. D., D. L. Simpson, J. E. Rose & S. H. Barondes. 1974. Nature (London) **252:** 148.
5. Simpson, D. L., D. R. Thorne & H. H. Loh. 1977. Nature (London) **266:** 367.

COLD-INSOLUBLE GLOBULIN IS NOT ABSOLUTELY REQUIRED FOR CELLULAR ADHESION TO NATIVE COLLAGEN SUBSTRATA*

Frederick Grinnell and Dianne Minter

Department of Cell Biology
University of Texas Health Science Center
Dallas, Texas 75235

Baby hamster kidney (BHK) cells require a specific serum factor in order to spread on tissue culture substrata.[1] The factor has been purified from fetal calf serum; it appears to function by adsorbing onto the substratum and forming the sites to which the cells attach and spread.[2] More recent studies performed with human serum indicate that the cell-spreading factor is identical to cold-insoluble globulin (CIg) by electrophoretic, chromatographic, and immunologic criteria.[3]

Cold-insoluble globulin is also thought to be required for cellular adhesion and spreading to occur with dried collagen substrata.[4,5] However, the physiologic significance of this dependence is unclear, because the dried collagen substrata are probably denatured.[6] To clarify this point, we have undertaken a systematic study of CIg involvement in BHK cellular adhesion and spreading with a variety of collagen-coated substrata, both denatured and native.

Dried collagen-coated substrata were prepared in tissue culture dishes as described by Klebe,[7] except that the urea step was omitted. Gelatin-coated substrata were prepared by incubating tissue culture dishes with a solution of gelatin (30 mg/ml) for 10 min at 22°C. Native collagen-coated substrata were prepared by incubating tissue culture dishes with a solution of freshly prepared rat tail collagen (2 mg/ml) for 10 min at 22°C. Native collagen-coated gels were prepared in tissue culture dishes as described by Elsdale and Bard.[8] Gelatin gels were prepared in tissue culture dishes by gelling a solution of gelatin (100 mg/ml) at 22°C for about 30 min.

Dried collagen-coated substrata and gelatin-coated substrata reacted identically. In the absence of serum or CIg, there was little adhesion and no spreading of BHK cells after a 30-min incubation at 37°C. Addition of purified CIg to the incubation medium at a concentration of 10 μg/ml or pretreatment of the substratum for 10 min at 20°C with a CIg-containing solution permitted complete cellular adhesion and spreading. In marked contrast to the denatured collagen-coated substrata, native collagen-coated substrata did not require CIg for cellular adhesion to occur within 30 min, and after a 2-hr incubation, complete spreading of BHK cells was observed. It should be noted that the addition of CIg increased the rate of cellular spreading on these substrata. Native collagen gels reacted in the same manner, regardless of the presence or absence of CIg in the incubation medium. Cellular adhesion to native collagen gels occurred within 30 min. Unlike the other substrata, cellular spreading was bipolar and not typically triangular.[8] Within 2 hr, about half of the cells spread into a bipolar morphology. Longer incubations (e.g., 24 hr) did not result in increased spreading. Cellular adhesion and spreading were not observed to occur with gelatin gels under any circumstances. In this regard, gelatin gels behave similarly to agar gels.

*Supported by Grant CA 14609 from the National Cancer Institute.

The data indicate that addition of CIg is required for cellular adhesion and spreading with denatured collagen-coated substrata but not with native collagen-coated substrata. This difference may be due to the presence of endogenous CIg as a contaminant in the rat tail collagen. Several kinds of experiments were performed to test this hypothesis.

Antibodies against the fetal calf serum-spreading factor and against human CIg inhibited BHK cellular spreading with rat serum, indicating that these antibodies cross-react with the rat factor. Neither of these antisera was found to inhibit cellular adhesion and spreading with native collagen-coated substrata prepared from rat tail collagen.

Biochemical analysis has indicated that tyrosine hydroxyl groups, tryptophan residues, and free carboxyl groups are required for CIg activity. This conclusion is based on inhibition studies with lactoperoxidase, *N*-bromosuccinimide, and a soluble carbodiimide, respectively.[9] Similar studies were conducted with native collagen-coated substrata. With adsorbed collagen, lactoperoxidase was slightly inhibitory, *N*-bromosuccinimide was without effect, and carbodiimide treatment alone stimulated cellular spreading, while carbodiimide treatment in the presence of an amino acceptor was slightly inhibitory. With native collagen gels, lactoperoxidase was without effect. The other treatments could not be carried out with native gels because of washing and handling problems.

The results of the immunologic and biochemical analyses indicate that adhesion and spreading of BHK cells on native collagen-coated substrata are probably not mediated by endogenous CIg, which is present as a contaminant in the collagen preparations.

In conclusion, we find that CIg is required for cellular adhesion and spreading with denatured collagen-coated substrata; with native collagen-coated substrata, CIg is not required. Therefore, it seems unlikely that CIg mediates fibroblast adhesion and spreading on native collagen fibrils *in situ*. Nevertheless, in areas where collagen is denatured *in situ*, collagen-adsorbed CIg may form the sites of fibroblast adhesion. The simplest explanation of our findings with native collagen is that BHK cells bear a cell surface collagen receptor and interact directly with specific regions of native collagen.

References

1. Grinnell, F. 1976. Exp. Cell Res. **102:** 51–62.
2. Grinnell, F., D. Hays & D. Minter. 1977. Exp. Cell Res. **110:** 175–190.
3. Grinnell, F. & D. Hays. 1978. Exp. Cell Res. In press.
4. Pearlstein, E. 1976. Nature (London) **262:** 497–499.
5. Jilek, F. & H. Hörmann. 1977. Hoppe-Seyler's Z. Physiol. Chem. **358:** 133–136.
6. Maroudas, N. G. 1977. Nature (London) **267:** 183.
7. Klebe, R. 1974. Nature (London) **250:** 248–251.
8. Elsdale, T. & J. Bard. 1972. J. Cell Biol. **54:** 626–637.
9. Grinnell, F. & D. Minter. 1978. Exp. Cell Res. Submitted for publication.

BINDING OF CELL ATTACHMENT PROTEIN TO COLLAGEN: EFFECT OF CHEMICAL MODIFICATIONS

Hynda K. Kleinman, Ermona B. McGoodwin, and George R. Martin

National Institute of Dental Research
Bethesda, Maryland 20014

Robert J. Klebe

Department of Human Biological Chemistry and Genetics
University of Texas Medical Branch
Galveston, Texas 77550

Klebe[1] showed in 1974 that cells attach to collagen substrata through a serum glycoprotein, cell attachment protein (c-CAP). This glycoprotein is produced by cells but destroyed by trypsin. Both whole serum and isolated c-CAP will promote the attachment of cells to collagen-coated dishes. Pearlstein[2] has suggested that cell surface-associated proteins, such as large, external, transformation-sensitive (LETS) and cell surface protein (CSP),[4] are identical to c-CAP. We have observed that CSP isolated from chick cells will promote the attachment of cells to collagen. c-CAP is secreted as a 440,000-dalton protein with a collagen-binding site(s) and a different cell-binding site(s).

Previously, we have demonstrated that preincubation of serum with collagenous peptides reduces its cellular attachment activity. This effect has been utilized to develop an assay for measuring the interaction of c-CAP with peptides from collagen for localization of the binding site. The major active peptide produced from the α_1(I) chain by cyanogen bromide cleavage was α_1(I)-CB7, which comprises residues 551–822.[5] This peptide, as illustrated in FIGURE 1, is capable of promoting cellular attachment when coated on a bacteriologic plate, as are the α_1(I) chains. Some adhesion of cells to plastic occurs but may be due to the direct binding of c-CAP in the serum to the plastic or to an unrelated reaction. Collagen from the cuticle of *Ascaris* did not support cellular attachment.

Further localization of the binding site was achieved by assay of the products of various proteolytic enzyme treatments of α_1(I)-CB7, and a peptide that is composed of residues 757–791 was found to be active.[6] This peptide contains a sequence that may be less helical[7] than other portions of the molecule. The integrity of the collagenase-sensitive bond (residues 775 and 776) and a nearby chymotrypsin-sensitive bond (residues 779 and 780) was found necessary for retention of c-CAP-binding activity.

Chemical modification was employed to establish the importance of charge to activity in the binding region. Modification of arginyl residues[8] with 1,2-cyclohexanedione resulted in a fivefold stimulation of the binding of c-CAP to either the α_1(I) chain or α_1(I)-CB7 (see TABLE 1). However, when lysyl residues were modified with maleic anhydride,[9] c-CAP did not bind to collagen. Removal of this group restored activity. Acetic anhydride modification of lysyl residues[10] suppressed the binding by 50%, while isourea[11] was without effect.

The reason for the enhanced binding activity after modification of the arginyl residues is unclear. Arginine is contained in or near the c-CAP-binding site, because binding activity is lost when chymotrypsin cleaves residues 779 and 780, a

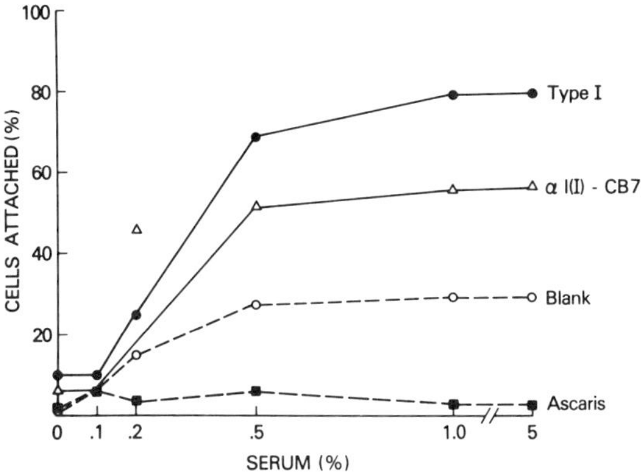

FIGURE 1. Attachment of Chinese hamster ovary (CHO) cells to α_1(I), α_1(I)-CB7, and *Ascaris* collagen in the presence of increasing concentrations of serum. The collagen was dissolved in dilute acetic acid, and 10 μg in 1 ml were added to each 35-mm bacteriologic plate and allowed to air dry. The attachment assay was performed as described previously.[1]

glutamine-arginine bond. Cyclohexanedione contributes a large hydrophobic, positively charged group to an already hydrophobic region of the collagen chain.

It is known that lysine is not contained in the 35-amino acid-active binding fragment. This fact is supported by the data shown in TABLE 1, because modification with isourea did not change the binding of c-CAP to collagen. Together, the lysyl modification data suggest the collagen-c-CAP interaction is in part ionic and that a reduction of the positive charge on collagen outside of the binding region will inhibit c-CAP binding.

In conclusion, the interaction between collagen and c-CAP appears to be quite strong. For example, many preparations of collagen contain c-CAP, and it is necessary to wash plates coated with these preparations with 8 M urea to demonstrate the c-CAP requirement. In addition, 1 M KBr has also been found to release c-CAP from collagen.[12] Chemical modifications of the collagen chain suggest that the reaction is in part ionic and strongly inhibited by negatively charged groups. This fact may be important in cell-matrix interactions, because the matrix, like cartilage, is rich in anionic collagen-binding components, such as proteoglycans. Such an inhibitory effect is suggested by the results of Linsenmeyer *et al.*,[13] who found that chondroitin sulfate blocked the binding of c-CAP to collagen, while a less negatively charged polysaccharide, hyaluronic acid, did not.

TABLE 1

CHEMICAL MODIFICATIONS OF COLLAGEN

Reagent	Group Modified	c-CAP-Binding Activity
1,2-Cyclohexanedione	arginine	fivefold increase*
Maleic anhydride	lysine	abolished
Acetic anhydride	lysine	reduced by 50%
O-methylisourea	lysine	unchanged

*c-CAP-binding activity was determined as described previously.[5]

References

1. Klebe, R. J. 1974. Nature (London) **250:** 248–251.
2. Pearlstein, E. 1975. Nature (London) **262:** 497–500.
3. Hynes, R. O. 1973. Proc. Nat. Acad. Sci. USA **70:** 3170–3174.
4. Yamada, K. M. & J. A. Weston. 1974. Proc. Nat. Acad. Sci. USA **71:** 3492–3496.
5. Kleinman, H. K., E. B. McGoodwin & R. J. Klebe. 1976. Biochem. Biophys. Res. Commun. **72:** 426–432.
6. Kleinman, H. K., E. B. McGoodwin, G. R. Martin, R. J. Klebe, P. P. Fietzek & D. E. Woolley. 1978. J. Biol. Chem. In press.
7. Ramachandran, G. N. & C. Ramakrishan. 1977. *In* Biochemistry of Collagen. G. N. Ramakrishan & A. H. Reddi, Eds.: 45–81. Plenum Publishing Corporation. New York, N.Y.
8. Patthy, L. & E. L. Smith. 1975. J. Biol. Chem. **250:** 557–564.
9. Butler, P. J. G., J. T. Harris, B. S. Hartley & P. Leverman. 1969. Biochem. J. **112:** 679–689.
10. Means, J. E. & R. E. Feeney. *In* Chemical Modification of Protein.: 69–70. Holden-Day, Inc. San Francisco, Calif.
11. Kimmel, J. 1967. *In* Methods in Enzymology. C. H. W. Hirs, Ed. Vol. XI: 584–589.
12. Dessau, W., B. Adelmann, R. Timpl & G. R. Martin. 1978. Biochem. J. **169:** 55–59.
13. Linsenmeyer, T. F., B. P. Toole & J. Gross. 1976. J. Cell Biol. **70:** 59(Abst.).

FIBROBLASTS RELEASE SURFACE PROTEINS THAT FUNCTION AS CONCANAVALIN A RECEPTORS*

Albert J. T. Millis,† Marian Hoyle, and Paulette J. McCormick‡

Department of Biological Sciences
State University of New York at Albany
and †Albany Medical College
Albany, New York 12222

Proliferation of diploid fibroblasts is limited to cells that are spread on a substrate. Previously, we demonstrated that cellular attachment and spreading, the first steps in the initiation of cellular growth, were density-dependent processes. We reported that human diploid fibroblasts (HDF), plated at greater than 2×10^3 cells/cm^2 of surface area (10^4 cells/ml of medium), attached and spread within several hours, whereas cells plated at lower densities did not spread.[1] We suggested that the ability to spread was dependent on the presence of factors, released by cells, in the culture medium.[2] The presence of growth-promoting factors in conditioned medium has been known for many years.[3,4] Therefore, we examined the conditioned medium produced by HDF and found that the factors responsible for cellular spreading were fibroblast surface proteins ("released proteins") that were released into the culture medium[2] and that those proteins were necessary for concanavalin A (con A) -mediated cellular agglutination.[5] In this report, we will summarize those findings.

The fact that conditioned medium contained proteins that originated at the cell surface was established by examining both released proteins and proteins that were extracted from the cell surface[6] on 7.5% polyacrylamide gels in the presence of sodium dodecyl sulfate (SDS).[7] Cells were first labeled with either [^{14}C]leucine, [^{14}C]glucosamine, or [^{125}I]NaI.[8] The released proteins and the extracted proteins both contained four components that were labeled with [^{14}C]precursors, and three of those components were labeled with ^{125}I. Those four components, from released and extracted materials, comigrated on SDS gels and had apparent molecular weights that ranged from 96,000 to 220,000 daltons. Those components bound [^{3}H]con A and promoted cellular spreading. In addition, when low-density cells were mixed with released proteins and then washed to remove excess released proteins, their ability to agglutinate in the presence of 100 μg/ml of con A was increased from 29 to 85%.

To determine whether released proteins are transferred intracellularly, we examined the effects of released and extracted proteins on con A-mediated agglutination of both fibroblasts and red blood cells. We found that when released proteins were mixed with con A before exposure to cells, the cells did not agglutinate. This finding may indicate that once the released proteins were bound by con A, neither component could interact with the cell surface. However, the con A-mediated agglutination of HDF was enhanced after fibroblasts were first exposed to released protein and subsequently to con A. Further, the spreading ac-

*Supported by Grant HD-09515 from the National Institute of Child Health and Human Development and by Grant AG-00697 from the National Institute on Aging.
‡Supported by a National Science Foundation Predoctoral Fellowship.

tivity of released protein was reduced by previous absorption of the released protein solution to HDF.

We do not know which of the released proteins are responsible for the activities tested, because that determination must await separation of the proteins from each other. Evidence indicates that the four proteins exhibit differential affinities for con A and that their separation may be achieved by elution from a con A-Sepharose affinity column.[11] However, one of the released proteins has an apparent molecular weight similar to that reported for the surface glycoprotein known as large, external, transformation-sensitive protein, fibronectin, or cell surface protein[9] and some of the same biologic properties.[10]

It is likely that alterations of surface architecture, achieved through the release and absorption of surface proteins, result in detectable variations in surface-related phenomena. Further, the release and absorption of membrane components are major factors in the density-dependent regulation of cellular growth.

References

1. Millis, A. J. T., M. Hoyle & B. Field. 1977. J. Cell. Physiol. **93:** 17–24.
2. Millis, A. J. T. & M. Hoyle. 1978. Nature (London) **271:** 668–669.
3. Puck, T. T. & P. I. Marcus. 1955. Proc. Nat. Acad. Sci. USA **44:** 432–437.
4. Rubin, H. 1966. Exp. Cell Res. **41:** 138–148.
5. McCormick, P. J. & A. J. T. Millis. 1978. Exp. Cell Res. **113:** 211–215.
6. Yamada, K. M. & J. A. Weston. 1974. Proc. Nat. Acad. Sci. USA **71:** 3492–3496.
7. Laemmli, U. 1970. Nature (London) **227:** 680–685.
8. Kennel, S. J., B. C. DeVillano & R. A. Lerner. 1973. Methods Mol. Biol. **6:** 1–38.
9. Hynes, R. 1976. Biochim. Biophys. Acta **458:** 73–107.
10. Yamada, K. M., S. S. Yamada & I. Pastan. 1975. Proc. Nat. Acad. Sci. USA **72:** 3158–3162.
11. McCormick, P. J. & A. J. T. Millis. 1978. Submitted for publication.

MODELS OF CELL SURFACE EVENTS DURING NORMAL AND TRANSFORMED HUMAN CELLULAR ADHESION *IN VITRO*

R. Rajaraman,*† R. A. Fox,*† and J. M. MacSween†

*Departments of *Microbiology and †Medicine*
Faculty of Medicine
Dalhousie University
Halifax, Nova Scotia, Canada

Coordination among cellular adhesion, detachment, locomotion, proliferation, and differentiation regulated by the microenvironment guides the destiny of the developing embryo through morphogenesis and the maintenance of the adult organism. A loss of coordination among these phenomena appears to characterize neoplastic conditions. Neoplastic cells do not respond to controls on cellular division and cell-cell interaction at the level of the cell surface. Studies on cellular adhesion have demonstrated that neoplastic cells are weakly adherent[1] and defective in locomotion.[2]

Recent studies on cellular adhesion indicate that several phenomena, such as capping, endocytosis, locomotion, and adhesion, may be interrelated.[3–5] Based on these and other observations,[6–9] we propose that patching, capping, endocytosis, and cellular locomotion are manifestations of a single process: the cell discarding foreign and unwanted materials. Capping is caused by abortive adhesion and does not result in locomotion. Successful adhesion to the substratum renders capping and endocytosis impossible; the cell detaches itself from the substratum by spurning it in a manner analogous to capping.

The interpretation of capping as a result of abortive adhesion permits construction of molecular models of normal and neoplastic cellular adhesion in the presence and absence of serum. In this model, normal cells have A molecules (adhesion sites) on their surface; they can spread on the substratum in the absence of serum. In the presence of serum, A molecules on the normal cell surface bind to B molecules in serum, which may be substratum bound or free in suspension. Binding of free B molecules with cell surface A molecules results in blockage of adhesion sites; these blocked sites are cleared via capping. New adhesion sites (A molecules) are produced at the active edges of the cell. Binding of cell surface A molecules with substratum-bound B molecules results in cellular adhesion. Transformed cells do not bear A molecules on their surfaces; they cannot spread in the absence of serum. Transformed cells may recruit A molecules from serum to attain deformability and spreading (FIGURE 1). The transformed cell may also be defective in capping surface-bound ligands.

Thus, the degree of dependence on serum for deformation and spreading may be determined by the rate at which the cell can synthesize endogenously or "recruit" exogenously "A" molecules or by a combination of both rates. The major differences between normal and transformed cellular adhesion and spreading appear to be the lack of adhesion sites and the lower rate of formation of adhesion sites at the active edges in the latter.

The nature of molecules A and B is not known at present. Several recent

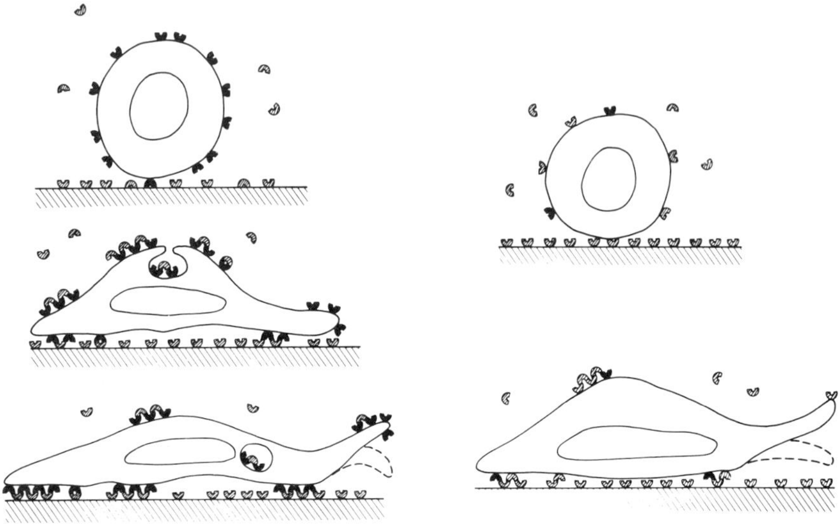

FIGURE 1. *Left:* Schematic model of normal human fibroblast adhesion to a substratum in the presence of serum. Serum factors, including B molecules, are quickly adsorbed onto the substratum surface. These factors are also in suspension in the medium (*top*).

When the cell settles down on the serum protein layer, an A molecule may bind with B molecules adsorbed on the substratum, effecting cellular adhesion; some of the cell surface A molecules may also bind with free B molecules. The latter event will not result in adhesion; these A-B complexes will be subjected to capping and endocytosis (*middle*).

New adhesion sites (A molecules) produced at the active edges during spreading and locomotion effect adhesion by binding with B molecules adsorbed on the substratum (*bottom*).

Serum factors other than B molecules that may be adsorbed onto the substratum are omitted for clarity.

Right: Model of transformed fibroblast adhesion to a substratum in the presence of serum. Serum factors, including B molecules, are quickly adsorbed onto the substratum surface. These factors are also in suspension in the medium (*top*). Transformed cells bear little or no adhesion sites and seem to adsorb serum factors from the medium (*top*).

The transformed cell may synthesize new adhesion sites or "recruit" adhesion sites from the medium and begin to spread. It can also spread by sticking directly to substrate-adsorbed serum factors. The cell surface-adsorbed serum factors may be blocked by identical or nonidentical complementary factors that may cause capping (*bottom*). For the sake of simplicity, it is assumed here that the A molecule (in the cell surface) and the B molecule (in the serum) are identical.

studies on cellular adhesion[8,10–15] are compatible with the hypothesis that fibronectin is a probable candidate for molecule A. Molecule B may be identical to molecule A or it may be different and complementary to molecule A. Since fibronectin may occur as a dimer,[16–18] it is likely that the cell-spreading factor (CSF) found in serum[19] is a dimer of fibronectin. Other likely candidates for molecule B would be fibrin, fibrinogen,[20,21] and collagen,[12,22] which are known to bind to the cell surface, probably via fibronectin.

References

1. Coman, D. E. 1944. Cancer Res. **4:** 625.
2. Vasiliev, J. M. & I. M. Gelfand. 1976. *In* Proceedings of Symposium on the Fundamental Aspects of Metastasis. L. Weiss, Ed.: 71.
3. Edelman, G. M., J. L. Wang & I. Yahara. 1976. *In* Cell Mobility. R. Goldman, T. Pollard & J. Rosenbaum, Eds.: 305. Cold Spring Harbor Laboratory. Cold Spring Harbor, N.Y.
4. Vasiliev, J. M., I. M. Gelfand, L. V. Domnina, N. A. Dorfman & O. Y. Pletyushkima. 1976. Proc. Nat. Acad. Sci. USA **73:** 4085.
5. Albrecht-Buehler, G. & L. B. Chen. 1977. Nature (London) **266:** 454.
6. Rajaraman, R., D. E. Rounds, S. P. S. Yen & A. Rembaum. 1974. Exp. Cell Res. **88:** 327.
7. Rajaraman, R., J. M. MacSween, V. G. Vethamany & R. A. Fox. 1978. Exp. Cell Res. Submitted for publication.
8. Rajaraman, R., K. Westermark, A. Vaheri & J. Pontén. This monograph.
9. Rajaraman, R., B. Westermark, J. Pontén, A. Vaheri, R. A. Fox & J. M. MacSween. 1978. J. Supramol. Struct. Suppl. **2:** 333.
10. Yamada, K. M., S. S. Yamada & I. Pastan. 1975. Proc. Nat. Acad. Sci. USA **72:** 3158.
11. Yamada, K. M., S. S. Yamada & I. Pastan. 1976. Proc. Nat. Acad. Sci. USA **73:** 1217.
12. Pearlstein, E. 1976. Nature (London) **262:** 497.
13. Noonan, K. D., D. S. Lindberg & J. A. McClure. 1971. Progr. Clin. Biol. Res. **9:** 245.
14. Brown, S., W. Levinson & J. A. Spudich. 1976. J. Supramol. Struct. **5:** 119.
15. Chen, L. B., N. Maitland, P. H. Gallimore & J. K. McDougall. 1977. Exp. Cell Res. **106:** 39.
16. Sear, C. H. J., M. E. Grant & D. S. Jackson. 1976. Biochem. Biophys. Res. Commun. **71:** 379.
17. Keski-Oja, J., D. F. Mosher & A. Vaheri. 1977. Biochem. Biophys. Res. Commun. **74:** 699.
18. Hynes, R. O. & A. Destree. 1977. Proc. Nat. Acad. Sci. USA **74:** 2855.
19. Grinnell, F. 1976. Exp. Cell Res. **102:** 51.
20. Ruoslahti, E. & A. Vaheri. 1975. J. Exp. Med. **141:** 497.
21. Nozawa, R. T. & R. L. Guerrant, 1977. Exp. Cell Res. **107:** 25.
22. Bornstein, P. & J. F. Ash. 1977. Proc. Nat. Acad. Sci. USA **74:** 2480.

IMMUNOFLUORESCENCE STUDIES ON FIBRONECTIN DISTRIBUTION PATTERNS DURING ADHESION, DEFORMATION, AND SPREADING OF HUMAN GLIAL AND GLIOMA CELLS*

R. Rajaraman,† B. Westermark,‡ A. Vaheri,§ and J. Pontén‡

†Departments of Microbiology and Medicine
Faculty of Medicine
Dalhousie University
Halifax, Nova Scotia, Canada
‡ Wallenburg Laboratory
University of Uppsala
Uppsala, Sweden
and §Department of Virology
University of Helsinki
00290 Helsinki 29, Finland

Cells *in vitro* adhere, deform, and spread via progressive stages, such as adhesion, filopodial growth, cytoplasmic webbing (spreading), and flattening of the nucleus.[1] Normal human fibroblasts can adhere and spread on the substratum irrespective of the presence or absence of serum proteins in the medium, whereas transformed cells cannot deform and spread unless serum proteins are present.[2,3] Since the discovery of the absence of fibronectin in transformed cells,[4-8] much attention has been directed to its role in the altered social behavior of these cells, including their reduced adhesiveness. It has been proposed that fibronectin is involved in cellular adhesion to the substratum.[9] This proposal predicts that the transformed cells dependent on serum proteins for deformation and spreading may recruit serum fibronectin to attain deformability. We, therefore, studied the fibronectin distribution pattern in spreading glial and glioma cells.

Human glial and glioma cells were grown in Eagle's basal medium supplemented with 10% fetal calf serum and antibiotics under standard culture conditions. Cells were harvested when the cultures were 4–5 days old with 0.25% trypsin, rinsed in serum-free medium, and layered on 35-mm petri plates that contained glass coverslips in serum-free medium or medium supplemented with 10% fetal calf serum. At varying intervals, cells were fixed in 2% paraformaldehyde and scored for percentage of well-spread cells[3]; cells were also stained for fibronectin by use of rabbit antihuman fibronectin, followed by fluorescein isothiocyanate-conjugated swine antirabbit IgG. "Footprints" of spreading cells were prepared by exposure of unfixed cells to 0.1% Triton® X-100, 0.01% EDTA, or by incubating cells at 4°C for 30–60 min with antifibronectin; these footprints were also stained for fibronectin. The specificity of antifibronectin has been described previously.[10]

The glial cells spread to the extent of 30–35% within 3 hr after layering in the absence of serum, whereas in serum supplemented medium, almost 100% cellular spreading was observed. Trypsin-dispersed glial cells were spherical in shape and exhibited diffuse surface distributions of fibronectin within 30–60 min after layer-

*Supported in part by the International Cancer Data Bank Programme of the National Cancer Institute under Contract N01-C0-65341 with the International Union Against Cancer.

0077-8923/78/0312-0444 $01.75/0

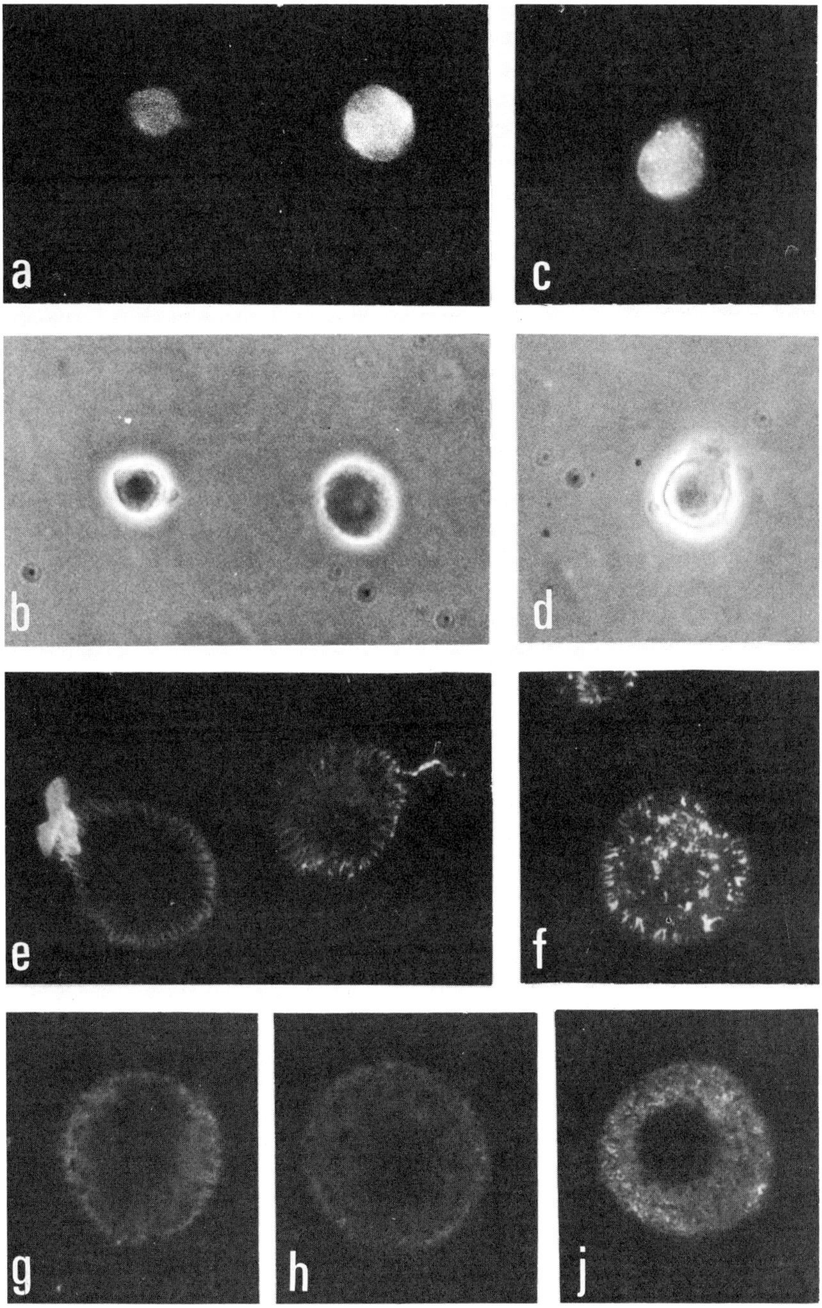

FIGURE 1. Fibronectin distribution during glial cell spreading *in vitro*. a, Diffuse surface distribution of fibronectin in nonspread cells 30 min after layering; b, phase-contrast appearance of the same field shown in a; c, glial cell begins to form a lamellipod on one side

(Continued on next page)

ing (FIGURE 1, a & b). Some cells displayed discrete patches of fibronectin on their dorsal surfaces, even when they were still more or less spherical, but these patches were preferentially oriented on the sides that faced the beginnings of filopodia or lamellipodia (FIGURE 1, c & d). When the formation of filopodia and lamellipodia was prominent, fibronectin was preferentially oriented around the cellular periphery and did not show a diffuse pattern on the dorsal surface of the cell. Fibronectin deposition between the centrifugally advancing cellular periphery and the substratum became evident only at the early cytoplasmic webbing stage. Fibronectin deposition was initiated in the form of radial filaments along the cellular periphery, probably along the ventral side of the advancing filopodia, and, therefore, whole-cell staining was not an efficient method for studying fibronectin deposition between the cell and the substratum. Footprints prepared by different techniques were ideal for this purpose; the Triton X-100 and EDTA treatments were the best techniques. Analysis of the stained footprints revealed a sequential progression of fibronectin deposition by the spreading cell. Following radial deposition of fibronectin at the cellular periphery, fibronectin was deposited in a centripetal direction in the form of discontinuous spots when cells were spread in the presence of serum. When the cells were well spread (about 3–6 hr after layering), peripheral radial filamentous deposition was less prominent, and the mosaic pattern was dominant (FIGURE 1, e & f). In the absence of serum, peripheral radial filamentous deposition of fibronectin was followed by centripetal deposition in a contiguous mosaic pattern throughout the underside of the cell, leaving small fibronectin-free islands (FIGURE 1, g–j). Fibronectin staining was very bright along the margins of cells in contact with each other (FIGURE 2, a & b).

Thus, the initial adhesive contact between the spherical normal glial cell and the substratum is not mediated by fibronectin. Further, this initial contact also does not trigger microfilament formation in the cytoplasm[11] and thus is probably of different chemical nature. However, this initial cell surface contact with the substratum triggers the surrounding surface membrane area to form filopodia and lamellipodia for cellular spreading.[1] Our studies demonstrate that contact of the filopodia and lamellipodia with the substratum triggers deposition of fibronectin at these sites and that more fibronectin-mediated sites are established in a centripetal direction. That cell surface contact is the stimulant for fibronectin deposition is also demonstrated by increased deposition of fibronectin in cell-cell contact areas (FIGURE 2, a & b). Microfilaments are in a diassembled form in recently harvested cells in suspension,[11] and cellular spreading is accompanied by cytoplasmic assembly of microfilaments.[11-13] Therefore, it is reasonable to assume that cell surface contact triggers fibronectin deposition, which, in turn, triggers microfilament assembly in the cytoplasm. Further evidence for this sequence of events

with a speckled distribution of fibronectin superimposed on the diffuse pattern; d, the same cell shown in c visualized by phase-contrast microscopy; e & f, footprints of spreading glial cells in the presence of serum show changing patterns of fibronectin; e, the footprint (*left*) exhibits a radial filamentous distribution of fibronectin at the cellular periphery; note the absence of fibronectin in the center of the cell where the initial contact between the cell surface and the substratum occurred. The bright object on this footprint (*upper left*) is noncellular debris; another footprint (*right*) is in a later stage, where centripetal fibronectin deposition is in progress; f, later stage of fibronectin deposition displays a spotty distribution; note that peripheral radial filaments are not very prominent; g–j, footprints of glial cells spread in the absence of serum. The peripheral radial distribution of fibronectin as shown in the left footprint in e is quickly masked by more diffuse and centripetal progressive deposition of fibronectin. ×1700.

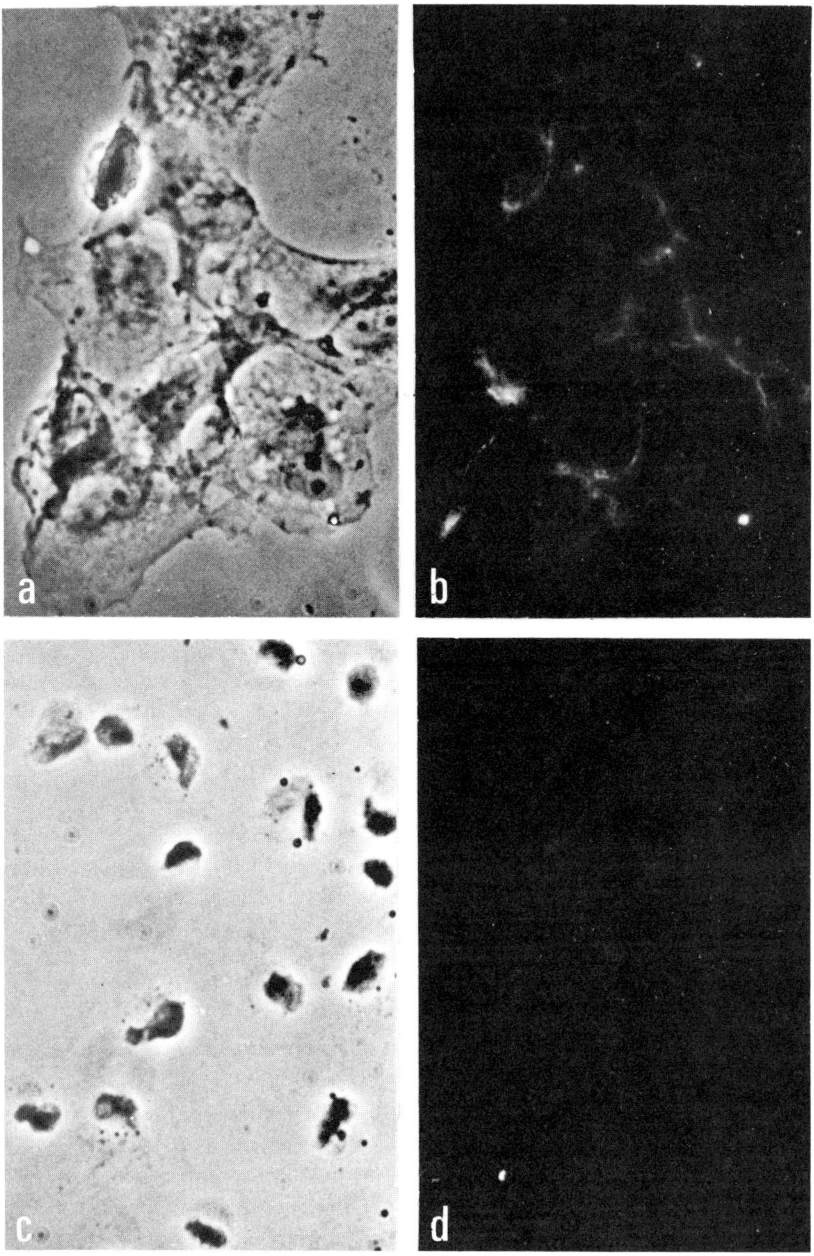

FIGURE 2. a, Glial cells spread for 3 hr; b, the cells illustrated in a show fibronectin at cell-cell contact areas; c, glioma cells spread for 3 hr and extracted with Triton X-100; d, the footprints of the cells illustrated in c display no fibronectin deposition by the immunofluorescent technique. × 1500.

derives from the fact that in transformed cells, addition of exogenous fibronectin induces a well-spread morphology that is also accompanied by the formation of cytoplasmic actin cables (microfilaments).[14,15] On the other hand, cell surface contact with the substratum may initiate secretion of fibronectin and polymerization of actin cables simultaneously, but the transmembrane link between external fibronectin and the submembranous cables would be effected only when fibronectin binds with the substratum or other ligands. The nature of the transmembrane link is not known at present, although a postulated protein-X has been implicated.[16]

In the absence of serum, fibronectin is deposited more or less throughout the undersurface; this distribution pattern may cause the undersurface of the cell membrane to be closely apposed to the substratum[17–19] and thus may render these cells highly resistant to trypsin or EDTA.[18–21]

Glioma cells did not spread in the absence of serum, even though spherical cells exhibited a diffuse surface distribution of fibronectin. Thus, the initial contact of the glioma cell surface with the substratum failed to initiate filopodial and lamellipodial formation in the surrounding surface membrane in the absence of serum. These cells were able to form filopodia and lamellipodia only in the presence of serum proteins. However, no fibronectin was detectable in the footprints of glioma cells spread in the presence of serum (FIGURE 2, c & d). It appears that either glioma cell spreading in the presence of serum occurs independent of the presence of fibronectin or that the amount of fibronectin deposited by spread glioma cells was too low to be detected by the immunofluorescent technique. Other experiments with ^{125}I-labeled fibronectin indicate that neoplastic cells can deform and spread only on fibronectin-coated substrata or after recruiting fibronectin from the medium.[22] Neoplastic cells can synthesize fibronectin, although at a lower rate than normal cells, but they are unable to deploy fibronectin for efficient adhesion or to retain fibronectin on their surfaces. This finding suggests a possible defect in fibronectin-anchoring protein (fibronectin receptor or protein-X?) in the transformed cell membrane.

REFERENCES

1. RAJARAMAN, R., D. E. ROUNDS, S. P. S. YEN & A. REMBAUM. 1974. Exp. Cell Res. **88:** 327.
2. RAJARAMAN, R., D. E. ROUNDS, S. P. S. YEN & A. REMBAUM. 1975. *In* Polyelectrolytes and Their Applications. A. Rembaum & E. Selegny, Eds.: 163. D. Reidel Publishing Company. Dordrecht, The Netherlands.
3. RAJARAMAN, R., J. M. MACSWEEN, V. G. VETHAMANY & R. A. FOX. 1978. Exp. Cell Res. Submitted for publication.
4. HYNES, R. O. 1973. Proc. Nat. Acad. Sci. USA **70:** 3170.
5. HYNES, R. O. 1976. Biochim. Biophys. Acta **458:** 73.
6. GAHMBERG, C. G. & S. HAKOMORI. 1973. Proc. Nat. Acad. Sci. USA **70:** 3329.
7. VAHERI, A. & E. RUOSLAHTI. 1974. Int. J. Cancer **13:** 579.
8. VAHERI, A. 1977. *In* Virus Transformed Cell Membranes. C. Nicolan, Ed. Academic Press. New York and London. In press.
9. YAMADA, K. M., S. S. YAMADA & I. PASTAN. 1976. Proc. Nat. Acad. Sci. USA **73:** 1217.
10. STENMAN, S., J. WARTIOVAARA & A. VAHERI. 1977. J. Cell Biol. **74:** 453.
11. BRAGINA, E. E., J. U. M. VASILIEV & I. M. GELFAND. 1976. Exp. Cell Res. **97:** 241.
12. FOLLETT, E. A. C. & R. D. GOLDMAN. 1970. Exp. Cell Res. **59:** 124.

13. GOLDMAN, R. D. & D. M. KNIPE. 1973. Cold Spring Harbor Symp. Quant. Biol. **37:** 523.
14. HYNES, R. O., A. T. DESTREE, V. M. MAUTNER & I. U. ALI. 1977. J. Supramol. Struct. **7:** 397.
15. ALI, I. U., V. MAUTNER, R. LANZA & R. O. HYNES. 1977. Cell **11:** 115.
16. ASH, J. F., D. LOUVARD & S. J. SINGER. 1977. Proc. Nat. Acad. Sci. USA **74:** 5584.
17. PEGRUM, S. M. & N. G. MARONDAS. 1975. Exp. Cell Res. **96:** 416.
18. RAJARAMAN, R., J. M. MACSWEEN & R. A. FOX. This monograph.
19. RAJARAMAN, R., J. M. MACSWEEN & R. A. FOX. 1978. J. Theor. Biol. In press.
20. TAKEICHI, M. 1971. Exp. Cell Res. **68:** 88.
21. CULP, L. A. & J. F. BUNIEL. 1976. J. Cell. Physiol. **88:** 89.
22. RAJARAMAN, R., J. M. MACSWEEN & R. A. FOX. 1978. Biochem. Biophys. Res. Commun. Submitted for publication.

A COLLAGENOUS COMPONENT OF THE MICROEXUDATE CARPET SECRETED BY ATTACHING HUMAN FIBROBLASTS

C. E. Schwartz, C. G. Hellerqvist, and L. W. Cunningham

Department of Biochemistry
Vanderbilt University
Nashville, Tennessee 37232

Trypsinized fibroblasts, in the absence of serum, attach to plastic substrates and then rapidly secrete onto the substrate a layer of organic material referred to as microexudate carpet.[1] The cells then attach to this material in a manner that permits them to assume their characteristic morphology. The chemical nature of the carpet has been examined by many laboratories.[2-4] One group[5] showed that both leucine and glucosamine are incorporated into the carpet under serum-free conditions. Recently, we reported the presence of hydroxyproline in the carpet secreted by attaching fibroblasts in the absence of serum.[6] In view of the great interest in the role of collagen[7,8] or cell-derived attachment factors[9-11] in cell-substratum adhesion, we have attempted to assess, in a preliminary way, the functional participation in cell-substratum adhesion of this hydroxyproline-containing material.

Human skin fibroblasts were established from an explant from newborn foreskin,[6] and the cells were maintained in Dulbecco's modified Eagle's medium supplemented with 10% fetal calf serum. The cells were grown at 37°C in a humidified atmosphere of 10% CO_2, 90% air. For experiments, cultured fibroblasts, suspended in serum-free Hepes-Hanks' medium (pH 7.3) supplemented with D-glucose (1.00 g/liter), ascorbic acid (50 μg/ml), and amino acids, were plated at 5×10^4 cells/cm^2 and incubated for 4 hr at 37°C. Labeled microexudate carpet was obtained by introducing 10 μC of L-[U-^{14}C]proline. Under these conditions, the cells attached rapidly and were flattened within 30 min.

After 4 hr, the medium was removed. The cells were released with 1 M urea as described previously.[6] The remaining residue was then solubilized by adding a 1% sodium dodecyl sulfate (SDS) solution, 1 mM in EDTA and buffered with 0.1 M Tris HCl (pH 8.0). The dish was sonicated for 10 min at 40°C and then incubated for 15 hr at 37°C. After an additional 10-min sonication, the solubilized carpet was collected, and the dish was washed with the 1% SDS solution, which was combined with the solubilized residue. This mixture was placed in a boiling water bath for 2 min, and 50 μg of rat skin salt-soluble collagen were added as the carrier. The sample was then exhaustively dialyzed against 3% aqueous *n*-butanol, 1 mM in pyridine acetate (pH 5.5). The retentate was lyophilized and stored at −20°C until used.

Amino acid analyses of the solubilized carpet demonstrated the presence of both 3- and 4-hydroxyproline, with ratios of 3-/4-hydroxyproline and hydroxyproline/proline of 0.08 and 0.11, respectively. Gel filtration of the L-[U-^{14}C]-proline-labeled carpet on BioGel A5M agarose, equilibrated by use of 0.1% SDS buffered with 0.1 M Tris HCl (pH 7.5) at 28°C, resulted in a complex profile, with a large portion of the radioactivity in high-molecular-weight molecules. Analyses of the void volume peak produced 3-/4-hydroxyproline and hydroxyproline/proline ratios of 0.07 and 0.12, respectively.

TABLE 1

EFFECT OF METHOD OF CELLULAR RELEASE ON L-[U-^{14}C]PROLINE-LABELED RESIDUE*

Residue	Total Counts per Minute	Hydroxyproline/Proline Ratio
Urea	3420	0.118
Collagenase	3450	0.024
Trypsin	3255	0.114

*Cells were incubated in 10-cm dishes with 10 μCi of [^{14}C]proline for 4 hr in Hepes-Hanks' buffer. Cells were removed by 1 M urea, 0.1 M in Tris-HCl (pH 8.0; 30 min at 37°C), 0.1% Worthington CLSPA collagenase in Hepes-Hanks' buffer (30 min at 37°C), or 0.1% crystalline trypsin in phosphate-buffered saline (5 min at 37°C). The residue was solubilized as described.

These results indicate that the carpet is partly collagenous in nature and suggest that collagenase might be effective in releasing such fibroblasts. Indeed, fibroblasts incubated for 4 hr in serum-free medium are readily removed by collagenase (0.1% in Hepes-Hanks' buffer for 30 min at 37°C). The collagenease-released cells were more self-adherent than the trypsin-released cells and had a tendency to form aggregates at 25–37°C. Aggregation could be reduced if the cell suspension was kept at 4°C. Ovalbumin added with the collagenase did not inhibit the release, demonstrating indirectly that collagenase causes release of the cells. Radioactive amino acid analyses of the microcarpet that remained after release of cells by various procedures reveal that regardless of treatment, the total counts per minute that remained are essentially the same. However, the hydroxy-

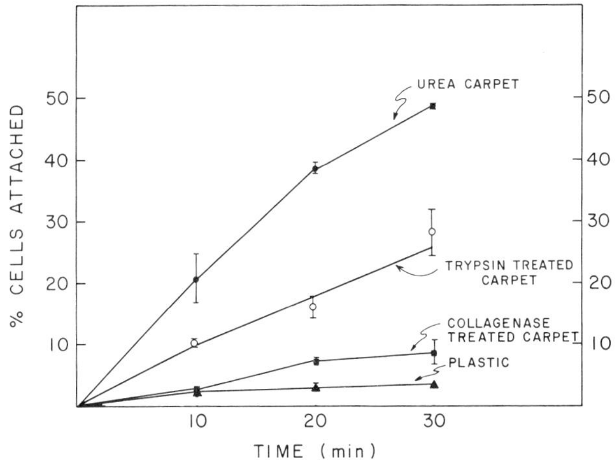

FIGURE 1. Adhesion of collagenase-released single cells. Collagenase-released [^{3}H]-leucine-labeled single cells were added to unlabeled urea carpets at 100,000 cells/Linbro well. Collagenase-treated carpets were subjected to 100 U/ml of collagenase solution for 15 min at 37°C. Trypsin carpets were treated with 0.1% crystalline trypsin for 5 min at 37°C. All assay media were 0.3% in bovine serum albumin.

proline/proline ratios indicate, as anticipated, that collagenase preferentially solubilizes collagen-like material (TABLE 1).

The involvement of the collagen-like component in attachment and growth was also studied. The rate of adhesion was monitored by an established assay[12] in which the urea residue is used as the attaching substrate. The addition of 3 mg/ml of bovine serum albumin to the single-cell suspension greatly reduced nonspecific attachment to plastic. FIGURE 1 illustrates the existence of a linear increase in cellular attachment to the urea carpet over a 20-min period. Pretreatment of the carpet with collagenase reduced this linear attachment by 80%. The rates of growth of both trypsinized and collagenase-released cells were measured as total [*methyl*-^{3}H]thymidine incorporation during continuous labeling for 7 days in the presence of serum. The growth rate was stimulated by the urea carpet, and pretreatment of the microcarpet preparation with collagenase eliminated this growth stimulation.

These preliminary results clearly demonstrate the presence of a collagenous component in the microcarpet. Cellular release and inhibition of both cellular attachment and growth by collagenase strongly suggest an important role for this component in cell-matrix interactions in this system. However, at this time, it is not possible to determine whether the mechanism responsible for this phenomenon involves the disruption of a specific cell receptor-collagen interaction or of an attachment complex in the carpet of which collagen is an integral part.

REFERENCES

1. ROSENBERG, M. D. 1960. Biophys. J. **1:** 138.
2. WEISS, L., G. POSTE, A. MACKEARIN & K. WILLET. 1975. J. Cell Biol. **64:** 135.
3. CULP, L. A. 1974. J. Cell Biol. **63:** 71.
4. CULP, L. A. 1976. Biochemistry **15:** 4094.
5. YAOI, Y. & T. KANASEKI. 1972. Nature (London) **237:** 283.
6. LEMBACH, J. K., R. E. BRANSON, P. B. HEWGLEY & L. W. CUNNINGHAM. 1977. Eur. J. Biochem. **72:** 379.
7. KLEBE, R. 1974. Nature (London) **250:** 248.
8. BORNSTEIN, P. & J. F. ASH. 1977. Proc. Nat. Acad. Sci. USA **74:** 2480.
9. KLEINMAN, H. K., E. B. MCGOODWIN & R. J. KLEBE. 1976. Biochem. Biophys. Res. Commun. **72:** 426.
10. PEARLSTEIN, E. 1976. Nature (London) **262:** 497.
11. MOORE, E. 1976. J. Cell Biol. **70:** 634.
12. WALTER, B. T., P. OHMAN & S. ROSEMAN. 1973. Proc. Nat. Acad. Sci. USA **70:** 1469.

SERUM GLYCOPROTEINS IN FIBROBLAST SPREADING ON GLASS

David Thom, Andrew J. Powell, R. Andrew Badley,
Anne Woods, Colin G. Smith, and David A. Rees

Colworth Laboratory
Unilever Research
Sharnbrook, Bedford MK44 1LQ, England

Our aim is to investigate the mechanism of the transition from the round shape of suspended fibroblasts to their characteristic spread shape on an inert substratum in the presence of serum. Spreading is accompanied by the formation[1-3] of discrete adhesion zones and by the concomitant development of a fibrillar protein cytoskeleton with at least some components organized in relation to focal contacts.[4-6] Following evidence[7-9] that the transition requires prior adsorption of serum macromolecules whose presence in the solution phase is not essential and that the stimulatory components were related[10,11] to cold-insoluble globulin (CIg) and large, external, transformation-sensitive (LETS) protein, we isolated two such fractions from chicken plasma.

Initially, a chick CIg preparation[12] was further resolved into two distinct fractions, a closely spaced 225,000/215,000-dalton protein doublet and a homogeneous 140,000-dalton fraction, by gel chromatography (Sepharose 4B) in the presence of sodium dodecyl sulfate (SDS). Routine isolation of biologically active material was achieved by use of affinity columns prepared from antibodies to the denatured preparations. The unretarded material ("CIg-depleted" serum) contained the other serum proteins. Specifically eluted fractions contained only a mixture of the components used to raise the antisera, which were then partially separated by gel chromatography (BioGel A1.5). Two types of immunologic test confirmed their close relationship to each other and to CIg/LETS protein: the distinct immunodiffusion arc formed against an antiserum against human CIg and the fibrillar distribution characteristic of LETS protein on chick embryo fibroblasts, shown by indirect immunofluorescence by means of antibodies against either preparation.

Cold-insoluble globulin from other sources has convincingly been shown[13-16] to be a protease-sensitive, disulfide-bridged dimer of molecular weight 400,000–450,000 daltons ($S^{\circ}_{20,w}$ = 12–14.5). Sodium dodecyl sulfate polyacrylamide gel electrophoresis (SDS-PAGE) and sedimentation analysis confirmed the purity of our 140,000-dalton preparation ($S^{\circ}_{20,w}$ = 7.3), but the 225,000/215,000-dalton doublet ($S_{20,w}$ = 14.5) was consistently contaminated by residual traces of 140,000-dalton material. Reduction with 2-mercaptoethanol invariably produced five or six prominent bands of lower molecular weight (by SDS-PAGE) from the doublet, suggesting that this preparation from chicken plasma had been proteolytically cleaved. Similar reduction of the 140,000-dalton material yielded two distinct bands with molecular weights of 65,000 and 30,000 daltons, but we are not yet able to distinguish whether this material is a proteolytically derived fragment of CIg or a distinct but closely related analog.

The amino acid compositions of the chicken CIg preparations (TABLE 1) were distinguished only by their striking similarity to each other and to previously reported CIg/LETS preparations.[13,14,17] Carbohydrate residue analysis (TABLE 1) confirmed the glycoprotein nature of the preparations (< 5% carbohydrate) and revealed more distinctive features. We were unable to detect *N*-acetylneuraminic acid in either preparation, and the absence of *N*-acetylgalactosamine also indicated that the side chains are likely to be of the *N*-glycosidically linked type. Characterization of secondary structure by means of an iterative computer method

TABLE 1

AMINO ACID AND CARBOHYDRATE RESIDUE ANALYSES OF CHICKEN COLD-INSOLUBLE GLOBULIN PREPARATIONS

	Composition (moles/100 moles of amino acids)	
Amino Acid Residue	225,000/215,000-Dalton Glycoprotein	140,000-Dalton Glycoprotein
Aspartic acid	10.2	7.9
Threonine	7.1	7.2
Serine	8.1	10.7
Glutamic acid	10.3	9.7
Proline	6.2	7.4
Glycine	9.4	9.2
Alanine	7.6	8.3
Cystine	4.0	2.5
Valine	6.8	8.0
Methionine	1.7	1.3
Isoleucine	2.5	2.4
Leucine	5.0	6.0
Tyrosine	4.1	3.9
Phenylalanine	3.7	3.1
Histidine	2.3	2.3
Lysine	5.6	4.9
Arginine	5.3	5.4
	Composition (moles/100 moles of amino acids in the protein*)	
Carbohydrate Residue	**225,000/215,000-Dalton Glycoprotein†**	**140,000-Dalton Glycoprotein†**
Mannose	0.95	1.16
Galactose	0.43	0.38
N-acetylglucosamine	0.48	0.19
N-acetylgalactosamine	—	—
N-acetylneuraminic acid	—	—

*Protein concentration was determined by the Lowry method and was converted to moles of amino acids by use of average residue molecular weights (116 and 107 for the 225,000/215,000- and 140,000-dalton glycoprotein, respectively) calculated from amino acid analysis.

†Analyzed essentially according to the procedure described by Clamp *et al.*[23,24]

to match observed circular dichroism (CD) spectra in terms of linear combinations of model spectra also provided very similar conclusions for both preparations.[18] a) Both displayed a strong positive band centered below 200 nm. This band is absent in the usual model spectra,[19,20] and the observed CD could not be matched below 215 nm by combinations of these spectra. b) For both preparations, the optimal overall match gave 35–40% β-sheet conformation, with the remainder in disordered segments, and no α-helix could be detected. This observation agrees with current interpretations of the CD spectrum of human CIg.[21]

Indirect immunofluorescent staining of isolated focal adhesions by use of antibodies to CIg provided evidence that a form of this glycoprotein was present at the specialized adhesion zones, because at least some of the cross-reacting spots corresponded to focal contacts, as visualized by interference reflection microscopy.[6,22] Electron micrographs of vertical sections from such preparations and of cells captured in the act of detachment revealed[6] a distinct external organization of surface macromolecules, with marked side-to-side periodicity mediating this interaction with the serum layer, and also the presence of a tightly associated and similarly ordered cytoplasmic plaque almost certainly associated with actin stress fibers.

Experiments with living cells also confirmed that cold-insoluble globulins are probable candidates for the serum components that facilitate cellular spreading. In serum-free medium, glass preadsorbed with 10% (v/v) chick serum initiated a process of cellular spreading and organization for as long as 4 hr that was directly analogous to that observed for fibroblasts (rat 16C, Colworth strain) cultured in complete medium. Complete inhibition of this spreading by subsequent treatment of this serum-coated glass with antibodies to our CIg preparations indicated that they are the essential components; this conclusion was confirmed by the spreading observed with purified glycoproteins. Preadsorption of glass with the 140,000-dalton glycoprotein (25–200 μg/ml) resulted in fully spread fibroblasts, and spreading activity was also observed for the 225,000/215,000-dalton preparation. In addition, "CIg-depleted" chick serum that has been passed twice through the affinity column completely loses this ability to promote cellular spreading.

Although preadsorption of CIg onto glass was a simple condition for spreading, without the stimulus of additional serum molecules in the solution phase, this process progressed normally for only a limited period (4 hr). At this stage, the cells had completely established their characteristic fibroblastic shape, but interference reflection microscopy revealed no subsequent development of the pattern of adhesions beyond the mottled, semiorganized state that normally precedes a fully developed arrangement of focal contacts.

In contrast, cells cultured for 24 hr in medium that contained "CIg-depleted" serum (10% v/v) had developed completely organized arrays of focal contacts, which represent appropriate development of a cytoskeletal organizatic . of actomyosin and other fibrillar protein systems, whether the glass had been pretreated with CIg or "CIg-depleted" serum. Our evidence therefore indicates that although adsorption of CIg onto glass is a sufficient condition for cellular spreading, it is not an essential one. Our interpreation is that given the stimulus of the requisite serum molecules, cells can produce all of the components necessary for highly organized contacts, including endogenous CIg/LETS.

Acknowledgments

We thank Miss H. M. Webb and Mrs. T. P. Ogden for skilled technical assistance, Mr. A. J. Marks and Mr. C. Allcock for the amino acid and carbohydrate residue analyses, respectively, and Professor M. W. Mosesson for kindly providing the antiserum to human CIg.

References

1. Curtis, A. S. G. 1964. J. Cell Biol. **20:** 199–215.
2. Abercrombie, M., J. E. M. Heaysman & S. M. Pegrum. 1971. Exp. Cell Res. **67:** 359–367.
3. Abercrombie, M. & G. A. Dunn. 1975. Exp. Cell Res. **92:** 57–62.
4. Izzard, C. S. & L. R. Lochner. 1976. J. Cell Sci. **21:** 129–159.
5. Heath, J. P. & G. A. Dunn. 1978. J. Cell Sci. **29:** 197–212.
6. Rees, D. A., R. A. Badley, C. W. Lloyd, D. Thom & C. G. Smith. 1978. *In* Society for Experimental Biology Symposium 32, Cell-Cell Recognition: 241–260.
7. Klebe, R. J. 1974. Nature (London) **250:** 248–251.
8. Grinnell, F. 1976a. Exp. Cell Res. **97:** 265–274.
9. Grinnell, F. 1976b. Exp. Cell Res. **102:** 51–62.
10. Pearlstein, E. 1976. Nature (London) **262:** 497–500.
11. Chiquet, M., E. C. Puri, D. C. Turner & H. M. Eppenberger. 1977. Experientia **33:** 815.
12. Mosesson, M. W. & R. A. Umfleet. 1970. J. Biol. Chem. **245:** 5728–5736.
13. Mosher, D. F. 1975. J. Biol. Chem. **250:** 6614–6621.
14. Mosesson, M. W., A. B. Chen & R. M. Huseby. 1975. Biochim. Biophys. Acta **386:** 509–524.
15. Chen, A. B., D. L. Amrani & M. W. Mosesson. 1977. Biochim. Biophys. Acta **493:** 310–322.
16. Jilek, F. & H. Hörmann. 1977. Hoppe Seyler's Z. Physiol. Chem. **358:** 133–136.
17. Carter, W. G. & S.-I. Hakomori. 1977. Biochem. Biophys. Res. Commun. **76:** 299–308.
18. Thom, D., A. J. Powell & D. A. Rees. 1978. Exp. Cell Res. Submitted for publication.
19. Greenfield, N. & G. D. Fasman. 1969. Biochemistry **8:** 4108–4115.
20. Chen, Y. H., J. T. Yang & K. H. Chau. 1974. Biochemistry **13:** 3350–3359.
21. Mosesson, M. W. 1977. Personal communication.
22. Badley, R. A., C. W. Lloyd, C. G. Smith, A. Woods, L. Carruthers, C. Allcock & D. A. Rees. 1978. Exp. Cell Res. Submitted for publication.
23. Chambers, R. E. & J. R. Clamp. 1971. Biochem. J. **125:** 1009–1018.
24. Bhatti, T., R. E. Chambers & J. R. Clamp. 1970. Biochim. Biophys. Acta **222:** 339–347.